CAMBRIDGE STUDIES IN
ADVANCED MATHEMATICS 32
EDITORIAL BOARD
D.J.H. GARLING, T. TOM DIECK, P. WALTERS

COHOMOLOGICAL METHODS IN TRANSFORMATION GROUPS

Already published

1 W.M.L. Holcombe *Algebraic automata theory*
2 K. Petersen *Ergodic theory*
3 P.T. Johnstone *Stone spaces*
4 W.H. Schikhof *Ultrametric calculus*
5 J.-P. Kahane *Some random series of functions, 2nd edition*
6 H. Cohn *Introduction to the construction of class fields*
7 J. Lambek & P.J. Scott *Introduction to higher-order categorical logic*
8 H. Matsumura *Commutative ring theory*
9 C.B. Thomas *Characteristic classes and the cohomology of finite groups*
10 M. Aschbacher *Finite group theory*
11 J.L. Alperin *Local representation theory*
12 P. Koosis *The logarithmic integral I*
13 A. Pietsch *Eigenvalues and s-numbers*
14 S.J. Patterson *An introduction to the theory of the Riemann zeta-function*
15 H.J. Baues *Algebraic homotopy*
16 V.S. Varadarajan *Introduction to harmonic analysis on semisimple Lie groups*
17 W. Dicks & M. Dunwoody *Groups acting on graphs*
18 L.J. Corwin & F.P. Greenleaf *Representations of nilpotent Lie groups and their applications*
19 R. Fritsch & R. Piccinini *Cellular structures in topology*
20 H Klingen *Introductory lectures on Siegel modular forms*
22 M.J. Collins *Representations and characters of finite groups*
24 H. Kunita *Stochastic flows and stochastic differential equations*
25 P. Wojtaszczyk *Banach spaces for analysts*
26 J.E. Gilbert & M.A.M. Murray *Clifford algebras and Dirac operators in harmonic analysis*
27 A. Frohlich & M.J. Taylor *Algebraic number theory*
28 K. Goebel & W.A. Kirk *Topics in metric fixed point theory*
29 J.F. Humphreys *Reflection groups and Coxeter groups*
30 D.J. Benson *Representations and cohomology I*
31 D.J. Benson *Representations and cohomology II*
32 C. Allday & V. Puppe *Cohomological methods in transformation groups*
33 C. Soulé et al *Lectures on Arakelov geometry*
34 A. Ambrosetti & G. Prodi *A primer of nonlinear analysis*
35 J. Palis & F. Takens *Hyperbolicity, stability and chaos at homoclinic bifurcations*
37 Y. Meyer *Wavelets and operators I*

COHOMOLOGICAL METHODS IN TRANSFORMATION GROUPS

C. Allday
Professor of Mathematics, University of Hawaii

V. Puppe
Professor of Mathematics, University of Konstanz

Published by the Press Syndicate of the University of Cambridge
The Pitt Building, Trumpington Street, Cambridge CB2 1RP
40 West 20th Street, New York, NY 10011-4211, USA
10 Stamford Road, Melbourne, Victoria 3166, Australia

© Cambridge University Press 1993

First published 1993

Printed in Great Britain at the University Press, Cambridge

Library of Congress cataloguing in publication data available
A catalogue record for this book is available from the British Library

ISBN 0 521 35022 0 hardback

Contents

Preface ix

Chapter 1 1
Equivariant cohomology of G-CW-complexes and the Borel construction
1.1 G-CW-complexes and a comparison theorem for equivariant cohomology theories 3
1.2 The Borel construction 10
1.3 The Borel construction for 2-tori 22
1.4 The Borel construction for p-tori 61
Exercises 85

Chapter 2 92
Summary of some aspects of rational homotopy theory
2.1 The Sullivan-de Rham algebra 92
2.2 Minimal models 96
2.3 Rational homotopy theory 99
2.4 Finite-dimensional rational homotopy 104
2.5 The Grivel-Halperin-Thomas theorem 106
2.6 Sullivan-de Rham theory for rational Alexander-Spanier cohomology 111
2.7 Formal spaces, formal maps and the Eilenberg-Moore spectral sequence 116

Chapter 3   129
Localization

3.1	The Localization Theorem	130
3.2	The Localization Theorem for general G-spaces	141
3.3	Equivariant rational homotopy	148
3.4	Equivariant rational homotopy for general G-spaces	158
3.5	The Evaluation Theorem for torus actions	160
3.6	The ideals $\mathfrak{p}(K,c)$	166
3.7	Chang-Skjelbred modules and Hsiang-Serre ideals	175
3.8	Hsiang's Fundamental Fixed Point Theorem	189
3.9	Remarks on the Weyl group	205
3.10	The cohomology inequalities and other basic results	208
3.11	The Hirsch-Brown model and the Evaluation Theorem for p-torus actions on general spaces	223
	Exercises	244

Chapter 4   253
General results on torus and p-torus actions

4.1	Rank and Poincaré series	254
4.2	Generalities on torus actions	264
4.3	Almost-free torus actions and the rank of a space	269
4.4	More results on almost-free torus actions	280
4.5	The method of Browder and Gottlieb	286
4.6	Equivariant Tate cohomology	301
4.7	Two theorems on topological symmetry	326
4.8	Localization and the Steenrod algebra	332
4.9	The rational homotopy Lie algebra of a fixed point set	336
	Exercises	338

Chapter 5   342
Actions on Poincaré duality spaces

5.1	Algebraic preliminaries	343
5.2	Poincaré duality for the fixed point set	347
5.3	Equivariant Gysin homomorphism, Euler classes and a formula of A.Borel	360
5.4	Torus actions and Pontryagin classes	383
5.5	Golber formulas and other results	386
	Exercises	394

	Appendix A	**398**
	Commutative algebra	
A.1	Krull dimension	399
A.2	Modules	401
A.3	Primary decomposition	405
A.4	Homological dimension	408
A.5	Regular sequences	413
A.6	Cohen-Macaulay rings and modules	418
A.7	Evaluations and presentations	425
	Appendix B	**435**
	Some homotopy theory of differential modules	
B.1	Basic notions and elementary results	436
B.2	Applications to cochain complexes over graded algebras	448
	Exercises	454
	References	**456**
	Index	**466**
	Index of Notation	**469**

Preface

It is our aim to give a contemporary account of a small, but well-developed and useful, part of the immense mathematical field of (compact) transformation groups - namely that in which the main tools are ordinary cohomology theory and rational homotopy theory. Furthermore, except for occasional excursions, we shall restrict our attention to those groups for which these methods work best: these are tori and elementary abelian p-groups. (An elementary abelian p-group, or p-torus, where p is a prime number, is just a product of finitely many copies of the cyclic group of order p.) Torus and p-torus actions are of more than mere intrinsic interest, however: one can extrapolate to gain much useful information about more general group actions, often in a way similar to that in which the classification of compact connected Lie groups was achieved by studying the roots and weights associated with representations of maximal tori. Two important references where the reader can see such extrapolation at work are [Quillen, 1971a, b] and [Hsiang, 1975].

Our subject began with the work of P.A.Smith in the 1930s and 1940s; and consequently, it is often called P.A. Smith theory. Important developments were brought together in the Princeton seminar [Borel *et al.*, 1960]. Later the subject received substantial clarification and inspiration, when, prompted by the work of Atiyah and Segal in equivariant K-theory, some ideas, which were implicit in the work of Borel, were reformulated in the succinct Localization Theorem proven independently by W.-Y. Hsiang and Quillen. (Versions of the theorem for generalized cohomology theories were also given by tom Dieck.) Since then one of the most powerful methods to be introduced into the subject has been Sullivan's theory of minimal models, which is particularly useful for studying torus actions by means of rational cohomology and rational homotopy.

Essential contributions, which can now be considered as a classical part of the subject initiated by P.A. Smith, have been made by A. Borel, G. Bredon, P. Conner, E. Floyd, W.-Y. Hsiang, R. Oliver, D. Quillen, J.C. Su, T. Chang, T. Skjelbred and many others. We apologize if we sometimes fail to give the original reference for these results. Recent developments in the area are due to A. Adem, W. Browder, G. Carlsson, D. Gottlieb, S. Halperin and others.

In our mode of presentation we have a dual purpose. On the one hand, part of the book is written as an introduction to the subject for a person with relatively little background who wishes to get the gist of things without undue pain. Thus we have tried to simplify many parts of the book by treating there only G-CW-complexes. This has the advantage that the cohomology theory used is ordinary singular cohomology. We have added exercises in the hope that the book can be used as a textbook as well as for self-study. On the other hand, we want the book to be a useful reference for anyone working on transformation groups; and for this reason in many parts of the book we treat some more general classes of G-spaces. In these more general cases Čech or, equivalently, Alexander-Spanier, cohomology is used.

For most of the book we have tried to keep the prerequisites minimal, assuming only results which can be found easily in standard textbooks. For algebraic topology we refer mainly to [Dold, 1980] and [Spanier, 1966], for homological algebra to [Cartan, Eilenberg, 1956] and [MacLane, 1967], for the general theory of compact transformation groups to [Bredon, 1972] and [tom Dieck, 1987], and for group rings and group cohomology to [Brown, K.S. 1982]. In each case only a small part of the book is required: the reader is not expected to have read them all from cover to cover. We have included an appendix, which summarizes many of the results from commutative algebra which we need; and we have included another appendix, which outlines the homotopy theory of chain complexes. In Chapter 2 we have summarized the main points from rational homotopy theory which are used in the book.

As usual the presentation becomes a little more terse and the prerequisites increase as the book wears on. In Chapter 1, for example, spectral sequences are not used; but from Chapter 3 onward they are used frequently. The spectral sequence which occurs most often is the Leray-Serre spectral sequence of a fibration. For singular theory this can be found in [Spanier, 1966]. For sheaf cohomology, which is equivalent to Čech cohomology for paracompact spaces, it can be found in [Bredon, 1967a]. A useful version can also be found in [Brown, K.S. 1982]. The basic properties of localization of rings and modules are not included in Appendix A: useful references are [Bourbaki, 1961] or [Atiyah, MacDonald, 1969]. The Steenrod algebra, which crops up on a few occasions, can be found in [Spanier, 1966] and [Steenrod, Epstein, 1962].

Since it is our purpose to supplement and not to supersede such references as [Borel, et al., 1960], [Quillen, 1971a, b], [Bredon, 1972], [Hsiang, 1975] and [tom Dieck, 1987], we sometimes quote results from these works without proof. We always include proofs, however, whenever they are essential to the continuity of the text, whenever they are particularly revealing, whenever we can achieve some generalization of existing proofs, or whenever, in our context, existing proofs can be simplified.

A brief first reading of the book might consist of the following: Chapter 1 and Sections 3.1, 3.10, 4.5 and 5.1-5.4. The cohomological part of Section 3.5 could be added easily, and so could be Section 4.6, which is quite self-contained. Armed with the results of Sections 2.1-2.5, one could add Section 3.3, the homotopical part of Section 3.5, and Sections 4.2 and 4.3. For cohomological results involving more general G-spaces one needs Sections 2.6 and 3.4. The main theorem of Section 3.8 can be found quickly and easily in [Hsiang, 1975]; our presentation is more complicated since we treat a non-localized version of the theorem, and we rely on the rather technical Section 3.6. Some of the shorter sections, such as 3.9, 4.7 and 4.8, are not particularly demanding. (In some places we quote results from Sections 3.6 and 3.7 because they are on hand: the reader seeking to avoid these sections can often fill in the arguments without too much difficulty.)

We would like to thank all the people who have helped us with the writing of the book; and, in particular, our heartfelt thanks go to those who have typed the manuscript, mainly Mrs. A. Giese and Mrs. P. Goldstein. We wish to thank also the University of Konstanz, the University of Hawaii and the Deutsche Forschungsgemeinschaft for their support. And we thank Mr. D. Tranah of Cambridge University Press, especially for his much appreciated patience.

Honolulu,   C. Allday
Konstanz   V. Puppe

1

Equivariant cohomology of G-CW-complexes and the Borel construction

By analogy with the notion of a (co)homology theory on some category of topological spaces one can define the notion of an equivariant or G-(co)homology theory on G-spaces, where G is a compact Lie group. Depending on the given frame and intended purpose one actually might impose different sets of axioms for such a definition (see, e.g., [Bredon, 1967b], [tom Dieck, 1987], [Lee, 1968], [May, 1982], [Seymour, 1982]), but the minimal request would be the G-homotopy invariance of the (co)homology functor and a suitable Mayer-Vietoris long exact sequence. These two requirements suffice to get an elementary comparison theorem for G-(co)homology theories similar to the usual non-equivariant case; i.e. if $\tau : h_G \to k_G$ is a natural transformation between G-(co)homology theories, which is an isomorphism on 'G-points' (i.e. homogeneous spaces G/K, K a closed subgroup of G) then $\tau(X)$ is an isomorphism for all G-spaces X, which can be obtained from (finitely many) 'G-points' by a finite number of the following steps (in any order):

(1) replacing a G-space by a G-homotopy equivalent G-space;
(2) taking finite coproducts (topological sums) of G-spaces;
(3) taking homotopy pushouts (double mapping cylinders of G-maps between G-spaces);

(see [Seymour, 1982] for more sophisticated versions of the comparison theorem). The category obtained this way is just the category of G-spaces, which are G-homotopy equivalent to finite G-CW-complexes (cf. [Puppe, D., 1983] for a discussion of this and related questions in the non-equivariant case, not restricted to finite CW-complexes). To extend the comparison theorem to arbitrary G-CW-complexes one needs further additivity properties of the (co)homology theories involved. It is important to realize that these additional properties are not always fulfilled for those G-theories in which we are interested.

The G-theories we are going to consider are variants of the 'Borel (co)homology' of a G-space X, which is the ordinary (co)homology of the Borel construction $EG \times_G X$ on X, where $EG \times_G X$ denotes the total space of the bundle associated with the universal principal G-bundle $G \to EG \to BG$, i.e. the orbit space of $EG \times X$ with respect to the diagonal action (see [Borel et al., 1960]). For G a finite group we will actually use a more algebraic description of this theory, namely as the (co)homology of the group G with coefficients in the (co)chain complex $C(X)$ over $\mathbb{Z}[G]$, where $\mathbb{Z}[G]$ denotes the group ring of G over $\mathbb{Z}$ (see, e.g., [Brown, K.S., 1982]). This provides (co)chain models for the equivariant (co)homology of X, which are very suitable for certain algebraic constructions and for calculations. If G is a p-torus, i.e. an elementary abelian p-group, $G = (\mathbb{Z}_p)^r$, p prime, one can describe the ordinary cohomology of the fixed point set X^K for a subgroup $K < G$ roughly as the Borel cohomology of X with certain coefficients (which depend on K) at least for $p = 2$, the case p an odd prime being somewhat more complicated. In the case of a torus $G = S^1 \times \ldots \times S^1$ the theory of minimal models (see Chapter 2) leads to a formally very similar situation (cf. Section 3.5). This description can be used effectively in place of the Localization Theorem for the Borel cohomology to relate the ordinary cohomology of the G-space X and of the fixed point sets X^K of subgroups $K < G$, which, in fact, is our main theme. The relation to the Localization Theorem is discussed in Chapter 3 in a more general setting.

Other important G-theories like equivariant K-theory (see, e.g. [Segal, 1968a], [Petrie, Randall, 1984]) or equivariant bordism (see, e.g., [Conner, Floyd, 1964]) and cobordism (see, e.g., [tom Dieck, 1970]) will not be considered in this book.

The reader to whom the brief exposition given in the following section, which assumes some familiarity with the very basic definitions and elementary properties of G-spaces, seems unsatisfactory is advised to consult, e.g., [tom Dieck, 1987], Chapter I, as a general reference for basic notions and results in equivariant topology, in particular for the fundamentals in equivariant homotopy theory.

1.1 G-CW-complexes and a comparison theorem for equivariant cohomology theories

The category of G-spaces which are G-homotopy equivalent to G-CW-complexes forms - very similar to the non-equivariant case - a convenient setting for equivariant topology (see [Illman, 1972a, b, 1975], [Matumoto, 1971]). It is closed under homotopy pushouts (double mapping cylinders) and contains, for example, differentiable G-manifolds, which, in fact, admit an actual G-CW-decomposition (see [Illman, 1978, 1983]).

We prove in this section an elementary comparison theorem for equivariant cohomology theories, which provides a useful tool in studying the relations amongst the ordinary cohomologies of the different fixed point sets X^K, $K < G$, where X is a G-space, which is G-homotopy equivalent to a finite (resp. finite-dimensional) G-CW-complex. We will be mainly considering the case where G is a torus or a p-torus (i.e. an elementary abelian p-group, p a prime), but for this section there is no essential difficulty in treating the more general case where G is any compact Lie group and $K < G$ denotes a closed subgroup.

(1.1.1) Definition

(1) By the n-dimensional G-cell of type G/K we mean the G-space $G/K \times D^n$, where G acts by left translation on G/K and trivially on the n-dimensional ball D^n. We call $G/K \times S^{n-1}$ the G-boundary of the G-cell $G/K \times D^n$.

(2) We say that the G-space X is obtained from the G-space Y by attaching a disjoint union $\coprod_\nu G/K_\nu \times D^n$ of n-dimensional G-cells $G/K_\nu \times D^n$, $K_\nu < G$, along the G-maps $\phi_\nu \colon G/K_\nu \times S^{n-1} \to Y$ if $X = Y \cup_{\phi_\nu} (\coprod_\nu G/K_\nu \times D^n) := Y \sqcup (\coprod_\nu G/K_\nu \times D^n)/\sim$ where $\phi_\nu([g]_\nu, s) \sim ([g]_\nu, s)$ for $([g]_\nu, s) \in G/K_\nu \times S^{n-1}$. (Observe that G-maps $\phi \colon G/K \times S^{n-1} \to Y$ correspond one-to-one to (non-equivariant) maps
$$\tilde\phi \colon S^{n-1} \cong \{1\} \times S^{n-1} \to Y^K,$$
(1 = unit element of G).) There is an obvious canonical G-inclusion $Y \subset X$.

(3) A G-CW-complex X is a G-space which is obtained as the colimit of a sequence of G-inclusions $X^0 \subset X^1 \subset \ldots \subset X^{n-1} \subset X^n \subset \ldots$, where X^0 is a disjoint union of homogeneous spaces G/K_ν (0-dimensional G-cells) and X^n is obtained from X^{n-1} by attaching a disjoint union of n-dimensional G-cells.

A finite (resp. finite-dimensional) G-CW-complex is, of course, a G-

CW-complex which is constructed from finitely many G-cells (resp. from G-cells of bounded dimension).

It should be noted that one might have to distinguish between the equivariant and the non-equivariant dimension since the dimsension of G/K might be positive, if G is not a finite group.

The orbit space X/G of a G-CW-complex inherits in the obvious way a CW-structure from the G-CW-structure of X. The (non-equivariant) cells of X/G are just the orbit spaces of the G-cells of X and the attaching maps of X/G are those induced from the equivariant maps of X by dividing out the G-action.

The above definition is modelled after the usual definition of CW-complexes, replacing the 'points' by 'G-points', i.e. homogeneous spaces, and many topological and homological properties of CW-complexes carry over or have their 'natural' counterparts in the equivariant setting (see [Bredon, 1967b], [Illman, 1972a, b, 1975], [Matumoto, 1971], [May, 1982], [Seymour, 1982], [tom Dieck, 1987], Lück [1989]). Here we shall only need a few rather simple facts about G-CW-complexes. The main use we are making of them is to get a simple comparison theorem for G-cohomology theories (see Theorem (1.1.3)).

(1.1.2) Examples
(1) For $G = \mathbb{Z}_2$ the unit sphere $S^{n-1} = \{x \in \mathbb{R}^n : \|x\| = 1\}$ together with the G-action given by scalar multiplication,

$$G \times S^{n-1} \to S^{n-1}, (\lambda, x) \mapsto \lambda x, \lambda \in G = \{\pm 1\} \subset \mathbb{R}, x \in S^{n-1} \subset \mathbb{R}^n$$

is a free G-space, denoted by X, which has the following decomposition as G-CW-complex:

$$X^0 := G \cong S^0$$
$$X^m := X^{m-1} \cup_{\phi^m} G \times D^m, \text{ where } \phi^m \colon G \times S^{m-1} \to X^{m-1} = S^{m-1}$$
is given by $\phi^m(\lambda, x) = \lambda x$ $(1 \leq m < n)$.

The orbit space X/G is the real projective space $\mathbb{R}P^{n-1}$, and the G-CW-structure of X induces the standard (non-equivariant) cell decomposition of $\mathbb{R}P^{n-1}$.

(2) For $G = S^1$ the unit sphere $S^{2n-1} = \{z \in \mathbb{C}^n : \|z\| = 1\}$ together with the G-action given by complex scalar multiplication

$$G \times S^{2n-1} \to S^{2n-1}, (\lambda, z) \mapsto \lambda z, \lambda \in G = S^1 \subset \mathbb{C}, z \in S^{2n-1} \subset \mathbb{C}^n$$

is a free G-space X with an analogous decomposition as a G-CW-complex:

$$X^0 := G \cong S^1, X^1 := X^0$$

$$X^{2m} = X^{2m-1} \cup_{\phi^{2m}} G \times D^{2m}, \text{ where}$$
$$\phi^{2m} : G \times S^{2m-1} \to X^{2m-1} = X^{2m-2} = S^{2m-1}$$

is given by
$$\phi^{2m}(\lambda, z) = \lambda z \quad (1 \le m < n).$$

(Note that the equivariant $2m$-skeleton X^{2m} of X is not a $2m$-skeleton of a corresponding non-equivariant cell decomposition of X due to the fact that $G = S^1$ has dimension 1.) The orbit space X/G is the complex projective space $\mathbb{C}P^{n-1}$, and again the given G-CW-structure of X induces the standard non-equivariant CW-decomposition of $\mathbb{C}P^{n-1}$.

(3) Restricting the S^1-action on S^{2n-1} in Example (2) to a subgroup $G = \mathbb{Z}_n < S^1$ clearly gives a free G-action on S^{2n-1}. But the S^1-CW-structure of S^{2n-1} does not give a G-CW-structure of S^{2n-1} for $G = \mathbb{Z}_n$ 'on the nose', since $S^1 \times D^{2m}$ is obviously not a G-cell in the sense of Definition (1.1.1). Of course $S^1 \times D^{2m}$ can be decomposed into G-cells and this leads to the following G-CW-structure of the G-space S^{2n-1}

$X^0 := G$
$X^1 := G \cup_{\phi^1} G \times D^1 \cong S^1$,
$\phi^1 : G \times S^0 \to G$ is the unique equivariant extension of
$$\tilde{\phi} : S^0 \to G,$$
$\tilde{\phi}(1) = e$, $\tilde{\phi}(-1) = g \in G$, where g denotes a fixed generator of G
$X^{2m} := X^{2m-1} \cup_{\phi^{2m}} G \times D^{2m}$,
$\phi^{2m} : G \times S^{2m-1} \to X^{2m-1} \cong S^{2m-1}$ is given by $\phi^{2m}(\lambda, x) = \lambda x (1 \le m < n)$
$X^{2m+1} := X^{2m} \cup_{\phi^{2m+1}} G \times D^{2m+1}$,
$\phi^{2m+1} : G \times S^{2m} \to X^{2m}$ is the unique equivariant extension of $\tilde{\phi}^{2m+1} : S^{2m} \to X^{2m}$, where $\tilde{\phi}^{2m+1}$ is defined as follows: S^{2m} can be written as the union of the right (D^{2m}_+) and the left (D^{2m}_-) hemisphere glued together along the boundary S^{2m-1}, $S^{2m} = D^{2m}_+ \cup_{S^{2m-1}} D^{2m}_-$; and X^{2m} is the quotient of $G \times D^{2m}$ by an equivalence relation given by the attaching map ϕ^{2m}. With this notation we put $\tilde{\phi}^{2m+1}(x_+) = [e, x]$, $x_+ \in D^{2m}_+$ and x the corresponding element in D^{2m}, $\tilde{\phi}^{2m+1}(x_-) = [g, g^{-1}x]$, $x_- \in D^{2m}_-$, $x \in D^{2m}$, g the chosen generator of G. Note that the equivalence relation on $G \times D^{2m+1}$ given by ϕ^{2m+1} is such

that the above definition gives a well-defined map $\tilde{\phi}^{2m+1}$: $S^{2m} \to X^{2m}$ ($[e,x] = [g,g^{-1}x]$ in X^{2m} if $x \in S^{2m-1}$).

(4) The universal free G-space EG (contractible, disregarding the G-action), i.e. the total space of the universal principal G-bundle can be considered an (infinite) G-CW-complex. The orbit space $BG := EG/G$, i.e. the classifying space of the group G, inherits a CW-structure from the G-CW-structure of EG.

The space EG is well defined only up to G-homotopy equivalence. For concise treatment of the relevant homotopy theory see, e.g., [tom Dieck, 1987]. For the groups $G = \mathbb{Z}_2, \mathbb{Z}_n$ or $G = S^1$ one can just take the colimit (over the canonical inclusions given by the G-CW-structure) of the free G-space in (1), (3) or (2), respectively, to get a G-CW-structure on $EG = \operatorname{colim} S^n = S^\infty$. (By standard homotopy arguments this colimit is contractible since $S^n \subset S^{n+1}$ is a cofibration and is homotopic to the constant map.) Note that in case $G = \mathbb{Z}_n$ (including $n = 2$), the cellular chain complex $W_*(EG)$ considered as a complex over the group ring $\mathbb{Z}[G]$ together with the natural augmentation $\varepsilon : W_*(EG) \to W_*(pt) = \mathbb{Z}$ is the standard minimal free acyclic resolution of $\mathbb{Z}$ as a trivial $\mathbb{Z}[G]$-module, which is used to define the homology (and cohomology) of the group G in a purely algebraic context (see, e.g., [Brown, K.S., 1982]).

To get EG as a G-CW-complex for the groups we mainly consider (i.e. $G = (\mathbb{Z}_p)^r$, p prime, or $G = (S^1)^r = T^r$) one can take the product of the universal free G-spaces of the single factors defining the G-CW-structure (and the 'weak' topology) on the product in a similar way as in the non-equivariant situation.

In general, the following construction gives an explicit G-CW-structure on EG. Note that it is not a generalization of the construction described above, in fact for $G = \mathbb{Z}_p$ the number of cells used in a fixed positive dimension is bigger than in the 'minimal' construction above.

$EG_0 := G$

$EG_n := EG_{n-1} \cup_{\phi_n} G \times (G^n \times \triangle_n)$, where $\triangle_n \cong D^n$ is the standard n-simplex, $\dot{\triangle}_n \cong S^{n-1}$ its boundary and $\phi_n : G \times (G^n \times \dot{\triangle}_n) \to EG_{n-1}$ given by

$$\phi_n(g_0, \ldots, g_n, x_0, \ldots, x_n) := \begin{cases} [g_0, \ldots, g_{n-1}, x_0, \ldots, x_{n-1}] \\ \quad \text{if } x_n = 0 \\ [g_0, \ldots, g_i g_{i+1}, \ldots, g_n, x_0, \\ \quad \ldots, \hat{x}_i, \ldots, x_n] \\ \quad \text{if } x_i = 0 \text{ and } 0 \leq i < n. \end{cases}$$

1.1. G-CW-complexes and a comparison theorem

Here $(x_0, \ldots, x_n)$ are the barycentric coordinates of a point $x \in \triangle_n$; $\hat{}$ means omit the corresponding term and $[\ldots]$ denotes the equivalence class in EG_{n-1}. It is straightforward to check that ϕ_n is a well-defined G-map. $EG := \operatorname{colim} EG_n$, where the colimit is taken with respect to the canonical inclusions $EG_{n-1} \subset EG_n$. By construction these inclusions are cofibrations (in fact they are G-cofibrations, i.e. cofibrations in the category of G-spaces). Hence to show that EG is contractible (disregarding the G-action) it suffices to prove that the inclusions $EG_{n-1} \subset EG_n$ are null-homotopic. The map

$$\phi : G \times (G^{n-1} \times \triangle_{n-1}) \times I \to G \times (G^n \times \triangle_n)$$

given by

$$\phi(g_0, \ldots, g_{n-1}, x_0, \ldots, x_{n-1}, t)$$
$$= (g_0, \ldots, g_{n-1}, g_{n-1}^{-1} \cdots g_0^{-1}, (1-t)x_0, \ldots, (1-t)x_{n-1}, t)$$

induces the desired homotopy between the inclusion $EG_{n-1} \subset EG_n$ and the constant map to $[e] \in EG_0 \subset EG_n$ (here e denotes the unit element in $G = EG_0$) since ϕ is compatible with the necessary identifications to obtain a map from $EG_{n-1} \times I$ to EG_n and

$$\phi(g_0, \ldots, g_{n-1}, x_0, \ldots, x_{n-1}, 0) = [g_0, \ldots, g_{n-1}, x_0, \ldots, x_{n-1}]$$
$$\phi(g_0, \ldots, g_{n-1}, x_0, \ldots, x_{n-1}, 1) = [g_0, \ldots, g_{n-1}, g_{n-1}^{-1} \cdots g_0^{-1}, 0, \ldots 0, 1]$$
$$= [e, 1] = [e].$$

If G is a finite group, then $EG_0 \subset EG_1 \subset \ldots EG_{n-1} \subset \ldots$ together with the attaching maps ϕ_n is a G-CW-structure for EG since $G \times (G^n \times \triangle_n)$ is a disjoint union of n-dimensional G-cells. In case G is not finite (i.e. the dimension of G as a manifold is greater than 0) the attaching of $G \times (G^n \times \triangle_n)$ to EG_{n-1} can be decomposed into successively attaching free G-cells of different dimensions to EG_{n-1} corresponding to a non-equivariant cell decomposition of $G^n \times \triangle_n$.

For a very conceptual construction of EG and its orbit space $BG = EG/G$ from a categorical viewpoint and a discussion of the relation to Milnor's construction of the universal principal G-bundle see [Segal, 1968b].

(5) If X is a (compact) differentiable manifold with a differentiable G-action, then X can be viewed as a (finite) G-CW-complex (see [Illman, 1978, 1983]).

(6) For G a finite group, a (finite) simplicial G-complex in the sense of Bredon [1972] can be viewed - after barycentric subdivision - as a (finite) G-CW-complex.

Many of the basic concepts and results (e.g. Whitehead Theorem, cellular approximation, obstruction theory) carry over to G-CW-complexes (see [May, 1982] for a resumé and the references given there for details). In particular any G-space X can be approximated by a G-CW-complex X_{CW} in the following sense: there exists a G-map

$$\phi : X_{CW} \to X$$

which induces a weak equivalence for the fixed point sets of all closed subgroups $K \subset G$, i.e. $\phi^K : X_{CW}^K \to X^K$ induces isomorphisms of the (non-equivariant) homotopy groups. This is quite obvious in the case of a finite group G (using the canonical G-structure on the singular simplicial complex of X and on its realization), but holds, more generally, for any compact Lie group G (see [Waner, 1980], [Seymour, 1983], [Matumoto, 1984]).

In certain situations it is of interest to decide whether a G-space X can be approximated by a finite G-CW-complex in the above sense. This is a very difficult question in general (see, e.g., [Lück, 1989] for a comprehensive treatment of this and related topics). It is shown in [Petrie, Randall, 1986] that a non-singular affine variety, on which a finite group G acts algebraically, has the G-homotopy type of a finite G-CW-complex.

The following result, which is, in fact, a rather straightforward G-analogue of the well-known comparison theorem for (non-equivariant) cohomology theories, will be applied in the next section.

Equivariant cohomology theories are considered on suitable categories of G-spaces or pairs of G-spaces (see, e.g., [Bredon, 1967b], [tom Dieck, 1987], [May, 1982], [Lee, 1968], [Segal, 1968a], [Seymour, 1982]). Although the axioms required for an equivariant cohomology theory depend to some extent on the context, a $\mathbb{Z}$-graded equivariant cohomology theory $h_G^* = \{h_G^q\}_{q \in \mathbb{Z}}$ on the category of (finite) G-CW-complexes taking values in a category of modules over a ring R should always fulfil:

GCT I: h_G^* is G-homotopy invariant, i.e. $h_G^*(f_0) = h_G^*(f_1)$ if $f_0, f_1 : X \to Y$ are G-homotopic.

GCT II: If X is obtained from X_1 by attaching X_2 along a cellular G-map $\phi : X_0 \to X_1$, where X_0 is a subcomplex of X_2, then there is a long exact Mayer-Vietoris sequence

$$\ldots h_G^q(X_0) \xrightarrow{\delta} h_G^{q+1}(X) \to h_G^{q+1}(X_1) \times h_G^{q+1}(X_2) \to h_G^{q+1}(X_0) \to \ldots$$

The properties GCT I and GCT II suffice to prove the following result:

(1.1.3) Theorem

If $\tau : h_G^* \to k_G^*$ is a natural transformation between G-cohomology theories, which fulfil GCT I and GCT II above and if $\tau(G/K)$ is an isomor-

phism for each $K < G$ (i.e. for each 'G-point'), then τ is an isomorphism for each finite G-CW-complex.

Proof.
(a) That $\tau(G/K \times S^{n-1})$ is an isomorphism follows from GCT I and GCT II and the five lemma by decomposing $X = G/K \times S^{n-1} = X_1 \cup_\phi X_2$ with $X_1 = G/K \times D^0$, $X_2 = G/K \times D^{n-1}$ and
$$\phi: X_0 = G/K \times S^{n-2} \to X_1$$
given by collapsing S^{n-2} to a point ($n \geq 1$, $S^{-1} := \emptyset$).

(b) That $\tau(X)$ is an isomorphism for any finite G-CW-complex X now follows by induction on the skeletons of X using again GCT I and GCT II, the five lemma, and part (a). □

(1.1.4) Remarks
(1) If one assumes that the cohomology theories considered in (1.1.3) are strongly additive (i.e. $h_G^*(\coprod_\nu X_\nu) = \prod_\nu h_G^*(X_\nu)$ for any disjoint union $\coprod_\nu X_\nu$, and similarly for k_G^*), then Milnor's $\lim^1$-argument (see [Milnor, 1962]) together with the above gives that $\tau(X)$ is an isomorphism for any G-CW-complex. But although many equivariant cohomology theories are strongly additive, some of those we are going to consider in the next paragraph are not. As we shall point out this reflects the fact that P.A. Smith theory needs some kind of finiteness assumption.

(2) Using the same argument as above one obtains that $\tau(X)$ is an isomorphism for any finite-dimensional G-CW-complex X if $\tau(\coprod_\nu G/K_\nu)$ is an isomorphism for any disjoint union $\coprod_\nu G/K_\nu$ of G-points. If only finitely many orbit types G/K occur in X it suffices that $\tau(\coprod G/K)$ is an isomorphism, where $\coprod G/K$ denotes an arbitrary disjoint union of G-points of the same type G/K (see, e.g., (1.3.5), (1.4.5) and (3.1.6)).

(3) The above results obviously extend to G-spaces that are G-homotopy equivalent to finite, resp. finite-dimensional, G-CW-complexes. This holds, in fact, for almost all the results we are going to present, i.e. the assumption 'X a G-CW-complex' can usually be replaced by 'X G-homotopy equivalent to a G-CW-complex' for obvious reasons.

(1.1.5) Remark
It is left to the interested reader to formulate and prove the analogous results for equivariant homology theories.

For more sophisticated versions of the comparison theorem in the equivariant context see [Seymour, 1982].

1.2 The Borel Construction

A. Borel [Borel et al., 1960] introduced the following method to study the cohomology of G-spaces. It has become a basic tool in the study of transformation groups.

Let X be a G-space and EG the universal free G-space (G a compact Lie group) then $X_G := EG \times_G X$ - the orbit space of the diagonal action on the product $EG \times X$ (where EG and X are assumed to be left G-spaces) - is the total space of the bundle $X \to X_G \to BG$ associated to the universal principle bundle $G \to EG \to BG$ ($BG := EG/G$ the classifying space of G). The cohomology $H^*(X_G)$ of X_G, viewed as a functor in the variable X can be considered as an equivariant cohomology theory in the sense of Section 1.1. For certain groups G, in particular for tori and p-tori, and appropriate coefficients, the $H^*(BG)$-module structure of $H^*(X_G)$, given by the map $H^*(BG) = H^*(*_G) \to H^*(X_G)$ induced by the projection $X_G = EG \times_G X \to EG \times_G * = BG$ ($*$ denotes a one-point space) and by the cup-product, strongly reflects relations between the (ordinary) cohomology of the G-space X and its fixed point set. Most of what we discuss in this book is concerned with exploiting this connection.

In Chapter 1 we restrict ourselves to finite groups if not explicitly noted otherwise; in fact we mainly consider p-tori, and we give a more algebraic description of the above equivariant cohomology which provides certain cochain models that can be used very effectively to study the relation between the cohomology $H^*(X^K)$ of the different fixed-point sets X^K, $K < G$, of the G-space X (including in particular the fixed point set X^G and the whole space $X = X^{\{1\}}$, $\{1\} < G$). For a concise treatment of the fundamental topological properties of the Borel construction see, e.g., [tom Dieck, 1987]. (The reader should realize that the Borel construction is well-defined only up to homotopy, since the universal free G-space is only well-defined up to G-homotopy. But applying the homotopy invariant functor $H^*(-)$ to the Borel construction X_G on the G-space X gives (up to natural equivalence) a well-defined equivariant cohomology theory $H^*_G(X) = H^*(X_G)$.)

Let $\mathcal{E}_*(G) := W_*(EG; k)$ be the cellular chain complex of the universal free G-CW-complex (see Section 1.1) with coefficients in a principal ideal domain k. (We are mainly concerned with $k = \mathbb{Z}$ or k a field and we are going to suppress the coefficients from the notation when there is no danger of confusion. If necessary we will indicate them by writing $\mathcal{E}_*(G; k)$ instead of $\mathcal{E}_*(G)$.) The complex $\mathcal{E}_*(G)$ inherits a G-structure from EG, i.e. can be considered as a complex over the group ring $k[G]$ of

1.2. The Borel construction

G (recall that G is assumed to be finite). In fact $\mathcal{E}_*(G)$ is a free acyclic resolution of the trivial G-module k over the ring $k[G]$ (see, e.g., [Brown, K.S., 1982]).

(1.2.1) Definition
For a chain complex C_* (resp. a cochain complex C^*) over $k[G]$, where G is a finite group, we define:
(1) $\beta_*^G(C_*) := \mathcal{E}_*(G) \otimes_{k[G]} C_*$, $(\beta_*^G(C_*))_n := \bigoplus_i \mathcal{E}_i(G) \otimes_{k[G]} C_{n-i}$)
(2) $\beta_G^*(C^*) := \text{Hom}_{k[G]}(\mathcal{E}_*(G), C^*)$, $(\beta_G^*(C^*)^n$
$:= \prod_i \text{Hom}_{k[G]}(E_i(G), C^{n-i}))$
Here we assume that C_* is a left $k[G]$-module and C^* is a right $k[G]$-module and we transform the left G-action on EG to a right G-action by defining $eg := g^{-1}e$ for $e \in EG$ and $g \in G$, and take the induced right $k[G]$-module structure on $\mathcal{E}_*(G)$.

Note that
$$\beta_G^*(C^*) \cong \text{Hom}_k(\beta_*^G(C_*), k) \quad \text{if } C^* = \text{Hom}_k(C_*, k).$$

$H_*^G(C_*) := H(\beta_*^G(C_*))$, (resp. $H_G^*(C^*) := H(\beta_G^*(C^*))$ are called the homology (resp. cohomology) of G with coefficients C_* (resp. C^*) (see, e.g., [Brown, K.S., 1982], Chapter VII). In particular, if C_* (resp. C^*) is just a $k[G]$-module (concentrated in degree zero) then $H(\beta_*^G(C_*))$ (resp. $H(\beta_G^*(C^*))$) is the usual algebraically defined homology (resp. cohomology) of the group G with coefficients in the respective module.

For a $k[G]$-complex C^* which is bounded below (i.e. $C^n = 0$ for $n \leq n_0$) we give another description of $\beta_G^*(C^*)$. Let
$$\mathcal{E}^*(G) := \text{Hom}_{k[G]}(\mathcal{E}_*(G), k[G])$$
denote the dual of $\mathcal{E}_*(G)$ over $k[G]$. We consider $\mathcal{E}_*(G)$ as a right $k[G]$-module as before and $\mathcal{E}^*(G)$ as a left $k[G]$-module via $(g\phi)(e) := \phi(eg)$.

(1.2.2) Proposition
For any cochain complex C^* over $k[G]$, which is bounded below, there is a natural isomorphism of cochain complexes over k
$$\Phi : C^* \otimes_{k[G]} \mathcal{E}^*(G) \to \text{Hom}_{k[G]}(\mathcal{E}_*(G), C^*)$$
given by $c \otimes \phi \mapsto \Phi_{c \otimes \phi}$, where $\Phi_{c \otimes \phi}(e) := c\phi(e)$,
$$c \in C^*, \phi \in \mathcal{E}^*(G), e \in \mathcal{E}_*(G).$$

Proof.
It is easy to check (left to the reader) that the map Φ is compatible with the respective coboundaries (defined in the usual way; see, e.g., [Dold, 1980] or [Brown, K.S., 1982]). Since $\mathcal{E}^*(G)$ is a free $k[G]$-module,

and C^* and $\mathcal{E}_*(G)$ are bounded below it suffices to show that the corresponding map is an isomorphism if $\mathcal{E}^*(G)$ is replaced by $k[G]$. But this is immediate. □

It is clear from the definition that the construction $\beta_*^G(-)$ (resp. $\beta_G^*(-)$) gives an additive functor $\beta_*^G : \partial gk[G]$-Mod $\to \partial gk$-Mod (resp. $\beta_G^* : \delta gk[G]$-Mod $\to \delta gk$-Mod) from the category of chain complexes (resp. cochain complexes) over $k[G]$ to the category of chain complexes (resp. cochain complexes) over k.

(1.2.3) Proposition
The functor $\beta_*^G : \partial gk[G]$-Mod $\to \partial gk$-Mod has the following properties:

(1) *β_*^G preserves homotopies, if $f_0, f_1 : C_* \to C'_*$ are chain homotopic in $\partial gk[G]$-Mod then $\beta_*^G(f_0)$, $\beta_*^G(f_1)$ are chain homotopic in ∂gk-Mod.*

(2) *β_*^G is exact, i.e. if $C' \to C \to C''$ is an exact sequence in $\partial gk[G]$-Mod, then $\beta_*^G(C') \to \beta_*^G(C) \to \beta_*^G(C'')$ is exact in ∂gk-Mod.*

(3) *If $f : C \to C'$ is a morphism in $\partial gk[G]$-Mod which is a chain homotopy equivalence in ∂gk-Mod (forgetting the G-structure), then $\beta_*^G(f) : \beta_*^G(C) \to \beta_*^G(C')$ is a homotopy equivalence in ∂gk-Mod.*

Proof.
(1) if $s : C_* \to C_{*+1}$ is a chain homotopy between f_0 and f_1 in $\partial gk[G]$-Mod, i.e. s is a $k[G]$-module morphism such that $\partial s + s \partial = f_0 - f_1$, then $\beta_*^G(s) = \mathrm{id}_{\mathcal{E}_*(G)} \otimes_{k[G]} s$ is a homotopy (in ∂gk-Mod) between $\beta_*^G(f_0)$ and $\beta_*^G(f_1)$.

(2) Since $\mathcal{E}_*(G)$ is free as a $k[G]$-module the functor $\mathcal{E}_*(G) \otimes_{k[G]} -$ is exact (i.e. $\mathcal{E}_*(G)$ is a flat $k[G]$-module).

(3) Since the mapping cone of $\beta_*^G(f)$ is given by $\beta_*^G(C(f))$, where $C(f)$ is the mapping cone of f, it suffices to show (see (B.1.7)) that $\beta_*^C(N)$ is ht-trivial if N is. If we filter $\mathcal{E}_*(G)$ by degree, i.e. $\mathcal{F}_q(\mathcal{E}_*(G)) = \bigoplus_{i=0}^q \mathcal{E}_i(G)$ we get a filtration of $\mathcal{E}_*(G) \otimes_{k[G]} N$, namely

$$0 = \mathcal{F}_{-1}(\mathcal{E}_*(G)) \otimes_{k[G]} N \subset \ldots \subset \mathcal{F}_q(\mathcal{E}_*(G)) \otimes_{k(G)} N$$
$$\subset \mathcal{F}_{q+1}(\mathcal{E}_*(G)) \otimes_{k[G]} N \subset \ldots$$

such that

$$\mathcal{E}_*(G) \otimes_{k(G)} N = \varinjlim_q \mathcal{F}_q(\mathcal{E}_*(G)) \otimes_{k[G]} N,$$

and

$$\mathcal{F}_{q+1}(\mathcal{E}_*(G)) \otimes_{k[G]} N / \mathcal{F}_q(\mathcal{E}_*(G)) \otimes_{k[G]} N \cong \mathcal{E}_q(G) \otimes_{k[G]} N$$

1.2. The Borel construction

is ht-trivial (since it is just a direct sum of copies of N). Corollary (B.1.18) therefore implies that $\mathcal{E}_*(G) \otimes N$ is ht-trivial. □

(1.2.4) Remarks
(1) The above proposition just relies on standard properties of the tensor product functor. The analogous result for the functor $\beta_G^* : \delta gk[G]$-Mod $\to \delta gk$-Mod is proved similarly using standard properties of the Hom-functor.
(2) For an object $C \in \partial gk$-Mod, k a principal ideal domain, which is free as a k-module the conditions 'C ht-trivial' and 'C hl-trivial' are equivalent (see (B.1.11)). Thus, for (1.2.3)(3), it suffices to assume that $H(f)$ is an isomorphism (instead of f being a homotopy equivalence in ∂gk-Mod) if C and C' are free as k-modules.
The following consequence of (1.2.3) will be needed in Chapter 5.

(1.2.5) Corollary
Let $A \xrightarrow{f} B \xleftarrow{g} C \xrightarrow{h} D$ be a sequence of morphisms in $\partial gk[G]$-Mod such that g and $hg^{-1}f$ are homotopy equivalences in ∂gk-Mod, where g^{-1} is a homotopy inverse of g in ∂gk-Mod. Then $\beta_*^G(h)(\beta_*^G(g))^{-1}\beta_*^G(f)$ and $\beta_*^G(g)$ are also homotopy equivalences in ∂gk-Mod. The analogous result holds for β_G^*.
(Note that g might not have a homotopy inverse in $\partial gk[G]$-Mod. Otherwise the corollary would not need any extra argument.)

Proof.
We consider Diagram 1.1 in $\partial gk[G]$-Mod in which $Z(g,h)$ is the

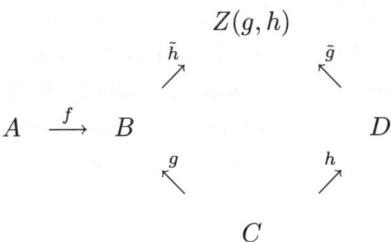

Diagram 1.1

double mapping cylinder (homotopy pushout) of $B \xleftarrow{g} C \xrightarrow{h} D$ and $\tilde{g}$, $\tilde{h}$ are the canonical inclusions. One has that $\tilde{h}g$ is homotopic to $\tilde{g}h$ in $\partial gk[G]$-Mod and $\tilde{g}$ is a homotopy equivalence in ∂gk-Mod (since $C(g) \cong Z(g,h)/D$, (see Appendix B.1)). Therefore the assumption implies that

$\tilde{h}f$ is a homotopy equivalence (and $\tilde{g}^{-1}\tilde{h}f \simeq hg^{-1}f$) in ∂gk-Mod. Applying (1.2.3) gives that $\beta_*^G(g)$, $\beta_*^G(\tilde{g})$ and $\beta_*^G(\tilde{h}f) = \beta_*^G(\tilde{h})\beta_*^G(f)$ are homotopy equivalences in ∂gk-Mod. Since $\tilde{h}g$ is homotopic to $\tilde{g}h$ in $\partial gk[G]$-Mod one gets $\beta_*^G(\tilde{h})\beta_*^G(g) = \beta_*^G(\tilde{h}g) \simeq \beta_*^G(\tilde{g})\beta_*^G(h) = \beta_*^G(\tilde{g}h)$ in ∂gk-Mod. Therefore $(\beta_*^G(\tilde{g}))^{-1}\beta_*^G(\tilde{h})\beta_*^G(f) \simeq \beta_*^G(h)(\beta_*^G(g))^{-1}\beta_*^G(f)$ is a homotopy equivalence in ∂gk-Mod. □

If X is a G-space we can apply the above construction to the singular chain complex $S_*(X;k)$ (resp. cochain complex $S^*(X;k)$) with coefficients in k. We use the following notation:

$$\beta_*^G(X;k) := \beta_*^G(S_*(X;k)), \quad H_*^G(X;k) := H(\beta_*^G(X;k))$$
$$\beta_G^*(X;k) := \beta_G^*(S^*(X;k)), \quad H_G^*(X;k) := H(\beta_G^*(X;k)).$$

(1.2.6) Theorem

The functor $H_G^*(-;k)$ (resp. $H_*^G(-;k)$) is an equivariant cohomology (resp. homology) theory in the sense of Section 1.1.

Proof.
GCT I: Since a G-homotopy between two G-maps $f_0, f_1 : X \to Y$ induces a chain homotopy over $k[G]$ between $S^*(f_0;k)$ and $S^*(f_1;k)$ and since $\beta_G^*(-)$ preserves homotopies (see (1.2.3)(1)) the functor

$$H_G^*(-;k) = H(\beta_G^*(S^*(-;k)))$$

is homotopy invariant.

GCT II: If $X = X_1 \cup_\phi X_2$, $\phi : X_0 \to X_1$ a cellular G-map, X_1, X_2 G-CW-complexes with $X_0 \subset X_2$ a G-subcomplex, then there is a short exact sequence of $k[G]$-cochain complexes

$$0 \to S^*(\{X_1, X_2\};k) \to S^*(X_1;k) \times S^*(X_2;k) \to S^*(X_0;k) \to 0$$

and a morphism $S^*(X;k) \to S^*(\{X_1,X_2\};k)$ in $\delta gk[G]$-Mod, which is a homotopy equivalence in δgk-Mod (cf. [Dold, 1980], III.8 and VI.7.6). The desired long exact sequence for $H_G^*(-;k)$ is now obtained by applying (1.2.3)(2), (3). The proof for homology is similar. □

(1.2.7) Remarks
(1) For the proof of GCT I we do not need to assume that the G-spaces involved are G-CW-complexes. Analogous to the non-equivariant situation the long exact sequence of GCT II also exists under much more general assumptions, e.g., it would suffice to assume that $X_0 \to X_2$ is a G-cofibration (and X_0, X_1, X_2 are arbitrary G-spaces). Equivariant cohomology theories on more general spaces than G-CW-complexes are discussed in Chapter 3.

1.2. The Borel construction

(2) For a G-CW-complex X the cellular $k[G]$-complex $W_*(X;k)$ may be used instead of the singular $k[G]$-complex $S_*(X;k)$ to define $H_*^G(X;k)$ (resp. $H_G^*(X;k)$), i.e. $H(\beta_*^G(S_*(X;k)))$ and $H(\beta_*^G(W_*(X;k)))$ (resp. $H(\beta_G^*(S^*(X;k)))$ and $H(\beta_G^*(W^*(X;k)))$) are naturally isomorphic.

Proof of part (2).

Let $X^0 \subset X^1 \subset \ldots X^{n-1} \subset X^n \subset \ldots$ be the skeletal filtration of X. Define the subcomplex $C_*(X;k)$ of $S_*(X;k)$ by $C_n(X;k) := \{c \in S_n(X^n;k); \partial c \in S_{n-1}(X^{n-1};k)\}$, where ∂ is the boundary in $S_*(X;k)$. Then $C_*(X;k)$ is a $k[G]$-subcomplex of $S_*(X;k)$, and one also has a $k[G]$-chain map $C_*(X;k) \xrightarrow{p} W_*(X;k)$, which maps the element $c \in C_n(X;k)$ to its class in $H_n(X^n, X^{n-1};k) \cong W_n(X;k)$. We claim that both maps - the inclusion $C_*(X;k) \xrightarrow{i} S_*(X;k)$ and the projection $C_*(X;k) \xrightarrow{p} W_*(X;k)$ - are homotopy equivalences over k, i.e. induce isomorphism in homology.

Since any class in $H_n(S_*(X;k))$ can be represented by a cycle in $S_n(X^n;k)$ (and hence in $C_n(X;k)$), i_* is surjective. On the other hand if $c \in C_n(X;k)$ is a cycle in $C_*(X;k)$ such that there exists $b \in S_{n+1}(X;k)$ with $\partial b = c$ (i.e. $i_*[c] = 0 \in H_n(S_*(X;k))$), then one can find an element $b' \in S_{n+1}(X^{n+1};k)$ such that $\partial b' = \partial b = c$. Clearly $b' \in C_{n+1}(X;k)$ and hence $[c] = 0$ in $H_n(C_*(X;k))$. So i_* is also injective.

We now show that p_* is an isomorphism, too. If $c \in S_n(X^n;k)$ represents a cycle in

$$W_n(X,k) = H_n(X^n, X^{n-1};k) = H_n(S_*(X^n;k)/S_*(X^{n-1};k)),$$

then $\partial c \in S_{n-1}(X^{n-2};k) + \partial(S_n(X^{n-1};k))$. Let $\partial c = a' + \partial b''$, with $a' \in S_{n-1}(X^{n-2};k)$, $b'' \in S_n(X^{n-1};k)$. Since $H_{n-1}(S_*(X^{n-2};k)) = 0$ and $\partial a' = 0$, there exists $b' \in S_n(X^{n-2};k)$ such that $\partial b' = a'$. The element $\tilde{c} = c - b' - b'' \in S_n(X^n;k)$ is a cycle in $C_n(X^n;k)$ and represents the same element as c in $W_n(X;k)$, i.e. p_* is surjective. Now assume that $c \in C_n(X^n;k)$ is a cycle in $C_*(X;k)$ which is mapped to zero in $H(W_*(X;k))$, i.e. there exist elements $b' \in S_{n+1}(X^{n+1};k)$ and $a'' \in S_n(X^{n-1};k)$, such that $\partial b' = c + a''$. Since $\partial c = 0$ one also has $\partial a'' = 0$. But $H_n(S_*(X^{n-1};k)) = 0$, and hence there exists an element $b'' \in S_{n+1}(X^{n-1};k)$ such that $\partial b'' = a''$. Therefore $\partial(b' - b'') = 0$, and $b' - b'' \in C_{n+1}(X;k)$, i.e. $[c] = 0$ in $H_n(C_*(X;k))$, and p_* is injective.

This proves the above claim. Since i and p are morphisms in $\partial_g k[G]$-Mod, part (2) now follows from Proposition (1.2.3)(3). $\square$

As already noted Borel's method of studying transformation groups relies essentially on the multiplicative structure of the equivariant cohomology, in particular on the $H^*(BG)$-module structure of $H^*(EG\times_G X)$. We now describe this structure for the algebraic version of the Borel construction.

The complex $\mathcal{E}_*(G)$ can be equipped with a diagonal
$$\triangle: \mathcal{E}_*(G) \to \mathcal{E}_*(G) \otimes \mathcal{E}_*(G),$$
which is a morphism in $\partial gk[G]$-Mod, where $\mathcal{E}_*(G) \otimes \mathcal{E}_*(G)$ carries the diagonal action. This diagonal can be thought of either as induced (on the cellular complex $\mathcal{E}_*(G) = W_*(EG;k)$) by the (topological) diagonal of the space EG or as a lifting of the isomorphism $k \to k \otimes k$ to the respective acyclic free resolutions $\mathcal{E}_*(G) \to k$, resp. $\mathcal{E}_*(G) \otimes \mathcal{E}_*(G) \to k \otimes k$. It is determined uniquely up to homotopy in $\partial gk[G]$-Mod.

For any cochain complex C^* over $k[G]$ this diagonal induces a product
$$\beta_G^*(C^*) \otimes \beta_G^*(k) \to \beta_G^*(C^*)$$
defined by
$$\phi \otimes \psi \mapsto \phi \cup \psi,$$
where
$$(\phi \cup \psi)(e) = \eta((\phi \otimes \psi)(\triangle e)) \text{ for } e \in \mathcal{E}_*(G),$$
$\phi \in \beta_G^*(C^*) = \text{Hom}_{k[G]}(\mathcal{E}_*(G), C^*), \psi \in \beta_G^*(k) = \text{Hom}_{k[G]}(\mathcal{E}_*(G), k)$
and
$$\eta: C^* \otimes k \to C^*$$
the k-module structure of C^*. This product induces a map
$$H_G^*(C^*) \otimes H_G^*(k) \to H_G^*(C^*)$$
which gives a (right) $H_G^*(k)$-module structure on $H_G^*(C^*)$. Of course, one gets a left $H_G^*(k)$-module structure on $H_G^*(C^*)$ in an analogous way. (Note that - since $\triangle: \mathcal{E}_*(G) \to \mathcal{E}_*(G) \otimes \mathcal{E}_*(G)$ is commutative only up to homotopy - the left and right $H_G^*(k)$-module structures on $H_G^*(C^*)$ coincide (up to the usual sign) on the cohomology level, but only up to homotopy on the cochain level.) In particular, for $C^* = k$ we get the usual cup-product on $H_G^*(k) \cong H^*(EG \times_G *; k) = H^*(BG; k)$.

If C^* has a product $\mu: C^* \otimes C^* \to C^*$ (where μ is a morphism in $\delta gk[G]$-Mod, with $C^* \otimes C^*$ carrying the diagonal action), then in a similar way to that used above we get a product
$$\beta_G^*(C^*) \otimes \beta_G^*(C^*) \to \beta_G^*(C^*),$$
where
$$\phi_1 \otimes \phi_2 \mapsto \phi_1 \cup \phi_2$$
$$(\phi_1 \cup \phi_2)(e) = \mu((\phi_1 \otimes \phi_2)(\triangle e)).$$

If $C^* = S^*(X;k)$ is the singular cochain complex of a G-space X with the product $\mu : S^*(X;k) \otimes S^*(X;k) \to S^*(X;k)$ induced by the diagonal $X \to X \times X$ of X, then one gets an induced multiplication on $H(\beta_G^*(S^*(X;k))) = H_G^*(X;k)$.

Although in the case of a finite group G we mainly work with the more algebraic description of equivariant (co)homology given above, it is useful to know that this theory coincides with the one defined by the (topological) Borel construction. The latter has to be used in case of a compact Lie group G which is not finite.

(1.2.8) Theorem
The (co)homology theories $H_*^G(-;k)$ and $H_*(EG \times_G -;k)$ (resp. $H_G^*(-;k)$ and $H^*(EG \times_G -;k)$) are naturally isomorphic on the category of G-spaces, when G is a finite group.

(Note that for a point, as G-space, one has the classical result, that the algebraically defined (co)homology of the group G, using resolutions of the trivial $k[G]$-module k over $k[G]$, is equivalent to the (co)homology of the classifying space BG.)

Proof.
We first consider the homology theories. For a G-space X, $H_*^G(X;k)$ is defined as $H(\beta_*^G(X;k))$, where $\beta_*^G(X;k) := \mathcal{E}_*(G;k) \otimes_{k[G]} S_*(X;k)$. Since $\mathcal{E}_*(G;k) = W_*(EG;k)$ one can replace the cellular chain complex $W_*(EG;k)$ of EG by the singular chain complex $S_*(EG;k)$; both can be considered as free resolutions of the trivial $k[G]$-module k and hence are chain homotopy equivalent over $k[G]$. One can choose natural Eilenberg-Zilber maps
$$S_*(EG;k) \otimes S_*(X;k) \overset{\Phi}{\underset{\Psi}{\rightleftarrows}} S_*(EG \times X;k)$$
such that $\Psi\Phi$ and $\Phi\Psi$ are naturally homotopic to the (respective) identities (see, e.g., [Dold, 1980]VI. 12). Hence Φ, Ψ and the homotopies are compatible with the diagonal action of G, and thus induce a pair of chain homotopy equivalences over k
$$S_*(EG;k) \otimes_{k[G]} S_*(X;k) \overset{\bar\Phi}{\underset{\bar\Psi}{\rightleftarrows}} k \otimes_{k[G]} S_*(EG \times X;k).$$
On the other hand, since $\pi : EG \times X \to EG \times_G X$ is a covering with deck transformation group G, the map
$$\bar\Theta : k \otimes_{k[G]} S_*(EG \times X;k) \to S_*(EG \times_G X;k),$$
induced by $S_*(\pi)$, is an isomorphism. ($\bar\Theta$ is injective since two singular simplices $\sigma_q^1, \sigma_q^2 : \triangle_q \to EG \times X$ coincide when composed with π if and

only if there exists a $g \in G$ such that $g\sigma_q^1 = \sigma_q^2$. The map $\bar\Theta$ is surjective since any singular simplex $\sigma_q : \triangle_q \to EG \times_G X$ can be lifted to $EG \times X$, for $\triangle_q$ is contractible and hence the induced covering of $\triangle_q$ is trivial.)

The composition $\bar\Theta\bar\Phi$ induces the desired natural isomorphism

$$H_*^G(X;k) \cong H_*(EG \times_G X; k).$$

The argument for cohomology is similar. We only need to check in addition that the isomorphism is compatible with the respective products.

In the definition of $\beta_G^*(S^*(X;k)) = \operatorname{Hom}_{k[G]}(\mathcal{E}_*(G;k), S^*(X;k))$ we can again replace $\mathcal{E}_*(G;k) = W_*(EG;k)$ by $S_*(EG;k)$, including the diagonal, which on $S_*(EG;k)$ is given by the geometric diagonal and an Eilenberg-Zilber map $S_*(EG \times EG; k) \to S_*(EG;k) \otimes S_*(EG;k)$. One has a natural isomorphism

$$\operatorname{Hom}_{k[G]}(S_*(EG;k), S^*(X;k))$$
$$= \operatorname{Hom}_{k[G]}(S_*(EG;k), \operatorname{Hom}(S_*(X;k), k))$$
$$\cong \operatorname{Hom}(S_*(EG;k) \otimes_{k[G]} S_*(X;k), k).$$

The product on $H_G^*(X;k) = H(\beta_G^*(S^*(X;k)))$ is induced by the 'tensor product diagonal' on $S_*(EG;k) \otimes S_*(X;k)$, i.e. taking the diagonal on each factor, followed by the appropriate twist of the middle factors in the image, while the product on $H^*(EG \times_G X; k)$ is induced by the diagonal of $S_*(EG \times X; k)$. Since Diagram 1.2, where we abbreviate $S_*(EG \times X)$ by S_*^{EGX},

$$\begin{array}{ccc}
S_*(EG) \otimes S_*(X) & \xrightarrow{\Phi} & S_*^{EGX} \\
\downarrow{\scriptstyle \triangle_{EG} \otimes \triangle_X} & & \downarrow{\scriptstyle \triangle_{EGX}} \\
S_*(EG) \otimes S_*(EG) \otimes S_*(X) \otimes S_*(X) & & \\
\downarrow{\scriptstyle \tau} & & \\
(S_*(EG) \otimes S_*(X)) \otimes (S_*(EG) \otimes S_*(X)) & \xrightarrow{\Phi \otimes \Phi} & S_*^{EGX} \otimes S_*^{EGX}
\end{array}$$

Diagram 1.2

is naturally homotopy commutative, one gets the induced homotopy commutative Diagram 1.3, where $S_*(EG;k) \otimes_{k[G]} S_*(X;k)$ is abbreviated by S_*^2,

1.2. The Borel construction

$$\begin{array}{ccc}
S_*^2 & \xrightarrow{\bar{\Theta}\bar{\Phi}} & S_*(EG \times_G X; k) \\
\downarrow & & \downarrow \triangle_{EG \times_G X} \\
(S_*^2) \otimes (S_*^2) & \xrightarrow{(\bar{\Theta}\bar{\Phi}) \otimes (\bar{\Theta}\bar{\Phi})} & S_*(EG \times_G X; k) \otimes S_*(EG \times_G X; k),
\end{array}$$

Diagram 1.3

where the left vertical map gives the product in $H_G^*(X; k)$. Hence the natural isomorphism $H^*(EG \times_G X; k) \cong H_G^*(X; k)$, induced by the dual of $\bar{\Theta}\bar{\Phi}$, is compatible with the respective multiplications. □

(1.2.9) Remarks

(1) So far we have viewed the construction $H_*^G(X)$ (resp. $H_G^*(X)$) as a functor in the variable X. It is possible to consider these constructions also as functors in the variable G or even the pair (G, X), more precisely: If $a : K \to G$ is a group homomorphism and $f : Y \to X$ is a map of a K-space Y into a G-space X such that $f(ky) = a(k)f(y)$ for all $y \in Y$, $k \in K$, then there is a (well-defined) induced morphism $H_*^K(Y) \to H_*^G(X)$ with the expected functorial properties. For the algebraic version of the equivariant (co)homology this morphism is obtained from a map of chain complexes (unique up to homotopy)

$$\mathcal{E}_*(a) \otimes S_*(f) : \mathcal{E}_*(K) \otimes_{\mathbf{Z}[K]} S_*(Y) \to \mathcal{E}_*(G) \otimes_{\mathbf{Z}[G]} S_*(X),$$

where the map $\mathcal{E}_*(a) : \mathcal{E}_*(K) \to \mathcal{E}_*(G)$ is a map of resolutions corresponding to $a : K \to G$.

One gets a corresponding map $E(a) \times f : EK \times_K Y \to EG \times_G X$ of the Borel constructions, where $E(a) : EK \to EG$ is a map on the universal free spaces induced by $a : K \to G$. The map $H_*^K(Y) \to H_*^G(X)$ can be written as a composition

$$H_*^K(Y) \to H_*^K(\text{res}_a X) \to H_*^G(X),$$

where $\text{res}_a X$ denotes the space X, viewed as a K-space via a. The first map is induced by the K-map $f : Y \to \text{res}_a X$, and the second by the pair $a : K \to G$, id: $\text{res}_a X \to X$.

The most interesting special case is the situation where $a : K \to G$ is the inclusion of a subgroup and f is the identity, i.e. the G-space X is viewed as a K-space by restricting the action to the subgroup K.

On the other hand, if X is K-space and $K < G$ a subgroup then $G \times_K X$ is a G-space. It is easy to see that $H^G_*(G \times_K X; k)$ is naturally isomorphic to $H^K_*(X; k)$; for the algebraic version one uses the chain homotopy equivalences

$$\mathcal{E}_*(G; k) \otimes_{k[G]} S_*(G \times_K X; k) \simeq \mathcal{E}_*(G; k) \otimes_{k[G]} (k[G] \otimes_{k[K]} S_*(X; k))$$
$$\simeq \mathcal{E}_*(G; k) \otimes_{k[K]} S_*(X; k)$$
$$\simeq \mathcal{E}_*(K; k) \otimes_{k[K]} S_*(X; k).$$

Note that $\mathcal{E}_*(G; k)$ viewed as $k[K]$-complex by restriction is a resolution over $k[K]$ of the trivial $k[K]$-module k; hence $\mathcal{E}_*(G; k)$ and $\mathcal{E}_*(K; k)$ are homotopy equivalent over $k[K]$. Similarly for the topological construction one has homotopy equivalences $EG \times_G (G \times_K X) \simeq EG \times_K X \simeq EK \times_K X$, using the fact that EG, considered as a K-space by restriction, is a universal free K-space, i.e. EG and EK are K-homotopy equivalent.

In particular one has: $H^G_*((G/K) \times X; k) \cong H^K_*(X; k)$ if X is a G-space, since $(g, x) \mapsto (g, gx)$ induces a homeomorphism $G \times_K X \to (G/K) \times X$. The corresponding fact in cohomology will be used later in certain places, e.g. in case X is a trivial G-space. For X a point, one gets

$$H^G_*(G/K; k) \cong H^K_*(*; k) \cong H_*(BK; k).$$

(2) If X is a paracompact free G-space then the projection $p : X \to X/G$ is a principal G-bundle and $\psi : EG \times_G X \to X/G$ is a fibre bundle with contractible fibre EG (see, e.g., [tom Dieck, 1987]). Hence by the long exact homotopy sequence, ψ is a weak homotopy equivalence; in fact, it is a homotopy equivalence (see [Dold, 1963] or [tom Dieck, 1987]). In particular $\psi_* : H_*(EG \times_G X) \to H_*(X/G)$ is an isomorphism (cf. Exercise (1.6) in the case X is a G-CW-complex).

In the case of a finite group G one can use a more algebraic argument to obtain the isomorphism $H^G_*(X) \cong H_*(X/G)$ for a paracompact free G-space X: as in the proof of (1.2.8) one has that $S_*(X/G; k)$ is isomorphic to $k \otimes_{k[G]} S_*(X; k)$. The augmentation $\varepsilon : \mathcal{E}_*(G; k) \to k$ induces a morphism $\phi : \mathcal{E}_*(G; k) \otimes_{k[G]} S_*(X; k) \to k \otimes S_*(X; k)$, which actually corresponds to the map $\psi : EG \times_G X \to X/G$ above. We claim that ϕ is a homotopy equivalence of chain complexes over k. Define a filtration on $S_*(X; k)$ by $\mathcal{F}_q(S_*(X; k)) := \bigoplus_{i \leq q} S_i(X; k)$. Then the $\mathcal{F}_q(S_*(X; k))$ are free $k[G]$-subcomplexes of $S_*(X; k)$. The induced map on the successive quotients

$$\bar{\phi} : \mathcal{E}_*(G; k) \otimes_{k[G]} \mathcal{F}_q(S_*(X; k))/\mathcal{E}_*(G; k) \otimes_{k[G]} \mathcal{F}_{q-1}(S_*(X; k))$$
$$\to k \otimes_{k[G]} \mathcal{F}_q(S_*(X; k))/k \otimes_{k[G]} \mathcal{F}_{q-1}(S_*(X; k))$$

is a homotopy equivalence over k, since $\mathcal{F}_q(S_*(X;k))/\mathcal{F}_{q-1}(S_*(X;k))$ is free over $k[G]$ (with trivial boundary). Hence ϕ itself is a homotopy equivalence over k (cf. (B.1.18)). (A reader who is familiar with spectral sequences - which we do not want to use in this chapter - might prefer the corresponding spectral sequence argument.) See also (3.10.9) for generalizations. Analogous results hold for cohomology.

(1.2.10) Remark
With the usual techniques we can extend the equivariant (co)homology $H^G_*(-)$ (resp. $H^*_G(-)$) to pairs of G-spaces, e.g, for a pair of G-spaces (X, A) we define: $H^G_*(X, A) := H(\mathcal{E}_*(G) \otimes_{\mathbb{Z}[G]} (S_*(X)/S_*(A)))$. The formal properties in the non-equivariant situation (see, e.g., [Dold, 1980] for a comprehensive treatment) can be copied; in particular one gets boundary maps, long exact sequences of pairs, (relative) Mayer-Vietoris sequences, excision etc..

As a k-module (i.e. disregarding differentials) $\beta^G_*(C_*) = \mathcal{E}_*(G) \otimes_{k[G]} C_*$ ($C_* \in \partial g k[G]$-Mod) is isomorphic to $W_*(BG) \otimes C_*$, since $\mathcal{E}_*(G) = W_*(EG) \cong W_*(BG) \otimes k[G]$ (as $k[G]$-modules). But the differential on $\mathcal{E}_*(G) \otimes_{k[G]} C_*$ does not correspond, in general, to the usual tensor product differential on $W_*(BG) \otimes C_*$. We therefore denote the complex $W_*(BG) \otimes C_*$ with the differential inherited from $\beta^G_*(C_*)$ by $W_*(BG)\tilde{\otimes}C_*$. In particular, if X is a G-space and $C_* = S_*(X;k)$ we get a 'twisted tensor product' $W_*(BG)\tilde{\otimes}S_*(X;k)$ which gives the homology of the Borel construction X_G. (Co)chain complexes of this type have been introduced by G.Hirsch [1954] and E.H.Brown [1959] to study the (co)homology of certain fibre spaces. We therefore call the complex $W_*(BG)\tilde{\otimes}S_*(X;k)$ a Hirsch-Brown model for the homology of the Borel construction on X.

The situation in cohomology is similar, i.e. one gets a Hirsch-Brown model of the form $S^*(X;k)\tilde{\otimes}W^*(BG)$ for the cohomology of the Borel construction on X (cf. (1.2.2)). For G a p-torus these models are studied in detail in the course of the book (see, in particular, Sections 1.3, 1.4, 3.11, cf. Sections 3.5, 5.2). We will sometimes change the notation to $W^*(BG)\tilde{\otimes}S^*(X;k)$, switching the factors in the tensor product (cf. also Section B.2), to have it closer to the commonly used notation for the Leray-Serre spectral sequence of the Borel fibration $X \to X_G \to BG$. But if we want to emphasize the 'deformation theoretical' aspects of this construction, as in the case, for example, in this chapter, we keep the first notation.

1.3 The Borel construction for 2-tori

In this section we describe in detail the algebraic version of the Borel construction, i.e. the Hirsch-Brown model, in the case where G is a 2-torus of rank r, i.e. $G \cong (\mathbb{Z}_2)^r$. It turns out that in this special situation the $H_G^*(k)$-module structure of $H_G^*(C^*)$ can already be represented by a ('strict', i.e. not only 'up to homotopy') $H_G^*(k)$-module structure on $\beta_G^*(C^*)$ if k is a field of characteristic 2. This gives the possibility of describing the (non-equivariant) cohomology

$$H^{(*)}(X^K; k) := \bigoplus_{i=0}^{\infty} H^i(X^K; k)$$

of the fixed point set $X^K := \{x \in X; gx = x \text{ for all } g \in K\}$ of any subgroup $K < G$ as the cohomology of $\beta_G^*(X; k)$ with certain $H_G^*(k)$-modules (depending on K) as coefficients, where X is a suitable G-space. It is easy to derive several results of 'P.A. Smith type' from this description of the cohomology of the fixed point set.

We first consider the Borel construction for $G = \mathbb{Z}_2$. The cellular chain complex $\mathcal{E}_*(G) := W_*(EG)$ of the G-CW-complex EG described in Section 1.1 is a free chain complex over the group ring $\mathbb{Z}[G]$ generated by one element w_n in each dimension $n = 0, 1, \ldots$, i.e. $\mathcal{E}_*(G) \cong W_* \otimes \mathbb{Z}[G]$ as $\mathbb{Z}[G]$-modules, where W_* is freely generated by $\{w_n, n = 0, 1, \ldots\}$ as a $\mathbb{Z}$-module. In terms of the left action of $\mathbb{Z}[G]$ on $\mathcal{E}_*(G)$ the boundary is given by

$$\partial w_n = \begin{cases} (1+g)w_{n-1} & \text{if } n \text{ is even } (\partial w_0 = 0) \\ (1-g)w_{n-1} & \text{if } n \text{ is odd} \end{cases}$$

where 1 denotes the unit element and g the generator in $G = \mathbb{Z}_2$.

We therefore get the following result:

(1.3.1) Proposition

(1)
$$H_n(BG; \mathbb{Z}) \cong H_n(\mathcal{E}_*(G) \otimes_{\mathbb{Z}[G]} \mathbb{Z}) \cong \begin{cases} \mathbb{Z} & \text{if } n = 0 \\ \mathbb{Z}_2 & \text{if } n \text{ is odd} \\ 0 & \text{if } n \text{ is even}, n \neq 0 \end{cases}$$

(2) *If k is a field of characteristic 2, then*

$$H_n(BG; k) \cong H_n(\mathcal{E}_*(G; k) \otimes_{k[G]} k) \cong k$$

for all $n \geq 0$.

The cohomology of BG as a k-module can easily be calculated by taking the dual complexes (or by using the Universal Coefficient Formula). But

1.3. The Borel construction for 2-tori

to get the multiplicative structure on $H^*(BG;k)$ we need a diagonal $\Delta : \mathcal{E}_*(G) \to \mathcal{E}_*(G) \otimes \mathcal{E}_*(G)$ on $\mathcal{E}_*(G)$.

This can be given by

$$\Delta(w_n) = \sum_{p+q=n} \Delta_{p,q},$$

where

$$\Delta_{p,q} = \begin{cases} w_p \otimes w_q & \text{if } p \text{ is even} \\ w_p \otimes gw_q & \text{if } p \text{ is odd and } q \text{ is even} \\ -w_p \otimes gw_q & \text{if p and q are odd} \end{cases}$$

(To see this one only needs to check that Δ is a morphism of chain complexes over $\mathbb{Z}[G]$, which is a lifting of the canonical isomorphism $\mathbb{Z} \to \mathbb{Z} \otimes \mathbb{Z}$ to the respective resolutions.)

(1.3.2) Proposition
(1) $H^*(BG;\mathbb{Z}) \cong H(\text{Hom}_{\mathbb{Z}[G]}(\mathcal{E}_*(G),\mathbb{Z})) \cong \mathbb{Z}[t]/(2t)$ with $\deg(t) = 2$
(2) If k is a field of characteristic 2 then

$$H^*(BG;k) \cong H(\text{Hom}_{k[G]}(\mathcal{E}_*(G;k),k) \cong k[t]$$

with $\deg(t) = 1$.

Proof.
Let $\bar{w}^n \in \text{Hom}_{\mathbb{Z}[G]}(\mathcal{E}_*(G),\mathbb{Z})$ be the element which is defined by

$$\bar{w}^n(w_m) = \begin{cases} 1 & \text{if } n = m \\ 0 & \text{if } n \neq m \end{cases}$$

(extended to all of $\mathcal{E}_*(G) \cong W_* \otimes \mathbb{Z}[G]$ as a $\mathbb{Z}[G]$-module morphism). Then

$$\bar{w}^p \cup \bar{w}^q(w_m) = (\bar{w}^p \otimes \bar{w}^q)(\Delta w_m) = \begin{cases} 1 & \text{if } p+q = m \\ 0 & \text{if } p+q \neq m \end{cases}$$

(Note the standard sign convention is applied when evaluating $\bar{w}^p \otimes \bar{w}^q$ on $(-w_p \otimes gw_q)$.) Hence $\bar{w}^p \cup \bar{w}^q = \bar{w}^{p+q}$. In case (1) $\bar{w}^n$, for n even, represents the generator of $H^n(BG;\mathbb{Z})$ and $2\bar{w}^n \sim 0$ for $n > 0$ (n even). In case (2) $\bar{w}^n$ represents the generator of $H^n(BG;k)$ for any $n = 0, 1, \ldots$. Therefore $\bar{w}^p \cup \bar{w}^q = \bar{w}^{p+q}$ implies the desired result. $\square$

(1.3.3) Remark
The proof of (1.3.2) shows that,

$$\beta_G^*(k) = \text{Hom}_{k[G]}(\mathcal{E}_*(G),k)$$

24   1. Equivariant cohomology and the Borel construction

with the cup-product induced by the above diagonal on the cochain level is isomorphic to $k[t]$, the polynomial algebra in one variable of degree 1 over k ($\bar{w}^i \leftrightarrow t^i$). The corresponding coboundary on $k[t]$ is the derivation given by $\delta(t) = 2t^2$. This holds for any coefficient ring k. If k is a field of characteristic 2 then the coboundary on $k[t]$ is zero.

If C^* is an object in $\delta g k[G]$-Mod, bounded below, then (by (1.2.2)) $\beta_G^*(C^*) \cong C^* \otimes_{k[G]} \mathcal{E}^*(G)$ with the usual coboundary on the tensor product.

Since
$$\mathcal{E}^*(G) = \mathrm{Hom}_{k[G]}(\mathcal{E}_*(G), k[G]) \cong \mathrm{Hom}_{k[G]}(W_* \otimes k[G], k[G])$$
$$\cong k[G] \otimes W^*$$

as $k[G]$-modules, where $W^* := \mathrm{Hom}_k(W_*, k)$, one gets an isomorphism of graded k-modules $\beta_G^*(C^*) \cong C^* \otimes_{k[G]} \mathcal{E}^*(G) \cong C^* \otimes W^*$. (Tensor products which are not specified otherwise are taken over k.) The coboundary $\tilde{\delta}$ which is induced on $C^* \otimes W^*$ through this isomorphism is given by

$$\tilde{\delta}(c \otimes w^n) = \begin{cases} (\delta c) \otimes w^n + (-1)^{|c|} c(1-g) \otimes w^{n+1} & \text{if } n \text{ is even} \\ (\delta c) \otimes w^n + (-1)^{|c|} c(1+g) \otimes w^{n+1} & \text{if } n \text{ is odd,} \end{cases}$$

where w^n is the dual of w_n, δ denotes the coboundary of C^* and $|c|$ the degree of $c \in C^*$. Note that the coboundary of $\mathcal{E}^*(G) \cong k[G] \otimes W^*$ is given by

$$\delta(1 \otimes w^n) = \begin{cases} (1-g) \otimes w^{n+1} & \text{if } n \text{ is even} \\ (1+g) \otimes w^{n+1} & \text{if } n \text{ is odd} \end{cases}$$

The above formula for $\tilde{\delta}$ then follows from the fact that $C^* \otimes_{k[G]} \mathcal{E}^*(G)$ carries the usual tensor product differential and that

$$cg \otimes (1 \otimes w) = c \otimes (g \otimes w) \text{ in } C^* \otimes_{k[G]} \mathcal{E}^*(G) = C^* \otimes_{k[G]} (k[G] \otimes W^*).$$

To indicate that $C^* \otimes W^*$ together with the above coboundary $\tilde{\delta}$ is not a tensor product of two complexes C^* and W^* we use $C^* \tilde{\otimes} W^*$ as a notation for the complex $(C^* \otimes W^*, \tilde{\delta})$.

(1.3.4) Proposition
$\beta_G^*(C^*) \cong C^* \tilde{\otimes} W^*$ is naturally isomorphic to the 'extended module' $C^* \otimes k[t]$ as a right module over $\beta_G^*(k) \cong k[t]$.

Proof.
We have to calculate the product $(c \otimes w^p) \cup \bar{w}^q$ for $c \otimes w^p \in C^* \otimes W^* \cong$

1.3. The Borel construction for 2-tori

$\beta_G^*(C^*)$ and $\bar{w}^q \in \beta_G^*(k)$. By definition

$$((c \otimes w^p) \cup \bar{w}^q)(w_m) = \eta((c \otimes w^p \otimes \bar{w}^q)(\triangle w_m))$$
$$= \begin{cases} c & \text{if } p+q = m \\ 0 & \text{if } p+q \neq m, \end{cases}$$

where $\eta : C^* \otimes k \to C^*$ is the k-module structure on C^*. Hence $(c \otimes w^p)\bar{w}^q = c \otimes w^{p+q}$, and $\Psi(c \otimes w^n) := c \otimes t^n$ gives the desired isomorphism $\Psi : C^* \otimes W^* \to C^* \otimes k[t]$ of $k[t]$-modules. $\square$

Again we denote by $C^* \tilde{\otimes} k[t]$ the complex whose underlying module is $C^* \otimes k[t]$ and which inherits the coboundary from $C^* \tilde{\otimes} W^*$ via the isomorphism Ψ. $C^* \tilde{\otimes} k[t]$ can be thought of as a differential graded module over the differential graded ring $k[t]$ (see Remark (1.3.3)), in particular the product $\beta_G^*(C^*) \otimes \beta_G^*(k) \to \beta_G^*(C^*)$ is already associative on the cochain level in the case at hand.

Note that the analogous left multiplication $\beta_G^*(k) \otimes \beta_G^*(C^*) \to \beta_G^*(C^*)$ defined by using the same diagonal on $\mathcal{E}_*(G)$ is associative only up to homotopy.

For the rest of this section we assume that k is a field of characteristic 2 if not explicitly stated otherwise. Then the above consideration shows that $C^* \tilde{\otimes} k[t]$ is an object in $\delta g k[t]$-Mod, where $k[t]$ has trivial coboundary; in particular the coboundary $\tilde{\delta} : C^* \tilde{\otimes} k[t] \to C^* \tilde{\otimes} k[t]$ is $k[t]$-linear. The elements $\tilde{c} \in C^* \otimes k[t]$ of total degree n can be thought of as polynomials in t with coefficients in C^*, which are homogeneous with respect to the total degree n, i.e. $\tilde{c} = \sum_{i=0}^{n} c^{n-i} \otimes t^i$, where $c^{n-i} \in C^{n-i}$ ('$n-i$' is just an index for c indicating the degree in C^*, while t^i should be interpreted as the i-th power of t in $k[t]$). Since $\tilde{\delta}$ is $k[t]$-linear it is determined by the restriction to the constant polynomials $C^* \otimes 1 \subset C^* \tilde{\otimes} k[t]$, and $\tilde{\delta}(c \otimes 1) = (\delta c) \otimes 1 + c(1+g) \otimes t$ is just a linear polynomial in t whose constant term is determined by the coboundary δc of c in C^*, the coefficient of t being given by the $k[G]$-structure of C^*. (Note that we assume char$(k) = 2$).

If we evaluate the element $\tilde{c} = \sum_{i=0}^{n} c^{n-i} \otimes t^i$ at a point $\alpha \in k$ we get the element

$$\tilde{c}(\alpha) = \sum_{i=0}^{n} c^{n-i} \otimes \alpha^i \in (\bigoplus_i C^i) \otimes k \cong \bigoplus_i C^i =: C^{(*)}.$$

This evaluation map $C^* \tilde{\otimes} k[t] \to C^{(*)}$ can also be described as the projection

$$C^* \otimes k[t] \cong (C^* \tilde{\otimes} k[t]) \otimes k_\alpha \to (C^* \tilde{\otimes} k[t]) \otimes_{k[t]} k_\alpha := C^{(*)} \tilde{\otimes} k[t] \otimes_{k[t]} k_\alpha,$$

where k_α denotes k together with the $k[t]$-module structure given by extending the map $t \mapsto \alpha$ to a morphism of k-algebras $\alpha : k[t] \to k$. (In more algebraic terms: k is considered as the residue field of $k[t]$ with respect to the maximal ideal $(t - \alpha) \subset k[t]$ and $\alpha : k[t] \to k$ is the corresponding projection.)

The differential $\tilde{\delta}$ on $C^* \tilde{\otimes} k[t]$ can be viewed as a perturbation of the 'trivial' extension of the differential δ on C^* to $C^* \otimes k[t]$, or as a one-parameter family of differentials on $C^{(*)}$, parameterized by t, which at the point $t = 0$ coincides with the differential δ on C^*.

We use the following notation:

$$H_{G,\alpha}(C^*) := H(\beta_G^*(C^*) \otimes_{k[t]} k_\alpha) \text{ for } C^* \in \delta gk[G]\text{- Mod}, \alpha \in k.$$

It should be realized that $\beta_G^*(C^*) \otimes_{k[t]} k_\alpha$ does not inherit a grading if $\alpha \neq 0$, since $\alpha : k[t] \to k$ maps the element t of degree 1 to the element α of degree 0. As a k-module $\beta_{G,\alpha}(C^*) := \beta_G^*(C^*) \otimes_{k[t]} k_\alpha$ is isomorphic to $\bigoplus_{i=0}^\infty C^i$ since

$$\beta_G^*(C^*) \otimes_{k[t]} k_\alpha \cong (C^* \tilde{\otimes} k[t]) \otimes_{k[t]} k_\alpha \cong \bigoplus_{i=0}^\infty C^i.$$

and the coboundary δ_α of $\beta_{G,\alpha}(C^*)$ is given by $\tilde{\delta}$ evaluated at $\alpha \in k$.

For $\alpha \neq 0$ the evaluation at $t = \alpha$, i.e. the functor $- \otimes_{k[t]} k_\alpha$, is exact on $\delta gk[t]$-Mod (see (A.7.2)), and hence $H_{G,\alpha}(C^*) = H_G^*(C^*) \otimes_{k[t]} k_\alpha$.

If $C^* = S^*(X; k)$, where X is a G-space, then

$$\beta_G^*(X; k) = \beta_G^*(S^*(X; k))$$

carries a multiplication coming from the diagonal of X and of $\mathcal{E}_*(G)$, which induces the cup-product on $H_G^*(X; k) \cong H^*(EG \times_G X; k)$ (see (1.2.8)). This product is compatible with the $k[t]$-module structure in the sense that $H_G^*(X; k)$ can be considered as a $k[t]$-algebra. On the cochain level, however, the multiplication on $\beta_G^*(X; k)$ is in general not $k[t]$-bilinear and does not induce a multiplication on $\beta_G^*(X; k) \otimes_{k[t]} k_\alpha$ for $\alpha \neq 0$ in an obvious way. In case $\alpha = 0$, however, it is easy to check that there is an induced multiplication on $\beta_G^*(X; k) \otimes_{k[t]} k_0$ which - under the isomorphism $\beta_G^*(X; k) \otimes_{k[t]} k_0 \cong S^*(X; k)$ - corresponds to the multiplication on $S^*(X; k)$.

We therefore can consider $H_{G,\alpha}(X; k)$ as a k-algebra, where the product structure is induced by the multiplication on $\beta_G^*(X; k)$ as follows:

for $\alpha \neq 0$ via the isomorphism $H_{G,\alpha}(X; k) \cong H_G^*(X; k) \otimes_{k[t]} k_\alpha$;
for $\alpha = 0$ via the isomorphism $\beta_G^*(X; k) \otimes_{k[t]} k_0 \cong S^*(X; k)$.

(Note that in general: $H_{G,0}(X; k) \not\cong H_G^*(X; k) \otimes_{k[t]} k_0$).

The following theorem is the least complicated of a number of results similar in spirit that we discuss later.

(1.3.5) Theorem
Let X be a finite-dimensional G-CW-complex, $G = \mathbb{Z}_2$. Then there are natural isomorphisms of k-algebras
$$H_{G,\alpha}(X;k) := H(\beta_G^*(X;k) \otimes_{k[t]} k_\alpha) \cong \begin{cases} H^{(*)}(X;k) & \text{for } \alpha = 0 \\ H^{(*)}(X^G;k) & \text{for } \alpha \neq 0. \end{cases}$$

Proof.
We first consider the special case where X has trivial G-action. Then the coboundary $\tilde{\delta}$ on $S^*(X;k)\tilde{\otimes}k[t]$ is given by
$$\tilde{\delta}(c \otimes 1) = (\delta c) \otimes 1 + c(1+g) \otimes t$$
$$= (\delta c) \otimes 1$$
since g acts trivially and $\operatorname{char}(k) = 2$. Hence in this case $\tilde{\delta}$ is just the trivial extension of the differential δ of $S^*(X;k)$ to $S^*(X;k) \otimes k[t]$. Therefore the evaluation $\tilde{\delta}_\alpha$ of $\tilde{\delta}$ at α coincides with δ for any $\alpha \in k$ and we get that $\beta_{G,\alpha}(S^*(X;k))$ is naturally (in the variable X) isomorphic to $S^{(*)}(X;k) = \bigoplus_{i=0}^{\infty} S^i(X;k)$ as a differential k-module. Moreover it is straightforward to check - using the Künneth formula
$$H_G^*(X;k) = H(S^*(X;k) \otimes k[t]) \cong H^*(X;k) \otimes k[t]$$
- that this isomorphism is compatible with the respective multiplicative structures. In particular we get that $H_{G,\alpha}(X;k) = H(\beta_{G,\alpha}(S^*(X;k)))$ is naturally isomorphic to $H(S^{(*)}(X;k)) = H^{(*)}(X;k)$ as a k-algebra.

We now come to the general case and assume at first that $\alpha \neq 0$. We define
$$h_G^q(X) := H_{G,\alpha}(X;k)$$
and
$$k_G^q(X) := H_{G,\alpha}(X^G;k)$$
for $q \in \mathbb{Z}$. Both $\{h_G^q(-)\}_{q\in\mathbb{Z}}$ and $\{k_G^q(-)\}_{q\in\mathbb{Z}}$ can be viewed as $\mathbb{Z}$-graded equivariant cohomology theories on the category of G-spaces (see (1.2.3), (1.2.4)(1) and the proof of (1.2.6); one uses here the fact that $\beta_G^*(C^*)$ is free as a module over $k[t]$ (see (1.3.4))). The inclusion $X^G \subset X$ induces a natural transformation $\tau(X) : h_G^*(X) \to k_G^*(X)$, and in view of (1.1.3) and (1.1.4)(2) it suffices to check that $\tau(X)$ is an isomorphism in the following cases:
(i) $X = \coprod *$, i.e. X is a disjoint union of points (with trivial G-action);
(ii) $X = \coprod G$, i.e. X is a disjoint union of free G-orbits.
In case (i) it is obvious that $\tau(X)$ is an isomorphism since $X^G = X$. In case (ii) the fixed point set X^G of G is empty and therefore it remains to show that $H_{G,\alpha}(X;k) = 0$ in that case (for $\alpha \neq 0$). Using the cellular

cochain complex
$$W^*(X;k) = W^*(\coprod G;k) = \prod W^*(G;k) \cong \prod k[G]$$
instead of the singular complex $S^*(X;k)$ we get
$$\beta_{G,\alpha}(S^*(X;k)) \simeq \beta_{G,\alpha}(W^*(X;k))$$
$$\cong ((\prod k[G])\tilde{\otimes}k[t]) \otimes_{k[t]} k_\alpha$$
$$\cong \prod ((k[G]\tilde{\otimes}k[t]) \otimes_{k[t]} k_\alpha).$$

(Warning: Tensor products do not commute in general with (categorical) products. This is the reason behind the fact that $H_{G,\alpha}(-;k)$ is not strongly additive (cf. Remark (1.3.6)).)

To finish the argument we need only show $H((k[G])\tilde{\otimes}k[t]) \otimes_{k[t]} k_\alpha) = 0$; but $(k[G]\tilde{\otimes}k[t]) \otimes_{k[t]} k_\alpha \cong k[G]$ as $k[G]$-modules, and the induced boundary $\tilde{\delta}_\alpha$ is given by
$$\tilde{\delta}_\alpha(1) = \tilde{\delta}_\alpha(g) = \alpha(1+g),$$
where $\{1,g\}$ is the canonical k-basis of $k[G] \cong k \oplus k$. Since α is non-zero the cohomology vanishes.

For $\alpha = 0$, one can use an analogous argument (checking that
$$H_{G,0}(G;k) \cong H^*(G;k)$$
where again G is considered as a discrete topological space); however, it is much simpler to observe that
$$\beta_{G,0}(S^*(X;k)) = (S^*(X;k)\tilde{\otimes}k[t]) \otimes_{k[t]} k_0$$
is naturally isomorphic to $S^*(X;k)$ including the respective differential and multiplication (even if the action of G on X is non-trivial) since the perturbation of the differential δ of $S^*(X;k)$ which gives the twisted differential $\tilde{\delta}$ of $S^*(X;k)\tilde{\otimes}k[t]$ consists of terms which contain the factor t, and hence disappear if evaluated at $t = 0$. A similar argument applies to the product structure. □

(1.3.6) Remark

For the part of the above theorem which concerns the evaluation at $\alpha = 0$ the assumption that X is finite-dimensional is not necessary. But for the result about evaluating at $\alpha \neq 0$ some finiteness condition is definitely unavoidable, since for $X = EG$ one gets $H_{G,\alpha}(EG;k) \cong k$ for all α (see Exercise (1.9)) although $(EG)^G = \emptyset$.

Before we start with the discussion of variations and generalizations of Theorem (1.3.5) we give some typical applications including certain classical results by P.A. Smith. One basic question underlying P.A. Smith's work was to find out to what extent topological group actions on

1.3. The Borel construction for 2-tori

spaces like discs, euclidean spaces, spheres, resemble linear group actions (i.e. those given by representations of the group) (see [Bredon, 1977] for some remarks on the history of the subject). Clearly the fixed point set of a linear action on a disc, euclidean space or sphere will again be a disc, euclidean space or sphere respectively. P.A. Smith showed that as far as the (co)homology of the fixed point set is concerned an analogous result holds even for topological actions of certain groups if the coefficients for the cohomology are chosen appropriately. The following corollary gives the precise statement for $G = \mathbb{Z}_2$. See (1.4.7) and (3.1.10) for the cases $G = \mathbb{Z}_p$, p odd, and $G = S^1$.

(1.3.7) Corollary
Let X be a finite-dimensional G-CW-complex $(G = \mathbb{Z}_2)$, and let k be a field of characterstic 2 (as usual for this section).
(1) If $H^*(X;k) \cong k$ then $H^*(X^G;k) \cong k$
(2) If $H^*(X;k) \cong H^*(S^n;k)$ then $H^*(X^G;k) \cong H^*(S^m;k)$ with $-1 \leq m \leq n$ ($S^{-1} := \emptyset$ by definition, hence $H^*(S^{-1};k) = 0$) cf. Exercise (1.12).)

Proof.
Using (1.3.5) we want to calculate $H_{G,\alpha}(X;k)$ for $\alpha \neq 0$. We can replace $\beta_G^*(S^*(X;k))$ by the minimal Hirsch-Brown model $H^*(X;k) \tilde{\otimes} k[t]$ up to chain homotopy equivalence over $k[t]$ (see (B.2.4)). We therefore only need to calculate the cohomology of this model with coefficients in k_α. The constant (with respect to t) part of the differential of the minimal Hirsch-Brown model applied to an element in $H^*(X;k) \otimes 1 \subset H^*(X;k) \otimes k[t]$ is zero, while the linear part in t is given by the induced action in cohomology (see Section B.2).
Part (1): Since $G = \mathbb{Z}_2$ cannot act non-trivially on $\mathbb{Z}_2$ by module homomorphisms (the only module automorphism of $\mathbb{Z}_2$ is the identity) we get that the differential of the minimal Hirsch-Brown model must vanish altogether because terms of order higher than 1 cannot occur in the differential of elements $H^*(X;k) \otimes 1 \subset H^*(X;k) \otimes k[t]$ for degree reasons. Hence $H^*(X;k) \tilde{\otimes} k[t] \cong k[t]$ with trivial differential in this case, and therefore $H_{G,\alpha}(X;k) \cong k[t] \otimes_{k[t]} k_\alpha \cong k$ for all α.
Part (2): Again we make use of the minimal Hirsch-Brown model $H^*(X;k) \tilde{\otimes} k[t] \cong k[t] \oplus k[t]$ equipped with a $k[t]$-linear differential $\tilde{\delta}$. If $n > 0$ then the two summands $k[t]$ are in two different dimensions 0 and n. Let a and b be respective generators of the two summands. By the same argument as above one sees that the only possibly non-zero term

of $\tilde{\delta}(b)$ must be of the form $\tilde{\delta}(b) = \lambda a t^{n+1}$ where $\lambda \in k$. If $\lambda \neq 0$ then, for $a \neq 0$, $\tilde{\delta}_\alpha \neq 0$ and $H_{G,\alpha}(X;k) = 0$.

On the other hand if $\lambda = 0$ then $\tilde{\delta}_\alpha = 0$ for all α and
$$H_{G,\alpha}(X;k) \cong k \oplus k.$$
If $n = 0$ then the only possibly non-zero part of $\tilde{\delta}$ is given by the induced action of G on $H^*(X;k) \cong k \oplus k$. If the action is non-trivial it must interchange the two summands and hence $X^G = \emptyset$ (it is also clear algebraically that $H_{G,\alpha}(X;k) = 0$ for $\alpha \neq 0$ in this case). For a trivial action we obviously get $H_{G,\alpha}(X;k) = k \oplus k$. In all possible cases one therefore has $H^*(X^G;k) \cong H^*(S^m;k)$ for some $m \geq -1$.

It remains to show that m cannot exceed n. It clearly suffices to handle the case where $\tilde{\delta} = 0$ (since otherwise $X^G = \emptyset$ and $m = -1$). In this case the inclusion $j : X^G \hookrightarrow X$ induces a morphism
$$j^* : H^*(X) \tilde{\otimes} k[t] \to H^*(X^G) \otimes k[t]$$
of the minimal Hirsch-Brown models of X and X^G (considered as a G-space with trivial action), which gives an isomorphism when evaluated at $\alpha \neq 0$.

If m were larger than n the map j^* would factor through
$$H^0(X^G) \otimes k[t] \cong k[t]$$
for degree reasons, and therefore the evaluation at $\alpha \neq 0$ could not become an isomorphism. □

The next corollary can be considered as a generalization of the above. Although related to the argument we used to show (1.3.7), the proof of (1.3.8) does not use Corollary (1.3.7) (or its proof) and hence one could also obtain (1.3.7) immediately from (1.3.8) as a special case (cf. Section 3.10 (in particular (3.10.6), (3.10.7)) and (4.2.2)).

(1.3.8) Corollary
Let X be a finite-dimensional G-CW-complex ($G = \mathbb{Z}_2$) such that $\sum_{i=0}^{\infty} \dim_k H^i(X;k)$ is finite (k is a field of characteristic 2), then
(1) $\sum_{i=0}^{\infty} \dim_k H^{m+i}(X^G;k) \leq \sum_{i=0}^{\infty} \dim_k H^{m+i}(X;k)$ for all $m \geq 0$; moreover:
(2)
$$\sum_{i=0}^{\infty} \dim_k H^i(X;k) - \sum_{i=0}^{\infty} \dim_k H^i(X^G;k)$$
$$= 2 \dim_k \operatorname{Tor}^{k[t]}(H_G^*(X;k), k_0);$$

1.3. The Borel construction for 2-tori

in particular:
(3) $\sum_{i=0}^{\infty} \dim_k H^i(X;k) \equiv \sum_{i=0}^{\infty} \dim_k H^i(X^G;k)$ modulo 2.

Proof.
We consider the minimal Hirsch-Brown model for X_G and
$$(X^G)_G = X^G \times BG$$
and the morphism $j^* : H^*(X;k)\tilde{\otimes}k[t] \to H^*(X^G;k) \otimes k[t]$, between them (defined up to homotopy in $\delta gk[t]$-Mod) which is induced by the inclusion $j : X^G \to X$. With respect to the filtration of $H^*(X;k)\tilde{\otimes}k[t]$ by subcomplexes
$$\mathcal{F}_m(H^*(X;k)\tilde{\otimes}k[t]) := (\bigoplus_{i=0}^{m} H^i(X;k))\tilde{\otimes}k[t]$$
(similarly for X^G) the morphism j^* is filtration preserving since it preserves the total degree and is $k[t]$-linear. We get Diagram 1.4 in $\delta gk[t]$-Mod, where $\tilde{H}$ abbreviates $H^*(X;k)\tilde{\otimes}k[t]$, H^G abbreviates $H^*(X^G;k)\otimes k[t]$, and $\tilde{H}^G$ abbreviates $H^*(X^G;k)\tilde{\otimes}k[t]$.

$$\begin{array}{ccccccccc}
0 & \to & \mathcal{F}_{m-1}(\tilde{H}) & \to & \tilde{H} & \to & \tilde{H}/\mathcal{F}_{m-1}(\tilde{H}) & \to & 0 \\
& & \downarrow \bar{j}^* & & \downarrow \bar{j}^* & & \downarrow \bar{j}^* & & \\
0 & \to & \mathcal{F}_{m-1}(H^G) & \to & H^G & \to & \tilde{H}^G/\mathcal{F}_{m-1}(H^G) & \to & 0
\end{array}$$

Diagram 1.4

Applying cohomology with coefficients k_α (over $k[t]$) gives a ladder of two long exact sequences where the vertical maps induced by j^* are isomorphisms if $\alpha \neq 0$. But the differential of $H^*(X^G;k) \otimes k[t]$ is trivial and hence the corresponding long exact sequence decomposes into short exact sequences. It therefore follows that $\bar{j}^*$ induces a surjection in cohomology with coefficients k_α. As a $k[t]$-module
$$H^*(X;k)\tilde{\otimes}k[t]/\mathcal{F}_{m-1}(H^*(X;k)\tilde{\otimes}k[t])$$
is isomorphic to $(\bigoplus_{i=m}^{\infty} H^i(X;k)) \otimes k[t]$; similarly for X^G instead of X, but here the isomorphism is even compatible with the - actually trivial - differential on both sides. The map $\bar{j}^* \otimes_{k[t]} k_\alpha$ must be surjective already (since the target is a complex with trivial differential and $\bar{j}^* \otimes_{k[t]} k_\alpha$ induces a surjection in cohomology). Hence the dimension of the domain as a k-vector space is at least as large as the dimension of the range; thus
$$\dim_k \bigoplus_{i=m}^{\infty} H^i(X^G;k) \leq \dim_k \bigoplus_{i=m}^{\infty} H^i(X;k),$$

i.e.
$$\sum_{i=0}^{\infty} \dim_k H^{m+i}(X^G; k) \leq \sum_{i=0}^{\infty} \dim_k H^{m+i}(X; k) \text{ for } m \geq 0.$$

To prove (2) of (1.3.8) we use the fact that $- \otimes_{k[t]} k_\alpha$ is exact on the category of graded $k[t]$-modules (see A.7.2) for $\alpha \neq 0$, in particular
$$H_{G,\alpha}(X; k) = H_G^*(X; k) \otimes_{k[t]} k_\alpha \text{ for } \alpha \neq 0.$$
We therefore get
$$(*) \quad \dim_k H_G^{(*)}(X; k) \otimes_{k[t]} k_0$$
$$= \dim_k H_{G,\alpha}(X; k) + \dim_k \operatorname{Tor}^{k[t]}(H_G^{(*)}(X; k), k_0)$$

(see A.7.4). On the other hand by the Universal Coefficients Formula one has
$$(**) \quad \dim_k H((H^{(*)}(X) \tilde{\otimes} k[t]) \otimes_{k[t]} k_0)$$
$$= \dim_k (H_G^{(*)}(X) \otimes_{k[t]} k_0) + \dim_k \operatorname{Tor}^{k[t]}(H_G^{(*)}(X), k_0).$$

Since $(H^{(*)}(X) \tilde{\otimes} k[t]) \otimes_{k[t]} k_0 \cong H^{(*)}(X)$ (with trivial differential), the equalities (*) and (**) (and (1.3.5)) imply
$$\dim_k H^{(*)}(X; k) - \dim_k H^{(*)}(X^G; k) = 2 \dim_k \operatorname{Tor}^{k[t]}(H_G^{(*)}(X), k_0).$$
$\square$

The next three corollaries demonstrate how Theorem (1.3.5) can be used to detect fixed points. Similar and more general results will be given later (see, in particular, Chapter 3).

(1.3.9) Corollary
Let X be a finite-dimensional G-CW-complex $(G = \mathbb{Z}_2)$. Then X has a fixed point if and only if the map
$$H_G^*(*; k) \to H_G^*(X; k)$$
induced by the projection $p : X \to *$ is injective.

Proof.
If X has a fixed point then p has a right inverse and hence the induced map is injective. On the other hand if $p_G^* : H_G^*(*; k) \to H_G^*(X; k)$ is injective then so is
$$p_{G,\alpha}^* : H_{G,\alpha}(*; k) \cong k \to H_{G,\alpha}(X; k) \cong H^{(*)}(X^G; k)$$
for $\alpha \neq 0$, since $- \otimes_{k[t]} k_\alpha$ is exact on $gk[t]$-Mod. Hence $H^{(*)}(X^G; k) \neq 0$ and $X^G \neq \emptyset$.
$\square$

We want to remark at this point that Corollary (1.3.9) is the classical prototype of a number of similar criteria for the existence of fixed points for actions of more general groups. The Borel-Smith theory using equivariant ordinary cohomology works very well for elementary abelian groups (see (3.1.12)) while for more general abelian groups equivariant cobordism is appropriate (see [tom Dieck, 1972]), and, more recently, formally similar results have been derived for arbitrary finite groups using equivariant cohomotopy (see [Jackowski, 1988]). In this book we only deal with equivariant ordinary cohomology.

The next corollary (see [Assadi, 1988], cf. Exercise (3.12)) can be considered as a generalization of the classical Borsuk-Ulam Theorem.

(1.3.10) Corollary
Let $f : X \to Y$ be a G-map between connected G-spaces ($G = \mathbb{Z}_2$). Assume that $H^i(X;k) = 0$ for $i = 1, \ldots, m-1$, and that Y is a finite-dimensional G-CW-complex with free G-action and $H^j(Y;k) = 0$ for $j > n$. Then $n > m$, or $n = m$ and $H^n(f;k) \neq 0$ (k a field of characteristic 2).

Proof.
We look again at the morphism $\tilde{f}^* : H^*(Y;k)\tilde{\otimes}k[t] \to H^*(X;k)\tilde{\otimes}k[t]$ induced by f on the minimal Hirsch-Brown model. (Note that we do not need X to be a finite-dimensional G-CW-complex in order to obtain the minimal Hirsch-Brown model. This additional assumption is needed when one wants to establish the connection to the fixed point set.) The map $\tilde{f}^*$ is compatible with the inclusion of $k[t] \cong H_G^*(*;k)$ into $H^*(X;k)\tilde{\otimes}k[t]$ and $H^*(Y;k)\tilde{\otimes}k[t]$ respectively induced by the projections to a point, i.e. one has the commutative triangle shown in Diagram 1.5.

$$H^*(Y;k)\tilde{\otimes}k[t] \xrightarrow{\tilde{f}^*} H^*(X;k)\tilde{\otimes}k[t]$$
$$\tilde{q}^* \nwarrow \quad \nearrow \tilde{p}^*$$
$$k[t]$$

Diagram 1.5

Assume $n < m$, or $n = m$ and $H^n(f) = 0$. Then for degree reasons $\tilde{f}^*$ factors through $H^0(X;k) \otimes k[t] \cong k[t]$ (since $\tilde{f}^* \otimes_{k[t]} k_0 = H^*(f;k)$).

The image of $\tilde{p}^* : k[t] \to H^*(X;k)\tilde{\otimes}k[t]$ is just $H^0(X;k) \otimes k[t]$, similarly for Y instead of X. Hence under the above assumptions the dia-

gram would imply that $\tilde{q}^*$ has a left inverse, and in particular
$$q_G^* : k[t] \to H_G^*(Y;k)$$
would be injective. By (1.3.9) this means that Y has a fixed point which contradicts the hypothesis that Y is a free G-space. $\square$

(1.3.11) Examples
(1) (Borsuk-Ulam). If $f : S^m \to S^n$ is a (continuous) map which fulfils $f(-x) = -f(x)$ for all $x \in S^m$ then $m \leq n$. This follows immediately from (1.3.10) if one interprets f as a $\mathbb{Z}_2$-map, where the $\mathbb{Z}_2$-actions on S^m and S^n are given by the antipodal map.
(2) If $f : EG \to Y$ is a G-map ($G = \mathbb{Z}_2$) and Y is a finite-dimensional G-CW-complex then Y must have a fixed point.

If $f : X \to Y$ is a G-map and X has a fixed point then obviously Y must have a fixed point. The following result gives a partial converse to this statement.

(1.3.12) Corollary
Let X be a finite-dimensional G-CW-complex, $G = \mathbb{Z}_2$, such that $H^i(X;k) = 0$ for $i > n$ (k a field of characteristic 2). Let S^n be the n-dimensional sphere equipped with a G-action which has fixed points. If $f : X \to S^n$ is a G-map which induces an isomorphism $H^n(f) : H^n(S^n;k) \to H^n(X;k)$, then $X^G \neq \emptyset$.

Proof.
We consider the (homotopy) commutative diagram shown as Diagram 1.6, where H_{S_k} (resp. H_{X_k}) is an abbreviation for $H^n(S^n;k)$ (resp. $H^n(X;k)$, and $H^*_{S_k}$ (resp. $H^*_{X_k}$) is one for $H^*(S^n;k)$ (resp. $H^*(X;k)$);

$$\begin{array}{ccccccccc}
R & \cong & H_{S_k} \otimes R \subset H^*_{S_k} \tilde{\otimes} R & \xrightarrow{\tilde{f}^*} & H^*_{X_k} \tilde{\otimes} R & \xrightarrow{\tilde{p}} & H_{X_k} \otimes R & \cong & R \\
\downarrow & & \downarrow & & \downarrow & & \downarrow & & \downarrow \\
k & \cong & H_{S_k} \subset H^*_{S_k} & \xrightarrow{f^*} & H^*_{X_k} & \xrightarrow{p} & H_{X_k} & \cong & k
\end{array}$$

Diagram 1.6

where $\tilde{f}^*$ is the morphism between the minimal Hirsch-Brown models of S^n and X, respectively, induced by $f : X \to S^n$, $R = k[t]$, and $\tilde{p}$ is the projection
$$H^*(X;k)\tilde{\otimes}R \to H^*(X;k)\tilde{\otimes}R/(\bigoplus_{i<n} H^i(X;k) \otimes R)) \cong H^n(X;k) \otimes R.$$

Since $\bigoplus_{i<n} H^i(X;k) \otimes R$ is an R-subcomplex of $H^*(X;k)\tilde{\otimes}R$ the map $\tilde{p}$ is a morphism of R-complexes, where the induced R-linear differential on $H^n(X;k) \otimes R \cong R$ clearly must be trivial.

Since - by assumption - $(S^n)^G \neq \emptyset$, the differential on $H^*(S^n;k)\tilde{\otimes}R$ vanishes (cf. (1.3.7)(2)), and therefore $H^n(S^n;k) \otimes R$ (with trivial differential) is an R-subcomplex of $H^*(S^n;k)\tilde{\otimes}R$.

The vertical maps in Diagram 1.6 are given by evaluating at $t = 0$. Because of the vanishing of the differential of $H^*(S^n;k)\tilde{\otimes}R$ we actually get a strictly commutative diagram.

By assumption the composition of the maps in the bottom row is an isomorphism, and hence the same holds for the top row. Therefore $H_G^*(X;k) \cong H(H^*(X;k)\tilde{\otimes}R)$ splits off a copy of R as a direct summand. In particular $H_G^*(X;k)$ evaluated at $\alpha = 1$ is non-trivial, which implies $X^G \neq \emptyset$. $\square$

(1.3.13) Example

The assumptions in (1.3.12) may look somewhat technical but they are actually fulfilled quite naturally in the following situation.

Let $f' : M \to N$ be a G-map between differentiable n-dimensional compact, connected G-manifolds, $G = \mathbb{Z}_2$. Assume that $N^G \neq \emptyset$, and that the (non-equivariant) degree of f' is non-zero modulo 2. Choosing a G-invariant neighbourhood V of a fixed point $y \in N$ such that

$$(\overline{V}, V) \cong (D^n, D^n - S^{n-1})$$

one gets a G-map

$$f'' : N \to N/(N-V) \cong D^n/S^{n-1} \cong S^n$$

such that the composition $f = f''f'$ fulfils the assumption in (1.3.12). Hence $M^G \neq \emptyset$. Generalizations of (1.3.12) and (1.3.13) to p-tori and tori can be obtained easily using the results we derive in Chapters 1 and 3 (see, in particular, (1.4.11), (3.1.6) and (3.5.7), cf. Exercise (5.8)); but see [Browder, 1987] for more general and sophisticated results in this direction.

The next proposition characterizes the situations in which actual equality holds in (1.3.8) (1), cf. also (3.10.4).

(1.3.14) Proposition

Let X be a finite-dimensional G-CW-complex $(G = \mathbb{Z}_2)$ such that $\sum_{i=0}^{\infty} \dim_k(H^i(X;k))$ is finite (k a field of characteristic 2). Then the following conditions are equivalent:

(1) $\sum_{i=0}^{\infty} \dim_k H^i(X;k) = \sum_{i=0}^{\infty} \dim_k H^i(X^G;k)$
(2) The differential $\tilde{\delta}$ of the minimal Hirsch-Brown model
$$H^*(X;k)\tilde{\otimes}k[t]$$
vanishes.
(3) The map $j_G^* : H_G^*(X;k) \to H_G^*(X^G;k)$, induced by the inclusion $j : X^G \subset X$, is injective.
(4) $\mathrm{Tor}^{k[t]}(H_G^*(X), k_0) = 0$
(5) $H_G^*(X;k) \otimes_{k[t]} k_0 \cong H^*(X;k)$ (induced by the projection
$$p_0 : H^*(X;k) \otimes k[t] \to H^*(X;k),\ t \mapsto 0)$$
(6) $p_0 : H^*(X;k)\tilde{\otimes}k[t] \to H^*(X;k)$ induces a surjection
$$p_0^* : H_G^*(X;k) \to H^*(X;k).$$

Proof.

(1) *implies* (2): Since $H^*(X^G;k) \cong H_{G,\alpha}(X;k)$ for $\alpha \neq 0$ (see (1.3.5)), the condition (1) is equivalent to $\tilde{\delta}_\alpha = 0$ for $\alpha \neq 0$; note that $H_{G,\alpha}(X;k)$ is the cohomology of a complex whose underlying k-vector space is just $H^{(*)}(X;k)$, the differential being given by the differential $\tilde{\delta}$ of the minimal Hirsch-Brown model $H^*(X;k)\tilde{\otimes}k[t]$ evaluated at α. In Diagram 1.7 we factorize $\tilde{\delta}$ into a surjection followed by an injection. Since the evaluation at $\alpha \neq 0$ (i.e. - $\otimes_{k[t]} k_\alpha$) is exact we get a similar diagram for $\tilde{\delta}_\alpha$.

$$H^*(X;k) \otimes k[t] \xrightarrow{\tilde{\delta}} H^*(X;k) \otimes k[t]$$
$$\searrow \qquad \nearrow$$
$$\tilde{B}$$

Diagram 1.7

But $\tilde{B}_\alpha = \tilde{\delta}_\alpha((H^*(X;k)\tilde{\otimes}k[t]) \otimes_{k[t]} k_\alpha) = 0$; therefore $\tilde{B}$ must be zero since it is a free $k[t]$-module, being a submodule of the free $k[t]$-module $H^*(X;k)\tilde{\otimes}k[t]$. Hence $\tilde{\delta} = 0$.

(2) *implies* (3): The map $j : X^G \to X$ induces a morphism of minimal Hirsch-Brown models $j^* : H^*(X;k)\tilde{\otimes}k[t] \to H^*(X^G;k) \otimes k[t]$. Since the differentials on both complexes vanish, j^* is already equal to the map $j_G^* : H_G^*(X;k) \to H_G^*(X^G;k)$. Let $\tilde{K} :=$ kernel of j_G^*, then
$$\tilde{K}_\alpha = \tilde{K} \otimes_{k[t]} k_\alpha = 0$$
since $j_G^* \otimes_{k[t]} k_\alpha$ is an isomorphism; see (1.3.5). But $\tilde{K}$ is a free $k[t]$-module and hence $\tilde{K}_\alpha = 0$ implies $\tilde{K} = 0$.

(3) *implies* (4): Since $H_G^*(X;k)$ is isomorphic to a submodule of $H_G^*(X^G;k) \cong H^*(X^G;k) \otimes k[t]$ it is a free $k[t]$-module; and in particular $\mathrm{Tor}^{k[t]}(H_G^*(X;k), k_0) = 0$.

(4) *is equivalent to* (5): By the universal coefficients formula there is a short exact sequence

$$0 \to H_G^*(X;k) \otimes_{k[t]} k_0 \xrightarrow{\phi} H(H^*(X;k)\tilde{\otimes}k[t] \otimes_{k[t]} k_0)$$
$$\xrightarrow{\psi} \mathrm{Tor}^{k[t]}(H_G^*(X;k), k_0) \to 0$$

The middle term of this sequence is nothing but $H^*(X;k)$. Hence the vanishing of the Tor-term is equivalent to (5).

(5) *is equivalent to* (6): The map $p_0 : H^*(X;k)\tilde{\otimes}k[t] \to H^*(X;k)$, $t \mapsto 0$, induces a map $p_0^* : H_G^*(X;k) \to H^*(X;k)$ which factors through $H_G^*(X;k) \otimes_{k[t]} k_0$, i.e. $p_0^* = \phi\rho$, where

$$\rho : H_G^*(X;k) \to H_G^*(X;k) \otimes_{k[t]} k_0 \cong H_G^*(X;k)/tH_G^*(X;k)$$

is the canonical projection and ϕ is as above. Since ρ is always surjective and ϕ always injective, (5) and (6) are equivalent.

(4) *implies* (1) by Corollary (1.3.8)(2). □

(1.3.15) Remark
If one interprets $H_G^*(X;k)$ as the cohomology

$$H^*(EG \times_G X; k)$$

of the Borel construction $EG \times_G X$, then condition (1.3.14)(2) is equivalent to the condition that the induced action of G on $H^*(X;k)$ is trivial and the Serre spectral sequence of the fibration $X \to EG \times_G X \to BG$ for cohomology with coefficients in k collapses from the E_2-term on. Condition (6) means that X is totally non-homologous to zero (TNHZ) in $EG \times_G X$ with respect to $H^*(-;k)$. The Leray-Hirsch theorem (see, e.g., [Spanier, 1966]) then also shows that $H_G^*(X;k)$ is isomorphic to $H^*(X) \otimes k[t]$ as a $k[t]$-module.

The multiplicative structure on $\beta_G^*(X;k) = S^*(X;k)\tilde{\otimes}k[t]$ induces one on the minimal Hirsch-Brown model $H^*(X;k)\tilde{\otimes}k[t]$ using the homotopy equivalence between $\beta_G^*(X;k)$ and $H^*(X;k)\tilde{\otimes}k[t]$ in $\delta gk[t]$-Mod. Similar to the product on $\beta_G^*(X;k)$ this multiplication is not $k[t]$-bilinear but it is compatible (up to homotopy) with the projection $H^*(X;k)\tilde{\otimes}k[t] \to H^*(X;k)$, $t \mapsto 0$ ('evaluation at $t = 0$'). It also induces the 'correct' $k[t]$-algebra structure on $H(H^*(X;k)\tilde{\otimes}k[t]) \cong H_G^*(X;k)$. (In a sense $\beta_G^*(X;k)$ and $H^*(X;k)\tilde{\otimes}k[t]$ are '$k[t]$-algebras up to homotopy'; but we will not elaborate on that statement.) In particular, if all the conditions in Proposition (1.3.14) are fulfilled $H^*(X;k)\tilde{\otimes}k[t] \cong H_G^*(X;k)$ itself is

a (strict) $k[t]$-algebra. The multiplication is given by a $k[t]$-bilinear map
$$\tilde{\mu} : (H^*(X;k)\tilde{\otimes}k[t]) \otimes (H^*(X;k)\tilde{\otimes}k[t]) \to H^*(X;k)\tilde{\otimes}k[t]$$
which induces the cup-product on $H^*(X;k)$ when evaluated at $t = 0$. In general $\tilde{\mu}$ is not just the componentwise multiplication on $H^*(X;k) \otimes k[t]$, using the cup-product on the first factor and the multiplication in the polynomial ring $k[t]$ on the second. (We therefore still use the notation $H^*(X;k)\tilde{\otimes}k[t]$, even though the differential $\tilde{\delta}$ on $H^*(X;k)\tilde{\otimes}k[t]$ vanishes.)

But $\tilde{\mu}$ can be considered as a perturbation (or deformation) of this componentwise multiplication by higher order terms in t, more precisely:
$$\tilde{\mu}(a,b) = \mu_0(a,b) + \mu_1(a,b)t + \ldots + \mu_i(a,b)t^i + \ldots \in H^*(X;k)\tilde{\otimes}k[t]$$
where $a \in H^p(X;k)$, $b \in H^q(X;k)$, $\mu_0(a,b) = a \cup b \in H^{p+q}(X;k)$ (since $(H^*(X;k)\tilde{\otimes}k[t]) \otimes_{k[t]} k_0 = H^*(X;k))$, $\mu_i(a,b) \in H^{p+q-i}(X;k)$. (Here we write the elements of $H^*(X,k)\tilde{\otimes}k[t]$ as 'polynomials' in t with coefficients in $H^*(X;k)$. Note though, that $\tilde{\mu}$ is not the standard product on $H^*(X)[t]$.) Since $\tilde{\mu}$ is $k[t]$-bilinear it is already determined by its restriction to $H^*(X;k) \otimes H^*(X;k)$.

The multiplication $\tilde{\mu}$ can be thought of as a family of k-algebra structures on the k-vector space $H^{(*)}(X)$ parameterized by t. Theorem (1.3.5) implies that $\tilde{\mu}$ evaluated at $\alpha \in k$ gives the cup-product on $H^*(X;k)$ if $\alpha = 0$ and the cup-product on $H^*(X^G;k)$ if $\alpha \neq 0$. The algebraic structure that comes up in the above topological situation has been studied in a purely algebraic context by several authors (see, e.g., [Gerstenhaber, 1964, 1966, 1974], [Iarrobino, Emsalem, 1978], [Gerstenhaber, Schack, 1988]).

The set up of Gerstenhaber seems particularly suitable to be applied to the above situation. Here we only give a few results in this direction which require very little technical apparatus. More extensive use of the relevant theory of deformations of algebraic structures is made in [Puppe, V., 1978, 1988], [Hauschild, 1983].

The reader should note the formal similarity between the deformation (twisting, perturbation) of the (trivial) differential on $H^*(X;k)$ to obtain the differential $\tilde{\delta}$ on $H^*(X;k)\tilde{\otimes}k[t]$ in the minimal Hirsch-Brown model and the deformation (twisting, perturbation) of the cup-product on $H^*(X;k)$ to obtain the multiplication on $H_G^*(X;k) \cong H^*(X;k)\tilde{\otimes}k[t]$ in case $\tilde{\delta} = 0$. In fact, the reason behind the inequalities (1.3.8)(1) is the 'semicontinuity' of the rank of the k-linear map $\tilde{\delta}_\alpha$ as a function of $\alpha \in k$, i.e. rank $(\tilde{\delta}_0) \leq$ rank $(\tilde{\delta}_\alpha)$ for $\alpha \in k$ (see (A.7.4)); and this kind of semicontinuity under deformation is typical for certain invariants of algebraic structures. We are going to discuss some simple examples of

1.3. The Borel construction for 2-tori

this phenomenon for deformations of commutative graded, associative algebras. This has immediate applications to the cohomology of transformation groups. Later we will also consider deformations of graded Lie algebras from a similar point of view and apply the results to the rational homotopy of G-spaces for $G \cong (S^1)^r$ a (real) torus. A systematic discussion of deformations of algebraic structures can be found in [Gerstenhaber, 1964, 1966, 1974] for rings and algebras and in [Nijenhuis, Richardson, 1966] for graded Lie algebras (see also [Hazewinkel, Gerstenhaber, 1988]).

Let A^* be a connected (i.e. $A^0 \cong k$), commutative graded algebra (over the field k) which fulfils $\sum_{i=0}^{\infty} \dim_k A^i < \infty$. In view of the applications in this section we assume that the characteristic of k is 2, and hence the multiplication is strictly commutative. But the reader should not have serious difficulties with the necessary modifications for the following algebraic concepts and results in case $\text{char}(k) \neq 2$ and A^* is commutative in the graded sense.

There are several interesting invariants of the algebra A^* which can be defined as the rank (dimension) of certain k-vector spaces constructed from the multiplication of A^*. For example the minimal number $\mathfrak{g}(A^*)$ of elements necessary to generate A^* as an algebra (with unit) is characterized as $\mathfrak{g}(A^*) = \dim_k(\overline{A^*}/(\overline{A^*})^2)$, where $\overline{A^*} := \bigoplus_{i=1}^{\infty} A^i$ is the augmentation ideal of the canonical augmentation $\varepsilon : A^* \to k$ given by $\varepsilon|A^0 : A^0 \cong k$ (see (A.7.7)). Of course, for any $q \in \mathbb{N}$, $\dim_k(\overline{A^*}/(\overline{A^*})^q)$ is a reasonable invariant of the algebra A^*, and one can characterize the product length $\ell(A^*)$ (cup length if A^* is the cohomology algebra of a topological space) as

$$\ell(A^*) = \max\{q; \dim_k(\overline{A^*}/(\overline{A^*})^q) < \dim_k(\overline{A^*})\}.$$

Let $\mu : A^* \otimes A^* \to A^*$ denote the multiplication of A^*, $\overline{\mu} : \overline{A^*} \otimes \overline{A^*} \to \overline{A^*}$ its restriction to $\overline{A^*} \otimes \overline{A^*} \subset A^* \otimes A^*$ and $\hat{\mu} : A^* \to \text{Hom}_k(\overline{A^*}, \overline{A^*})$ the adjoint k-linear map. The dimension of the kernel of $\hat{\mu}$ is an interesting invariant of A^*. It is called the type of A^* (see, e.g., [Kunz, 1985] for a discussion of this invariant in a more general algebraic context). For example it is easy to see that A^* is a Poincaré algebra (i.e. fulfils Poincaré duality) if and only if $\text{type}(A^*) = 1$ (see (5.1.4)).

We use the semicontinuity properties of the above invariants under deformation to prove the following result.

(1.3.16) Theorem
Let X be a connected finite-dimensional G-CW-complex ($G = \mathbb{Z}_2$) such that $\dim_k H^{(*)}(X;k) = \dim_k H^{(*)}(X^G;k) < \infty$ (k a field of character-

istic 2, e.g. $k = \mathbb{F}_2$), and let $F \subset X^G$ be a component of the fixed point set, then:

(1)
$$\dim_k \overline{(H^*(F;k)/H^*(F;k)^q)} \leq \dim_k \overline{(H^*(X;k)/H^*(X;k)^q)}$$
for $q = 2, 3, \ldots$, in particular, $\mathfrak{g}(H^*(F;k)) \leq \mathfrak{g}(H^*(X;k))$;

(2) if X^G is connected (i.e. $F = X^G$), then
$$\ell(H^*(F;k)) \geq \ell(H^*(X;k));$$

(3) $\text{type}(H^*(F;k)) \leq \text{type}(H^*(X;k))$; in particular, $H^*(F;k)$ fulfils Poincaré duality if $H^*(X;k)$ does.

Proof.

Since by assumption all the equivalent conditions in (1.3.14) are fulfilled one has a $k[t]$-algebra structure
$$\tilde{\mu} : (H^*(X;k) \tilde{\otimes} k[t]) \otimes (H^*(X;k) \tilde{\otimes} k[t]) \to H^*(X;k) \tilde{\otimes} k[t]$$
on $H^*(X) \tilde{\otimes} k[t]$ such that the evaluation of $\tilde{\mu}$ at $t = 0$ gives the cup-product on $H^*(X;k)$ and the evaluation at $t = \alpha \neq 0$ (e.g. $\alpha = 1$) gives the cup-product on $H^{(*)}(X^G;k)$ (which by assumption is isomorphic to $H^{(*)}(X;k)$ as an ungraded k-vector space) (see (1.3.5)). We choose a component $F \subset X^G$; fixing a base point $* \in F$ gives an augmentation
$$\tilde{\varepsilon} : H^*(X;k) \tilde{\otimes} k[t] \to k[t] \cong H^*(*,k) \tilde{\otimes} k[t]$$
of the algebra $H^*(X;k) \tilde{\otimes} k[t]$. Note that $\tilde{\varepsilon}$, in general, is not the 'obvious' extension of $\varepsilon : H^*(X;k) \to k$ to $H^*(X;k) \tilde{\otimes} k[t]$ (the latter would not be compatible with a twisted multiplication on $H^*(X;k) \tilde{\otimes} k[t]$). But the evaluation of $\tilde{\varepsilon}$ at $t = 0$ gives $\varepsilon : H^*(X;k) \to k$ and the evaluation at $t = \alpha \neq 0$ gives the augmentation of $H^*(X^G;k)$ corresponding to the inclusion $* \in F \subset X^G$. The multiplication $\tilde{\mu}$ induces a multiplication $\overline{\tilde{\mu}}$ on $\ker \tilde{\varepsilon}$ which can be iterated to give $k[t]$-multilinear maps
$$\overline{\tilde{\mu}}^q : \ker \tilde{\varepsilon} \otimes \ldots \otimes \ker \tilde{\varepsilon} \to \ker \tilde{\varepsilon} \quad \text{for } q = 2, 3, \ldots.$$

By (1.3.5) one has
$$\text{coker}(\overline{\tilde{\mu}}_\alpha^q) \cong \begin{cases} \overline{H^*(X;k)/H^*(X;k)^q} & \text{for } \alpha = 0 \\ \overline{H^*(X^G;k)/H^*(X^G;k)^q} & \text{for } \alpha = 1, \end{cases}$$
where $\overline{H^*(X^G;k)} := \ker \tilde{\varepsilon}_1$, $\tilde{\varepsilon}_1 : H^*(X^G;k) \to k$ being the augmentation induced by $* \in F \subset X^G$. The semicontinuity of rank $(\overline{\tilde{\mu}}_\alpha^q)$ (as a function of $\alpha \in k$) (see (A.7.4)) now implies that
$$\dim_k \overline{(H^*(X;k)/H^*(X;k)^q)} \geq \dim_k \overline{(H^*(X^G;k)/H^*(X^G;k)^q)},$$
and we finish the proof of part (1) by observing that
$$\overline{H^*(X^G;k)/H^*(X^G;k)^q} \cong \overline{H^*(F;k)/H^*(F;k)^q}.$$

1.3. The Borel construction for 2-tori

One has that $H^*(X^G; k) = H^*(F; k) \times H^*(F'; k)$, where $F' := X^G - F$, and the augmentation $\tilde{\varepsilon}_1 : H^*(X^G; k) \to k$ is the projection to $H^*(F; k)$ followed by the canonical augmentation of $H^*(F; k)$. Hence
$$\ker \tilde{\varepsilon}_1 = \overline{H^*(X^G; k)} = \overline{H^*(F; k)} \times H^*(F'; k),$$
and since $H^*(F'; k)$ has a unit,
$$\overline{H^*(X^G; k)}^q = \overline{H^*(F; k)}^q \times H^*(F'; k).$$
Therefore
$$\overline{H^*(X^G; k)}/\overline{H^*(X^G; k)}^q$$
$$= \overline{H^*(F; k)} \times H^*(F'; k)/\overline{H^*(F; k)}^q \times H^*(F'; k)$$
$$\cong \overline{H^*(F; k)}/\overline{H^*(F; k)}^q.$$

Part (2) follows from (1) since
$$\dim_k(\overline{H^*(X; k)}/\overline{H^*(X; k)}^q) < \dim_k \overline{H^*(X; k)}$$
implies
$$\dim_k(\overline{H^*(X^G; k)}/\overline{H^*(X^G; k)}^q) \leq \dim_k(\overline{H^*(X; k)}/\overline{H^*(X; k)}^q)$$
$$< \dim_k \overline{H^*(X; k)}$$
$$= \dim_k \overline{H^*(X^G; k)}.$$
(Note that the assumption 'X^G connected' is essential.)

For the proof of part (3) one considers the map
$$\hat{\tilde{\mu}} : \ker \tilde{\varepsilon} \to \mathrm{Hom}(\ker \tilde{\varepsilon}, \ker \tilde{\varepsilon})$$
adjoint to
$$\overline{\tilde{\mu}} : \ker \tilde{\varepsilon} \otimes \ker \tilde{\varepsilon} \to \ker \tilde{\varepsilon}.$$
Again the semicontinuity of rank $(\hat{\tilde{\mu}}_\alpha)$ implies that
$$\dim_k(\ker \hat{\tilde{\mu}}_0) \geq \dim_k(\ker \hat{\tilde{\mu}}_1),$$
where
$$\hat{\tilde{\mu}}_1 : \overline{H^*(X^G; k)} \to \mathrm{Hom}(\overline{H^*(X^G; k)}, \overline{H^*(X^G; k)})$$
is adjoint to the multiplication in $\overline{H^*(X^G; k)}$. (Note that evaluation at $t = \alpha$ commutes with taking adjoint maps.)

Since $\mathrm{Hom}(\overline{H^*(F; k)}, \overline{H^*(F; k)})$ maps injectively into
$$\mathrm{Hom}(\overline{H^*(F; k)} \times H^*(F'; k), \overline{H^*(F; k)} \times H^*(F'; k))$$
$$= \mathrm{Hom}(\overline{H^*(X^G; k)}, \overline{H^*(X^G; k)})$$
under the canonical map, the dimension of the kernel of the map
$$\overline{H^*(F; k)} \to \mathrm{Hom}(\overline{H^*(F; k)}, \overline{H^*(F; k)}),$$
adjoint to multiplication in $\overline{H^*(F, k)}$, is bounded above by $\dim_k(\ker \hat{\tilde{\mu}}_1)$. Hence
$$\mathrm{type}(H^*(F; k)) \leq \mathrm{type}(H^*(X; k)).$$

42   1. Equivariant cohomology and the Borel construction

The rest of part (3) now follows from the characterization of Poincaré duality algebras A^* as having type$(A^*) = 1$ and the obvious fact that already for degree reasons $\dim_k(\ker \hat{\mu}) \geq 1$ if $\sum_{i=0}^{\infty} \dim_k A^i$ is finite. □

We will give a generalization of the second part of (3) which does not need the assumption '$\dim_k H^{(*)}(X;k) = \dim_k H^{(*)}(X^G;k)$' in Chapter 5 (see (5.2.1), cf. Exercise (1.16)). Here we are going to discuss an extension of the result on the minimal number of generators (see (1.3.16)(1)) to the minimal number of relations necessary to present $H^*(X;k)$ resp. $H^*(F;k)$ (cf. (3.8.12) for generalizations).

For A^* a connected graded commutative k-algebra such that $\dim_k A^{(*)} < \infty$ we denote the minimal number of relations by $\mathfrak{r}(A^*)$, and we define

$$\mathrm{cid}(A^*) := \mathfrak{r}(A^*) - \mathfrak{g}(A^*).$$

(See Section A.7 for some basic results on presentations of connected graded algebras.)

(1.3.17) Theorem
Let X be a connected finite-dimensional G-CW-complex ($G = \mathbb{Z}_2$) such that $\dim_k H^*(X;k) = \dim_k H^{(*)}(X^G;k) < \infty$ (k a field of characteristic 2), and let $F \subset X^G$ be a component of the fixed point set, then:

(1) $\mathfrak{g}(H^*(F;k)) \leq \mathfrak{g}(H^*(X;k))$,
(2) $\mathfrak{r}(H^*(F;k)) \leq \mathfrak{r}(H^*(X;k))$,
(3) $\mathrm{cid}(H^*(F;k)) \leq \mathrm{cid}(H^*(X;k))$.

Proof.
Let $k[W] \xrightarrow{\psi} k[V] \xrightarrow{\phi} H^*(X;k)$ be a graded minimal presentation of the graded k-algebra $H^*(X;k)$, i.e. $\dim_k V = \mathfrak{g}(H^*(X;k))$ and $\dim_k W = \mathfrak{r}(H^*(X;k))$ (see (A.7.17)(2)), where $k[V]$ (resp. $k[W]$) denotes the free graded k-algebra over the graded k-vector space V (resp. W).

Since $\dim_k H^{(*)}(X;k) = \dim_k H^{(*)}(X^G;k)$ one has that

$$H_G^*(X;k) \cong H^*(X;k) \tilde{\otimes} k[t]$$

(see (1.3.14)) is a 'one-parameter family of deformations of $H^*(X;k)$' in the sense described above, i.e. $H^*(X;k)\tilde{\otimes}k[t]$ evaluated at $t = 0$ gives $H^*(X;k)$ ($H_G^*(X;k) \otimes_{k[t]} k_0 \cong H^*(X;k)$). We show that one can 'lift' ϕ and ψ to morphisms of augmented, graded $k[t]$-algebras

$$\tilde{\phi}: k[V] \otimes k[t] \to H_G^*(X) \cong H^*(X;k)\tilde{\otimes}k[t]$$

and

$$\tilde{\psi}: k[W] \otimes k[t] \to k[V] \otimes k[t],$$

1.3. The Borel construction for 2-tori

where $k[V] \otimes k[t]$ and $k[W] \otimes k[t]$ are the usual tensor products considered as $k[t]$-algebras and equipped with the canonical augmentations mapping all $v \in V$, resp. $w \in W$, to zero, and the (grading preserving) augmentation $\tilde{\varepsilon} : H^*_G(X;k) \to k[t]$ of $H^*_G(X;k)$ is induced by choosing a base point $* \in F \subset X$, such that

(i) $\tilde{\phi} \otimes_{k[t]} k_0 = \phi$, $\tilde{\psi} \otimes_{k[t]} k_0 = \psi$
(ii) $\tilde{\phi}\tilde{\psi}(k[W] \otimes k[t]) = 0$
(iii) $\tilde{\phi} : k[V] \otimes k[t] \to H^*_G(X;k)$ (hence also the induced map between the $k[t]$-modules of indecomposables

$$Q(\tilde{\phi}) : Q(k[V] \otimes k[t]) \cong V \otimes k[t] \to Q(H^*_G(X)) = \ker \tilde{\varepsilon}/(\ker \tilde{\varepsilon})^2)$$

and

$$Q(\tilde{\psi}) : Q(k[W] \otimes k[t]) \cong W \otimes k[t] \to \ker \tilde{\phi}/(\ker \tilde{\phi})(\overline{k[V]} \otimes k[t])$$

are surjective.
(Note that the above conditions imply that

$$k[W] \otimes k[t] \xrightarrow{\tilde{\psi}} k[V] \otimes k[t] \xrightarrow{\tilde{\phi}} H^*_G(X;k)$$

is an augmented (graded) presentation of the $k[t]$-algebra $H^*_G(X;k)$ (see Section A.7).)

To construct $\tilde{\phi}$ one chooses a grading preserving k-linear map

$$V \to \ker \tilde{\varepsilon} \subset H^*_G(X;k)$$

such that the composition with the projection

$$H^*_G(X;k) \to H^*_G(X;k) \otimes_{k[t]} k_0 = H^*(X;k)$$

coincides with the restriction $\phi|V$ of ϕ to $V \subset k[V]$. There then is a unique extension of $V \to \ker \tilde{\varepsilon}$ to a morphism

$$\tilde{\phi} : k[V] \otimes k[t] \to H^*_G(X;k)$$

of augmented, graded $k[t]$-algebras which, by construction, fulfils

$$\tilde{\phi} \otimes_{k[t]} k_0 = \phi.$$

Since ϕ is surjective by assumption, so is $\tilde{\phi}$ (and hence $Q(\tilde{\phi})$) (see (A.7.3)(2) and (A.7.17)(3)). The construction of $\tilde{\psi}$ is similar. One chooses a k-linear map $W \to \ker \tilde{\phi} \subset k[V] \otimes k[t]$ such that the composition with the projection $k[V] \otimes k[t] \to (k[V] \otimes k[t]) \otimes_{k[t]} k_0 = k[V]$ is equal to $\psi|W$. This is possible since $\tilde{\phi}$ has a $k[t]$-linear splitting (because $H^*_G(X)$ is a free $k[t]$-module). It remains to show that

$$Q(\tilde{\psi}) : W \otimes k[t] \to \ker \tilde{\phi}/(\ker \tilde{\phi}) \cdot (\overline{k[V]} \otimes k[t]) =: \tilde{K}$$

is surjective; and, as above, it suffices to observe that

$$Q(\tilde{\psi}) \otimes_{k[t]} k_0 : W \to \tilde{K} \otimes_{k[t]} k_0$$

is surjective; but by construction,
$$Q(\tilde{\psi}) \otimes_{k[t]} k_0 : W \to \tilde{K} \otimes_{k[t]} k_0$$
coincides with
$$Q(\tilde{\psi}) : W \to \ker \phi/(\ker \phi)(\overline{k[V]}) \cong \tilde{K} \otimes_{k[t]} k_0.$$

We now evaluate the sequence
$$k[W] \otimes k[t] \xrightarrow{\tilde{\psi}} k[V] \otimes k[t] \xrightarrow{\tilde{\phi}} H^*_G(X;k)$$
of augmented, graded $k[t]$-algebras at $t = \alpha \neq 0$ to get a sequence
$$k[W] \xrightarrow{\tilde{\psi}_\alpha} k[V] \xrightarrow{\tilde{\phi}_\alpha} H^*_G(X;k)_\alpha \cong H^{(*)}(X^G;k)$$
(see (1.3.5)) of augmented k-algebras and augmentation preserving algebra morphisms. (Note that though the algebras involved can be equipped with a grading in a rather natural way, the morphisms between them may not respect the grading.) We compose $\tilde{\phi}_\alpha$ with the morphism $H^{(*)}(X^G;k) \to H^{(*)}(F;k)$ induced by the inclusion $F \subset X^G$ to get $\tilde{\phi}_F : k[V] \to H^{(*)}(F,k)$, and consider the sequence
$$k[W] \xrightarrow{\tilde{\psi}_\alpha} k[V] \xrightarrow{\tilde{\phi}_F} H^{(*)}(F;k)$$
of augmented k-algebras. Since $\tilde{\phi}$ is surjective, so is $\tilde{\phi}_\alpha$, and hence $\tilde{\phi}_F$ and $Q(\tilde{\phi}_F) : V \to Q(H^{(*)}(F;k))$. Therefore
$$\mathfrak{g}(H^*(X;k)) = \dim_k V \geq \dim Q(H^{(*)}(F;k)) = \mathfrak{g}(H^{(*)}(F:k)),$$
which proves (1.3.17)(1). Also $Q(\tilde{\psi}_\alpha) : W \to \ker \tilde{\phi}_\alpha/(\ker \alpha)\overline{k[V]}$ is surjective since $Q(\tilde{\psi})$ is (and $Q(\tilde{\psi})_\alpha = Q(\tilde{\psi}_\alpha)$). We claim that the inclusion $\ker \tilde{\phi}_\alpha \subset \ker \tilde{\phi}_F$ induces a surjection
$$\ker \tilde{\phi}_\alpha/(\ker \tilde{\phi}_\alpha)\overline{k[V]} \to \ker \tilde{\phi}_F/(\ker \tilde{\phi}_F)\overline{k[V]}.$$

The algebra $H^*(X^G;k)$ is the direct product $H^*(F;k) \times H^*(F';k)$ where $F' := X^G - F$ and $\ker \tilde{\phi}_F = \tilde{\phi}_\alpha^{-1}(0, H^*(F';k))$. Choose $e \in \overline{k[V]}$ such that $\tilde{\phi}_\alpha(e) = (0,1)$. If $x \in \ker \tilde{\phi}_F$ then the equation $x = (x - xe) + xe$ shows that $x \in \ker \tilde{\phi}_\alpha + (\ker \tilde{\phi}_F) \cdot \overline{K[V]}$, hence the above claim holds.

What we have proved so far shows that the sequence
$$k[W] \xrightarrow{\tilde{\psi}_\alpha} k[V] \xrightarrow{\tilde{\phi}_F} H^{(*)}(F;k)$$
is very close to being an augmented presentation of $H^{(*)}(F;k)$; in fact the only difference is that instead of having
$$\tilde{\psi}_\alpha(\overline{k[W]}) \cdot k[V] = \ker \tilde{\phi}_F$$
we only have proved the weaker fact that
$$Q(\tilde{\psi}_\alpha) : W \to \ker \tilde{\phi}_F/(\ker \tilde{\phi}_F) \cdot k[V]$$
is surjective.

Nevertheless (see Section A.7, in particular (A.7.16) and (A.7.17)) we get

$$\begin{aligned}\operatorname{cid}(H^*(F;k)) &= \mathfrak{r}(H^*(F;k)) - \mathfrak{g}(H^*(F;k)) \\ &\leq \dim_k \ker \tilde{\phi}_F / (\ker \tilde{\phi}_F)\overline{k[V]} - \dim_k V \\ &\leq \dim_k W - \dim_k V \\ &\leq \mathfrak{r}(H^*(X;k)) - \mathfrak{g}(H^*(X;k)) \\ &= \operatorname{cid}(H^*(X;k)).\end{aligned}$$

This proves part (3); and part (2) is, of course, an immediate consequence of (1) and (3). □

(1.3.18) Remark

The proof of (1.3.17) shows in particular that (under the above assumptions) for any presentation of the k-algebra $H^*(X;k)$ as a quotient $H^*(X;k) = k[x_1,\ldots,x_n]/(f_1,\ldots,f_m)$ of a polynomial ring $k[x_1,\ldots,x_n]$ in n (graded) variables divided by a homogeneous ideal $(f_1,\ldots,f_m)$ generated by the elements $f_1,\ldots,f_m$ one can find a presentation of $H_G^*(X;k) = k[x_1,\ldots,x_n,t]/(F_1,\ldots,F_m)$ where F_i, $i=1,\ldots,m$ are homogeneous polynomials in $k[x_1,\ldots,x_n,t]$ such that the evaluation of F_i at $t=0$ gives f_i for $i=1,\ldots,m$ (i.e. F_i can be written as $F_i = f_i + tG_i$, $i=1,\ldots,m$, for some $G_i \in k[x_1,\ldots,x_n,t]$.

(1.3.19) Example

Let X be a finite-dimensional G-CW-complex ($G = \mathbb{Z}_2$) such that $H^*(X;k) \cong k[x]/(x^{n+1})$, where $\deg(x) = 1$, i.e. the cohomology with coefficients in k looks like that of the n-dimensional real projective space (k a field of charcteristic 2). We assume that $X^G \neq \emptyset$. Then the projection $\pi : X \to *$ and the inclusion $\iota : * \to X$ of one fixed point induce morphisms

$$\pi^* : k[t] \to H^*(X;k)\tilde{\otimes}k[t]$$

and

$$\iota^* : H^*(X;k)\tilde{\otimes}k[t] \to k[t]$$

such that $\iota^*\pi^* = \operatorname{id}_{k[t]}$. It follows that

$$\tilde{\delta}(x \otimes 1) = 0$$

($\tilde{\delta}(x \otimes 1)$ cannot have a non-trivial component in $1 \otimes k[t] = \pi^*(k[t])$ since this is a direct summand of $H^*(X;k)\tilde{\otimes}k[t]$ as a $k[t]$-complex; and it cannot have a non-trivial component in $H^*(X;k) \otimes t$ since this would correspond to a non-trivial action of G on $H^1(X;k) \cong k$ which is impossible (since this action of G on $k \cong \mathbb{Z}_2 \otimes k$ is induced by an action on $\mathbb{Z}_2$);

and other components in $\tilde{\delta}(x \otimes 1)$ cannot occur for degree reasons). The product on $S^*(X;k)\tilde{\otimes}k[t]$ induces a product $\tilde{\mu}$ on $H^*(X;k)\tilde{\otimes}k[t]$ which evaluated at $t = 0$ gives the cup-product in $H^*(X;k)$ (up to homotopy). Hence, on cocycles in $H^*(X;k)\tilde{\otimes}k[t]$, the evaluation at $t = 0$ coincides precisely with the product in $H^*(X;k)$.

Since $\tilde{\mu}$ must be compatible with the differential one has
$$\tilde{\delta}\tilde{\mu}((x \otimes 1) \otimes (x \otimes 1)) = \tilde{\mu}(\tilde{\delta}(x \otimes 1) \otimes (x \otimes 1) + (x \otimes 1) \otimes \tilde{\delta}(x \otimes 1)) = 0$$
(since $\tilde{\delta}(x \otimes 1) = 0$). But
$$\tilde{\mu}((x \otimes 1) \otimes (x \otimes 1)) = x^2 \otimes 1 + a(x \otimes t) + b(1 \otimes t^2)$$
for some $a,\, b \in k$. Since $\tilde{\delta}(x \otimes t) = \tilde{\delta}(x \otimes 1)t$ and $\tilde{\delta}(1 \otimes t^2) = 0$, we get $\tilde{\delta}(x^2 \otimes 1) = 0$. Inductively one shows that $\tilde{\delta}(x^i \otimes 1) = 0$ for $i = 1, \ldots, n$ and hence $\tilde{\delta}$ must be trivial on $H^*(X;k)\tilde{\otimes}k[t]$. (This could also be concluded using the multiplicative properties of the Serre spectral sequence for the fibration $X \to EG \times_G X \to BG$.)

The algebra $H_G^*(X;k) \cong H^*(X;k)\tilde{\otimes}k[t]$ can be presented as
$$H^*(X;k)\tilde{\otimes}k[t] \cong k[x,t]/(x^{n+1} + b_n x^n t + \ldots b_0 t^{n+1})$$
with $b_i \in k$ for $i = 0, \ldots, n$ (see (1.3.18)) and the evaluation at $\alpha \neq 0$, e.g., $\alpha = 1$, gives $H^*(X^G;k)$ as a k-algebra (see (1.3.5)). If k is algebraically closed then
$$F_1(x) := x^{n+1} + b_n x^n + \ldots + b_0 = \prod_{i=1}^{\nu}(x - a_i)^{n_i+1}$$
where $a_i \in k$ are the (pairwise disjoint) zeros of $F_1(x)$ and $n_i + 1$ the corresponding multiplicities ($\sum_{i=1}^{\nu}(n_i + 1) = n + 1$). By the Chinese Remainder Theorem
$$k[x,1]/(F_1(x)) \cong \prod_{i=1}^{\nu} k[x_i]/(x_i^{n_i+1}).$$
This decomposition must correspond to the decomposition of
$$H^*(X^G;k) \cong \prod_{i=1}^{\nu} H^*(F_i;k),$$
where F_i, $i = 1, \ldots, \nu$ are the components of X^G. Since the latter decomposition must hold already over $\mathbb{F}_2$ if it holds over k, so does the former; and hence there are only two possible values for the a_i, namely 0 and 1; and $F_1(x) = x^j(x-1)^{n+1-j}$ for some $0 \leq j \leq n+1$.

In particular we get under the above assumption, that the fixed point set X^G consists of at most two components and each component has the k-cohomology of a real projective space of dimension $\leq n$. In Chapter 3 we discuss the phenomenon, which occurs in this example, in greater generality using more algebraic machinery.

(1.3.20) Remark

The invariant $\mathrm{cid}(A^*) = \mathrm{r}(A^*) - \mathrm{g}(A^*)$ for A^* a connected graded commutative algebra with $\dim_k A^* < \infty$ is called the 'complete intersection defect of A^*', since the condition '$\mathrm{cid}(A^*) = 0$' is equivalent to 'A^* is a complete intersection' (see, e.g., [Kunz, 1985] for algebraic background on complete intersections, and note that in our case the Krull dimension of A^* is zero (because of the condition $\dim_k A^* < \infty$))(cf. Appendix A). For an algebra A^* as above $\mathrm{cid}(A^*)$ can never be negative, i.e. to present such an algebra one needs at least as many relations as one needs generators (see, e.g., [Kunz, 1985]). Part (3) of (1.3.17) therefore implies in particular that $H^*(F; k)$ is a complete intersection if $H^*(X; k)$ is.

By (1.3.5) and (1.3.14) many results from the theory of deformations of algebras can be translated to results in transformation groups, Theorems (1.3.16) and (1.3.17) being fairly simple examples. This could be done in a more thorough and systematic way (see [Hauschild, 1983] and [Puppe, V., 1984] for some steps in this direction), and leads to results of quite a different nature than (1.3.16) and (1.3.17): e.g., one can exhibit the existence of simply connected manifolds with little symmetry (see [Puppe, V. 1988]). But one would need some more machinery from deformation theory, which we are not going to present in this book.

Some of the results we have derived so far can easily be generalized to finite 2-groups (i.e. the order of the group is a power of 2) by a simple induction argument. Let G be a 2-group and X a G-CW-complex. Since the centre of G is non-trivial (G is solvable, cf. Exercise (1.5)) one can find a normal subgroup $\mathbb{Z}_2 < G$. The quotient $\overline{G} = G/(\mathbb{Z}_2)$ acts on the fixed point set $F(\mathbb{Z}_2, X) := X^{\mathbb{Z}_2}$ of the subgroup $\mathbb{Z}_2$ and the fixed point set $F(\overline{G}, F(\mathbb{Z}_2, X))$ of this action clearly coincides with $F(G, X) = X^G$. The order of $\overline{G}$ is of course smaller than the order of G. If one has a result of the following type: 'If X is a finite-dimensional $\mathbb{Z}_2$-CW-complex which (as a topological space disregarding the action) has the property $\mathcal{P}$, then the fixed point set $X^{\mathbb{Z}_2}$ also has this property', then an induction argument with respect to the order of the group G gives the analogous result for G-CW-complexes X where G is a finite 2-group. (The beginning of the induction is clear and the induction step uses the fact that $F(G, X) = F(\overline{G}, (F(\mathbb{Z}_2, X)))$ (see above), i.e. if X has property $\mathcal{P}$ then $F(\mathbb{Z}_2, X) = X^{\mathbb{Z}_2}$ has property $\mathcal{P}$ by the original result, and hence $F(\overline{G}, X^{\mathbb{Z}_2})$ has property $\mathcal{P}$ by the induction hypothesis).

If, for example, property $\mathcal{P}$ means that $\sum_{i=0}^{\infty} \dim_k H^{m+i}(X)$ is smaller than or equal to a constant N_m for $m \in \mathbb{N}$, then (1.3.8)(1) is a result of the type in question (choosing $N_m := \sum_{i=0}^{\infty} \dim_k H^{m+i}(X; k)$ for $m \in \mathbb{N}$).

The above argument gives:

(1.3.21) Remark
The results (1.3.8)(1), (3); (1.3.16)(1), (2), (3) and (1.3.17)(1), (2), (3) hold for G being any finite 2-group instead of $\mathbb{Z}_2$.

This is immediate for (1.3.8)(1), (3), (cf. (4.1.8) for a generalization of (1.3.8)(2) to 2-tori) whereas for (1.3.16) and (1.3.17) one has to make sure that the necessary assumptions for the induction step are fulfilled. Let G be a finite 2-group, $\mathbb{Z}_2 < G$ a normal subgroup, $\overline{G} = G/(\mathbb{Z}_2)$ and X a G-CW-complex such that
$$\dim_k H^{(*)}(X;k) = \dim_k H^{(*)}(X^G;k),$$
then the inequalities
$$\dim_k H^{(*)}(X;k) \geq \dim_k H^{(*)}(X^{\mathbb{Z}_2};k) \geq \dim_k H^{(*)}(F(\overline{G}, X^{\mathbb{Z}_2});k)$$
(see (1.3.17)(1)) actually must be equalities. Of course, X need not be connected and therefore the assumptions of (1.3.16) resp. (1.3.17), may actually not be satisfied for $X^{\mathbb{Z}_2}$. But one can choose a component Y of $X^{\mathbb{Z}_2}$ which contains the component F of X^G one is interested in, and the equality
$$\dim_k H^{(*)}(X^{\mathbb{Z}_2};k) = \dim_k H^{(*)}(F(\overline{G}, X^{\mathbb{Z}_2});k)$$
implies that Y is invariant under $\overline{G}$ and again fulfils
$$\dim_k H^{(*)}(Y;k) = \dim_k H^{(*)}(F(\overline{G}, Y);k).$$

If one wants to discuss free G-actions and, for example, determine necessary conditions on a space X to allow a free action of a 2-torus, an inductive procedure to reduce the problem to the $\mathbb{Z}_2$-case as above does not seem adequate. We therefore now extend the constructions given for $G = \mathbb{Z}_2$ in the beginning of this section to $G = (\mathbb{Z}_2)^r$. As a universal free G-space EG we can take the r-fold product of the $\mathbb{Z}_2$-CW-complex $E\mathbb{Z}_2$. The corresponding cellular chain complex $W_*(EG;k) = \mathcal{E}_*(G)$ is just the r-fold tensor product of $W_*(E\mathbb{Z}_2;k) = \mathcal{E}_*(\mathbb{Z}_2)$ considered as a complex over the group ring $k[G] = k[\mathbb{Z}_2] \otimes \ldots \otimes k[\mathbb{Z}_2]$ with the componentwise action. Many of the constructions and calculations in the first part of this section can be carried through 'componentwise' in an analagous fashion. For example, the Künneth formula and Proposition (1.3.2)(3) give:

(1.3.22) Proposition
If k is a field of characteristic 2 and $G = (\mathbb{Z}_2)^r$ a 2-torus of rank r, then
$$H^*(BG;k) = H^*_G(*;k) \cong k[t_1, \ldots, t_r],$$
with $\deg(t_i) = 1$ for $i = 1, \ldots, r$.

Proposition (1.3.4) leads to the following result:

(1.3.23) Proposition
If C^* is an object in $\delta gk[G]$-Mod, then $\beta_G^*(C^*)$ is isomorphic to $C^* \otimes k[t_1, \ldots, t_r]$ as a right $k[t_1, \ldots, t_r]$-module.

The differential of $\beta_G^*(C^*)$ corresponds to a 'twisted' differential on $C^* \otimes k[t_1, \ldots, t_r]$, and - as before - we denote the last complex together with this differential by $C^* \tilde{\otimes} k[t_1, \ldots, t_r]$. To shorten the notation we put $R := H^*(BG; k) \cong k[t_1, \ldots, t_r]$.

We want now to get a generalization of the basic Theorem (1.3.5). For any morphism $\alpha : R \to k$ of k-algebras we consider the evaluation $\beta_{G,\alpha}(C^*) := \beta_G^*(C^*) \otimes_R k_\alpha$ of $\beta_G^*(C^*)$ at α, where $C^* \in \delta gk[G]$-Mod and k_α is the field k considered as an R-module via $\alpha : R \to k$.

Note that after a choice of coordinates for R, i.e. an isomorphism $R \cong k[t_1, \ldots, t_r]$, the morphism α corresponds to an n-tuple $(\alpha_1, \ldots, \alpha_r) \in k^r$, where $\alpha_i := \alpha(t_i)$, $i = 1, \ldots, r$. We denote by $\mathcal{L}(G)$ the set of all k-algebra morphisms $\alpha : R \to k$. A choice of coordinates for R gives a bijection $\mathcal{L}(G) \cong k^r$; in particular, $\mathcal{L}(G)$ can be considered as a k-vector space.

If X is a G-space we define the subset $X_\alpha \subset X$ by
$$X_\alpha := \{x \in X : \alpha \in \mathcal{L}(G_x)\}$$
where $G_x = \{g \in G : gx = x\}$ is the isotropy group of x, and $\mathcal{L}(G_x)$ is considered as a linear subspace of $\mathcal{L}(G)$ via the inclusion $\mathcal{L}(G_x) \to \mathcal{L}(G)$ induced by $G_x < G$, i.e. by
$$H^*(BG; k) \to H^*(BG_x; k).$$

(1.3.24) Theorem
Let X be a finite-dimensional G-CW-complex. For any $\alpha \in \mathcal{L}(G)$ there is a natural isomorphism
$$\tau(X) : H_{G,\alpha}(X; k) := H(\beta_{G,\alpha}(S^*(X; k))) \to H^{(*)}(X_\alpha; k).$$

We will use the following lemma to prove (1.3.24).

(1.3.25) Lemma
Let X' (resp. X'') be a finite G' (resp. G'')-CW-complex where G', G''' are 2-tori. Then for the $G = G' \times G''$-CW-complex $X := X' \times X''$ one has:

(1) $\beta_G^*(W^*(X;k)) \cong \beta_{G'}^*(W^*(X';k)) \otimes \beta_{G''}^*(W^*(X'';k))$; in particular, $H_G^*(X;k) \cong H_{G'}^*(X';k) \otimes H_{G''}^*(X'';k)$;

50   1. Equivariant cohomology and the Borel construction

(2) $\beta^*_{G,\alpha}(W^*(X;k)) \cong \beta^*_{G',\alpha'}(W^*(X';k)) \otimes \beta^*_{G'',\alpha''}(W^*(X'';k))$, where $\alpha = (\alpha', \alpha'') \in \mathcal{L}(G') \times \mathcal{L}(G'') = \mathcal{L}(G)$; in particular,
$$H^*_{G,\alpha}(X;k) \cong H^*_{G',\alpha'}(X';k) \otimes H^*_{G'',\alpha''}(X'';k).$$

Proof of (1.3.25).
(1) The cellular cochain complex $W^*(X;k)$ can be written as the tensor product $W^*(X;k) \cong W^*(X';k) \otimes W^*(X'';k)$ including the $k[G]$-structure, which on the tensor product is given by the $k[G']$ (resp. $k[G'']$)-structure of the factors. (Note that $k[G] \cong k[G'] \otimes k[G'']$.) Therefore

$\beta^*_G(W^*(X;k))$
$\cong (W^*(X';k) \otimes W^*(X'';k)) \otimes_{k[G']\otimes k[G'']} (\mathcal{E}^*(G') \otimes \mathcal{E}^*(G''))$
$\cong (W^*(X';k) \otimes_{k[G']} \mathcal{E}^*(G')) \otimes (W^*(X'';k) \otimes_{k[G'']} \mathcal{E}^*(G''))$
$\cong \beta^*_{G'}(W^*(X';k)) \otimes \beta^*_{G''}(W^*(X'';k))$.

(Note that this argument does not use the assumption that G', G'' are 2-tori; compare Exercise (1.17).)

(2) Part (2) follows from part (1) by applying the functor $- \otimes_R k_\alpha$. i.e.

$\beta_{G,\alpha}(W^*(X;k))$
$\cong \beta^*_G(W^*(X;k))) \otimes_R k_\alpha$
$\cong (\beta^*_{G'}(W^*(X';k)) \otimes \beta^*_{G''}(W^*(X'';k)) \otimes_R k_\alpha$
$\cong (\beta^*_{G'}(W^*(X';k)) \otimes_{R'} k_{\alpha'}) \otimes (\beta^*_{G''}(W^*(X'';k)) \otimes_{R''} k_{\alpha''})$
$\cong \beta_{G',\alpha'}(W^*(X';k)) \otimes \beta_{G'',\alpha''}(W^*(X'';k))$,

where
$$R = H^*(BG;k) \cong H^*(BG',k) \otimes H^*(BG'',k) = R' \otimes R''$$
and $\alpha : R \to k$ corresponds to $\alpha' \otimes \alpha'' : R' \otimes R'' \to k \otimes k = k$. The statements about the (co)homology follow, of course, from the Künneth formula since we work with coefficients in a field (of characteristic 2). □

Proof of (1.3.24).
This is similar to the proof of (1.3.5), i.e. we view $H_{G,\alpha}(X;k)$ and $H^{(*)}(X_\alpha;k)$ as equivariant cohomology theories (in the variable X) and consider the natural transformation $\tau(X)$ between them, which is induced essentially by the inclusion $X_\alpha \subset X$. We first show $H_{G,\alpha}(X_\alpha;k)$ is naturally isomorphic to $H^{(*)}(X_\alpha;k)$. Clearly $\beta_{G,\alpha}(W^*(X_\alpha;k)) = (W^*(X_\alpha;k)\tilde\otimes R) \otimes_R k_\alpha$ is naturally isomorphic to $W^{(*)}(X_\alpha;k)$ as a k-vector space. We claim that this isomorphism is compatible with the respective coboundaries; i.e. the twisting (or perturbation) of the coboundary on $W^*(X_\alpha;k)\tilde\otimes R$, which comes from the G-action on X_α, vanishes

1.3. The Borel construction for 2-tori

as one evaluates at α. Since this perturbation depends only on the $k[G]$-module structure of $W^*(X_\alpha; k)$, it suffices to check the above claim for each G-point G/G_x with $\alpha \in \mathcal{L}(G_x)$ separately. We choose a splitting $G \cong G' \times G''$ with $G' = G/G_x$, $G'' = G_x$. One gets corresponding decompositions

$$R = H^*(BG; k) \cong H^*(BG'; k) \otimes H^*(BG''; k) = R' \otimes R''$$

and

$$\mathcal{L}(G) = \mathcal{L}(G') \times \mathcal{L}(G'').$$

We now apply (1.3.25)(2) to the $(G' \times G'')$-space $G/G_x \times \{*\}$ and obtain

$$\beta_{G,\alpha}(W^*(G/G_x \times \{*\})) \cong \beta_{G',\alpha'}(W^*(G/G_x; k)) \otimes \beta_{G'',\alpha''}(W^*(*; k)).$$

The condition '$\alpha \in \mathcal{L}(G_x)$' is equivalent to '$\alpha|R'$: $R' \to k$ being trivial on $\overline{R}'$ (i.e. being the standard augmentation of R')'. Therefore

$$\beta_{G',\alpha'}(W^*(G/G_x; k)) \cong W^{(*)}(G/G_x; k).$$

Hence we obtain the desired isomorphism

$$\beta_{G,\alpha}(W^*(G/G_x; k)) \cong W^{(*)}(G/G_x; k)$$

(since clearly $\beta_{G'',\alpha''}(W^*(*; k)) \cong k$).

It remains to show that $X_\alpha \subset X$ induces an isomorphism

$$\tau(X) : H_{G,\alpha}(X; k) \to H_{G,\alpha}(X_\alpha; k).$$

As in the proof of (1.3.5) it suffices to check this for $X = \coprod G/K$, $K < G$, (X a possibly infinite disjoint union of orbits of the same type).

(i) If $\alpha \in \mathcal{L}(K)$ then $(\coprod G/K)_\alpha = \coprod G/K$ and there is nothing to prove.

(ii) If $\alpha \notin \mathcal{L}(K)$ then $(\coprod G/K)_\alpha = \emptyset$ and one has to show that $H_{G,\alpha}(\coprod G/K; k)$ vanishes.

Although $H_{G,\alpha}(-; k)$ is not strongly additive, it suffices to prove $H_{G,\alpha}(G/K; k) = 0$; see proof of (1.3.5). Lemma (1.3.25)(2) implies that $H_{G,\alpha}(G/K; k)$ is isomorphic to $H_{G/K,\overline{\alpha}}(G/K; k)$ with

$$\overline{\alpha} = \alpha | H^*(B(G/K); k).$$

Hence it suffices to consider the case $K = \{1\} < G$ and $\alpha \neq 0$. We can choose a splitting $G = G' \times G''$ ($\mathcal{L}(G) \cong \mathcal{L}(G') \times \mathcal{L}(G'')$) such that $G' = \mathbb{Z}_2$ and $\alpha \neq 0$ ($\alpha = (\alpha', \alpha'') \in \mathcal{L}(G') \times \mathcal{L}(G'')$). By (1.3.25)(2) we get

$$H_{G,\alpha}(G; k) \cong H_{G',\alpha'}(G'; k) \otimes H_{G'',\alpha''}(G''; k).$$

From (1.3.5) we know that $H_{G',\alpha'}(G'; k)$ vanishes, and hence so does $H_{G,\alpha}(G; k)$. $\square$

The construction $\beta_{G,\alpha}(X;k)$ can be obtained in two steps; one first takes the tensor product of $\beta_G(X;k)$ with $k[t]_{\tilde{\alpha}}$ over $R = k[t_1,\ldots,t_r]$, where $k[t]_{\tilde{\alpha}}$ is $k[t]$ considered as an R-module via the algebra morphism $\tilde{\alpha} : k[t_1,\ldots,t_r] \to k[t]$, $t_i \mapsto \alpha_i t$ ($\alpha = (\alpha_1,\ldots,\alpha_r) \in k^r$); and then one evaluates at $t = 1$ (i.e. applies the functor $- \otimes_{k[t]} k_1$).

If $\alpha = (\alpha_1,\ldots,\alpha_r) \in \mathbb{F}_2^r \subset k^r$ then the first step can be given a geometrical meaning. The elements $(\alpha_1,\ldots,\alpha_r) \in \mathbb{F}_2^r$ are in one-to-one correspondence with group homomorphisms (again denoted by α) of $\mathbb{Z}_2$ to $G = (\mathbb{Z}_2)^r$; one just defines $\alpha : \mathbb{Z}_2 \to (\mathbb{Z}_2)^r$ by $\alpha(1) := (\alpha_1,\ldots,\alpha_r)$ (or written multiplicatively: $\alpha(g) := g_1^{\alpha_1}\ldots g_r^{\alpha_r}$, where $g \in \mathbb{Z}_2$ is the generator and g_i is the generator of the ith component in $(\mathbb{Z}_2)^r$). Any such α defines corresponding maps for the group rings, $\mathcal{E}_*(-)$, $H_-^*(*)$ etc. If C^* is an object in $\delta gk[G]$-Mod then we can consider the restriction of the G-action to a $\mathbb{Z}_2$-action via α and obtain the object $\text{res}_\alpha C^*$ in $\delta gk[\mathbb{Z}_2]$-Mod. We want to compare $\beta_G^*(C^*) \cong C^*\tilde{\otimes}R$ and $\beta_{\mathbb{Z}_2}^*(\text{res}_\alpha C^*) \cong \text{res}_\alpha C^*\tilde{\otimes}k[t]$. We get a morphism

$$\rho_\alpha(C^*) : \beta_G^*(C^*) = \text{Hom}_{k[G]}(\mathcal{E}_*(G), C^*) \to \text{Hom}_{k[\mathbb{Z}_2]}(\mathcal{E}_*(\mathbb{Z}_2), \text{res}_\alpha C^*)$$
$$= \beta_{\mathbb{Z}_2}^*(\text{res}_\alpha C^*)$$

by composing $\phi \in \text{Hom}_{k[G]}(\mathcal{E}_*(G), C^*)$ with $\mathcal{E}_*(\alpha) : \mathcal{E}_*(\mathbb{Z}_2) \to \mathcal{E}_*(G)$. As a map of resolutions the map $\mathcal{E}_*(\alpha)$ is only determined up to appropriate homotopy, but one can choose $\mathcal{E}_*(\alpha)$ and the diagonal of $\mathcal{E}_*(G)$ and $\mathcal{E}_*(\mathbb{Z}_2)$ in such a way that $\rho_\alpha(C^*) : \beta_G^*(C^*) \to \beta_{\mathbb{Z}_2}^*(\text{res}_\alpha C^*)$ becomes a morphism over

$$\rho_\alpha(k) : R \cong \beta_G^*(k) = \text{Hom}_{k[G]}(\mathcal{E}_*(G), k) \to \text{Hom}_{k[\mathbb{Z}_2]}(\mathcal{E}_*(\mathbb{Z}_2), k)$$
$$= \beta_{\mathbb{Z}_2}^*(k) \cong k[t].$$

where $\beta_G^*(C^*)$ is considered as an R-module and $\beta_{\mathbb{Z}_2}^*(\text{res}_\alpha C^*)$ as a $k[t]$-module. We leave it to the reader to verify that $\rho_\alpha(k)$ coincides with the above $\tilde{\alpha} : R \cong k[t_1,\ldots,t_r] \to k[t]$, $t_i \mapsto \alpha_i t$. Up to a change of coordinates, i.e. an isomorphism $G \cong (\mathbb{Z}_2)^r$, one can assume $(\alpha_1,\ldots,\alpha_r) = (1,0,\ldots,0)$, and this special case is easily checked. (The similar - but somewhat more complicated - situation for $(\mathbb{Z}_p)^r$, where p is an odd prime, is considered in detail in Section 3.11.) Hence $\rho_\alpha(C^*)$ induces a morphism

$$\overline{\rho_\alpha(C^*)} : \beta_G^*(C^*) \otimes_R k[t]_{\tilde{\alpha}} \to \beta_{\mathbb{Z}_2}^*(\text{res}_\alpha C^*).$$

(1.3.26) Proposition

The map $\overline{\rho_\alpha(C^*)}$ is an isomorphism in $\delta gk[t]$-Mod.

1.3. The Borel construction for 2-tori

Proof.
As an R-module $\beta_G^*(C^*)$ is isomorphic to $C^* \otimes R$ (see (1.3.23)) and hence $\beta_G^*(C^*) \otimes_R k[t]_{\tilde{\alpha}}$ is isomorphic to $C^* \otimes k[t]$ as a $k[t]$-module. But $\beta_{\mathbb{Z}_2}^*(\text{res}_\alpha C^*)$ is isomorphic to $C^* \otimes k[t]$ as a $k[t]$-module, too. Using these isomorphisms the map $\overline{\rho_\alpha(C^*)}$ need not become the identity, but $\overline{\rho_\alpha(C^*)}$ evaluated at $t = 0$, i.e. $\overline{\rho_\alpha(C^*)} \otimes_{k[t]} k_0 : C^* \to C^*$ will be an isomorphism (in fact, the identity if one chooses the isomorphisms according to (1.3.23) and (1.3.4)). Hence (A.7.3) implies that $\overline{\rho_\alpha(C^*)}$ is itself an isomorphism. $\square$

If X is a G-space then by (1.3.26)
$$H_{G,\tilde{\alpha}}^*(X;k) := H(\beta_{G,\tilde{\alpha}}^*(X;k)) = H(\beta_G^*(X;k) \otimes_R k[t]_{\tilde{\alpha}})$$
$$= H(\beta_{\mathbb{Z}_2}^*(\text{res}_\alpha X;k))$$
$$= H_{\mathbb{Z}_2}^*(\text{res}_\alpha X;k)$$
is the equivariant cohomology of $\text{res}_\alpha X$ with coefficients in k, where $\text{res}_\alpha X$ denotes X as a $\mathbb{Z}_2$-space with the action given by $\alpha : \mathbb{Z}_2 \to G$. Since the evaluation at $t = 1$ is exact on graded $k[t]$-modules one gets for X a finite dimensional G-CW-complex (see (1.3.5))
$$H_{G,\alpha}(X;k) = H_{G,\tilde{\alpha}}^*(X;k) \otimes_{k[t]} k_1 \cong H_{\mathbb{Z}_2,1}(\text{res}_\alpha X;k) \cong H^{(*)}(X^\alpha,k),$$
where X^α denotes the fixed point set of the $\mathbb{Z}_2$-action on X given by α.

It is easy to check that X^α coincides with X_α defined above: $x \in X^\alpha$ if and only if $\alpha : \mathbb{Z}_2 \to G$ factors through G_x. The map $\alpha : \mathbb{Z}_2 \to G$ induces on the level of the cohomology of the classifying spaces the map
$$\tilde{\alpha} : H^*(BG;k) \cong k[t_1,\ldots,t_r] \to k[t] = H^*(B\mathbb{Z}_2;k).$$
The composition of $\tilde{\alpha}$ with the evaluation at $t = 1$ gives the map
$$\alpha : R = k[t_1,\ldots,t_r] \xrightarrow{\tilde{\alpha}} k[t] \to k.$$
(The reader should not get confused by the use of the symbol α for two different - though closely related - maps.) If $x \in X^\alpha$ then
$$\tilde{\alpha} : H^*(BG;k) \to H^*(B\mathbb{Z}_2;k)$$
factors through $H^*(BG_x;k)$ and hence $\alpha \in \mathcal{L}(G_x)$, i.e. $x \in X_\alpha$. On the other hand if $\alpha : \mathbb{Z}_2 \to G$ does not factor through G_x then the map induced by the composition
$$\mathbb{Z}_2 \xrightarrow{\alpha} G \to G/G_x,$$
$H^*(B(G/G_x);k) \to H^*(B\mathbb{Z}_2) \to k$, is non-trivial on $H^1(B(G/G_x);k)$ and hence $\alpha \notin \mathcal{L}(G_x)$, i.e. $x \notin X_\alpha$. (Note that
$$H^*(BG;k) \cong H^*(B(G/G_x);k) \otimes H^*(BG_x;k)$$
since
$$BG = B(G/G_x) \times BG_x;$$

54   1. Equivariant cohomology and the Borel construction

and in particular
$$H^1(BG;k) \cong H^1(B(G/G_x);k) \oplus H^1(BG_x;k)).$$
Hence (1.3.26) together with (1.3.5) give another proof of (1.3.24) for $\alpha \in \mathbb{F}_2^r$. But one essential aspect of (1.3.24) is the possibility of choosing any $\alpha \in k^r \cong \mathcal{L}(G)$. This becomes clear from the following relations between $\alpha \in \mathcal{L}(G)$ and X_α.

(1.3.27) Proposition
Let $\alpha \in \mathcal{L}(G) \cong k^r$ and let
$$X_\alpha := \{x \in X; \alpha \in \mathcal{L}(G_x)\},$$
where X is a G-space with $G = (\mathbb{Z}_2)^r$, then
(1) $X_0 = X$ ($\alpha = 0$).
(2) For any $\alpha \in \mathcal{L}(G)$ there exists a subgroup $K(\alpha) < G$, such that $X_\alpha = X^{K(\alpha)}$.
(3) If k is an infinite field then for any subgroup $K < G$ there exists an $\alpha(K) \in \mathcal{L}(G)$ such that $X^K = X_{\alpha(K)}$.
(4) If $\mathcal{K}(X) := \{K \lneq G,$ such that $K = G_x$ for some $x \in X\}$ is the (finite) set of proper isotropy subgroups that 'occur in X', then $X^G = X_\alpha$ if and only if $\alpha \notin \bigcup_{K \in \mathcal{K}(X)} \mathcal{L}(K)$.

Proof:
  Parts (1) and (4) follow directly from the definition of X_α. But it should be noted that for part (4) it could happen that
$$\mathcal{L}(G) = \bigcup_{K \in \mathcal{K}(X)} \mathcal{L}(K)$$
if k is a finite field and hence in that case there is no α such that $X^G = X_\alpha$. For part (2) we put $K(\alpha) := \bigcap_{x \in X_\alpha} G_x$. Then clearly $X^{K(\alpha)} \supseteq X_\alpha$. Assume on the other hand that $\tilde{x} \in X^{K(\alpha)}$, then $K(\alpha) \subseteq G_{\tilde{x}}$ and hence
$$\mathcal{L}(K(\alpha)) \subset \mathcal{L}(G_{\tilde{x}}).$$
But $\alpha \in \mathcal{L}(G_x)$ for all $x \in X_\alpha$ implies
$$\alpha \in \mathcal{L}(K(\alpha)) = \bigcap_{x \in X_\alpha} \mathcal{L}(G_x)$$
and therefore $\alpha \in \mathcal{L}(G_{\tilde{x}})$, i.e. $\tilde{x} \in X_\alpha$, which gives the inclusion $X^{K(\alpha)} \subseteq X_\alpha$. For part (3) the assumption that k be infinite is essential (cf. the comment concerning part (4)). For any
$$\alpha \in \mathcal{L}(K) - \left(\bigcup_{G_x \not\supseteq K} \mathcal{L}(G_x) \right)$$

it follows by definition of X_α that $X_\alpha = X^K$, since this choice of α implies:
$$\alpha \in \mathcal{L}(G_x) \iff G_x \not\supset K$$
(cf. part (4) for the special case $K = G$). It only remains to show that
$$\mathcal{L}(K) - \left(\bigcup_{G_x \not\supset K} \mathcal{L}(G_x) \right) \neq \emptyset.$$
For every $G_x \not\supset K$ the intersection $\mathcal{L}(K) \cap \mathcal{L}(G_x) = \mathcal{L}(K \cap G_x)$ is a linear subspace of $\mathcal{L}(K)$ of strictly lower dimension and there are only finitely many isotropy groups G_x. Hence $\mathcal{L}(K) - (\bigcup_{G_x \not\supset K} \mathcal{L}(G_x))$ is the complement of a finite union of subspaces of strictly lower dimension in $\mathcal{L}(K)$ and therefore it is a non-empty set if the ground field contains infinitely many elements. Note that this set is in fact infinite if $K \neq 1$ (more precisely has 'dimension' equal to $\dim_k \mathcal{L}(K)$). □

(1.3.28) Remark
It is obvious from the proof of (1.3.27)(3) that in the case of an infinite field k there are (infinitely) many $\alpha \in \mathcal{L}(G)$ for a given subgroup $K < G$ such that $X^K = X_\alpha$, and there is no canonical choice. Also, for a given $\alpha \in \mathcal{L}(G)$ there can be (finitely) many subgroups $K < G$ such that $X_\alpha = X^K$. But here we can choose a subgroup $K(\alpha) < G$ with $X_\alpha = X^{K(\alpha)}$ in a canonical way by the following procedure.

Any projection $G \to G/K$, where $K < G$ is a subgroup, induces an inclusion $H^*(B(G/K); \mathbb{F}_2) \to H^*(BG; \mathbb{F}_2)$, and in particular in degree one: $H^1(B(G/K); \mathbb{F}_2) \to H^1(BG; \mathbb{F}_2) \cong \mathbb{F}_2^r$. On the other hand any $\mathbb{F}_2$-subspace V of $H^1(BG; \mathbb{F}_2)$ can be realized as $H^1(B(G/K); \mathbb{F}_2)$ for an appropriately chosen $K < G$. In fact, there is a one-to-one correspondence between the linear subspaces of $H^1(BG; \mathbb{F}_2)$ and the subgroups of G. The dimension of the subspace is equal to the corank of the corresponding subgroup (cf. Exercise (1.18)). Let V be the kernel of the map $\alpha^1 : H^1(BG; \mathbb{F}_2) \to k$, where α^1 denotes the restriction of $\alpha : H^*(BG; \mathbb{F}_2) \to k$ to
$$H^1(BG; \mathbb{F}_2) \subset H^1(BG; k) \subset H^*(BG; k).$$
Then the corresponding $K < G$ has the desired properties, i.e. $X_\alpha = X^K$ since the following conditions are equivalent:
(i) $\alpha \in \mathcal{L}(G_x)$ (i.e. $x \in X_\alpha$);
(ii) $\alpha^1 | H^1(B(G/G_x); \mathbb{F}_2) = 0$;
(iii) $H^1(B(G/G_x); \mathbb{F}_2) \subset H^1(B(G/K); \mathbb{F}_2)$, as subspaces of $H^1(BG; \mathbb{F}_2)$;
(iv) $K < G_x$ (i.e. $x \in X^K$).

It is easy to check that the above conditions are equivalent (the implication '(iii)⇒(iv)' may not be completely obvious, but follows from Exercise (1.18)). The subgroup

$$K(\alpha) = \bigcap_{x \in X_\alpha} G_x$$

given in the proof of (1.3.27)(2) can be bigger than the one described in (1.3.28). While the second is independent of the G-space X under consideration, the first is maximal with the property $X_\alpha = X^{K(\alpha)}$ for a given G-space X. We will discuss these aspects in greater detail in Section 3.11 for the somewhat more complicated case of $(\mathbb{Z}_p)^r$-actions, where p is an odd prime. The arguments there can be easily specialized to handle the case of $(\mathbb{Z}_2)^r$-actions.

The following corollary demonstrates that it is sometimes useful to be able to choose $\alpha \in k^r$ in (1.3.24), where k is the algebraic closure of $\mathbb{F}_2$. More general and also stronger results in the same direction are discussed in Section 1.4 and Chapters 3 and 4.

(1.3.29) Corollary

Let X be a finite-dimensional G-CW-complex such that the G-action is free; $G \cong (\mathbb{Z}_2)^r$. Then

$$\dim_k H^{(*)}(X;k) \geq r+1,$$

where k is a field of characteristic 2.

Proof.

We may assume that X is connected and that k is the algebraic closure of $\mathbb{F}_2$. Since $X_\alpha = \emptyset$ for all $\alpha \in \mathcal{L}(G)$, $\alpha \neq 0$, ($G_x = \{1\}$ for all $x \in X$) Theorem (1.3.24) implies: $H^*_{G,\alpha}(X;k) = 0$ for all $\alpha \neq 0$.

Let $H^*(X;k) \tilde{\otimes} k[t_1, \ldots, t_r]$ be the minimal Hirsch-Brown model for the G-space X. Let $1, x_1, \ldots, x_s \in H^*(X;k)$ be a (homogeneous) basis of the k-vector space $\bigoplus_{i=0}^{\infty} H^i(X;k)$.

For degree reasons

$$1 \otimes 1 \in H^0(X;k) \otimes k[t_1, \ldots, t_r]$$

must be a cycle in

$$H^*(X;k) \tilde{\otimes} k[t_1, \ldots, t_r]$$

(here it is used that X is connected and hence the G-action on $H^0(X;k)$ is trivial). Modulo terms in

$$\bigoplus_{i \geq 1} H^i(X;k) \otimes k[t_1, \ldots, t_r]$$

1.3. The Borel construction for 2-tori

the coboundaries $\tilde{\delta}(x_i \otimes 1)$ of $x_i \otimes 1 \in H^*(X;k)\tilde{\otimes}k[t_1,\ldots,t_r]$ are given by polynomials

$$p_i(t_1,\ldots,t_r) \in k[t_1,\ldots,t_r]$$

($\tilde{\delta}(x_i \otimes 1)$ is a linear combination of $(1, x_1, \ldots, x_s)$ with coefficients in $k[t_1,\ldots,t_r]$, and $p_i(t_1,\ldots,t_r)$ is the coefficient of 1). Assume $s < r$, then there exists a non-zero

$$\alpha = (\alpha_1,\ldots,\alpha_r) \in k^r$$

such that $p_i(\alpha_1,\ldots,\alpha_r) = 0$ for all $i = 1,\ldots,s$. (Here we use that k is algebraically closed so that Hilbert's *Nullstellensatz* applies (see, e.g., [Matsumura, 1986]); and we use the Krull Height Theorem (A.1.5). See also Lemma (3.11.15).) But then $(1 \otimes 1)$ would represent a non-zero class in $H^*_{G,\alpha}(X;k)$, since it cannot become a boundary after evaluation at α, i.e. in $H^*(X;k)\tilde{\otimes}k[t_1,\ldots,t_r] \otimes_{k[t_1,\ldots,t_r]} k_\alpha$. Hence $s \geq r$. □

We want to finish this section with some remarks on the multiplicative structure of $H_{G,\alpha}(X;k)$. The arguments are only sketched here. A detailed discussion which also includes an alternative proof of Theorem (1.3.24) and similar results for more general spaces is given in Section 3.11 in case $G \cong (\mathbb{Z}_p)^r$, p prime, and in Section 3.5. in case $G \cong (S^1)^r$.

Of course, one could use Theorem (1.3.24) to 'pull-back' the cup-product on $H^{(*)}(X_\alpha;k)$ in order to get a product on $H_{G,\alpha}(X;k)$ and hence - obviously - an isomorphism of k-algebras

$$H_{G,\alpha}(X;k) \cong H^{(*)}(X_\alpha;k)$$

(as in (1.3.5)). But this multiplication on $H_{G,\alpha}(X;k)$ should relate to the multiplication on $\beta^*_G(X;k)$ in a reasonable way, and this relation is fairly complicated, especially in case $\alpha \in k^r - \mathbb{F}_2^r$. The problem is that the multiplication on $\beta^*_G(X;k)$ need not be R-bilinear, and does not induce a multiplication on $\beta^*_{G,\alpha}(X;k)$ in the obvious way. If $\alpha \in \mathbb{F}_2^r$, then up to a change of coordinates one may assume

$$\alpha = (\alpha_1,\ldots,\alpha_r) = (1,0,\ldots,0),$$

and it is possible to choose a multiplication on $\beta^*_G(X;k)$ which induces a multiplication on the quotient

$$\beta_{G,\tilde{\alpha}}(X;k) := \beta^*_G(X;k) \otimes_R k[t]_{\tilde{\alpha}}$$

(cf. (1.3.26)). In particular one gets an induced $k[t]$-algebra structure on $H(\beta^*_{G,\tilde{\alpha}}(X;k))$ and a k-algebra structure on

$$H(\beta^*_{G,\tilde{\alpha}}(X;k)) \otimes_{k[t]} k_1 = H_{G,\alpha}(X;k).$$

Using the naturality of the isomorphism

$$\tau(X) : H_{G,\alpha}(X;k) \to H^{(*)}(X_\alpha;k)$$

(with respect to the inclusion $X_\alpha \hookrightarrow X$) one sees that $\tau(X)$ becomes an isomorphism of k-algebras (with respect to the k-algebra structure on $H(\beta^*_{G,\tilde{\alpha}}(X;k)) \otimes_{k[t]} k_1$).

In the general case, i.e. $\alpha \in k^r$ let $K(\alpha) \subset G$ be the subgroup defined in (1.3.28), i.e. $H^1(B(G/K(\alpha));\mathbb{F}_2)$ is the kernel of the map
$$\alpha^1 : H^1(BG;\mathbb{F}_2) \to k, \qquad t_i \mapsto \alpha_i.$$
(Note that for $\alpha \in \mathbb{F}_2^r$ the group $K(\alpha)$ is just the image of $\mathbb{Z}_2$ under the group homomorphism $\mathbb{Z}_2 \to G$ given by $\alpha \in \mathbb{F}_2^r$, whereas in general the rank of $K(\alpha)$ may be larger than 1.) Again one can choose the multiplication on $\beta^*_G(X;k)$ such that it induces a multiplication on
$$\beta^*_G(X;k) \otimes_R H^*(BK(\alpha);k),$$
where the R-module structure of $H^*(BK(\alpha);k)$ is given by the map
$$R = H^*(BG;k) \to H^*(BK(\alpha);k)$$
which is induced by the inclusion $K(\alpha) \hookrightarrow G$ (choose a splitting
$$G \cong K(\alpha) \times G/K(\alpha)$$
and coordinates in $G \cong (\mathbb{Z}_2)^r$ which respect this splitting). One gets an $H^*(BK(\alpha);k)$-algebra structure on
$$H(\beta^*_G(X;k) \otimes_R H^*(BK(\alpha);k)),$$
and we claim that
$$H(\beta^*_G(X;k) \otimes_R H^*(BK(\alpha);k)) \otimes_{H^*(BK(\alpha);k)} k_{\bar{\alpha}}$$
is naturally isomorphic to $H_{G,\alpha}(X;k)$ as a k-algebra, where $k_{\bar{\alpha}}$ denotes k considered as an $H^*(BK(\alpha);k)$-module via the morphism
$$H^*(BK(\alpha);k) \to k$$
induced by $\alpha : H^*(BG;k) \to k$. Note that by construction
$$\bar{\alpha}^1 : H^1(BK(\alpha);\mathbb{F}_2) \to k$$
is injective. The $H^*(BK(\alpha);k)$-algebra
$$H(\beta^*_G(X;k) \otimes_R H^*(BK(\alpha);k))$$
is nothing but the equivariant cohomology
$$H^*_{K(\alpha)}(\mathrm{res}_{K(\alpha)} X;k)$$
of $\mathrm{res}_{K(\alpha)} X$, i.e. X considered as a $K(\alpha)$-space by restricting the G-action to the subgroup $K(\alpha)$. Hence to prove the above claim it suffices to show that for any finite-dimensional G-CW-complex X one has a natural isomorphism of k-algebras
$$H^*_G(X;k) \otimes_R k_\alpha \cong H^{(*)}(X^G;k)$$
if
$$\alpha^1 : H^1(BG;\mathbb{F}_2) \to k$$

1.3. The Borel construction for 2-tori

is injective (and to apply this to the $K(\alpha)$-CW-complex $\mathrm{res}_{K(\alpha)} X$). To establish the last isomorphism one considers the localized equivariant cohomology $S^{-1}H_G(X; \mathbb{F}_2)$, where

$$S \subset H^*(BG; \mathbb{F}_2) \cong \mathbb{F}_2[t_1, \ldots, t_r]$$

is the multiplicative subset generated by the non-zero elements (linear polynomials) in $H^1(BG; \mathbb{F}_2)$. Since localization is exact,

$$S^{-1}H_G^*(X; \mathbb{F}_2)$$

is an equivariant cohomology theory on finite-dimensional G-CW-complexes, and the inclusion $X^G \hookrightarrow X$ induces an isomorphism

$$S^{-1}H_G^*(X; \mathbb{F}_2) \xrightarrow{\cong} S^{-1}H_G^*(X^G; \mathbb{F}_2).$$

This follows from the comparison theorem (see (1.1.3) and (1.1.4)(2)). (It is easy to check the isomorphism for the special cases $X = G/K$, $K < G$.) See Chapter 3 for more on localization theorems of this kind.

Evaluating an element in $H^1(BG; \mathbb{F}_2)$ at $\alpha \in k^r$ always gives a non-zero element in k (since $\alpha^1 : H^1(BG; \mathbb{F}_2) \to k$ is assumed to be injective). Therefore

$$\alpha : H^*(BG; \mathbb{F}_2) \to k$$

extends to

$$\hat{\alpha} : S^{-1}H^*(BG; \mathbb{F}_2) \to k$$

and the evaluation of $H_G^*(X; \mathbb{F}_2)$ at α factors through $S^{-1}H_G^*(X; \mathbb{F}_2)$, i.e.

$$H_G^*(X; \mathbb{F}_2) \otimes_{H^*(BG; \mathbb{F}_2)} k_\alpha \cong S^{-1}H_G^*(X; \mathbb{F}_2) \otimes_{S^{-1}H^*(BG; \mathbb{F}_2)} k_{\hat{\alpha}}.$$

By the Künneth formula one has

$$S^{-1}H_G^*(X^G; \mathbb{F}_2) \cong H^*(X^G; \mathbb{F}_2) \otimes S^{-1}H^*(BG; \mathbb{F}_2)$$

as $S^{-1}H^*(BG; \mathbb{F}_2)$-algebras; therefore

$$S^{-1}H_G^*(X^G; \mathbb{F}_2) \otimes_{S^{-1}H^*(BG; \mathbb{F}_2)} k_{\hat{\alpha}} \cong H^{(*)}(X^G; \mathbb{F}_2) \otimes k \cong H^{(*)}(X^G; k)$$

as k-algebras. Clearly

$$H_G^*(X; \mathbb{F}_2) \otimes_{H^*(BG; \mathbb{F}_2)} k_\alpha = H_G^*(X; k) \otimes_R k_\alpha$$

and hence we get the desired isomorphism

$$H_G^*(X; k) \otimes_R k_\alpha \cong H^{(*)}(X^G; k)$$

of k-algebras.

(1.3.30) Corollary
Let X be a finite-dimensional G-CW-complex, $G \cong (\mathbb{Z}_2)^r$, and let

$$\dim_k H^*(X; k) = \dim_k H^*(X^G; k) < \infty$$

for k a field of characteristic 2. Then for any $\alpha \in k^r$ one has
$$H_G^*(X;k) \otimes_R k_\alpha \cong H^{(*)}(X_\alpha;k)$$
as k-algebras.

Proof.
To calculate $H_G^*(X;k)$ one can again replace $\beta_G^*(X;k)$ by the minimal Hirsch-Brown model $H^*(X;k)\tilde\otimes R$, which is homotopy equivalent to $\beta_G^*(X;k)$ in δgR-Mod (see (B.2.4)). By induction the condition
$$\dim_k H^*(X;k) = \dim_k H^*(X^G;k) < \infty$$
together with Corollary (1.3.8) implies:
$$\dim_k H^*(X;k) = \dim_k H^*(X^K;k) = \dim_k H^*(X^G:k) < \infty$$
for any subgroup $K < G$. It follows from Theorem (1.3.24) that the differential $\tilde\delta$ induced on $H^*(X;k)\tilde\otimes R$ must vanish (since
$$H(H^*(X;k),\tilde\delta_\alpha) = H^{(*)}(X_\alpha;k)$$
and
$$\dim_k H^{(*)}(X_\alpha;k) = \dim_k H^*(X;k)$$
for all α; cf. (1.3.14)). Therefore
$$H^*(X;k)\tilde\otimes R \cong H_G^*(X;k)$$
as R-algebras, and
$$H_G^*(X;k) \otimes_R k_\alpha \cong (H^*(X;k)\tilde\otimes R) \otimes_R k_\alpha \cong H_{G,\alpha}(X;k) \cong H^{(*)}(X_\alpha;k).$$
□

It should be noted that the product structure on $H^*(X;k)\tilde\otimes R$ in the above proof is - in general - not the componentwise product but can be considered as an r-parameter family (parametrized by $t_1,\ldots,t_r \in R$) of k-algebra structures on $H^*(X;k)$ (which at $(t_1,\ldots,t_r) = (0,\ldots,0)$ coincides with the cup-product on $H^*(X;k)$) (cf. the comments following (1.3.15)). Without the assumption
$$`\dim_k H^{(*)}(X;k) = \dim_k H^{(*)}(X^G;k) < \infty\text{'}$$
the differential $\tilde\delta$ on the minimal Hirsch-Brown model may, of course, be non-trivial and can be considered as an r-parameter family of differentials on $H^*(X;k)$ (which at $(t_1,\ldots,t_r) = (0,\ldots,0)$ gives the trivial differential).

1.4 The Borel construction for p-tori

We now describe the algebraic version of the Borel construction in the case where G is a p-torus of rank r, i.e. $G \cong (\mathbb{Z}_p)^r$, and p is an odd prime. In many respects the situation is similar to that in the preceding section, but certain additional complications arise which, for example, are reflected in the appearance of the exterior algebra as part of the cohomology $H^*(BG; k)$ of BG, if k is a field of characteristic p, p odd (see below). Although it is not difficult to 'specialize the description given below for the case (odd p) to the case ($p = 2$), we have chosen to present the simpler case ($p = 2$) separately (in the preceding section) in order to demonstrate the basic ideas and methods without too much technical framework. Later the different cases will often be treated simultaneously.

Again we consider the group $G = \mathbb{Z}_p$ (i.e. $r = 1$) first. As before the cellular chain complex $\mathcal{E}_*(G) := W_*(EG)$ of the G-CW-complex EG described in Section 1.1 is a free chain complex over the group ring of G, generated by one element w_n in each dimension $n = 0, 1, \ldots$. The boundary is given by

$$\partial w_n \begin{cases} \nu w_{n-1} & \text{for } n \text{ even } (\partial w_0 = 0) \\ \tau w_{n-1} & \text{for } n \text{ odd} \end{cases}$$

where, with 1 denoting the unit element and g a generator of $G = \{1, g, g^2, \ldots, g^{p-1}\} \cong \mathbb{Z}_p$, $\nu := 1 + g + \ldots + g^{p-1}$, and $\tau := 1 - g$.

(1.4.1) Proposition

(1)
$$H_n(BG; \mathbb{Z}) \cong H_n(\mathcal{E}_*(G) \otimes_{\mathbb{Z}[G]} \mathbb{Z}) \cong \begin{cases} \mathbb{Z} & \text{if } n = 0 \\ \mathbb{Z}_p & \text{if } n \text{ is odd} \\ 0 & \text{if } n \neq 0 \text{ is even} \end{cases}$$

(2) If k is a field of characteristic p, then

$$H_n(BG; k) \cong H_n(\mathcal{E}_*(G; k) \otimes_{k[G]} k) \cong k$$

for all $n \geq 0$.

Proof.
This follows immediately from the above description of $\mathcal{E}_*(G)$. $\square$

A diagonal
$$\Delta : \mathcal{E}_*(G) \to \mathcal{E}_*(G) \otimes \mathcal{E}_*(G)$$

on $\mathcal{E}_*(G)$ can be given by the following formula:
$$\Delta(w_n) = \bigoplus_{a+b=n} \Delta_{a,b}(w_n),$$
where
$$\Delta_{a,b}(w_n) = \begin{cases} w_a \otimes w_b & \text{if } a \text{ is even} \\ w_a \otimes gw_b & \text{if } a \text{ is odd and } b \text{ is even} \\ -\sum_{0 \le i < j \le p-1} g^i w_a \otimes g^j w_b & \text{if } a \text{ and } b \text{ are odd.} \end{cases}$$

It is rather tedious but straightforward to check that
$$\Delta : \mathcal{E}_*(G) \to \mathcal{E}_*(G) \otimes \mathcal{E}_*(G)$$
is indeed a diagonal, i.e. a morphism of chain complexes over $\mathbb{Z}[G]$ (where $\mathcal{E}_*(G) \otimes \mathcal{E}_*(G)$ is equipped with the diagonal action), which is a lifting of the canonical isomorphism $\mathbb{Z} \to \mathbb{Z} \otimes \mathbb{Z}$ to the respective resolutions. The product in cohomology induced by the diagonal is described in the following proposition.

(1.4.2) Proposition

(1) $H^*(BG; \mathbb{Z}) \cong H^*(\text{Hom}_{\mathbb{Z}[G]}(\mathcal{E}_*(G), \mathbb{Z})) \cong \mathbb{Z}[t]/(pt)$ with $\deg(t) = 2$
(2) If k is a field of characteristic p then
$$H^*(BG; k) \cong H^*(\text{Hom}_{k[G]}(\mathcal{E}_*(G; k), k)) \cong \Lambda(s) \otimes k[t],$$
where $\deg(s) = 1$, $\deg(t) = 2$, and $\Lambda(s)$ denotes the exterior algebra over k generated by the element s. If $k = \mathbb{F}_p = \mathbb{Z}_p$ then the Bockstein homomorphism induced by the short exact coefficient sequence
$$0 \to \mathbb{Z}_p \to \mathbb{Z}_{p^2} \to \mathbb{Z}_p \to 0$$
is given by $\beta(s) = t$, $\beta(t) = 0$ (more generally: $\beta(st^i) = t^{i+1}$, $\beta(t^i) = 0$).

Proof.
Let $\bar{w}^n \in \text{Hom}_{\mathbb{Z}[G]}(\mathcal{E}_*(G), \mathbb{Z})$ again denote the element defined by
$$\bar{w}^n(w_m) = \begin{cases} 1 & \text{if } n = m \\ 0 & \text{if } n \ne m. \end{cases}$$
Then, using the above formula for the diagonal, one gets
$$\bar{w}^a \cup \bar{w}^b(w_m) = (\bar{w}^a \otimes \bar{w}^b)(\Delta w_m) = \begin{cases} 1 & \text{if } a+b = m, \text{ and } a \text{ or } b \text{ even} \\ \frac{p(1-p)}{2} & \text{if } a+b = m, \text{ and } a \text{ and } b \text{ odd} \\ 0 & \text{if } a+b \ne m \end{cases}$$
Hence, in case (1), the cocycles in $\text{Hom}_{\mathbb{Z}[G]}(\mathcal{E}_*(G), \mathbb{Z})$ form a subalgebra

isomorphic to $\mathbb{Z}[t]$, where $\deg(t) = 2$ and $\bar{w}^{2i}$ corresponds to t^i, and the coboundaries correspond to the ideal generated by pt, since
$$\delta(\bar{w}^{2i+1}) = p\bar{w}^{2i+2}.$$
In case (2) the coboundary on $\operatorname{Hom}_{k[G]}(\mathcal{E}_*(G;k), k)$ vanishes and the above formula for the cup-product on the cochain level shows that $H^*(BG; k)$ is isomorphic to $\Lambda(s) \otimes k[t]$, where s corresponds to $\bar{w}^1$ and t corresponds to $\bar{w}^2$. (Note that $\frac{p(1-p)}{2} \equiv 0 \bmod p$, since p is odd.) The assertion about the Bockstein homomorphism follows immediately since the coboundary in $\operatorname{Hom}_{\mathbb{Z}[G]}(\mathcal{E}_*(G); \mathbb{Z})$ is given by
$$\delta \bar{w}^n = \begin{cases} p\bar{w}^{n-1} & \text{if } n \text{ is odd} \\ 0 & \text{if } n \text{ is even.} \end{cases}$$
$\square$

(1.4.3) Remark
Comparing the situation at hand with case $p = 2$ (see Section 1.3), one sees that here $\operatorname{Hom}_{\mathbb{Z}[G]}(\mathcal{E}_*(G), \mathbb{Z})$, with the cup-product, is no longer a polynomial algebra although in each case the cycles in $\operatorname{Hom}_{\mathbb{Z}[G]}(\mathcal{E}_*(G), \mathbb{Z})$ are isomorphic to a polynomial algebra over $\mathbb{Z}$ generated by an element of degree 2. Nevertheless $\operatorname{Hom}_{\mathbb{Z}[G]}(\mathcal{E}_*(G), \mathbb{Z})$ is a 'strictly' commutative differential algebra (the coboundary being a derivation) and $\operatorname{Hom}_{k[G]}(\mathcal{E}_*(G; k), k)$ is the free graded commutative algebra generated by two elements of degree 1 and 2 respectively, if k is a field of characteristic p.

We now describe $\beta_G^*(C^*)$ for an object C^* in $\delta gk[G]$-Mod in a way similar to Section 1.3. We assume for the rest of this section that k is a field of characteristic p if not explicitly stated otherwise; but the reader should notice that certain analogous results to those given below also hold, for example, in the case $k = \mathbb{Z}$. We have again an isomorphism of graded k-modules
$$\beta_G^*(C^*) \cong C^* \otimes_{k[G]} \mathcal{E}^*(G) \cong C^* \otimes W^*,$$
where
$$W^* := \operatorname{Hom}_k(W_*, k), \qquad \mathcal{E}_*(G) \cong W_* \otimes k[G]$$
(as $k[G]$-modules), W_* being generated (as a k-vector space) by one element w_n in each dimension $n = 0, 1, 2, \ldots$, and
$$\mathcal{E}^*(G) = \operatorname{Hom}_{k[G]}(\mathcal{E}_*(G), k[G]).$$
The boundary of $\mathcal{E}_*(G)$ induces the coboundary on $\mathcal{E}^*(G)$, which - together with the coboundary δ on C^* - gives the coboundary $\tilde{\delta}$ on
$$\beta_G^*(C^*) \cong C^* \otimes_{k[G]} \mathcal{E}^*(G)$$

by the usual formula for the tensor product of complexes. Using the isomorphism
$$\beta_G^*(C^*) \cong C^* \otimes W^*$$
the coboundary $\tilde{\tilde{\delta}}$ is given by the following formula
$$\tilde{\tilde{\delta}}(c \otimes w^n) = \begin{cases} (\delta c) \otimes w^n + (-1)^{|c|} c\tau \otimes w^{n+1} & \text{if } n \text{ is even} \\ (\delta c) \otimes w^n + (-1)^{|c|} c\nu \otimes w^{n+1} & \text{if } n \text{ is odd,} \end{cases}$$
where w^n is the dual of w_n. (We write $\tilde{\tilde{\delta}}$ instead of $\tilde{\delta}$ for a reason to be made clear below.)

We use the notation $C^* \tilde{\otimes} W^*$ for the complex $(C^* \otimes W^*, \tilde{\tilde{\delta}})$. We next calculate the $\beta_G^*(k)$-module structure of
$$\beta_G^*(C^*) \cong C^* \tilde{\otimes} W^*$$
induced by the diagonal
$$\Delta : \mathcal{E}_*(G) \to \mathcal{E}_*(G) \otimes \mathcal{E}_*(G)$$
(and the k-module structure η of C^*). We make use of the description of
$$\beta_G^*(k) \cong H^*(BG; k) \cong \Lambda(s) \otimes k[t]$$
given in (1.4.2). By definition
$$((c \otimes w^a) \cup t^b)(w_m) = \eta((c \otimes w^a) \otimes \bar{w}^{2b})(\Delta w_m) = \begin{cases} c & \text{if } a + 2b = m \\ 0 & \text{if } a + 2b \neq m. \end{cases}$$
Hence $(c \otimes w^a) \cup t^b = c \otimes w^{a+2b}$.

The formula for the cup-product with s (and st^q) looks more complicated:
$$((c \otimes w^a) \cup (st^b))(w_m) = \eta(c \otimes w^a \otimes \bar{w}^{2b+1})(\Delta w_m)$$
$$= \begin{cases} c & \text{if } a \text{ is even and } a + 2b + 1 = m \\ -\sum_{i=0}^{p-2}(p - i - 1)cg^i & \text{if } a \text{ is odd} \\ & \text{and } a + 2b + 1 = m \\ 0 & \text{if } a + 2b + 1 \neq m \end{cases}$$
Hence
$$(c \otimes w^a) \cup (st^b) = \begin{cases} c \otimes w^{a+2b+1} & \text{if } a \text{ is even} \\ -\sum_{i=0}^{p-2}(p - i - 1)cg^i \otimes w^{a+2b+1} & \text{if } a \text{ is odd} \end{cases}$$
The above formulas show that the product on the cochain level is not associative $((- \cup s) \cup s \neq 0$ in general, whereas $s \cup s = 0$ in $H^*(BG; k))$,

1.4. The Borel construction for p-tori

and therefore $\beta_G^*(C^*)$ is not a complex over $H^*(BG;k)$ in the usual sense. But if one restricts to
$$k[t] \subset H^*(BG;k)$$
one gets the following:

(1.4.4) Proposition
As a right $k[t]$-module $\beta_G^*(C^*)$ is isomorphic to $C^* \otimes \Lambda(s) \otimes k[t]$.

The differential $\tilde{\tilde{\delta}}$ on $\beta_G^*(C^*)$ can again be considered as a perturbation of the 'trivial' extension of the differential δ on C^* to $C^* \otimes \Lambda(s) \otimes k[t]$, and this time it is conveniently described by a two-stage process, first a perturbation over $\Lambda(s)$ and then a perturbation over $k[t]$. The details are as follows. We define the differential $\tilde{\delta}$ on $C^* \otimes \Lambda(s)$ by:
$$\tilde{\delta}(c \otimes 1) = (\delta c) \otimes 1 + (-1)^{|c|} c\tau \otimes s$$
$$\tilde{\delta}(c \otimes s) = (\delta c) \otimes s.$$
This gives $C^* \otimes \Lambda(s)$ a structure as a differential graded $\Lambda(s)$-module. We denote $(C^* \otimes \Lambda(s), \tilde{\delta})$ by $C^* \tilde{\otimes} \Lambda(s)$. Next we define a perturbation $\tilde{\tilde{\delta}}$ of $\tilde{\delta}$ over $k[t]$ by:
$$\tilde{\tilde{\delta}}(c \otimes 1 \otimes t^q) = \tilde{\delta}(c \otimes 1) \otimes t^q$$
$$\tilde{\tilde{\delta}}(c \otimes s \otimes t^q) = \tilde{\delta}(c \otimes s) \otimes t^q + (-1)^{|c|} c\nu \otimes t^{q+1}$$
and we denote by $C^* \tilde{\otimes} \Lambda(s) \tilde{\otimes} k[t]$ the complex $(C^* \tilde{\otimes} \Lambda(s) \otimes k[t], \tilde{\tilde{\delta}})$.

The above calculation shows that $\beta_G^*(C^*) \cong C^* \tilde{\otimes} W^*$ is isomorphic to $C^* \tilde{\otimes} \Lambda(s) \tilde{\otimes} k[t]$ in $\delta gk[t]$-Mod. Note that, in general, $\tilde{\tilde{\delta}}$ is not compatible with the 'obvious' multiplication with the element s, hence $C^* \tilde{\otimes} \Lambda(s) \tilde{\otimes} k[t]$ is not a complex over $\Lambda(s)$.

We point out again that the cup-product $- \cup s$ on $\beta_G^*(C^*)$,
$$s \in H^1(BG;k) = \beta_G^1(k),$$
is - in general - not the 'obvious' map on all of $C^* \tilde{\otimes} \Lambda(s) \tilde{\otimes} k[t]$. Note, though, that by the above formula for the product:
$$(c \otimes 1 \otimes t^q) \cup s = c \otimes s \otimes t^q.$$
The map
$$(- \cup s) \cup s : C^* \tilde{\otimes} \Lambda(s) \tilde{\otimes} k[t] \to C^* \tilde{\otimes} \Lambda(s) \tilde{\otimes} k[t]$$
may not vanish, but it is homotopic to zero as a morphism in $\delta gk[t]$-Mod. The map $f(-) := (- \cup s) \cup s$ is given by
$$f(c \otimes \lambda \otimes t^q) = c\rho \otimes \lambda \otimes t^q,$$

where $\lambda \in \Lambda(s)$ and
$$\rho := \sum_{i=0}^{p-2}(p-i-1)g^i \in k[G].$$

A homotopy
$$h: C^* \tilde{\otimes} \Lambda(s) \tilde{\otimes} k[t] \to C^* \tilde{\otimes} \Lambda(s) \tilde{\otimes} k[t]$$
with $\tilde{\delta}h + h\tilde{\delta} = f$ can be given by the following formulas:
$$h(c \otimes 1 \otimes t^q) := (-1)^{|c|} c\gamma \otimes s \otimes t^q$$
$$h(c \otimes s \otimes t^q) := (-1)^{|c|} c\beta \otimes 1 \otimes t^{q+1},$$
where γ and β are elements in $k[G]$ which have to be determined in such a way that the equation
$$\tilde{\delta}h + h\tilde{\delta} = f$$
holds. Now
$$\tilde{\delta}h(c \otimes 1 \otimes t^q) = \tilde{\delta}((-1)^{|c|} c\gamma \otimes s \otimes t^q)$$
$$= (-1)^{|c|}\delta c\gamma \otimes s \otimes t^q + c\gamma\nu \otimes 1 \otimes t^{q+1}$$
$$\tilde{\delta}h(c \otimes s \otimes t^q) = \tilde{\delta}((-1)^{|c|} c\beta \otimes 1 \otimes t^{q+1})$$
$$= (-1)^{|c|}\delta c\beta \otimes 1 \otimes t^{q+1} + c\beta\tau \otimes s \otimes t^{q+1}$$
$$h\tilde{\delta}(c \otimes 1 \otimes t^q) = h(\delta c \otimes 1 \otimes t^q + (-1)^{|c|} c\tau \otimes s \otimes t^q)$$
$$= (-1)^{|\delta c|}\delta c\gamma \otimes s \otimes t^q + c\tau\beta \otimes 1 \otimes t^{q+1}$$
$$h\tilde{\delta}(c \otimes s \otimes t^q) = h(\delta c \otimes s \otimes t^q + (-1)^{|c|} c\nu \otimes 1 \otimes t^{q+1})$$
$$= (-1)^{|\delta c|}\delta c\beta \otimes 1 \otimes t^{q+1} + c\nu\gamma \otimes s \otimes t^{q+1}.$$
Hence, if γ and β are chosen in such a way that
$$\gamma\nu + \tau\beta = \beta\tau + \nu\gamma = \rho$$
then $\tilde{\delta}h + h\tilde{\delta} = f$. A simple calculation shows that $\gamma := \frac{1}{2}(p-1)g^0$ and
$$\beta := \beta_0 g^0 + \beta_1 g^1 + \ldots + \beta_{p-1} g^{p-1}$$
with
$$\beta_i := \frac{1}{2}(p-(i+1))(i+1)$$
for $i = 0, \ldots, p-1$ have the desired property.

For an object C^* in δgk-Mod we define
$$H_{G,\alpha}(C^*) := \begin{cases} H((\beta_G^*(C^*) \otimes_{k[t]} k_0) \otimes_{\Lambda(s)} k_0) & \text{if } \alpha = 0 \in k \\ (H(\beta_G^*(C^*) \otimes_{k[t]} k_\alpha)) \otimes_{\Lambda(s)} k_0 & \text{if } \alpha \in k, \alpha \neq 0, \end{cases}$$
where the $k[t]$-module structure of $k_\alpha \cong k$ is given by
$$\alpha: k[t] \to k, \qquad t \mapsto \alpha,$$

1.4. The Borel construction for p-tori

and the $\Lambda(s)$-module structure of $k_0 \cong k$ by $\Lambda(s) \to k$, $s \mapsto 0$. Note that
$$(\beta_G^*(C^*) \otimes_{k[t]} k_0) \otimes_{\Lambda(s)} k_0$$
inherits a well-defined coboundary from $\beta_G^*(C^*)$ - although the latter is not a complex over $\Lambda(s) \otimes k[t]$ - in fact
$$(\beta_G^*(C^*) \otimes_{k[t]} k_0) \otimes_{\Lambda[s]} k_0$$
is canonically isomorphic to C^* as an object in δgk-Mod. The $\Lambda(s)$-structure on
$$H(\beta_G^*(C^*) \otimes_{k[t]} k_\alpha)$$
which is used in order to define $H_{G,\alpha}(C^*)$ for $\alpha \neq 0$ is induced by the cup-product with $s \in H^1(BG,k)$ on $\beta_G^*(C^*)$. This indeed gives a $\Lambda(s)$-module structure on
$$H(\beta_G^*(C^*) \otimes_{k[t]} k_\alpha)$$
since
$$(- \cup s) \cup s : \beta_G^*(C^*) \to \beta_G^*(C^*)$$
is homotopic to zero in $\delta gk[t]$-Mod.

In contrast to the case $p = 2$ the $\mathbb{Z}$-grading on $\beta_G^*(C^*)$ is not completely destroyed by the evaluation at $t = \alpha$, $s = 0$, i.e. $H_{G,\alpha}(C^*)$ inherits a $\mathbb{Z}_2$-grading by the even and odd dimensional parts of $\beta_G^*(C^*)$. In the following theorem $H^{(*)}(-;k)$ denotes the $\mathbb{Z}_2$-graded cohomology
$$H^{(*)}(-;k) := \left(\bigoplus_{i \text{ even}} H^i(-;k) \right) \oplus \left(\delta \bigoplus_{i \text{ odd}} H^i(-;k) \right).$$

(1.4.5) Theorem
Let X be a finite-dimensional G-CW-complex, $G = \mathbb{Z}_p$. Then there are natural $\mathbb{Z}_2$-graded isomorphisms
$$H_{G,\alpha}(X;k) := H_{G,\alpha}(S^*(X;k)) \cong \begin{cases} H^{(*)}(X;k) & \text{for } \alpha = 0 \\ H^{(*)}(X^G;k) & \text{for } \alpha \neq 0. \end{cases}$$

Proof.
This is similar to the proof of Theorem (1.3.5), and we therefore only stress those points which need some extra attention because of the more complicated structure of $\beta_G^*(C^*)$ over $H^*(BG;k)$ in the case at hand.

The proof for the special case where the action is trivial, and for the case $\alpha = 0$ (but non-trivial action) is analogous to that in (1.3.5). For $\alpha \neq 0$ we compute explicitly $H_{G,\alpha}(X;k)$ for $X = G$ (the case $X = *$ is already contained in the above special cases):
$$H_{G,\alpha}(G;k) = (H(\beta_G^*(G) \otimes_{k[t]} k_\alpha)) \otimes_{\Lambda(s)} k_0$$

and
$$\beta_G^*(G) \otimes_{k[t]} k_\alpha \cong (k[G] \tilde{\otimes} \Lambda(s) \tilde{\otimes} k[t]) \otimes_{k[t]} k_\alpha \cong k[G] \tilde{\tilde{\otimes}} \Lambda(s),$$
where the coboundary on the right side is given by
$$\tilde{\tilde{\delta}}_\alpha(c \otimes 1) = c\tau \otimes s$$
$$\tilde{\tilde{\delta}}_\alpha(c \otimes s) = c\nu \otimes \alpha, \qquad c \in k[G].$$
Since
$$\cdots \to k[G] \xrightarrow{\tau} k[G] \xrightarrow{\nu} k[G] \xrightarrow{\tau} k[G] \to \cdots$$
is exact and $\alpha \neq 0$ is a unit in k, the cohomology of
$$(k[G] \tilde{\tilde{\otimes}} \Lambda(s), \tilde{\tilde{\delta}}_\alpha)$$
vanishes. Hence $H_{G,\alpha}(G; k) = 0$ for $\alpha \neq 0$. It is not clear a priori that $H_{G,\alpha}(X; k)$ is a cohomology theory on G-spaces, since $- \otimes_{\Lambda(s)} k_0$ is not an exact functor. But
$$H(\beta_G^*(X; k) \otimes_{k[t]} k_\alpha)$$
and
$$H^*(X^G; k) \otimes \Lambda(s)$$
certainly are cohomology theories, and one can apply the comparison argument in the proof of (1.3.5) first to these theories to get a natural isomorphism
$$H(\beta_G^*(X; k) \otimes_{k[t]} k_\alpha) \cong H^{(*)}(X^G; k) \otimes \Lambda(s)$$
of $\Lambda(s)$-algebras. In particular,
$$H(\beta_G^*(X; k) \otimes_{k[t]} k_\alpha)$$
is a free $\Lambda(s)$-module and therefore $H_{G,\alpha}(X; k)$ is indeed a cohomology theory. □

(1.4.6) Remark
The product structure on $\beta_G^*(S^*(X; k))$ in the case at hand is even more delicate than in the $G = \mathbb{Z}_2$ case. Nevertheless one can check that the multiplication on $\beta_G^*(S^*(X; k))$ - given by the diagonal on X and on $\mathcal{E}_*(G)$ - is $k[t]$-bilinear and hence induces a multiplication on
$$\beta_G^*(S^*(X; k)) \otimes_{k[t]} k_\alpha$$
and in turn a $\Lambda(s)$-algebra structure on
$$H(\beta_G^*(S^*(X; k)) \otimes_{k[t]} k_\alpha).$$
Therefore $H_{G,\alpha}(X; k)$ inherits a k-algebra structure for $\alpha \neq 0$. For $\alpha = 0$ the projection
$$\beta_G^*(S^*(X; k)) \to (\beta_G^*(S^*(X; k)) \otimes_{k[t]} k_\alpha) \otimes_{\Lambda(s)} k_0 \cong S^*(X; k)$$

is compatible with the respective multiplications and hence $H_{G,0}(X;k)$ is isomorphic to $H^{(*)}(X;k)$ as a k-algebra. With respect to these k-algebra structures on $H_{G,\alpha}(X;k)$ the isomorphism in Theorem (1.4.5) becomes an isomorphism of k-algebras. (The two-stage procedure described above to define a k-algebra structure on $H_{G,\alpha}(X;k)$ for $\alpha \neq 0$ can also be defined with some modification in the case $G = \mathbb{Z}_2$. The multiplication on the cochain level is bilinear with respect to the subring

$$H^{\mathrm{ev}}(B\mathbb{Z}_2; \mathbb{F}_2) \subset H^*(B\mathbb{Z}_2; \mathbb{F}_2) \cong \mathbb{F}_2[t],$$

and the two-stage procedure to be applied for $\alpha \neq 0$ corresponds to first evaluating at $t^2 = \alpha^2$, and then - after taking homology - at $\bar{t} = \alpha$, where $\bar{t}$ denotes the equivalence class of t in $k[t]/(t^2 - \alpha^2)$.) Alternatively one can describe the product structure on $H_{G,\alpha}(X;k)$ for $\alpha \neq 0$ via the isomorphism

$$H_{G,\alpha}(X;k) \cong ((S^{-1}H_G^*(X;k)) \otimes_{S^{-1}k[t]} k_{\hat{\alpha}}) \otimes_{\Lambda(s)} k_0,$$

where S is the multiplicative set generated by $t \in k[t]$, $S^{-1}(-)$ denotes localization with respect to this set, and

$$\hat{\alpha} : S^{-1}k[t] \to k,$$

$\alpha(t) = \alpha$, gives the $S^{-1}k[t]$-module structure of $k_{\hat{\alpha}}$. We are not very explicit here since these questions are dealt with in detail in Section 3.11.

We derive some consequences of Theorem (1.4.5) which are to a certain extent - but not completely - analogous to results obtained from Theorem (1.3.5) in the preceding section for the case $p = 2$. A more comprehensive treatment is given in Chapter 3.

(1.4.7) Corollary
Let X be a finite-dimensional G-CW-complex ($G = \mathbb{Z}_p$, p odd) and let k be a field of characteristic p (as usual in this section).
(1) *If $H^*(X;k) \cong k$ then $H^*(X^G;k) \cong k$.*
(2) *If $H^*(X;k) \cong H^*(S^n;k)$ then*
$$H^*(X^G;k) \cong H^*(S^m;k)$$
with $-1 \leq m \leq n$ ($S^{-1} := \emptyset$) and $n \equiv m$ modulo 2.
(Cf. also Exercise (1.12) and Examples (3.1.10).)

Proof.
The argument is similar to that for (1.3.7), i.e. we make use of the minimal Hirsch-Brown model which is described in detail in Proposition (1.4.10) below (for the case $G = (\mathbb{Z}_p)^r$). But the situation in case (odd p) is more subtle since the Hirsch-Brown model is not a cochain complex

over $H^*(BG;k)$, in general. In particular, certain degree arguments which hold in case (2) do not go through in the case (odd p), and hence the filtration described in (1.4.10) plays an essential role.

Part (1): Since the action of G on $k \cong H^*(X;k)$ must be trivial Proposition (1.4.10) implies that the differential $\bar{\delta}$ of the minimal Hirsch-Brown model $H^*(X;k)\tilde{\otimes}\Lambda(s)\tilde{\otimes}k[t]$ must vanish. Hence, by (1.4.5),

$$H^{(*)}(X^G;k) \cong H_{G,\alpha}(X;k) = ((H^*(X;k)\tilde{\otimes}\Lambda(s)\tilde{\otimes}k[t]) \otimes_{k[t]} k_\alpha) \otimes_{\Lambda(s)} k_0$$
$$\cong k \quad \text{for } \alpha \neq 0.$$

Part (2): Using part (1) we may, of course, assume that $n > 0$. Again the induced action in cohomology must be trivial. Let us assume that $X^G \neq S^{-1} = \emptyset$, i.e. there exists a fixed point $x \in X^G$. Then the maps $* \to X \to *$ induce corresponding morphisms of the respective minimal Hirsch-Brown models

$$H^*(*;k)\tilde{\otimes}\Lambda(s)\tilde{\otimes}k[t] \leftarrow H^*(X;k)\tilde{\otimes}\Lambda(s)\tilde{\otimes}k[t] \leftarrow H^*(*;k)\tilde{\otimes}\Lambda(s)\tilde{\otimes}k[t]$$

such that the composition is homotopic to the identity in $\delta gk[t]$-Mod (see (1.4.11)).

In fact, since the differential of $H^*(*;k)\tilde{\otimes}\Lambda(s)\tilde{\otimes}k[t]$ vanishes this compositon must be an isomorphism of $H^*(BG;k)$-modules. It follows from this and the filtration properties of the minimal Hirsch-Brown model (see (1.4.10)) that the differential of $H^*(X;k)\tilde{\otimes}\Lambda(s)\tilde{\otimes}k[t]$ must vanish, too. Hence, again by (1.4.5), one has

$$H^{(*)}(X^G;k) \cong ((H^*(X;k)\tilde{\otimes}\Lambda(s)\tilde{\otimes}k[t]) \otimes_{k[t]} k_\alpha) \otimes_{\Lambda(s)} k_0 \cong H^{(*)}(X;k)$$

as $\mathbb{Z}_2$-graded k-vector spaces, i.e. $H^*(X;k) \cong H^*(S^m;k)$ for some m with $n \equiv m$ modulo 2. Since the map

$$H^*(X;k)\tilde{\otimes}\Lambda(s)\tilde{\otimes}k[t] \to H^*(X^G;k) \otimes \Lambda(s) \otimes k[t],$$

induced by the inclusion $j : X^G \to X$, becomes an isomorphism when evaluated at $\alpha \neq 0$, $\alpha \in k$, (see (1.4.5)), one must have $m \leq n$.

It remains to show that for n even the fixed set is not empty. This follows from the preceding results if one considers the suspension

$$\Sigma X := X \times I / \sim ((x,1) \sim (x',1), \ (x,0) \sim (x',0) \text{ for } x,x' \in X)$$

equipped with the action given by

$$g([x,t]) := [gx,t],$$

where $g \in G$ and $[x,t] \in \Sigma X$ denotes the equivalence class of (x,t) in ΣX. Clearly $(\Sigma X)^G \neq \emptyset$, and hence by the above argument

$$H^*((\Sigma X)^G;k) \cong H^*(S^{m+1};k)$$

with $n \equiv m$ modulo 2. Since n is even, by assumption, so is m. Therefore $(\Sigma X)^G$ is connected and hence $X^G \neq \emptyset$. □

1.4. The Borel construction for p-tori

(1.4.8) Corollary
Let X be a finite-dimensional G-CW-complex ($G = \mathbb{Z}_p$) such that $\sum_{i=0}^{\infty} \dim_k H^i(X;k)$ is finite (k a field of characteristic p); then, for any $m \geq 0$,

$$\sum_{i=0}^{\infty} \dim_k H^{m+i}(X^G;k) \leq \sum_{i=0}^{\infty} \dim_k H^{m+i}(X;k)^G$$
$$\leq \sum_{i=0}^{\infty} \dim_k H^{m+i}(X;k).$$

(For a G-module M the submodule of invariant elements is denoted by M^G).
(Compare with Exercises (1.13) and (1.14).)

Proof.
Again the argument is similar to that for (1.3.8). We consider the minimal Hirsch-Brown model for the G-map $j : X^G \to X$ (see (1.4.10) and (1.4.11)). The $k[t]$-cochain complex $\beta_G^*(X;k) = (C^*(X;k)\tilde\otimes\Lambda(s))\tilde\otimes k[t]$ can be replaced (up to homotopy equivalence over $k[t]$) by a cochain complex of the form $H^*(X;k)\tilde\otimes\Lambda(s)\tilde\otimes k[t]$, and the inclusion $j : X^G \to X$ induces a morphism in $\delta gk[t]$-Mod

$$j^* : H^*(X;k)\tilde\otimes\Lambda(s)\tilde\otimes k[t] \to H^*(X^G;k) \otimes \Lambda(s) \otimes k[t]$$

which respects the filtration

$$\mathcal{F}_m(H^*(-;k)\tilde\otimes\Lambda(s)\tilde\otimes k[t]) := (\bigoplus_{i \geq 0}^{m} H^i(-;k)\tilde\otimes\Lambda(s))\tilde\otimes k[t]$$
$$= (H^{\leq m}(-;k)\tilde\otimes\Lambda(s))\tilde\otimes k[t].$$

One now proceeds as in the proof of (1.3.8)(1) to obtain the inequality

$$\dim_k(\bigoplus_{i=m}^{\infty} H^i(X^G;k) \otimes \Lambda(s)) \leq \dim_k(\bigoplus_{i=m}^{\infty} H^i(X;k) \otimes \Lambda(s)) \text{ for } m \in \mathbb{N}$$

and this clearly implies

$$\sum_{i=0}^{\infty} \dim_k H^{m+i}(X^G;k) \leq \sum_{i=0}^{\infty} \dim_k H^{m+i}(X;k) \quad \text{for } m \in \mathbb{N}.$$

A non-trivial action of the group G on $H^*(X;k)$ is reflected in the boundary of $H^*(X;k)\tilde\otimes\Lambda(s)$ as follows. In the two-stage construction of $\beta_G^*(X;k) = (C^*(X;k)\tilde\otimes\Lambda(s))\tilde\otimes k[t]$ the first step $C^*(X;k)\tilde\otimes\Lambda(s)$ is isomorphic to the mapping cone of the morphism

$$\tau = (1-g) : C^*(X;k) \to C^*(X;k)$$

in δgk-Mod. Therefore $H^*(X;k)\tilde\otimes\Lambda(s)$ is homotopy equivalent to the

mapping cone of
$$\tau^* = (1-g)^* : H^*(X;k) \to H^*(X;k),$$
i.e. the differential on $H^*(X;k)\tilde{\otimes}\Lambda(s)$ is given by
$$\tilde{\delta}(x \otimes 1) = (-1)^{|x|} \cdot \tau^*(x) \otimes s$$
$$\tilde{\delta}(x \otimes s) = 0.$$
By (B.2.4) we can replace $(H^*(X;k)\tilde{\otimes}\Lambda(s))\tilde{\otimes}k[t]$ by a complex over $k[t]$ of the form
$$H(H^*(X;k)\tilde{\otimes}\Lambda(s),\tilde{\delta})\tilde{\otimes}k[t].$$
But
$$H^r(H^*(X;k)\tilde{\otimes}\Lambda(s),\tilde{\delta}) = H^r(X;k)^G \oplus H^{r-1}(X;k)/\tau^* H^{r-1}(X;k).$$
Since
$$0 \to H^{r-1}(X;k)^G \to H^{r-1}(X;k) \xrightarrow{\tau^*} H^{r-1}(X;k)$$
$$\to H^{r-1}(X;k)/\tau^* H^{r-1}(X;k) \to 0$$
is exact, one has
$$\dim_k(H^{r-1}(X;k)/\tau^* H^{r-1}(X;k)) = \dim_k H^{r-1}(X;k)^G$$
and hence
$$\dim_k H^*(H^{\leq m}(X;k)\tilde{\otimes}\Lambda(s),\tilde{\delta}) = 2\dim_k(H^{\leq m}(X;k))^G.$$
The same argument as above now gives the desired inequality
$$\sum_{i=0}^{\infty} \dim_k H^{m+i}(X^G;k) \leq \sum_{i=0}^{\infty} \dim_k H^{m+i}(X;k)^G.$$
$\square$

The inequality of (1.4.8) also holds for $p = 2$ (see Exercise (1.13)). Refined inequalities are discussed in a more general context in Section 3.10.

It should be clear from the above discussion that many of the results developed in Section 1.3 for $p = 2$ have analogues in case p an odd prime, and can be obtained in a similar way. As we have pointed out already one essential difference, though, is the fact that $\beta_G^*(X;k)$ is an object in $\delta g H^*(BG;k)$-Mod for $G = \mathbb{Z}_2$, $k = \mathbb{F}_2$; while in the case of an odd prime p the corresponding statement does not hold - in general $\beta_G^*(X;k)$ is a cochain complex only over the polynomial part $k[t]$ of $H^*(BG;k)$ (and only 'up to homotopy' over the exterior part $\Lambda(s)$). This causes considerable complications if one wants to get an analogue of Theorem (1.3.24) in the case where G is a p-torus and p is an odd prime. We take up this problem in Chapter 3, where results for actions of tori and p-tori, p (any) prime, are treated simultaneously as far as

1.4. The Borel construction for p-tori

possible, using more and heavier machinery from algebraic topology. We finish this section by deriving certain properties of the Borel construction for p-tori and its minimal Hirsch-Brown model, which then are applied to study free p-torus actions. A more comprehensive discussion of free torus and p-torus actions is contained in Chapter 4. Let $G = (\mathbb{Z}_p)^r$ be a p-torus of rank r (p an odd prime). Since $BG \simeq B\mathbb{Z}_p \times \ldots \times B\mathbb{Z}_p$ the Künneth formula and Proposition (1.4.2) give the following.

(1.4.9) Proposition
If k is a field of characteristic p and $G = (\mathbb{Z}_p)^r$ a p-torus of rank r (p an odd prime), then
$$H^*(BG; k) \cong \Lambda(s_1, \ldots, s_r) \otimes k[t_1 \ldots, t_r],$$
where $\Lambda(s_1, \ldots, s_r)$ denotes the exterior algebra over k generated by elements $s_1, \ldots, s_r$ of degree 1, and $k[t_1, \ldots, t_r]$ the polynomial algebra generated by elements $t_1, \ldots, t_r$ of degree 2.

The Borel construction
$$\beta_G^*(C^*) = \operatorname{Hom}_{k[G]}(\mathcal{E}_*(G), C^*)$$
for a $k[G]$-cochain complex C^* can be described using the isomorphism
$$\mathcal{E}_*(G) = \mathcal{E}_*(\mathbb{Z}_p) \otimes \ldots \otimes \mathcal{E}_*(\mathbb{Z}_p)$$
and one finds that $\beta_G^*(C^*)$ can be considered as a cochain complex over $k[t_1, \ldots, t_r]$. As a $k[t_1, \ldots, t_r]$-module $\beta_G^*(C^*)$ is isomorphic to $C^* \otimes \Lambda(s_1, \ldots, s_r) \otimes k[t_1, \ldots, t_r]$ and hence the latter inherits a $k[t_1, \ldots, t_r]$-linear differential from $\beta_G^*(C^*)$, but this differential is not $\Lambda(s_1, \ldots, s_r)$-linear, in general. Yet, as in the special case $r = 1$, one can obtain the differential of $\beta_G^*(C^*)$ from the differential of C^* by a two-stage twisting process, first over $A := \Lambda(s_1, \ldots, s_r)$ and then over $R = k[t_1, \ldots, t_r]$. We describe the differential of $\beta_G^*(C^*)$ in more detail; in particular the behaviour of the induced differential on the minimal Hirsch-Brown model with respect to a certain filtration will become important for the application. Since $\mathcal{E}_*((\mathbb{Z}_p)^r)$ is isomorphic to the r-fold tensor product $\mathcal{E}_*(\mathbb{Z}_p) \otimes \ldots \otimes \mathcal{E}_*(\mathbb{Z}_p)$ the same holds for the dual complexes:
$$\mathcal{E}^*((\mathbb{Z}_p)^r) \cong \mathcal{E}^*(\mathbb{Z}_p) \otimes \ldots \otimes \mathcal{E}^*(\mathbb{Z}_p)$$
(including the differentials). The differential $\widetilde{\delta}$ on
$$\beta_G^*(C^*) \cong C^* \otimes_{k[G]} \mathcal{E}^*(G), \qquad G = (\mathbb{Z}_p)^r$$
is just the tensor product coboundary, i.e.
$$\widetilde{\delta}(c \otimes e) = (\delta_C c) \otimes e + (-1)^{|c|} c \otimes \delta_{\mathcal{E}} e, \text{ for } c \otimes e \in C^* \otimes_{k[G]} \mathcal{E}^*(G),$$
where δ_C and $\delta_{\mathcal{E}}$ denote the coboundaries of C^* and $\mathcal{E}^*(G)$, respectively.

If one filters $\beta_G^*(C^*)$ by
$$\mathcal{F}_q(\beta_G^*(C^*)) := (\bigoplus_{i \leq q} C^i) \otimes_{k[G]} \mathcal{E}^*(G) = \bigoplus_{i \leq q} C^i \otimes A \otimes R,$$
then - on $\mathcal{F}_q(\beta_G^*(C^*))$ - $\tilde{\delta}$ coincides with
$$\delta_C \otimes \mathrm{id} : \mathcal{F}_q(\beta_G^*(C^*)) \to \mathcal{F}_{q+1}(\beta_G^*(C^*)),$$
modulo $\mathcal{F}_q(\beta_G^*(C^*))$. (Note that $\mathcal{F}_q(\beta_G^*(C^*))$ is not a subcomplex of $\beta_G^*(C^*)$.)

To obtain the minimal Hirsch-Brown model (cf. Section B.2) one splits the complex C^* into a direct sum $C^* = H^* \oplus B^* \oplus D^*$, where $H^* = H(C^*)$, $Z^* := H^* \oplus B^*$ are the cocycles of C^*, and $N^* := B^* \oplus D^*$ is homotopy trivial with $\delta_N : D^{q-1} \to B^q$ ($\delta_N := \delta_C|N$) an isomorphism. The standard inclusion $D^* \otimes A \otimes R \subset C^* \tilde{\otimes} A \tilde{\otimes} R$ extends uniquely to a morphism $\tilde{\phi}_N : N^* \otimes A \otimes R \to C^* \tilde{\otimes} A \tilde{\otimes} R$ of complexes over R, where the differential on $N^* \otimes A \otimes R$ is just $\delta_N \otimes \mathrm{id}_{A \otimes R}$. $\tilde{\phi}_N$ is defined so that Diagram 1.8 is commutative.

$$\begin{array}{ccc} D^{q-1} \otimes A \otimes R & \longrightarrow & C^{q-1} \otimes A \otimes R \\ \downarrow \delta_N \otimes \mathrm{id}_{A \otimes R} & & \downarrow \tilde{\delta} \\ B^q \otimes A \otimes R & \xrightarrow{\tilde{\phi}_N} & \mathcal{F}_q(\beta_G^*(C^*)) \end{array}$$

Diagram 1.8

As, on $C^{q-1} \otimes A \otimes R$, $\tilde{\delta}$ coincides with $\delta_C \otimes \mathrm{id}_{A \otimes R}$ modulo $\mathcal{F}_{q-1}(\beta_G^*(C^*))$, it follows that $\tilde{\phi}_N$ coincides with $\phi_N \otimes \mathrm{id}_{A \otimes R}$ modulo elements of lower filtration, where $\phi_N : N^* \to C^* = H^* \oplus N^*$ is the standard inclusion as a direct summand. Let $\phi_H : H^* \to C^* = H^* \oplus N^*$ be the corresponding map for the other summand, and define
$$\phi : H^* \otimes A \otimes R \oplus N^* \otimes A \otimes R \to C^* \otimes A \otimes R$$
by
$$\phi|H^* \otimes A \otimes R = \phi_H \otimes \mathrm{id}_{A \otimes R}$$
and
$$\phi|N^* \otimes A \otimes R = \tilde{\phi}_N.$$
Since ϕ coincides with the identity modulo elements of lower filtration the map ϕ is an isomorphism of R-modules and its inverse $\psi := \phi^{-1}$ also coincides with the identity up to elements of lower filtration. We

1.4. The Borel construction for p-tori

consider Diagram 1.9 which is commutative; we abbreviate $A \otimes R$ by AR, $A \tilde\otimes R$ by $\widetilde{AR}$ and $N^* \otimes AR$ (resp. $H^* \otimes AR$) by N^*AR (resp. H^*AR).

$$\begin{array}{ccccc}
N^*AR & \stackrel{\phi_N \otimes \mathrm{id}_{AR}}{\longrightarrow} & (H^*AR) \oplus (N^*AR) & \stackrel{\rho_H \otimes \mathrm{id}_{AR}}{\longrightarrow} & H^* \tilde\otimes \widetilde{AR} \\
\mathrm{id} \downarrow \uparrow \mathrm{id} & & \phi \downarrow \uparrow \psi & & \overline\phi \downarrow \uparrow \overline\psi \\
N^*AR & \stackrel{\tilde\phi_N}{\underset{\approx}{\longrightarrow}} & C^* \tilde\otimes \widetilde{AR} & \longrightarrow & C^* \tilde\otimes \widetilde{AR} / \operatorname{im} \tilde\phi_N
\end{array}$$

Diagram 1.9

The middle term in the top row inherits a differential from $C^* \tilde\otimes A \tilde\otimes R$ via the isomorphisms ϕ and ψ, which hence become isomorphisms of R-complexes and therefore induce isomorphisms $\overline\phi$ and $\overline\psi$ on the quotients.

Note that - although $H^* \otimes A \otimes R$ will, in general, not be a subcomplex of the middle term in the top row - one obtains a well-defined differential $\overline\delta$ on the quotient with respect to the subcomplex $N^* \otimes A \otimes R$. This quotient is the desired minimal Hirsch-Brown model $H^* \tilde\otimes A \tilde\otimes R$. It should be noted that $H^* \tilde\otimes A \tilde\otimes R$ is the R-free minimal model of $C^* \tilde\otimes A \tilde\otimes R$ (as an object in δgR-Mod) in the sense of (B.2.4) only if the twisted differential $\overline\delta$ vanishes when evaluated at $t_i = 0$, $i = 1, \ldots, r$. In general, one has to replace $H^* \tilde\otimes A$ by its homology (with respect to the evaluated differential $\overline\delta \otimes_R k_0$) to obtain the R-free minimal model of the form $H(H^* \tilde\otimes A) \tilde\otimes R$ (cf. (1.4.13) and (3.11.20)).

The differential $\overline\delta$ of $H^* \tilde\otimes A \tilde\otimes R$ applied to an element

$$h^q \otimes a \otimes r \in H^q \otimes A \otimes R$$

is then defined as

$$\overline\delta(h^q \otimes a \otimes r) := (\rho_H \otimes \mathrm{id}_{A \otimes R}) \psi \tilde{\tilde\delta}(\phi_H(h^q) \otimes a \otimes r).$$

Since $\delta_C \otimes \mathrm{id}$ vanishes on $\phi(h^q \otimes a \otimes r)$ one gets:

$$\tilde{\tilde\delta}(\phi_H(h^q) \otimes a \otimes r) = (-1)^q \mathrm{id}_C \otimes_{k[G]} \delta_\mathcal{E}(\phi_H(h^q) \otimes a \otimes r) \in C^q \otimes A \otimes R,$$

considering $\phi_H(h^q) \otimes a \otimes r$ as an element in

$$Z^* \otimes_{k[G]} \mathcal{E}^*(G) \subset C^* \otimes_{k[G]} \mathcal{E}^*(G) \cong C^* \tilde\otimes A \tilde\otimes R.$$

(B^* and $Z^* = H^* \oplus B^*$ are $k[G]$-submodules of C^*, but the decomposition $C^* = H^* \oplus B^* \oplus D^*$ need not be compatible with the $k[G]$-structure.)

Since ϕ and ψ are equal to the identity modulo elements of lower filtration the induced differential $\overline\delta$ on $H^q \tilde\otimes A \tilde\otimes R$ is given by $\mathrm{id}_{H^*} \otimes_{k[G]} \delta_\mathcal{E}$ modulo elements in $\bigoplus_{i<q} H^i \otimes A \otimes R$, using the isomorphism

$$H^* \otimes A \otimes R \cong H^* \otimes_{k[G]} \mathcal{E}^*(G)$$

and the induced action of G on $H(C^*)$. If we define the filtration on

$H^*\tilde{\otimes}A\tilde{\otimes}R$ by

$$\mathcal{F}_q(H^*\tilde{\otimes}A\tilde{\otimes}R) := \bigoplus_{i\leq q} H^i \otimes A \otimes R$$

then $\bar{\delta}$ is determined by the induced action of G on $H(C^*)$ modulo elements of lower filtration. Note that, in particular, $\mathcal{F}_q(H^*\tilde{\otimes}A\tilde{\otimes}R)$ is a subcomplex of $H^*\tilde{\otimes}A\tilde{\otimes}R$. It also follows from the construction that

$$\tilde{\tilde{\rho}} := (\rho_H \otimes \mathrm{id}_{A\otimes R})\psi : C^*\tilde{\otimes}A\tilde{\otimes}R \to H^*\tilde{\otimes}A\tilde{\otimes}R$$

preserves the filtration and coincides with $\rho_H \otimes \mathrm{id}_{A\otimes R}$ modulo elements of lower filtration. Since $N^* \otimes A \otimes R$ is homotopy trivial the map $\tilde{\tilde{\rho}}$ is a homotopy equivalence. A homotopy inverse

$$\tilde{\tilde{\iota}} : H^*\tilde{\otimes}A\tilde{\otimes}R \to C^*\tilde{\otimes}A\tilde{\otimes}R$$

can be chosen to be filtration preserving (in fact, to map $\mathcal{F}_q(H^*\tilde{\otimes}A\tilde{\otimes}R)$ to the subcomplex

$$\bigoplus_{i<q} C^i \otimes A \otimes R \oplus Z^q \otimes A \otimes R \subset C^*\tilde{\otimes}A\tilde{\otimes}R,$$

since ϕ restricted to this subcomplex gives a homotopy equivalence with $\mathcal{F}_q(H^*\tilde{\otimes}A\tilde{\otimes}R)$), and such that $\tilde{\tilde{\rho}}\tilde{\tilde{\iota}}$ is the identity modulo terms of lower filtration.

We summarize the main properties of the minimal Hirsch-Brown model derived above in the following proposition.

(1.4.10) Proposition
Let $C^* = \{C^q, q \geq 0\}$ be a $k[G]$-complex, $G = (\mathbb{Z}_p)^r$. Then the algebraic Borel construction $\beta_G^*(C^*) \cong C^*\tilde{\otimes}A\tilde{\otimes}R$ has a minimal Hirsch-Brown model $H^*\tilde{\otimes}A\tilde{\otimes}R$ with the following properties:

(i) $H^*\tilde{\otimes}A\tilde{\otimes}R$ is filtered by subcomplexes

$$\mathcal{F}_q(H^*\tilde{\otimes}A\tilde{\otimes}R) := \bigoplus_{i\leq q} H^i \otimes A \otimes R$$

such that the differential on the successive quotients is given by the induced action of G on $H^* = H(C^*)$: i.e. the differential $\bar{\delta}$ of

$$H^*\tilde{\otimes}A\tilde{\otimes}R \cong H^* \otimes_{k[G]} \mathcal{E}^*(G)$$

(as R-modules, but - in general - not as R-complexes) coincides with

$$\mathrm{id}_{H^*} \otimes_{k[G]} \delta_\mathcal{E}$$

modulo terms of lower filtration.

(ii) There exists a filtration preserving pair of inverse homotopy equivalences
$$C^*\tilde{\otimes}A\tilde{\otimes}R \underset{\tilde{\iota}}{\overset{\tilde{\rho}}{\rightleftarrows}} H^*\tilde{\otimes}A\tilde{\otimes}R$$
where the filtration on $C^*\tilde{\otimes}A\tilde{\otimes}R$ is given by
$$\mathcal{F}_q(C^*\tilde{\otimes}A\tilde{\otimes}R) := \bigoplus_{i\leq q} C^i \otimes A \otimes R,$$
such that - modulo elements of lower filtration - $\tilde{\rho}$ coincides with $\rho_H \otimes \mathrm{id}_{A\otimes R}$ (where
$$\rho_H : C^* = H^* \oplus B^* \oplus D^* \to H^*$$
is the standard projection onto the first summand) and $\tilde{\rho}\tilde{\iota}$ coincides with $\mathrm{id}_{H^*\tilde{\otimes}A\tilde{\otimes}R}$. Furthermore,
$$\tilde{\iota}(\mathcal{F}_q(H^*\tilde{\otimes}A\tilde{\otimes}R)) \subset \bigoplus_{i<q} C^i \otimes A \otimes R \oplus Z^q \otimes A \otimes R.$$

(1.4.11) Corollary

(1) Let X be a G-space, $G = (\mathbb{Z}_p)^r$. Then the equivariant cohomology $H_G^*(X;k) = H(\beta_G^*(X;k))$ can be obtained from a minimal Hirsch-Brown model $H^*(X;k)\tilde{\otimes}A\tilde{\otimes}R$ with the properties described in (1.4.10).
(2) If $X \to Y$ is a G-map of G-spaces, then the induced map
$$f_G^* : H_G^*(Y;k) \to H_G^*(X;k)$$
can be represented by a chain map between the respective minimal Hirsch-Brown models
$$\tilde{f} : H^*(Y;k)\tilde{\otimes}A\tilde{\otimes}R \to H^*(X;k)\tilde{\otimes}A\tilde{\otimes}R,$$
which is filtration preserving and which coincides with $f^*\otimes\mathrm{id}_{A\otimes R}$ modulo terms of lower filtration ($f^* := H^*(f) : H^*(Y;k) \to H^*(X;k)$).

Proof.
Part (1) is an immediate consequence of (1.4.10). For part (2) consider Diagram 1.10, where we abbreviate as in Diagram 1.9.

$$\begin{array}{ccccc}
\beta_G^*(Y;k) \cong S^*(Y,k)\widetilde{\tilde{\otimes}AR} & \xrightarrow{S^*(f;k)\otimes\mathrm{id}_{AR}} & S^*(X;k)\widetilde{\tilde{\otimes}AR} \cong \beta_G^*(X;k) \\
\tilde{\iota}_Y \uparrow \downarrow \tilde{\rho}_Y & & \tilde{\iota}_X \uparrow \downarrow \tilde{\rho}_X \\
H^*(Y;k)\widetilde{\tilde{\otimes}AR} & & H^*(X;k)\widetilde{\tilde{\otimes}AR}
\end{array}$$

Diagram 1.10

If one defines $\tilde{f}^* := \tilde{\rho}_X(S^*(f,k) \otimes \mathrm{id}_{A\otimes R})\tilde{\iota}_Y$, then the assertion follows from (1.4.10). (Note that $\beta_G^*(f;k) = S^*(f;k) \otimes \mathrm{id}_{A\otimes R}$.) $\square$

It is possible to refine the filtration of $H^*\tilde{\otimes}A\tilde{\otimes}R$ such that the successive quotients of the refined filtration inherit the trivial differential; cf. [Carlsson, 1983] for related constructions when $p = 2$. Because of (1.4.10)(i) it suffices to consider the differential $\mathrm{id}_{H^*} \otimes_{k[G]} \delta_{\mathcal{E}}$ on $H^* \otimes_{k[G]} \mathcal{E}^*(G)$. We will use the structure of H^* as a $k[G]$-module to obtain the desired refinement of the above filtration. The group ring $k[\mathbb{Z}_p]$ is isomorphic to $k[\tau]/(\tau^p)$, where τ corresponds to $1-g \in k[\mathbb{Z}_p]$, g a chosen generator of $\mathbb{Z}_p$. (Note that $(1-g)^p = 1-g^p$ in $k[\mathbb{Z}_p]$ since $\mathrm{char}(k) = p$.) Let $g_1, \ldots, g_r$ be a system of generators of $G \cong \mathbb{Z}_p \times \ldots \times \mathbb{Z}_p$; then $k[G] \cong k[\mathbb{Z}_p] \otimes \ldots \otimes k[\mathbb{Z}_p]$ is isomorphic to $k[\tau_1, \ldots, \tau_r]/(\tau_1^p, \ldots, \tau_r^p)$, where τ_i corresponds to $1 - g_i$, $i = 1, \ldots, r$. Let $\mathfrak{m} \subset k[G]$ be the augmentation ideal, generated by $\tau_1, \ldots, \tau_r$. We define a decreasing filtration on $\mathcal{E}^*(G) \cong k[G] \otimes A \otimes R$ by powers of the ideal $\mathfrak{m}$, i.e.

$$\mathcal{F}^i(\mathcal{E}^*(G)) := \mathfrak{m}^i \otimes A \otimes R, i = 0, 1, \ldots.$$

Clearly

$$\mathcal{E}^*(G) = \mathcal{F}^0(\mathcal{E}^*(G)) \supseteq \mathcal{F}^1(\mathcal{E}^*(G)) \supseteq \ldots \supseteq \mathcal{F}^i(\mathcal{E}^*(G))$$
$$\supseteq \mathcal{F}^{i+1}(\mathcal{E}^*(G)) \ldots.$$

Because of $\mathfrak{m}^{r(p-1)+1} = 0$ one has $\mathcal{F}^{r(p-1)+1}(\mathcal{E}^*(G)) = 0$. Since

$$1 + g + \ldots + g^{p-1} = (1-g)^{p-1}$$

in $k[G]$ (note that $\binom{p-1}{j} \equiv (-1)^j \mod p$) the differential on $\mathcal{E}^*(\mathbb{Z}_p)$ (see above) raises the filtration degree in case $r = 1$, and by the tensor product formula for the differential of

$$\mathcal{E}^*(G) \cong \mathcal{E}^*(\mathbb{Z}_p) \otimes \ldots \otimes \mathcal{E}^*(\mathbb{Z}_p)$$

the same holds in the general case. Moreover, the differential of $\mathcal{E}^*(G)$ is R-linear. We now define the filtration on $H^*\tilde{\otimes}A\tilde{\otimes}R$ by

$$\mathcal{F}_q^i(H^*\tilde{\otimes}A\tilde{\otimes}R) := (\bigoplus_{j<q} H^j \oplus H^q\mathfrak{m}^i) \otimes A \otimes R.$$

One clearly has

$$\mathcal{F}_q^{i-1}(H^*\tilde{\otimes}A\tilde{\otimes}R) \supseteq \mathcal{F}_q^i(H^*\tilde{\otimes}A\tilde{\otimes}R),$$

$$\mathcal{F}_q^0(H^*\tilde{\otimes}A\tilde{\otimes}R) = \mathcal{F}_q(H^*\tilde{\otimes}A\tilde{\otimes}R),$$

and

$$\mathcal{F}_q^{r(p-1)+1}(H^*\tilde{\otimes}A\tilde{\otimes}R) = \mathcal{F}_{q-1}(H^*\tilde{\otimes}A\tilde{\otimes}R).$$

Since
$$H^q\mathfrak{m}^i \otimes A \otimes R = H^q \otimes_{k[G]} \mathcal{F}^i(\mathcal{E}^*(G)),$$
and $\delta_\mathcal{E}$ raises the filtration degree on $\mathcal{E}^*(G)$, so does $\mathrm{id}_{H^*} \otimes_{k[G]} \delta_\mathcal{E}$ and hence $\bar\delta$ on $H^*\tilde\otimes A\tilde\otimes R$. We arrive at the following proposition.

(1.4.12) Proposition
Let $C^* = \{C^q, q \geq 0\}$ be a $k[G]$-complex, $G = (\mathbb{Z}_p)^r$. Then the minimal Hirsch-Brown model $H^*\tilde\otimes A\tilde\otimes R$ of the algebraic Borel construction $\beta_G^*(C^*)$ has a filtration by subcomplexes over R,
$$\mathcal{F}_q^i(H^*\tilde\otimes A\tilde\otimes R) := (\bigoplus_{j<q} H^j \oplus H^q\mathfrak{m}^i) \otimes A \otimes R, \quad q \geq 0, 0 \leq i \leq r(p-1),$$
such that the induced differential on the successive quotients is trivial.

(1.4.13) Remarks
(1) A more detailed discussion of the minimal Hirsch-Brown model in case (odd p) including aspects of the two-stage twisting over A and R, and the uniqueness of the minimal model up to isomorphism in the appropriate category, is contained in [Baumgartner, 1990].
(2) As already remarked in connection with the $p = 2$ case (see Section 1.3) the behaviour of the model with respect to the restriction of the action to subgroups and a possible analogue of Theorem (1.3.24) is much more complicated in case (odd p) then in case (2). This theme is taken up in Section 3.11. On the other hand the results we derive in this section can easily be 'specialized' to the case (2), which means vaguely that one forgets about the exterior algebra A and assigns the degree 1 to the generators $t_i \in R$. In fact, using just the R-linearity of the minimal Hirsch-Brown model $H^*\tilde\otimes R$ and some simple degree arguments leads to a considerable simplification of the reasoning in case (2). (See also [Carlsson, 1983] for the case (2).)

If M is a right $k[G]$-module we denote by $\ell(M)$ the minimum of all $i \in \{0, 1, 2, \ldots\}$ such that $M\mathfrak{m}^i = 0$. This means that we have a strictly decreasing filtration
$$M = M\mathfrak{m}^0 \supsetneq M\mathfrak{m}^1 \supsetneq M\mathfrak{m}^2 \supsetneq \cdots \supsetneq M\mathfrak{m}^{\ell(M)} = 0,$$
where $\mathfrak{m} \subset k[G]$ is the augmentation ideal.

We will apply the filtered minimal Hirsch-Brown model to obtain the following result (see [Carlsson, 1983] for the case $p = 2$ and [Baumgartner, 1990] for the odd p case).

(1.4.14) Theorem

Let X be a paracompact free G-space, $G = (\mathbb{Z}_p)^r$, such that

$$H^i(X/G; k) = 0$$

if $i > N$ for some $N \in \mathbb{N}$ (this holds, for example, if X is a finite-dimensional G-CW-complex). Then

$$\sum_{j=0}^{\infty} \ell(H^j(X; k)) \geq r + 1.$$

In particular, if G acts trivially on $H^*(X; k)$, then $H^j(X; k) \neq 0$ for at least $r + 1$ different values of j.

(1.4.15) Corollary (see [Carlsson 1980, 1982])

Let X be a finite-dimensional free G-CW-complex, $G = (\mathbb{Z}_p)^r$, such that

$$H^*(X; k) \cong H^*(S^m \times \ldots \times S^m; k)$$

as graded k-vector spaces and G acts trivially on $H^*(X; k)$; then the number of factors in $S^m \times \ldots \times S^m$ is at least r.

To prove (1.4.14) we may assume that X is connected. The general case then follows by considering the way in which G permutes the components of X. For X connected, Theorem (1.4.14) is an immediate consequence of the following purely algebraic result applied to the filtered minimal Hirsch-Brown model $H^* \tilde{\otimes} A \tilde{\otimes} R$ (see (1.4.12)) of $\beta_G^*(X; k)$, using the fact that for a paracompact free G-space X the equivariant cohomology $H_G^*(X; k)$ is isomorphic to $H^*(X/G; k)$ (see (1.2.9)(2)).

(1.4.16) Proposition

Let $C^* \tilde{\otimes} R$ be a filtered R-complex, $R = k[t_1, \ldots, t_r]$ the graded polynomial ring in r variables over a field k ($|t_i| = 2$ for $i = 1, \ldots, r$ if $\mathrm{char}(k) \neq 2$; $|t_i| = 1$ for $i = 1, \ldots, r$ if $\mathrm{char}(k) = 2$), where $C^* = \{C^q; q \geq 0\}$ is a connected (i.e. $C^0 = k$), graded k-vector space and the filtration $\mathcal{F}_q(C^* \tilde{\otimes} R)$ of $C^* \tilde{\otimes} R$ is given as the extension of a filtration $0 = \mathcal{F}_{-1}(C^*) \subset \mathcal{F}_0(C^*) \subset \mathcal{F}_1(C^*) \subset \ldots \subset \mathcal{F}_{\ell-1}(C^*) = C^*$ of C^*, i.e. $\mathcal{F}_q(C^* \tilde{\otimes} R) := \mathcal{F}_q(C^*) \otimes R$. The differential $\tilde{\delta}$ of $C^* \otimes R$ is assumed to lower the filtration degree (and to increase the total degree). Assume, furthermore, that $C^0 \subset \mathcal{F}_0(C^*)$, and that $H^i(C^* \tilde{\otimes} R) = 0$ for $i > N$, for some $N \in \mathbb{N}$. Then the length ℓ of the filtration of C^* is at least $r + 1$.

(Note that one obtains an R-complex of the above type if one takes the R-free resolution of a graded, connected R-module M^* which is bounded above. In fact, one then gets a bigraded complex. The resolution degree can be used to obtain a filtration with the above properties. The proof of (1.4.16) uses slight modifications of standard results in homological algebra, carried from a graded to a filtered situation.)

We split the proof of (1.4.16) into several lemmata.

(1.4.17) Lemma
Let $K_1 := \Lambda(s_1,\ldots,s_r)\tilde\otimes R$ be the Koszul complex corresponding to $(t_1^{N+1},\ldots,t_r^{N+1}) \subset R$, i.e. $\Lambda(s_1,\ldots,s_r)$ is the exterior algebra over k on generators $s_1,\ldots,s_r$ and K_1 is a differential graded algebra over R, where the differential is given by $\tilde\delta(s_i) := t_i^{N+1}$ for $i = 1,\ldots,r$ (cf. Section A.5), and let $C^*\tilde\otimes R$ be an R-complex as in (1.4.16). Then there exists a morphism of R-complexes $\alpha : K_1 \to C^* \otimes R$ such that $\alpha(1 \otimes 1) = 1 \otimes 1 \in k \otimes R = C^0 \otimes R \subset C^* \otimes R$.

Proof.
We construct $\alpha : K_1 \to C^*\tilde\otimes R$ by induction over the product length in the s_is. On $1 \otimes R \subset K_1$ the R-linear map α is defined by
$$\alpha(1 \otimes 1) := 1 \otimes 1 \in C^*\tilde\otimes R.$$
Assume that α has already been defined on
$$\Lambda^j(s_1,\ldots,s_r) \otimes R \subset K_1$$
for $j < q$, where $\Lambda^j(s_1,\ldots,s_r)$ denotes the k-linear subspace of $\Lambda(s_1,\ldots,s_r)$ generated by all products in the s_is of length j. Let
$$s_{i_1}\ldots s_{i_q} \in \Lambda^q(s_1,\ldots,s_r),$$
then
$$\tilde\delta(s_{i_1}\ldots s_{i_q}) \in \Lambda^{q-1}(s_1,\ldots,s_r) \otimes R.$$
So $\alpha\tilde\delta(s_{i_1}\ldots s_{i_q})$ is defined already and is a cycle in $C^*\tilde\otimes R$. For degree reasons ($H^i(C^* \otimes R) = 0$ for $i > N$ by assumption) the element $\alpha\tilde\delta(s_{i_1}\ldots s_{i_q})$ must be a boundary, i.e. we can choose an element $u \in C^*\tilde\otimes R$ such that
$$\alpha\tilde\delta(s_{i_1}\ldots s_{i_q}) = \tilde\delta u.$$
We define $\alpha(s_{i_1}\ldots s_{i_q}) := u$. Since the (non-zero) monomials of length q form an R-basis of $\Lambda^q(s_1,\ldots,s_r) \otimes R$ we get the desired extension of α to $\Lambda^q(s_1,\ldots,s_r) \otimes R$. □

The next lemma and its proof are due to Baumgartner.

(1.4.18) Lemma
Let $K_2 := \Lambda(\sigma_1,\ldots,\sigma_r)\tilde{\otimes} R$ be the Koszul complex corresponding to $(t_1,\ldots,t_r) \subset R$, i.e. $\tilde{\delta}(\sigma_i) := t_i$ for $i = 1,\ldots,r$, and let $C^*\tilde{\otimes} R$ be as in (1.4.16).

(1) There exists a filtration preserving morphism of R-complexes
$$\beta: C^*\tilde{\otimes} R \to K_2$$
with $\beta(1 \otimes 1) = 1 \otimes 1$, where the filtration on K_2 is defined by
$$\mathcal{F}_q(K_2) := \bigoplus_{j \leq q} \Lambda^j(\sigma_1,\ldots,\sigma_r) \otimes R.$$

(2) Let $\beta', \beta'' : C^*\tilde{\otimes} R \to K_2$ be two morphisms of R-complexes (not necessarily filtration preserving) which coincide on $C^0 \otimes R$; then β' and β'' are R-homotopic.

Proof.
(1) We construct $\beta : C^*\tilde{\otimes} R \to K_2$ by induction on the filtration degree. On $C^0 \otimes R = k \otimes R$ the map β is already defined by the condition $\beta(1 \otimes 1) = 1 \otimes 1 \subset K_2$. One obviously can extend this map to
$$\mathcal{F}_0(C^*\tilde{\otimes} R) = \mathcal{F}_0(C^*) \otimes R \supset C^0 \otimes R = k \otimes R$$
such that
$$\beta: \mathcal{F}_0(C^*\tilde{\otimes} R) \to \mathcal{F}_0(K_2) = k \otimes R$$
is R-linear and compatible with the respective differentials. Assume that β has been constructed already on $\mathcal{F}_{q-1}(C^*\tilde{\otimes} R)$ as a filtration preserving morphism of R-complexes. We choose elements $\{c_i \in \mathcal{F}_q(C^*); i \in I\}$ such that the residue classes in $\mathcal{F}_q(C^*)/\mathcal{F}_{q-1}(C^*)$ form a k-basis. Since $\tilde{\delta}(c_i \otimes 1)$ is contained in $\mathcal{F}_{q-1}(C^*\tilde{\otimes} R)$ one has that $\beta\tilde{\delta}(c_i \otimes 1)$ is already defined and is a cycle in $\mathcal{F}_{q-1}(K_2)$. For degree reasons ($|c_i \otimes 1| > 0$, and $H^*(K_2) = R/(t_1,\ldots,t_r) = k$, cf. (A.4.13)(4)), $\beta\tilde{\delta}(c_i \otimes 1)$ is a boundary and we can choose an element $u_i \in \mathcal{F}_q(K_2)$ such that $\beta\tilde{\delta}(c_i \otimes 1) = \tilde{\delta}(u_i)$, $i \in I$. We define $\beta(c_i \otimes 1) := u_i$ for $i \in I$, to get the desired extension of β to $\mathcal{F}_q(C^*\tilde{\otimes} R)$.

To prove part (2) it suffices to show that $\beta := \beta' - \beta''$ is R-homotopic to zero. The homotopy is again constructed by induction on the filtration degree by a similar argument. By assumption β vanishes on $C^0 \otimes R$. Let $c \otimes 1 \in \mathcal{F}_0(C^*\tilde{\otimes} R)$ with $|c| > 0$. Then $\beta(c \otimes 1)$ is a cycle in K_2 of degree > 0, and hence a boundary ($H^j(K_2) = 0$ for $j > 0$).

We therefore can choose an R-linear map $h : \mathcal{F}_0(C^*\tilde{\otimes} R) \to K_2$ such that $\tilde{\delta}h + h\tilde{\delta} = \beta$ on $\mathcal{F}_0(C^*\tilde{\otimes} k)$. (Note that $\tilde{\delta}$ is zero on $\mathcal{F}_0(C^*\tilde{\otimes} R)$

by assumption, so $h\tilde{\delta}$ actually vanishes on $\mathcal{F}_0(C^*\tilde{\otimes}R)$.) Assume now that $h : \mathcal{F}_{q-1}(C^*\tilde{\otimes}R) \to K_2$ has been constructed already such that $\tilde{\delta}h + h\tilde{\delta} = \beta$ on $\mathcal{F}_{q-1}(C^*\tilde{\otimes}R)$.

Let $\{c_i \in \mathcal{F}_q(C^*); i \in I\}$ be as above. The R-linear map $\beta - h\tilde{\delta}$ is defined on $\mathcal{F}_q(C^*\tilde{\otimes}R)$ (since $\tilde{\delta}(\mathcal{F}_q(C^*\tilde{\otimes}R)) \subset \mathcal{F}_{q-1}(C^*\tilde{\otimes}R)$) and

$$(\beta - h\tilde{\delta})(c_i \otimes 1)$$

are cycles in K_2 of degree > 0, and hence boundaries

$$(\tilde{\delta}(\beta - h\tilde{\delta}) = \tilde{\delta}\beta - \tilde{\delta}h\tilde{\delta} = \beta\tilde{\delta} - \tilde{\delta}h\tilde{\delta} = (\beta - \tilde{\delta}h)\tilde{\delta} = h\tilde{\delta}\tilde{\delta} = 0$$

by induction hypothesis).

We therefore can choose elements $w_i \in K_2$, $i \in I$ such that

$$(\beta - h\tilde{\delta})(c_i \otimes 1) = \tilde{\delta}(w_i).$$

Defining $h(c_i \otimes 1) := w_i$ for $i \in I$ gives the desired extension of the homotopy to $\mathcal{F}_q(C^*\tilde{\otimes}R)$ $((\tilde{\delta}h + h\tilde{\delta})(c_i \otimes 1) = \beta(c_i \otimes 1))$. □

(1.4.19) Lemma
Let $\gamma' : K_1 \to K_2$ be any morphism of the Koszul complexes above, considered as complexes over R (γ' need not be multiplicative), such that $\gamma'(1 \otimes 1) = 1 \otimes 1$. Then $\gamma'(K_1) \not\subset \mathcal{F}_{r-1}(K_2)$.

Proof.
By Lemma (1.4.18)(2) the morphism γ' is R-homotopic to the map $\gamma'' : K_1 \to K_2$ of differential graded R-algebras, given by $\gamma''(s_i) = \sigma_i t_i^N$. Hence γ' and γ'' induce the same map in cohomology with any coefficients over R. Let $\overline{R} := R/(t_1^{N+1}, \ldots, t_r^{N+1})$. Then

$$H(K_1; \overline{R}) = \Lambda(s_1, \ldots, s_r) \otimes \overline{R}$$

and

$$H(K_2; \overline{R}) = \Lambda(\sigma_1 t_1^N, \ldots, \sigma_r t_r^N).$$

The map $H(\gamma''; \overline{R})$ is given by

$$H(\gamma''; \overline{R})(s_i) = \sigma_i t_i^N.$$

In particular,

$$H(\gamma''; \overline{R})(s_1 \ldots s_r) = (\sigma_1 \ldots \sigma_r) t_1^N \ldots t_r^N.$$

The same must hold for $H(\gamma'; \overline{R})$. Since

$$(\sigma_1 \ldots \sigma_r) t_1^N \ldots t_r^N \in H(K_2; \overline{R})$$

cannot be represented by an element in $\mathcal{F}_{r-1}(K_2)$ the assertion follows.

□

Proof of (1.4.16).
We choose morphisms $\alpha : K_1 \to C^* \tilde{\otimes} R$ and $\beta : C^* \tilde{\otimes} R \to K_2$ according to (1.4.17) and (1.4.18)(1). In particular, β is filtration preserving. The composition $\gamma' : \beta\alpha : K_1 \to K_2$ fulfils the assumption of (1.4.19): and so $\gamma'(K_1) \not\subset \mathcal{F}_{r-1}(K_2)$. Hence $\mathcal{F}_{r-1}(C^* \tilde{\otimes} R) \neq C^* \otimes R$, and therefore $\ell - 1 \geq r$. □

(1.4.20) Remarks
(1) As pointed out already the above discussion can also be done for $p = 2$ and, in fact, simplifies considerably. In particular, the above results (1.4.14) and (1.4.15) hold for any prime p.
(2) In case ($p = 2$) the above Koszul complexes K_1 and K_2 can be viewed as the minimal Hirsch-Brown models of the $(\mathbb{Z}_2)^r$-spaces X_1 and X_2 respectively, where $X_1 = S^N \times \ldots \times S^N$ (r-fold product with the standard free $(\mathbb{Z}_2)^r$-action) and $X_2 = (\mathbb{Z}_2)^r$ (considered as a $(\mathbb{Z}_2)^r$-space by left multiplication) (see Exercise (1.7)).
The map $\alpha : K_1 \to C^* \tilde{\otimes} R$ then represents the factorization of the classifying map $f : X \to EG$ of the free G-space X through an appropriate finite-dimensional G-subspace

$$S^N \times \ldots \times S^N \subset E\mathbb{Z}_2 \times \ldots \times E\mathbb{Z}_2 = EG,$$

and $\beta : C^* \tilde{\otimes} R \to K_2$ just represents the inclusion of an orbit on the level of the minimal Hirsch-Brown models. The 'geometric' interpretation of α and β in case odd p is not quite as simple, due to the appearance of the exterior algebra in the cohomology of the classifying space. The Koszul complexes K_1, resp. K_2, are not equal to the minimal Hirsch-Brown models of $S^N \times \ldots \times S^N$, resp. $(\mathbb{Z}_p)^r$, in that case.
(3) Refined versions of Theorem (1.4.14), containing also the case of real torus actions ($G = (S^1)^r$), are given in Chapter 4, where the topic of free torus actions is discussed in more detail.
Obviously Theorem (1.4.14) implies that $\sum_{j=0}^{\infty} \dim_k H^j(X; k) \geq r+1$. In case (2) one easily obtains the following improved inequality.

(1.4.21) Corollary
Let X be a paracompact free G-space, $G = (\mathbb{Z}_2)^r$ such that $H^i(X/G; \mathbb{F}_2) = 0$ if $i > N$ for some $N \in \mathbb{N}$. Then

$$\sum_{j=0}^{\infty} \dim_k H^j(X; \mathbb{F}_2) \geq 2r.$$

Proof.

Again we may assume that X is connected. We consider the filtered minimal Hirsch-Brown model $H^*(X)\tilde{\otimes}R$ of $\beta_G^*(X;\mathbb{F}_2)$. Without loss of generality one can assume that the filtration

$$0 = \mathcal{F}_{-1}(H^*(X)\tilde{\otimes}R) \subset \mathcal{F}_0(H^*(X)\tilde{\otimes}R) \subset \ldots \subset \mathcal{F}_q(H^*(X)\tilde{\otimes}R)$$
$$= \mathcal{F}_q(H^*(X)) \otimes R \subset \ldots \mathcal{F}_{\ell-1}(H^*(X)\tilde{\otimes}R) = H^*(X)\tilde{\otimes}R$$

has the property that $\tilde{\delta} : \mathcal{F}_q(H^*(X)\tilde{\otimes}R) \to \mathcal{F}_{q-1}(H^*(X)\tilde{\otimes}R)$ induces a non-trivial map

$$\bar{\delta} : \mathcal{F}_q(H^*(X)\tilde{\otimes}R)/\mathcal{F}_{q-1}(H^*(X)\tilde{\otimes}R)$$
$$\to \mathcal{F}_{q-1}(H^*(X)\tilde{\otimes}R)/\mathcal{F}_{q-2}(H^*(X)\tilde{\otimes}R)$$

for $q = 1,\ldots,\ell-1$. (Otherwise we could shorten the filtration by omitting the term $\mathcal{F}_{q-1}(H^*(X)\tilde{\otimes}R)$.) One clearly has $\bar{\delta}\bar{\delta} = 0$. If

$$\operatorname{rank}_R(\mathcal{F}_q(H^*(X)\tilde{\otimes}R)/\mathcal{F}_{q-1}(H^*(X)\tilde{\otimes}R))$$
$$= \dim_{\mathbb{F}_2}(\mathcal{F}_q(H^*(X))/\mathcal{F}_{q-1}(H^*(X))) = 1$$

then

$$\bar{\delta} : \mathcal{F}_{q+1}(H^*(X)\tilde{\otimes}R)/\mathcal{F}_q(H^*(X)\tilde{\otimes}R)$$
$$\to \mathcal{F}_q(H^*(X)\tilde{\otimes}R)/\mathcal{F}_{q-1}(H^*(X)\tilde{\otimes}R)$$

or

$$\bar{\delta} : \mathcal{F}_q(H^*(X)\tilde{\otimes}R)/\mathcal{F}_{q-1}(H^*(X)\tilde{\otimes}R)$$
$$\to \mathcal{F}_{q-1}(H^*(X)\tilde{\otimes}R)/\mathcal{F}_{q-2}(H^*(X)\tilde{\otimes}R)$$

has to vanish. This cannot happen if $0 < q < \ell - 1$. Since, by (1.4.16), $\ell \geq r+1$ one gets the desired inequality. □

Note that the above argument does not carry through in case (odd p) because of the presence of the exterior algebra in $H^*(B(\mathbb{Z}_p)^r;\mathbb{F}_p)$ (which hence contains nilpotent elements).

Exercises

For the following exercises we will always assume that G is a compact Lie group and $K < G$ is a closed subgroup, if not explicitly stated otherwise.

(1.1) Show that the category of spaces having the G-homotopy type of a (finite) G-CW-complex (and G-maps as morphisms) is the smallest category of G-spaces (and G-maps) which contain all 'G-points' G/K, $K < G$, and which is closed under homotopy colimits of (finite) diagrams; cf. [Puppe, D. 1983] for a conceptual proof of this result in case $G = \{1\}$.

(1.2) For a G-space X we denote by $\text{res}_K X$ the space X viewed as a K-space by restricting the G-action. Prove the following:
(a) If G is finite and X is a (finite, finite-dimensional) G-CW-complex then $\text{res}_K X$ can be considered as a (finite, finite-dimensional) K-CW-complex.
(b) If X has the G-homotopy type of a (finite, finite-dimensional) G-CW-complex then $\text{res}_K X$ has the K-homotopy type of a (finite, finite dimensional) K-CW-complex.

(Note that any orbit G/H, $H < G$, can be viewed as a finite K-CW-complex, $K < G$, but, in general, not in a canonical way (see [Illman, 1983, 1990], cf. [Mayer, K.H., 1989]).)

(1.3) Let X be a G-CW-complex. Show that:
(a)
$$X^K := \{x \in X; gx = x \text{ for all } g \in K\}$$
$$= \{x \in X; K < G_x\}$$
can be considered as a $W(K)$-CW-complex, where
$$W(K) := N(K)/K$$
and $N(K)$ is the normalizer of K in G.

(b)
$$X^{(K)} := \{x \in X; K \text{ conjugate to a subgroup of } G_x\}$$
$$= GX^K := \{gx; g \in G \text{ and } x \in X^K\}$$
can be considered as a G-subcomplex of X.

(1.4) Let X (resp. X') be a G (resp. G')-CW-complex, G and G' compact Lie groups. Prove that $X \times X'$, equipped with the compactly generated topology, can be considered as a $G \times G'$-CW-complex. In particular, $X \times X$ together with the diagonal action is G-homotopy equivalent to a G-CW-complex (see Exercise (1.2)).

The Euler characteristic $\chi(X)$ of a finite CW-complex X can be computed as the alternating sum of the number of cells in the different dimensions. One easily obtains the following results (note, though, that analogous results for more general G-spaces are much more subtle, cf. e.g. (3.10.19)(3)).

(1.5) Let X be a finite G-CW-complex. Prove the following:
(a) If $G = \mathbb{Z}_p$ then: $\chi(X) - \chi(X^G) = p(\chi(X/G) - \chi(X^G))$.
(b) If G is a finite p-group then: $\chi(X) - \chi(X^G) \equiv 0 \bmod p$.
(c) If G is a torus ($G = S^1 \times \ldots \times S^1$) then: $\chi(X) - \chi(X^G) = 0$.
(d) If G is a finite group and X is a free, finite G-CW-complex then: $\chi(X) = |G|\chi(X/G)$, where $|G|$ denotes the order of G.

Note that (b), applied to the action by conjugation on $X = G$, implies that any finite p-group has a non-trivial centre and hence is solvable.

(1.6) Use the comparison Theorem (1.1.3) and Remarks (1.1.4) to show that $H^G_*(X)$ (resp. $H^*_G(X)$) and $H_*(X/G)$ (resp. $H^*(X/G)$) are naturally isomorphic for any free G-CW-complex X. In particular,
$$H^G_*(EG) \cong H_*(BG)$$
(resp. $H^*_G(EG) \cong H^*(BG)$).

(1.7)
(a) Let S^m be the m-dimensional sphere considered as a $\mathbb{Z}_2$-space via the antipodal map, k a field of characteristic 2. Show that the minimal Hirsch-Brown model $H^*(S^m; k) \tilde{\otimes} k[t]$ is isomorphic (as a $k[t]$-complex) to the Koszul complex (cf. (A.5)) for the regular sequence $t^{m+1} \in k[t]$, i.e. $H^*(S^m; k) \tilde{\otimes} k[t] \cong \Lambda(\sigma) \tilde{\otimes} k[t]$, where the differential on the right side is given by $\tilde{\delta}\sigma = t^{m+1}$.

(b) Show that the G-map $(G = \mathbb{Z}_2)$ $\mathbb{Z}_2 \xrightarrow{i} S^m$ given by the inclusion of an orbit induces the following morphism (unique up to homotopy in $\delta gk[t]$ – Mod) between the respective minimal Hirsch-Brown models
$$i^* : \Lambda(\sigma) \tilde{\otimes} k[t] \to \Lambda(\tau) \tilde{\otimes} k[t], \qquad \sigma \mapsto \tau t^m.$$

(c) Generalize parts (a) and (b) to $S^m \times \ldots \times S^m$ (instead of S^m) considered as a $\mathbb{Z}_2 \times \ldots \times \mathbb{Z}_2$-space via the componentwise antipodal action.

(1.8) Prove an analogue of Exercise (1.7) for the case (odd p), i.e. replace the free $\mathbb{Z}_2$-space S^m by the free $\mathbb{Z}_p$-space S^{2m+1}, where the action is given by complex scalar multiplication,
$$\mathbb{Z}_p \subset S^1 \subset \mathbb{C}, \qquad S^{2m+1} \subset \mathbb{C}^{m+1}.$$

(1.9) For $G = \mathbb{Z}_p$ and k a field characteristic p show that
$$H_{G,\alpha}(EG; k) \cong k$$
for any $\alpha \in k$ (but $(EG)^G = \emptyset$) (cf. (1.3.5) and (1.4.5)). More generally: show that $H_{G,\alpha}(EG \times X; k) \cong H^*_{G,\alpha}(X; k)$ for any G-CW-complex X and any $\alpha \in k$ (here $EG \times X$ is equipped with the diagonal G-action).

(1.10) Let (G, k) be as in Exercise (1.9). Prove the following. If C^* is a free $k[G]$-complex which is bounded, i.e. $C^i \neq 0$ only for finitely many dimensions i, then $H_{G,\alpha}(C^*) = 0$ for $\alpha \neq 0$; cf. Exercise (1.9) to see that the boundedness assumption is essential.

(1.11) Let (G,k) be as in Exercise (1.9). Define
$$\overleftarrow{H}_{G,\alpha}(X;k) := \varprojlim_n H_{G,\alpha}(X^n;k),$$
where X is a G-CW-complex with n-skeleton X^n. Prove that
$$\overleftarrow{H}_{G,\alpha}(X;k) = \begin{cases} \prod_{i=0}^{\infty} H^i(X;k) \text{ for } \alpha = 0 \\ \prod_{i=0}^{\infty} H^i(X^G;k) \text{ for } \alpha \neq 0. \end{cases}$$
Note, in particular, that $\overleftarrow{H}_{G,\alpha}(EG;k) \neq H_{G,\alpha}(EG;k)$ for $\alpha \neq 0$.

(1.12) Let (G,k) be as in Exercise (1.9). Show that if (X,A) is a pair of finite-dimensional G-CW-complexes such that
$$H^i(X,A;k) = \begin{cases} k \text{ if } i = n \\ 0 \text{ if } i \neq n \end{cases}$$
then for some $0 \leq m \leq n$
$$H^i(X^G, A^G;k) = \begin{cases} k \text{ if } i = m \\ 0 \text{ if } i \neq m. \end{cases}$$
Furthermore, in case (odd p), $n \equiv m$ modulo 2.

(1.13) Let (G,k) be as in Exercise (1.9). For a $k[G]$-module M with $\dim_k M < \infty$ let
$$a(M) := \dim_k(\ker \tau_M / \nu_M(M)) = \dim_k(\ker \nu_M / \tau_M(M)),$$
where $\tau_M : M \to M$ is given by the multiplication with $(1-g) \in k[G]$ and $\nu_M : M \to M$ is given by the multiplication with $1+g+\ldots+g^{p-1} \in k[G]$, g a generator of $G = \mathbb{Z}_p$. Prove that if X is a finite-dimensional G-CW-complex such that $\dim_k H^*(X;k) < \infty$, then
$$\sum_{i=0}^{\infty} \dim_k H^{m+i}(X^G;k) \leq \sum_{i=0}^{\infty} a(H^{m+i}(X;k)) \leq \sum_{i=0}^{\infty} \dim_k H^{m+i}(X;k)^G$$
for all $m \in \mathbb{N}$.

(1.14) Let (G,k) be as in Exercise (1.9), and X be a finite-dimensional G-CW-complex such that $\dim_k H^*(X;k) < \infty$. In case (odd p) show that
(a)
$$\dim_k H^*(X;k)^G = \frac{1}{2}(\dim_k(H_G^*(X;k) \otimes_R k)$$
$$+ \dim_k \text{Tor}^R(H_G^*(X;k), k)),$$
$R = k[t] \subset H^*(BG;k)$

(b)
$$\dim_k H^*(X^G;k) = \frac{1}{2}(\dim_k(H^*_G(X;k) \otimes_R k)$$
$$- \dim_k \mathrm{Tor}^R(H^*_G(X;k),k))$$

in particular,
$$\dim H^*(X^G;k) = \dim_k H^*(X;k)^G - \dim_k \mathrm{Tor}^R(H^*_G(X;k),k).$$

Compare with (1.3.8) for the case (2), and Sections 4.1 and 4.2, in particular (4.1.8) and (4.2.2).

The following two exercises are essentially due to Raussen, who considered the analogous situation for S^1-actions (and rational coefficients).

(1.15) Let (G,k) be as in Exercise (1.9) and X a finite-dimensional G-CW-complex.

(a) Show that in case $(p=2)$ the inclusion $X^G \to X$ induces an injective R-linear map, $R := k[t]$,
$$H^*_G(X;k)/T \to H^*_G(X^G;k) \cong H^*(X^G;k) \otimes R,$$
where T is the R-torsion submodule of $H^*_G(X;k)$. The co-kernel of this map is R-torsion. In particular, $H^*_G(X;k)/T$ can be viewed as a one-parameter family of deformations of the algebra
$$(H^*_G(X;k)/T) \otimes_R k_0$$
with
$$(H^*_G(X;k)/T) \otimes_R k_\alpha \cong H^{(*)}(X^G;k)$$
for $\alpha \neq 0$. Show further that the k-algebra
$$(H^*_G(X;k)/T) \otimes_R k_0$$
is isomorphic to the subquotient $I^*(X;k)$ of $H^*(X;k)^G$, defined by
$$I^*(X;k) := i^*(H^*_G(X;k))/i^*(T)$$
where $i : X \to X_G$ is the inclusion of X into the Borel construction. Note that T (resp. $i^*(T)$) is an ideal in the ring $H^*_G(X;k)$ (resp. $i^*(H^*_G(X;k)) \subset H^*(X;k)$). Furthermore there is a filtration on the algebra $H^{(*)}(X^G;k)$ such that associated graded algebra is isomorphic to $I^*(X;k)$. Cf. (1.3.14)-(1.3.17) for the case $T = 0$.

(b) Prove the analogous results in case (odd p), i.e. show that $X^G \to X$ induces an inclusion
$$H^*_G(X;k)/T \to H^*(X^G;k) \otimes \Lambda \otimes R$$
such that the cokernel is R-torsion, where
$$R := k[t] \subset H^*(BG;k),$$

$T := R$-torsion submodule of $H_G^*(X;k)$. In particular, $H^*(X^G;k) \otimes \Lambda$ can be viewed as a deformation of the k-algebra,

$$J^*(X;k) := j^*(H_G^*(X;k))/j^*(T) \cong (H_G^*(X;k)/T) \otimes_R k_0,$$

where j^* is induced by

$$j : \beta_G^*(X;k) \to \beta_G^*(X;k) \otimes_R k_0 \simeq H^*(X;k) \tilde{\otimes} \Lambda.$$

(b) Show that in case (odd p) the inclusion $X^G \to X$ also induces an injection $\overline{H_G^*(X;k)}/\overline{T} \to H^*(X^G;k) \otimes R$ such that the cokernel is R-torsion, where

$$\overline{H_G^*(X;k)} := H_G^*(X;k) \otimes_\Lambda k_0 = H_G^*(X;k)/H_G^*(X;k)s$$

and $\overline{T}$ is the R-torsion submodule of $\overline{H_G^*(X;k)}$. Hence $H^*(X^G)$ is a deformation of $(\overline{H_G^*(X;k)}/\overline{T}) \otimes_R k_0$. Does $i: X \to X_G$ induce an isomorphism of $(H_G^*(X;k)/\overline{T}) \otimes_R k_0$ and a subquotient of $H^*(X;k)$?

(1.16) Let X be a finite-dimensional $\mathbb{Z}_2$-CW-complex such that $H^*(X;k)$ fulfils Poincaré duality for k a field of characteristic 2.

(a) Show that $I^*(X;k)$ as defined in Exercise (1.15) fulfils Poincaré duality. (Hint: Prove that $i^*(T) = (i^*(H_G^*(X;k)))^\perp$ in $H^*(X;k)$, i.e. $i^*(T)$ is the orthogonal complement of $i^*(H_G^*(X;k))$ with respect to the Poincaré pairing.)

(b) Prove that $H^*(F;k)$ fulfils Poincaré duality for any component $F \subset X^G$. (Hint: Use part (a), Exercise (1.15)(a) and the argument given for (1.3.16)(3).)

(1.17) Let G and G' be finite groups, and X (resp. X') a G (resp. G')-CW-complex. Prove the following:

(a) $\beta_*^{G \times G'}(X \times X') \simeq \beta_*^G(X) \otimes \beta_*^{G'}(X')$;

(b) $\beta_{G \times G'}^*(X \times X') \simeq \beta_G^*(X) \otimes \beta_{G'}^*(X')$ if X or X' is of finite type, i.e. having finitely many cells in each dimension.

(1.18) Let $G \cong (\mathbb{Z}_2)^r$. Show that the map which assigns to any subgroup $K < G$ the subspace

$$H^1(B(G/K);\mathbb{F}_2) \subset H^1(BG;\mathbb{F}_2) \cong \mathbb{F}_2^r$$

gives a one-to-one correspondence between all subgroups of G and all linear subspaces of $H^1(BG;\mathbb{F}_2)$ such that

(i) $\operatorname{corank}(K) = \dim_{\mathbb{F}_2} H^1(B(G/K);\mathbb{F}_2)$

(ii) $K < H \Leftrightarrow H^1(B(G/K);\mathbb{F}_2) \supset H^1(B(G/H);\mathbb{F}_2)$.

(1.19) Let $G \cong (\mathbb{Z}_2)^r$. Show that the map which assigns to any subgroup $K < G$ the subspace

$$\mathcal{L}(K) := \{\alpha : H^*(BG;\mathbb{F}_2) \to \mathbb{F}_2, \ \alpha|H^1(B(G/K);\mathbb{F}_2) = 0\}$$

of
$$\mathcal{L}(G) := \{\alpha : H^*(BG; \mathbb{F}_2) \to \mathbb{F}_2, \ \alpha \text{ a morphism of } \mathbb{F}_2\text{-algebras}\}$$
gives a one-to-one correspondence between all subgroups of G and all linear subspaces of $\mathcal{L}(G)$ such that
(i) $\text{rank}(K) = \dim_{\mathbb{F}_2} \mathcal{L}(K)$
(ii) $K < H \Leftrightarrow \mathcal{L}(K) \subset \mathcal{L}(H)$.

2

Summary of some aspects of rational homotopy theory

In this chapter we shall summarize the main results from Sullivan's theory of minimal models, which we shall need subsequently. Since the proofs are too long to be included here, we state the results without proofs. The seminal paper on the subject is [Sullivan, 1977] which was inspired in part by [Quillen, 1969]; but we refer mostly to [Bousfield, Gugenheim, 1976] and [Halperin, 1977(1983)], where detailed expositions and proofs may be found.

2.1 The Sullivan-de Rham algebra

Let $\Delta^n \subseteq \mathbb{R}^{n+1}$ be the standard n-simplex. In terms of the standard coordinate functions $x_0, \ldots, x_n$ on $\mathbb{R}^{n+1}$, Δ^n is the set of all points such that $0 \leq x_i \leq 1$, for $0 \leq i \leq n$, and $\sum_{i=0}^{n} x_i = 1$. Now a p-form on $\mathbb{R}^{n+1}$ is an expression $\sum_i f_i dx_{i_1} \ldots dx_{i_p}$, where $i = (i_1, \ldots, i_p)$, each f_i is a smooth function of $x_0, \ldots, x_n$, and the summation runs over all multi-indices i, where $0 \leq i_1 < \ldots < i_p \leq n$. Multiplication of forms and exterior differentiation make the set of all p-forms on $\mathbb{R}^{n+1}$, for $0 \leq p \leq n+1$, into a commutative graded differential algebra, $\Omega(\mathbb{R}^{n+1})$, the de Rham algebra of $\mathbb{R}^{n+1}$. It is commutative in the graded sense: if $\omega \in \Omega^p(\mathbb{R}^{n+1})$, the subset of p-forms, and $\theta \in \Omega^q(\mathbb{R}^{n+1})$, then $\omega\theta = (-1)^{pq}\theta\omega$. The differential, d, has degree $+1$: if $\omega \in \Omega^p(\mathbb{R}^{n+1})$, then $d\omega \in \Omega^{p+1}(\mathbb{R}^{n+1})$. Also d is a derivation: if $\omega \in \Omega^p(\mathbb{R}^{n+1})$ and $\theta \in \Omega^q(\mathbb{R}^{n+1})$, then $d(\omega\theta) = (d\omega)\theta + (-1)^p\omega(d\theta)$. If R denotes the

2.1. The Sullivan-de Rham algebra

ring of smooth functions on $\mathbb{R}^{n+1}$, then $\Omega(\mathbb{R}^{n+1})$, as an algebra, is the exterior algebra over R generated by $dx_0, \ldots, dx_n$.

Now we could restrict $\Omega(\mathbb{R}^{n+1})$ to the affine subspace where $\sum_{i=0}^{n} x_i = 1$, and we could simplify it further by restricting the coefficient ring from R to the rational polynomial ring $\mathbb{Q}[x_0, \ldots, x_n]$ modulo the ideal generated by $1 - \sum_{i=0}^{n} x_i$. The resulting algebra, B, can be viewed as the exterior algebra over $\mathbb{Q}[x_1, \ldots, x_n]$ generated by $dx_1, \ldots, dx_n$. The exterior differentiation is now a formal algebraic process: e.g.

$$d(x_i^m) = m x_i^{m-1} dx_i.$$

Before making a definition we review some algebraic notation and terminology. From now on, unless stated otherwise, we shall take $\mathbb{Q}$ as our ground field, and a commutative graded differential algebra (CGDA) will be a $\mathbb{Q}$-algebra, graded over the non-negative integers, commutative in the graded sense, with a differential of degree $+1$, which is a derivation.

If V is a non-negatively graded $\mathbb{Q}$-vector space, then $S(V)$ will denote the free commutative graded algebra (CGA) generated by V: i.e. $S(V) = T(V)/I$, where $T(V)$ is the tensor algebra of V, graded by degrees from V, and I is the ideal generated by all elements of the form $ab - (-1)^{\alpha\beta} ba$, where a and b are homogeneous elements of $T(V)$ having degrees α and β, respectively. If E is any non-negatively graded set, then E generates a graded $\mathbb{Q}$-vector space, $\mathbb{Q}E$, and we shall write $S(E)$ instead of $S(\mathbb{Q}E)$. Thus $S(E)$ is the tensor product of the polynomial ring generated by E^{ev}, the set of elements of E of even degree, and the exterior algebra generated by E^{od}, the set of elements of E of odd degree: i.e. $S(E) = \mathbb{Q}[x; x \in E^{ev}] \otimes \Lambda(y; y \in E^{od})$.

For example, the algebra B, above, is $S(E')/I \cong S(E)$, where

$$E' = \{x_0 \ldots, x_n, dx_0, \ldots, dx_n\},$$

$E = \{x_1, \ldots, x_n, dx_1, \ldots, dx_n\}$, $\deg(x_i) = 0$ and $\deg(dx_i) = 1$, for $0 \leq i \leq n$, and I is the ideal generated by the elements $1 - \sum_{i=0}^{n} x_i$ and $\sum_{i=0}^{n} dx_i$. (Note that some authors write $\Lambda(V)$ and ΛE instead of $S(V)$ and $S(E)$.)

(2.1.1) Definition
Let $t_0, \ldots, t_n$ be indeterminates. Introduce other indeterminates $dt_0, \ldots, dt_n$. Let $\deg(t_i) = 0$ and $\deg(dt_i) = 1$ for $0 \leq i \leq n$. Let $F = \{t_0, \ldots, t_n, dt_0, \ldots, dt_n\}$. Define $\triangledown(t_0, \ldots, t_n)$ as follows. As a graded algebra $\triangledown(t_0, \ldots, t_n) = S(F)/I$, where

$$I = (1 - \bigoplus_{i=0}^{n} t_i, \bigoplus_{i=0}^{n} dt_i).$$

The differential, d, on $\triangledown(t_0,\ldots,t_n)$ is defined by $d(t_i) = dt_i$, $d(dt_i) = 0$, for $0 \leq i \leq n$, and extended as a derivation. Thus

$$\triangledown(t_0,\ldots,t_n) \cong S(E),$$

where $E = \{t_1,\ldots,t_n, dt_1,\ldots,dt_n\}$, and $d(t_i) = dt_i$, $d(dt_i) = 0$ for $1 \leq i \leq n$.

Now we introduce indeterminates t_{ij} for $i \geq 0$ and $0 \leq j \leq i$. Let $\triangledown_p = \triangledown(t_{po},\ldots,t_{pp})$. For $p > 0$, $0 \leq i \leq p$, let $\partial_i : \triangledown_p \to \triangledown_{p-1}$ be the CGDA homomorphism such that $\partial_i t_{pj} = t_{p-1,j-1}$ if $j > i$, $\partial_i t_{pi} = 0$, and $\partial_i t_{pj} = t_{p-1,j}$ if $j < i$. For $p \geq 0$, $0 \leq i \leq p$, let $s_i : \triangledown_p \to \triangledown_{p+1}$ be the CGDA homomorphism such that $s_i t_{pj} = t_{p+1,j+1}$ if $j > i$, $s_i t_{pi} = t_{p+1,i} + t_{p+1,i+1}$ and $s_i t_{pj} = t_{p+1,j}$ if $j < i$. (Note that $\triangledown_0 = \mathbb{Q}$ concentrated in degree 0 with trivial differential). The ∂_i and s_i are uniquely determined as homomorphisms of CGDAs by the above requirements.

For $q \geq 0$ let $\triangledown_p^q$ denote the $\mathbb{Q}$-vector space of elements of degree q in $\triangledown_p$. Then for each $q \geq 0$, $\triangledown_*^q = \{\triangledown_p^q; p \geq 0\}$ is a simplicial set with the ∂_i as face operators and the s_i as degeneracy operators (see, e.g., [May, 1967], [Lamotke, 1968]).

(2.1.2) Definition

Let K be a simplicial set. Let $A^q(K) = [K, \triangledown_*^q]$, the set of all simplicial set morphisms from K to $\triangledown_*^q$. $A^q(K)$ inherits a $\mathbb{Q}$-vector space structure from the $\triangledown_p^q$. Let $A(K) = \bigoplus_{q=0}^{\infty} A^q(K)$. Then $A(K)$ is a CGDA in the obvious way. For a topological space X, let $\text{sing}(X)$ be the simplicial set of singular simplexes on X, and let $A(X) = A(\text{sing}(X))$; $A(X)$ is the Sullivan-de Rham algebra of X. Thus A is a covariant functor from Top, the category of topological spaces, to CGDA, the category of CGDAs.

Now let $\sigma : \Delta^n \to X$ be a singular n-simplex on X. Let $\omega \in A^n(X)$. Then $\omega(\sigma) \in \triangledown_n^n$; and hence $\omega(\sigma) = f(t_{n1},\ldots,t_{nn})dt_{n1}\ldots dt_{nn}$, for some $f(t_{n1},\ldots,t_{nn}) \in \mathbb{Q}[t_{n1},\ldots,t_{nn}]$. If we think of $t_{n1},\ldots,t_{nn}$ as independent real variables, then we can integrate $\omega(\sigma)$ over Δ^n embedded in $\mathbb{R}^n$, namely the set given by $0 \leq t_{ni} \leq 1$ for $1 \leq i \leq n$, and $\sum_{i=1}^n t_{ni} \leq 1$. Since f is a rational polynomial, this integration is a formal algebraic procedure, and it yields a rational number. Denote the integral by $I(\omega(\sigma))$.

Let $S_*(X)$ be the singular chain complex of X. Then any $\sigma \in S_n(X)$ has the form $\sigma = \sum_i a_i \sigma_i$, where each $a_i \in \mathbb{Z}$, σ_i is a singular n-simplex on X, and the sum is finite. Hence for $\omega \in A^n(X)$, we may define $I(\omega, \sigma) = \sum_i a_i I(\omega(\sigma_i)) \in \mathbb{Q}$. So the map $\omega \mapsto I(\omega, -)$ is a $\mathbb{Q}$-vector space homomorphism from $A^n(X)$ to $S^n(X; \mathbb{Q}) = \text{Hom}(S_n(X), \mathbb{Q})$. Thus

2.2. Minimal models

we get a graded $\mathbf{Q}$-vector space homomorphism
$$\mathcal{I} : A(X) \to S^*(X;\mathbf{Q}).$$
Clearly $\mathcal{I}$ induces a natural transformation of functors from $A(-)$ to $S^*(-;\mathbf{Q})$.

The first main theorem of Sullivan's theory is the following.

(2.1.3) Theorem
For any topological space X, $\mathcal{I}$ induces a natural isomorphism of graded algebras
$$\mathcal{I}^* : H(A(X)) \to H^*(X;\mathbf{Q}).$$

(2.1.4) Remarks
(1) It follows from the definition that if $X = \{x\}$ consists of a single point, then $A(X) = \mathbf{Q}$ (concentrated in degree 0). In particular, for any space X, the inclusion of a base point $\{x_0\} \to X$ induces an augmentation $A(X) \to \mathbf{Q}$.

(2) Let L be a simplicial complex, V the set of vertices of L, and Σ the set of simplexes of L. (So Σ is a set of finite subsets of V). Suppose that V is partially ordered so that each simplex is well ordered (e.g. V may be well ordered). Let $\sigma = \langle v_0, \ldots, v_n \rangle$ be a n-simplex, where $v_0 < \ldots < v_n$, $v_i \in V$, $0 \leq i \leq n$. Then a m-simplex τ, for $m \leq n$, is a face of σ if $\tau = \langle v_{i_0}, \ldots, v_{i_m} \rangle$, where $\{i_0, \ldots, i_m\} \subset \{0, \ldots, n\}$ with $v_{i_0} < \ldots < v_{i_m}$. Let $A_S^q(L)$ be the set of all functions ω with domain Σ such that:

(i) If $\sigma = \langle v_0, \ldots, v_n \rangle$ is an n-simplex, as above, then
$$\omega(\sigma) \in \bigtriangledown(v_0, \ldots, v_n)^q :$$
i.e. an element of degree q in $\bigtriangledown(v_0, \ldots, v_n)$; and

(ii) If τ is any face of σ, as above, then $\omega(\tau) = r(\sigma,\tau)(\omega(\sigma))$, where $r(\sigma,\tau) : \bigtriangledown(v_0, \ldots, v_n) \to \bigtriangledown(v_{i_0}, \ldots, v_{i_m})$ is the homomorphism of CGDAs defined by $r(\sigma,\tau)(v_j) = v_j$ if $j \in \{i_0, \ldots, i_m\}$ and $r(\sigma,\tau)(v_j) = 0$ if $j \notin \{i_0, \ldots, i_m\}$.

Then $A_S(L) = \bigoplus_{q=0}^{\infty} A_S^q(L)$ becomes a CGDA in the obvious way. If $|L|$ denotes the geometric realization of L, then it can be shown ([Bousfield, Gugenheim, 1976], Section 13) that there is a CGDA homomorphism $f(L) : A(|L|) \to A_S(L)$, which is natural on simplicial complexes ordered as above, such that $f(L)^* : H(A(|L|)) \to H(A_S(L))$ is an isomorphism. Thus the Sullivan-de Rham theory can be viewed rather more geometrically on simplicial complexes. (Note that we may define the integration map directly from $A_S(L)$ to the rational cochain complex of L.)

2.2 Minimal models

One of the most important features of the Sullivan-de Rham theory is that it allows us to replace the CGDA $A(X)$ by a CGDA $\mathcal{M}(X)$. This is algebraically much simpler than $A(X)$, but it carries all the rational cohomology information of X, and, if X satisfies certain conditions (to be described in Section 2.3), all of the rational homotopy information of X.

(2.2.1) Definitions
(1) A CGDA A is said to be connected if $A^0 \cong \mathbb{Q}$ and A is said to be simply-connected if A is connected and $A^1 = 0$.
(2) A CGDA A is said to be c-connected if $H^0(A) \cong \mathbb{Q}$. Thus a space X is path-connected if and only if $A(X)$ is c-connected.
(3) Let A be a CGDA and let $\varepsilon : A \to \mathbb{Q}$ be an augmentation. Then $\overline{A}$ denotes the augmentation ideal, $\ker \varepsilon$. If A is connected, then

$$\overline{A} = \bigoplus_{i=1}^{\infty} A^i.$$

(2.2.2) Definition
Let A be a CGDA. Then A is called a KS-complex if there exists $E \subseteq A$ such that:
(i) $E = \{x_\alpha; \alpha \in \mathcal{A}\}$ and $\mathcal{A}$ is well-ordered (i.e. E is a well-ordered set of elements of A);
(ii) $A = S(E)$; and
(iii) for any $\alpha \in \mathcal{A}$, $dx_\alpha \in S(E_\alpha)$ where

$$E_\alpha = \{x_\beta; \beta \in \mathcal{A} \text{ and } \beta < \alpha\}.$$

If, in addition,
(iv) $\deg(x_\alpha) > 0$ for all $\alpha \in \mathcal{A}$; and
(v) $\deg(x_\beta) < \deg(x_\alpha) \Rightarrow \beta < \alpha$, for any $\alpha, \beta \in \mathcal{A}$,
then A is said to be minimal. Whenever we say 'let $A = S(E)$ be a KS-complex (resp. minimal CGDA)' we shall tacitly assume that E is a well-ordered basis satisfying conditions (i)-(iii) (resp. (i)-(v)) above.

(2.2.3) Remarks and examples
(1) K stands for Koszul, S for Sullivan.
(2) Let $A = \Lambda(x, y)$ where $\deg(x) = \deg(y) = 1$, $dx = 0$ and $dy = xy$. Then A is free, but it is not a KS-complex.
(3) The CGDAs, $\triangledown(t_0, \ldots, t_n)$, of Section 2.1 are KS-complexes, but they are not minimal. (Both conditions (iv) and (v) are violated.)

2.2. Minimal models

(4) If $A = S(E)$, $E = \{x_\alpha; \alpha \in \mathcal{A}\}$ is minimal, then $dx_\alpha \in \overline{A} \cdot \overline{A}$ for all $\alpha \in \mathcal{A}$. Thus $d(\overline{A}) \subseteq \overline{A} \cdot \overline{A}$. This follows from condition (v).
(5) If A is a simply connected CGDA then it is easy to see that A is minimal if and only if A is free and $d(\overline{A}) \subseteq \overline{A} \cdot \overline{A}$.
(6) So far we have always denoted the differential in a CGDA by d. If we need to be more precise we shall write (A, d_A), (B, δ), (C, ∂), etc., for the algebra and its differential.

(2.2.4) Definitions
(1) A homomorphism $f : A \to B$ of CGDAs is called a weak equivalence (or quasi-isomorphism) if the induced map on homology, $f^* : H(A) \to H(B)$, is an isomorphism.
(2) If A is a CGDA, then a model of A is a KS-complex, K, together with a weak equivalence $K \to A$.
(3) If A is a CGDA, then a minimal model of A is a minimal CGDA, $\mathcal{M}$, together with a weak equivalence $\mathcal{M} \to A$.
(4) Recall from Section 2.1 that the CGDA ∇_1 can be viewed as a free CGDA generated by an element t of degree 0 and its differential dt. Then the maps $\partial_0, \partial_1 : \nabla_1 \to \nabla_0 = \mathbb{Q}$ are given by $\partial_0(t) = 0$ and $\partial_1(t) = 1$. Now let $f, g : A \to B$ be two maps of CGDAs. Then f and g are said to be homotopic, denoted $f \simeq g$, if there exists a map

$$H : A \to B \otimes \nabla_1$$

such that $(1_B \otimes \partial_0)H = f$ and $(1_B \otimes \partial_1)H = g$. If A and B are augmented and f and g are augmentation preserving, then the appropriate notion of homotopy (i.e. based homotopy) is defined by replacing $B \otimes \nabla_1$ in the above definition by

$$B \overline{\otimes} \nabla_1 = \mathbb{Q} \otimes \mathbb{Q} + \overline{B} \otimes \nabla_1 \subseteq B \otimes \nabla_1,$$

where $\overline{B}$ is the augmentation ideal of B. (In the language of [Quillen, 1967] the homotopies defined here are right homotopies: see [Bousfield, Gugenheim, 1976] for more details. See [Halperin, 1977(1983)] for a treatment of left homotopy of maps of CGDAs and augmented CGDAs.)

(2.2.5) Theorem
If A is a c-connected CGDA, then there exists a minimal model $\mathcal{M} \to A$. Furthermore, if $f : \mathcal{M} \to A$ and $g : \mathcal{N} \to A$ are any two minimal models of A, then there exists an isomorphism $\phi : \mathcal{M} \to \mathcal{N}$ such that $g\phi \simeq f$.

(2.2.6) Corollary
If X is a path-connected topological space, then there exists a minimal model $\mathcal{M}(X) \to A(X)$.

(2.2.7) Remark
Although $\mathcal{M}(X)$, called a minimal model of X, is unique up to isomorphism as a CGDA, it is not functorial. Note that $H(\mathcal{M}(X)) \cong H^*(X; \mathbb{Q})$.

(2.2.8) Examples
(1) Let $G = T^n$, a torus group.
$$R = H^*(BG, \mathbb{Q}) = \mathbb{Q}[t_1, \ldots, t_n],$$
$\deg(t_i) = 2$, $1 \leq i \leq n$. Let $\tau_i \in A(BG)$ be a cycle representing t_i for $1 \leq i \leq n$. Define $f : R \to A(BG)$ by $f(t_i) = \tau_i$, $1 \leq i \leq n$. Then this is a minimal model of BG.

(2) Let $X = \mathbb{C}P^n$. $H^*(X; \mathbb{Q}) = \mathbb{Q}[x]/(x^{n+1})$, where $\deg(x) = 2$. Let $\xi \in A(X)$ be a cycle representing x. Let $\eta \in A^{2n+1}(X)$ be such that $d\eta = \xi^{n+1}$. Now let $M = \mathbb{Q}[u] \otimes \Lambda(v)$, where $\deg(u) = 2$, $\deg(v) = 2n+1$, $du = 0$, $dv = u^{n+1}$. Then we have a minimal model $f : M \to A(X)$ given by $f(u) = \xi$, $f(v) = \eta$.

(3) Since $H^i(\mathbb{R}P^2; \mathbb{Q}) = 0$ for $i > 0$, $\mathbb{Q} \to A(\mathbb{R}P^2)$ is a minimal model: i.e. $\mathcal{M}(\mathbb{R}P^2) = \mathbb{Q}$.

Although minimal models are not functorial they are so up to homotopy as the following theorem shows.

(2.2.9) Theorem
Let A and B be c-connected CGDAs; and let $\mu_A : \mathcal{M}(A) \to A$ and $\mu_B : \mathcal{M}(B) \to B$ be minimal models of A and B respectively: i.e $\mathcal{M}(A)$ and $\mathcal{M}(B)$ are minimal CGDAs and μ_A and μ_B are weak-equivalences. Let $f : A \to B$ be any CGDA homomorphism. Then there is a homomorphism $\overline{f} : \mathcal{M}(A) \to \mathcal{M}(B)$ such that
$$\mu_B \overline{f} \simeq f \mu_A.$$
In particular, $\mu_B^* \overline{f}^* = f^* \mu_A^*$.

Furthermore if $\overline{f}' : \mathcal{M}(A) \to \mathcal{M}(B)$ is any other homomorphism such that $\mu_B \overline{f}' \simeq f \mu_A$, then $\overline{f}' \simeq \overline{f}$.

Another useful property of minimal CGDAs is the following.

(2.2.10) Proposition
If M and M' are minimal CGDAs, and if
$$\phi : M \to M'$$
is a weak-equivalence, then ϕ is an isomorphism.

We finish this section with a generalization of Definition (2.2.2).

(2.2.11) Definitions
(1) Let (A, d_A), (B, d_B) be CGDAs, and let $f : A \to B$ be a homomorphism. Then f is said to be a KS-extension if there exists a subset $E \subset B$, $E = \{x_\alpha; \alpha \in \mathcal{A}\}$ with $\mathcal{A}$ well-ordered, such that:

(i) if $j : S(E) \to B$ is the commutative graded algebra (CGA) homomorphism induced by the inclusion $E \subset B$, and if $\phi : A \otimes S(E) \to B$ is the CGA direct sum homomorphism induced by f and j, then ϕ is an ismorphism; and

(ii) $d_B \phi(1 \otimes x_\alpha) \in \phi(A \otimes S(E_\alpha))$, where $E_\alpha = \{x_\beta; \beta < \alpha\}$, for all $\alpha \in \mathcal{A}$, as before.

Since $d_B \phi(a \otimes 1) = d_B f(a) = f(d_A a) = \phi(d_A a \otimes 1)$, for all $a \in A$, it follows that, if we use ϕ to identify B with $A \otimes S(E)$, then d_B satisfies the conditions:

(a) $d_B(a \otimes 1) = d_A a \otimes 1$, for all $a \in A$, and
(b) $d_B(x_\alpha) \in A \otimes S(E_\alpha)$, for all $\alpha \in \mathcal{A}$.

If $\varepsilon_A : A \to \mathbb{Q}$ is an augmentation, then there is a CGA homomorphism $\varepsilon_A \otimes \mathrm{id} : A \otimes S(E) \to S(E)$. If a differential d is defined on $S(E)$ by the rule $d(x_\alpha) = (\varepsilon_A \otimes \mathrm{id})(d_B(x_\alpha))$, then $\varepsilon_A \otimes \mathrm{id}$ is a homomorphism of CGDAs, and $(S(E), d)$ is a KS-complex.

If $\varepsilon_B : B \to \mathbb{Q}$ is an augmentation and f is augmentation preserving (i.e. $\varepsilon_B f = \varepsilon_A$), then, replacing x_α by $x_\alpha - \varepsilon_B(x_\alpha)$, if necessary, we may suppose that $\varepsilon_B(x_\alpha) = 0$ for all $\alpha \in \mathcal{A}$; and ϕ and $\varepsilon_A \otimes \mathrm{id}$ are augmentation preserving when $S(E)$ is augmented by $\varepsilon : S(E) \to \mathbb{Q}$, where $\varepsilon(x_\alpha) = 0$ for all $\alpha \in \mathcal{A}$, and $A \otimes S(E)$ is augmented by $\varepsilon_A \otimes \varepsilon$.

In the augmented case, we shall refer to the sequence

$$(A, d_A) \to (A \otimes S(E), d_B) \to (S(E), d)$$

as a KS-extension.

(2) A KS-extension $(A, d_A) \to (A \otimes S(E), d_B)$ as above is called a minimal KS-extension if E also satisfies conditions (iv) and (v) of Definition (2.2.2).

(3) A KS-extension $(A, d_A) \to (A \otimes S(E), d_B)$ in which (A, d_A) is a KS-complex is called a Λ-extension.

(4) A minimal KS-extension $(A, d_A) \to (A \otimes S(E), d_B)$ in which (A, d_A) is a minimal CGDA is called a Λ-minimal Λ-extension.

2.3 Rational homotopy theory

Let G be a group and A an abelian group on which G acts by automorphisms: i.e. A is a $\mathbb{Z}G$-module.

(2.3.1) Definition
For $n \geq 1$, define $\Gamma_n A$ inductively by letting
$$\Gamma_1 A = A$$
and letting $\Gamma_{n+1} A$ be the $\mathbb{Z}G$-submodule of A generated by
$$\{ga - a; g \in G \text{ and } a \in \Gamma_n A\}.$$
Thus
$$A = \Gamma_1 A \supset \Gamma_2 A \supset \ldots \supset \Gamma_n A \supset \Gamma_{n+1} A \ldots$$
G is said to act nilpotently on A (A is a nilpotent $\mathbb{Z}G$-module) if $\Gamma_n A = 0$ for some $n \geq 1$.

(2.3.2) Definition
A path-connected topological space, X, is said to be nilpotent if:
(i) $\pi_1 := \pi_1(X)$ is a nilpotent group, and
(ii) π_1 acts nilpotently on $\pi_n(X)$ for all $n \geq 2$, where the action is the standard one.

(2.3.3) Discussion
There is an interesting theory of topological localization and completion due to Sullivan [1970] (see also [Hilton, Mislin, Roitberg, 1975], and [Bousfield, Kan, 1972]). A consequence of the theory is that if W is a nilpotent CW-complex, then there is a nilpotent CW-complex W_0 and a map $f : W \to W_0$ such that
(i) $\pi_n(W_0) \cong \pi_n(W) \otimes \mathbb{Q}$ for all $n \geq 1$, and
(ii) $f_* : \pi_n(W) \to \pi_n(W_0)$ is a localization (at (0)) for all $n \geq 1$ (see below).

For an abelian group A (written additively), $A \otimes \mathbb{Q}$ has its usual meaning; and $f : A \to B$, B also abelian, is a localization if there is an isomorphism $\phi : B \to A \otimes \mathbb{Q}$ such that $\phi f : A \to A \otimes \mathbb{Q}$ is the standard homomorphism. If G is a non-abelian nilpotent group, as may be the case with $\pi_1(W)$ above, then $G \otimes \mathbb{Q}$ and localization $G \to G \otimes \mathbb{Q}$ must be interpreted as Mal'tsev completion. This is somewhat complicated: see [Quillen, 1969], Appendix A, for a detailed definition.

The above conditions on $\pi_n(W)$ and $\pi_n(W_0)$ are, however, equivalent to the following ([Hilton, Mislin, Roitberg, 1975], Theorem 38):
(i) $H_n(W_0; \mathbb{Z}) \cong H_n(W; \mathbb{Z}) \otimes \mathbb{Q}$, for all $n \geq 1$, and
(ii) $f_* : H_n(W; \mathbb{Z}) \to H_n(W_0; \mathbb{Z})$ is a localization for all $n \geq 1$.

Note that simply-connected spaces and (homotopy) simple spaces (i.e spaces X for which $\pi_1(X)$ is abelian and acts trivially on the higher homotopy groups) are nilpotent.

2.3. Rational homotopy theory

Now we define the homotopy groups of an augmented KS-complex; these will be related to the rational homotopy groups of a nilpotent space by means of its minimal model.

(2.3.4) Definition

Let K be an augmented KS-complex with augmentation ideal $\overline{K}$. The indecomposable quotient of K, $Q(K)$, is defined to be $\overline{K}/\overline{K}\cdot\overline{K}$; $Q(K)$ is a graded vector space, and the differential on K induces a differential, δ_K, on $Q(K)$, so that $Q(K)$ becomes a rational cochain complex. By definition,

$$\pi^n(K) = H^n(Q(K), \delta_K), \text{ for } n \geq 0.$$

π^* is functorial on augmented KS-complexes; it can be shown ([Bousfield, Gugenheim, 1976] or [Halperin, 1977(1983)]) that, if $f : K \to L$ is an augmentation preserving weak equivalence of augmented KS-complexes, then the induced map

$$f^* : \pi^*(K) \to \pi^*(L)$$

is an isomorphism; and the converse holds if K and L are c-connected.

If (M, d_M) is a minimal CGDA, then $d_M(\overline{M}) \subseteq \overline{M} \cdot \overline{M}$; and hence δ_M is zero on $Q(M)$. Thus $\pi^n(M) = Q(M)^n$, the rational vector space of elements of degree n in $Q(M)$, in this case. In particular, $\pi^n(M)$ is isomorphic to the rational vector space generated by the elements of degree n in any free algebra basis of M.

(2.3.5) Definition

Let X be a path-connected topological space, and let $\mathcal{M}(X)$ be a minimal model of X. By definition,

$$\Pi_\psi^n(X) = \pi^n(\mathcal{M}(X)).$$

These are called the pseudo-dual rational homotopy groups of X. Although $\mathcal{M}$ is not functorial, it is sufficiently close to being so (see Theorem (2.2.5)), that Π_ψ^n is functorial. That is, Π_ψ^* is a covariant functor from the category of path-connected topological spaces to the category of graded rational vector spaces. In fact, the target of Π_ψ^* may be taken to be the category of graded rational co-Lie algebras, as we shall describe below.

Before introducing the co-Lie algebra structure, however, we shall state the second main theorem of Sullivan-de Rham theory. First though, comes a definition.

(2.3.6) Definition
A path-connected topological space, X, is said to be of finite $\mathbb{Q}$-type if $H^n(X;\mathbb{Q})$ is finite-dimensional (over $\mathbb{Q}$) for all $n \geq 1$. If X is nilpotent, this is equivalent to $H_1(X;\mathbb{Q})$ and $\pi_n(X) \otimes \mathbb{Q}$ being finite-dimensional (over $\mathbb{Q}$) for all $n \geq 2$.

(2.3.7) Theorem
Let X be a nilpotent topological space of finite $\mathbb{Q}$-type. Then, for each $n \geq 2$, there is a natural isomorphism
$$\Pi_\psi^n(X) \cong \operatorname{Hom}_{\mathbb{Z}}(\pi_n(X), \mathbb{Q}).$$
Furthermore, this holds for $n = 1$ if $\pi_1(X)$ is abelian. (Note that $\operatorname{Hom}_{\mathbb{Z}}(A, \mathbb{Q}) \cong \operatorname{Hom}_{\mathbb{Q}}(A \otimes \mathbb{Q}, \mathbb{Q})$ for any abelian group A.)

(2.3.8) Examples
(1) The theorem holds clearly for $X = BG$, G a torus, and $X = \mathbb{C}P^n$, as in Examples (2.2.8)(1) and (2).
(2) If $X = \mathbb{R}P^2$, then $\Pi_\psi^n(X) = 0$ for $n \geq 1$, by Example (2.2.8)(3). But $\pi_2(X) = \mathbb{Z}$. So the theorem fails in this case, X being non-nilpotent.
(3) [Bousfield, Gugenheim, 1976, 11.5.] Let V be a rational vector space of countably infinite dimension, and let X be an Eilenberg-Maclane space $K(V, n)$ for some $n \geq 2$. Then $\pi_{2n}(X) = 0$; but it can be shown that $\Pi_\psi^{2n}(X) \neq 0$. Here, of course, X is nilpotent, but it is not of finite $\mathbb{Q}$-type.

(2.3.9) Definition
Let $V = \bigoplus_{n=0}^{\infty} V_n$ be a graded rational vector space, and let $\bigtriangledown : V \to V \otimes V$ be a homomorphism such that:
(i) $T\bigtriangledown = -\bigtriangledown$, and
(ii) $(I \otimes \bigtriangledown)\bigtriangledown - (\bigtriangledown \otimes I)\bigtriangledown = (I \otimes T)(T \otimes I)(\bigtriangledown \otimes I)\bigtriangledown$,
where $I : V \to V$ is the identity, and $T : V \otimes V \to V \otimes V$ is given by $T(x \otimes y) = (-1)^{pq} y \otimes x$, for any $x \in V_p$, $y \in V_q$. Then V, together with $\bigtriangledown$, is said to be a (rational) co-Lie algebra.

Note that (i) corresponds to the graded anti-symmetry of a graded Lie algebra, and (ii) corresponds to the graded Jacobi identity. The maps in (ii) are from V to $V \otimes V \otimes V$.

(2.3.10) Definition
Let (K, d_K) be a connected KS-complex, and $E = \{x_\alpha; \alpha \in \mathcal{A}\}$ be an algebra basis. For each $\alpha \in \mathcal{A}$, we may write
$$d_K(x_\alpha) = \bigoplus_\beta r_{\alpha\beta} x_\beta + \bigoplus_{\beta,\gamma} r_{\alpha\beta\gamma} x_\beta x_\gamma + \text{ higher order terms,}$$

2.3. Rational homotopy theory

where $r_{\alpha\beta}, r_{\alpha\beta\gamma} \in \mathbb{Q}$. Thus, if $\bar{x}_\alpha$ is the image of x_α in $Q(K)$, then $\delta_K(\bar{x}_\alpha) = \sum_\beta r_{\alpha\beta}\bar{x}_\beta$. Now define $\triangle : Q(K) \to Q(K) \otimes Q(K)$ by

$$\triangle(\bar{x}_\alpha) = \frac{1}{2}\bigoplus_{\beta,\gamma} r_{\alpha\beta\gamma}(\bar{x}_\beta \otimes \bar{x}_\gamma + (-1)^{p(\beta)p(\gamma)}\bar{x}_\gamma \otimes \bar{x}_\beta),$$

where $p(\alpha) = \deg(x_\alpha)$, for any $\alpha \in \mathcal{A}$.

Now let $\lambda(K)$ be the graded rational vector space with basis

$$\{y_\alpha; \alpha \in \mathcal{A}\},$$

where $\deg(y_\alpha) = \deg(x_\alpha) - 1$. Define $\tau : Q(K) \to \lambda(K)$, of degree -1, by $\tau(\bar{x}_\alpha) = y_\alpha$, for all $\alpha \in \mathcal{A}$. Define a differential ∂_K on $\lambda(K)$ by

$$\partial_K(y_\alpha) = -\bigoplus_\beta r_{\alpha\beta} y_\beta.$$

So $\partial_K \tau = -\tau \delta_K$.

Extend τ to maps $\tau : Q(K) \otimes Q(K) \to \lambda(K) \otimes \lambda(K)$, and

$$\tau : Q(K) \otimes Q(K) \otimes Q(K) \to \lambda(K) \otimes \lambda(K) \otimes \lambda(K)$$

in the expected graded way. So

$$\tau(x \otimes y) = (-1)^{\deg(x)} \tau x \otimes \tau y,$$

and

$$\tau(x \otimes y \otimes z) = (-1)^{\deg(y)} \tau x \otimes \tau y \otimes \tau z.$$

Also extend the differentials δ_K and ∂_K is the standard way. (Note that, o.g., $\partial_K \tau = \tau \delta_K$ on $Q(K) \otimes Q(K)$.) Define $\triangledown : \lambda(K) \to \lambda(K) \otimes \lambda(K)$ by $\triangledown \tau = \tau \triangle$. Then $\triangledown \partial_K = \partial_K \triangledown$, since $\triangle \delta_K = -\delta_K \triangle$. Let

$$\mathcal{L}^*(K) = H(\lambda(K), \partial_K),$$

and let

$$\triangledown_K : \mathcal{L}^*(K) \to \mathcal{L}^*(K) \otimes \mathcal{L}^*(K)$$

be induced by $\triangledown$ and Künneth isomorphism. Then straightforward, but cumbersome, computations show that:

(i) $\triangledown_K$ is a co-Lie algebra structure on $\mathcal{L}^*(K)$;
(ii) up to isomorphism, $\mathcal{L}^*(K)$ and $\triangledown_K$ are independent of the choice of algebra basis E; and
(iii) $\mathcal{L}^*$ is a functor from the category of connected KS-complexes to the category of non-negatively graded rational co-Lie algebras.

Note that if K is minimal, then $\partial_K = 0$, and so $\mathcal{L}^*(K) = \lambda(K)$, which is just $Q(K)$ with degrees lowered by 1.

If X is a path-connected topological space, and if $\mathcal{M}(X)$ is a minimal model of X, then we set $\mathcal{L}^*(X) = \mathcal{L}^*(\mathcal{M}(X)) = \lambda(\mathcal{M}(X))$.

Now recall the rational homotopy Lie algebra of a simply-connected space X. The Whitehead product is a bilinear pairing

$$[-,-] : \pi_p(X) \times \pi_q(X) \to \pi_{p+q-1}(X),$$

for all p, $q \geq 2$. It induces a linear map, also denoted by

$$[-,-] : (\pi_p(X) \otimes \mathbb{Q}) \otimes (\pi_q(X) \otimes \mathbb{Q}) \to \pi_{p+q-1}(X) \otimes \mathbb{Q}.$$

Let $\mathcal{L}_n(X) = \pi_{n+1}(X) \otimes \mathbb{Q}$, for all $n \geq 1$; and, for $a \subset \pi_{n+1}(X) \otimes \mathbb{Q}$, let τa be the corresponding element in $\mathcal{L}_n(X)$. Now define

$$[-,-] : \mathcal{L}_p(X) \otimes \mathcal{L}_q(X) \to \mathcal{L}_{p+q}(X),$$

for all p, $q \geq 1$, by $[\tau a, \tau b] = (-1)^p \tau[a, b]$, for all $a \in \pi_{p+1}(X) \otimes \mathbb{Q}$, $b \in \pi_{q+1}(X) \otimes \mathbb{Q}$. Then $\mathcal{L}_*(X)$ together with $[-,-]$ is a graded Lie algebra (over $\mathbb{Q}$), called a rational homotopy Lie algebra of X.

From [Sullivan, 1977] or [Andrews, Arkowitz, 1978] we have the following.

(2.3.11) Theorem

If X is a simply-connected topological space of finite $\mathbb{Q}$-type, then the graded rational co-Lie algebra $\mathcal{L}^*(X)$ is dual to the graded rational Lie algebra $\mathcal{L}_*(X)$.

Thus, for such spaces, the minimal model determines the rational Whitehead products: they are given by the quadratic part of the differential.

We conclude this section with one more fact.

(2.3.12) Theorem
Given a Λ-extension,

$$(A, d_A) \to (B, d_B) \to (S(E), d),$$

of augmented CGDAs, the induced sequence of rational cochain complexes

$$0 \to (Q(A), \delta_A) \to (Q(B), \delta_B) \to (Q(S(E)), \delta) \to 0$$

is exact. Hence there is a long exact sequence

$$0 \to \pi^0(A) \to \pi^0(B) \to \pi^0(S(E)) \to \pi^1(A) \to \ldots$$
$$\to \pi^n(A) \to \pi^n(B) \to \pi^n(S(E)) \to \pi^{n+1}(A) \to \ldots$$

2.4 Finite-dimensional rational homotopy

In this section we state some of the very interesting and important results in [Halperin, 1977].

(2.4.1) Definitions
Let X be a path-connected space.
(1) X is said to be c-finite if $\dim_{\mathbb{Q}} H^*(X;\mathbb{Q}) < \infty$. In this case the Euler characteristic $\chi(X) = \sum_{n=0}^{\infty}(-1)^n \dim_{\mathbb{Q}} H^n(X;\mathbb{Q})$ is defined.
(2) X is said to be π-finite, or X has finite-dimensional rational homotopy (FDRH), if $\dim_{\mathbb{Q}} \Pi^*_\psi(X) < \infty$. In this case we may define the Euler homotopy characteristic

$$\chi\pi(X) = \bigoplus_{n=1}^{\infty}(-1)^n \dim_{\mathbb{Q}} \Pi^n_\psi(X).$$

(2.4.2) Examples
From Examples (2.2.8), we have that BG, for $G = T^n$, a torus, is π-finite but not c-finite. $\chi\pi(BG) = n$. In fact, for any compact connected Lie group of rank n, $\mathcal{M}(G)$ is an exterior algebra on n generators (of odd degree) with zero differential, and $\mathcal{M}(BG)$ is a polynomial algebra on n generators (of even degree) with zero differential: the generators of $\mathcal{M}(BG)$ correspond to those of $\mathcal{M}(G)$ with degrees raised by 1. Thus $\chi\pi(G) = -n$, $\chi\pi(BG) = n$, and, of course, $\chi(G) = 0$.

From (2.2.8), $\mathbb{C}P^n$ is both π-finite and c-finite, $\chi(\mathbb{C}P^n) = n+1$ and $\chi\pi(\mathbb{C}P^n) = 0$.

The following three theorems are from [Halperin, 1977].

(2.4.3) Theorem
Let X be a c-finite π-finite path-connected space. Then X is a rational Poincaré duality space.

Furthermore, if $\{x_1,\ldots,x_r\}$ is a basis for $\Pi^{ev}_\psi(X)$, and $\{y_1,\ldots,y_s\}$ is a basis for $\Pi^{od}_\psi(X)$, then the formal dimension of X, $\mathrm{fd}(X)$, is given by the formula

$$\mathrm{fd}(X) = r + \bigoplus_{i=1}^{s} \deg(y_i) - \bigoplus_{i=1}^{r} \deg(x_i).$$

(2.4.4) Theorem
Let X be a c-finite π-finite path-connected space. Then $\chi\pi(X) \leq 0$.
Furthermore, $\chi(X) \geq 0$; and $\chi(X) > 0$ if and only if $\chi\pi(X) = 0$.

(2.4.5) Theorem
Let X be a c-finite π-finite path-connected space with $\chi\pi(X) = 0$. Then $\mathcal{M}(X)$ has the form $\mathbb{Q}[x_1,\ldots,x_n] \otimes \Lambda(y_1,\ldots,y_n)$ with $dx_i = 0$ for $1 \leq i \leq n$, and $dy_i = f_i \in \mathbb{Q}[x_1,\ldots,x_n]$ for $1 \leq i \leq n$. And $H^*(X;\mathbb{Q}) \cong \mathbb{Q}[x_1,\ldots,x_n]/(f_1,\ldots,f_n)$ as a CGA.

Furthermore

$$\chi(X) = \dim_{\mathbb{Q}} H^*(X;\mathbb{Q}) = \prod_{i=1}^{n}(a_i+1)/\prod_{i=1}^{n} b_i,$$

where $a_i = \deg(y_i)$ (and so $a_i + 1 = \deg(f_i)$) and $b_i = \deg(x_i)$, for $1 \leq i \leq n$.

The main theorems of [Halperin, 1977] are actually theorems about minimal CGDAs: the theorems stated here follow by applying them to minimal models. In fact much more can be said along these lines (see, e.g., [Friedlander, Halperin, 1979]); and there is now a very substantial literature on this subject.

2.5 The Grivel-Halperin-Thomas theorem

In this section we give a theorem which, for certain fibrations

$$F \xrightarrow{i} E \xrightarrow{\pi} B,$$

gives a model of E in terms of minimal models of B and F. We also include a useful application of this theorem to pull-backs. Versions of the theorem, which follows, were proven independently by Grivel [1979], Halperin [1977(1983)] and Thomas [1980].

(2.5.1) Theorem
Let $\pi : E \to B$ be a Serre fibration of path-connected spaces; let $b \in B$ be a base-point and let $F = \pi^{-1}(b)$ be the fibre over b. Suppose that:

(i) F is path-connected;

(ii) $\pi_1(B)$ acts nilpotently on $H^j(F;\mathbb{Q})$ for all $j \geq 1$; (see Definition (2.3.1); the action here is the standard one); and

(iii) either B or F has finite $\mathbb{Q}$-type.

Then there is a commutative diagram (Diagram 2.1) where $i : F \to E$ is the inclusion of the fibre, the middle row is a minimal KS-extension, the bottom row is a Λ-minimal Λ-extension, and all the vertical arrows are weak equivalences. In particular, α is a minimal model of B, γ_1 and $\gamma_1\gamma_2$ are minimal models of F, and γ_2 is an isomorphism. $\beta_1\beta_2$ is a model of E; but, owing to the possible twisting of the differential of $\mathcal{M}(B) \otimes \mathcal{M}(F)$, $\beta_1\beta_2$ is not necessarily a minimal model of E.

2.5. The Grivel-Halperin-Thomas theorem

$$
\begin{array}{ccccc}
A(B) & \xrightarrow{A(\pi)} & A(E) & \xrightarrow{A(i)} & A(F) \\
\uparrow \text{id} & & \uparrow \beta_1 & & \uparrow \gamma_1 \\
A(B) & \longrightarrow & A(B) \otimes \mathcal{M}(F) & \longrightarrow & \mathcal{M}(F) \\
\uparrow \alpha & & \uparrow \beta_2 & & \uparrow \gamma_2 \\
\mathcal{M}(B) & \longrightarrow & \mathcal{M}(B) \otimes \mathcal{M}(F) & \longrightarrow & \mathcal{M}(F)
\end{array}
$$

Diagram 2.1

(2.5.2) Corollary

Under the conditions of Theorem (2.5.1), there is a long exact sequence
$$0 \to \Pi^1_\psi(B) \to \Pi^1_\psi(E) \to \Pi^1_\psi(F) \to \Pi^2_\psi(B) \to \ldots$$
$$\to \Pi^n_\psi(B) \to \Pi^n_\psi(E) \to \Pi^n_\psi(F) \to \Pi^{n+1}_\psi(B) \to \ldots$$

This follows at once from Theorem (2.3.12).

(2.5.3) Corollary

Under the conditions of Theorem (2.5.1), if any two of F, E, B are π-finite, then so is the third. Furthermore, in this case, $\chi\pi(E) = \chi\pi(B) + \chi\pi(F)$.

(2.5.4) Example

Let X be a G-CW-complex, $G = T^n$ a torus. Consider the fibration $X \to X_G \to BG$. Since $\pi_1(BG)$ is trivial and BG has finite $\mathbb{Q}$-type, conditions (ii) and (iii) of Theorem (2.5.1) are satisfied. Thus, if X is connected, the theorem gives a commutative diagram (Diagram 2.2)

$$
\begin{array}{ccccc}
A(BG) & \longrightarrow & A(X_G) & \longrightarrow & A(X) \\
\uparrow & & \uparrow & & \uparrow \\
R & \longrightarrow & R \otimes \mathcal{M}(X) & \longrightarrow & \mathcal{M}(X)
\end{array}
$$

Diagram 2.2

where $R = \mathbb{Q}[t_1, \ldots, t_n]$, $\deg(t_i) = 2$ for $1 \leq i \leq n$.

The only way in which $R \otimes \mathcal{M}(X)$ can fail to be minimal is if there exists ξ of degree 1 in $\mathcal{M}(X)$ and $\lambda_1, \ldots, \lambda_n$ in $\mathbb{Q}$, not all zero, such that $d\xi = \sum_{i=1}^n \lambda_i t_i + d_X(\xi)$, where d is the differential on $R \otimes \mathcal{M}(X)$ and d_X is the differential on $\mathcal{M}(X)$. So $R \otimes \mathcal{M}(X)$ is always minimal if $\Pi^1_\psi(X) = 0$, or, if $X^G \neq \emptyset$ (see Example (2.5.7) (2)).

If X is π-finite, then $\chi\pi(X_G) = n + \chi\pi(X)$.

(2.5.5) Remark
In Theorem (2.5.1), α may be chosen to be any minimal model of $A(B)$.

Now let $F \to E \xrightarrow{\pi} B$ be a Serre fibration satisfying the conditions of Theorem (2.5.1). Let

$$\begin{array}{ccc} E' & \longrightarrow & E \\ \downarrow{\pi'} & & \downarrow{\pi} \\ B' & \xrightarrow{f} & B \end{array}$$

Diagram 2.3

there be a pull-back (Diagram 2.3) with B' path-connected. Thus $\pi_1(B')$ acts nilpotently on $H^j(F;\mathbb{Q})$ for all $j \geq 1$. Suppose that either F has finite $\mathbb{Q}$-type or both B and B' have finite $\mathbb{Q}$-type. So both π and π' satisfy the conditions of Theorem (2.5.1).

(2.5.6) Theorem
In the above situation there is a commutative diagram (Diagram 2.4)

$$\begin{array}{ccccc} A(B') & \xrightarrow{A(\pi')} & A(E') & \longrightarrow & A(F) \\ \uparrow{\text{id}} & & \uparrow{\beta'} & & \uparrow{\gamma'} \\ A(B') & \longrightarrow & A(B') \otimes \mathcal{M}(F) & \longrightarrow & \mathcal{M}(F) \end{array}$$

Diagram 2.4

where the bottom row is a minimal KS-extension, β' and γ' are weak equivalences (and so, in particular, γ' is a minimal model of F); and the differential on $A(B') \otimes \mathcal{M}(F)$ comes from the differential on $A(B) \otimes \mathcal{M}(F)$ given by Theorem (2.5.1) together with the isomorphism

$$A(B') \otimes \mathcal{M}(F) \cong A(B') \otimes_{A(B)} (A(B) \otimes \mathcal{M}(F)).$$

Thus if d is the differential on $A(B) \otimes \mathcal{M}(F)$ coming from Theorem (2.5.1), and if d' is the differential on $A(B') \otimes \mathcal{M}(F)$, then, for any $\xi \in \mathcal{M}(F)$,

$$d'(1 \otimes \xi) = (A(f) \otimes 1)(d(1 \otimes \xi)).$$

Suppose further that $A(f) : A(B) \to A(B')$ is surjective. Let $\alpha'' : \mathcal{M}(B') \to A(B')$ be a minimal model. Then there is a minimal model $\alpha : \mathcal{M}(B) \to A(B)$, and a homomorphism $\phi : \mathcal{M}(B) \to \mathcal{M}(B')$ such that $\alpha''\phi = A(f)\alpha$; and there is a commutative diagram (Diagram 2.5)

2.5. The Grivel-Halperin-Thomas theorem

$$
\begin{array}{ccccc}
A(B') & \xrightarrow{A(\pi')} & A(E') & \longrightarrow & A(E) \\
\uparrow{\alpha''} & & \uparrow{\beta''} & & \uparrow{\gamma''} \\
\mathcal{M}(B') & \longrightarrow & \mathcal{M}(B') \otimes \mathcal{M}(F) & \longrightarrow & \mathcal{M}(F)
\end{array}
$$

Diagram 2.5

where the bottom row is a Λ-minimal Λ-extension, α'', β'' and γ'' are weak equivalences; and the differential on $\mathcal{M}(B') \otimes \mathcal{M}(F)$, δ', say, comes from the differential on $\mathcal{M}(B) \otimes \mathcal{M}(F)$ given by Theorem (2.5.1), δ, say, together with the isomorphism

$$\mathcal{M}(B') \otimes \mathcal{M}(F) \cong \mathcal{M}(B') \otimes_{\mathcal{M}(B)} (\mathcal{M}(B) \otimes \mathcal{M}(F)).$$

Thus, for any $\xi \in \mathcal{M}(F)$,

$$\delta'(1 \otimes \xi) = (\phi \otimes 1)(\delta(1 \otimes \xi)).$$

(2.5.7) Examples

(1) Let T be a torus and X a T-space. Let K be a subtorus. Then there is a pull-back (Diagram 2.6)

$$
\begin{array}{ccc}
X_K & \longrightarrow & X_T \\
\downarrow & & \downarrow \\
BK & \xrightarrow{f} & BT
\end{array}
$$

Diagram 2.6

where f is induced by the inclusion $K < T$. Since $T \cong K \times T/K$, we may define BK and BT so that we have a map $BT \to BK$ which is left inverse to f. Hence we may assume that $A(f)$ is onto; and so (2.5.6) applies. Thus, if $(R \otimes \mathcal{M}(X), d)$, $R := H^*(BT; \mathbb{Q}) = \mathcal{M}(BT)$, is a model for X_T as in Theorem (2.5.1), then there is a model $(R(K) \otimes \mathcal{M}(X), d')$ for X_K, where $R(K) := H^*(BK; \mathbb{Q}) = \mathcal{M}(BK)$, and, for any $\xi \in \mathcal{M}(X)$, $d'(1 \otimes \xi)$ is found from $d(1 \otimes \xi)$ by reducing all elements in R in the expression for $d(1 \otimes \xi)$ modulo the ideal $PK := \ker[f^* : R \to R(K)]$.

(2) Let G be a compact connected Lie group, and let X be a path-connected G-space. Suppose that $X^G \neq \emptyset$, and let $x_0 \in X^G$. Then the map $EG \to EG \times X$, given by $z \mapsto (z, x_0)$, induces a map $s : BG \to X_G$, which is right inverse to $\pi : X_G \to BG$. From Theorem (2.5.1), there is a model

$$\beta : (R \otimes \mathcal{M}(X), d) \to A(X_G),$$

where $R := H^*(BG; \mathbb{Q}) = \mathcal{M}(BG)$. If $i : R \to A(BG)$ is the minimal model (such that $A(\pi)i = \beta j$, where $j : R \to R \otimes \mathcal{M}(X)$ is the inclusion), then $A(BG)$ and $A(X_G)$ may be viewed as R-algebras by means of i and $A(\pi)i$, respectively. Furthermore, it follows from the general theory of CGDAs that there exists $\varepsilon : R \otimes \mathcal{M}(X) \to R$, and that β may be chosen so that Diagram 2.7 is a commutative diagram of CGDAs and R-algebras.

$$\begin{array}{ccc} R \otimes \mathcal{M}(X) & \xrightarrow{\varepsilon} & R \\ \downarrow{\beta} & & \downarrow{i} \\ A(X_G) & \xrightarrow{A(s)} & A(BG) \end{array}$$

Diagram 2.7

(Note that $A(s)$ is surjective, since $\pi s =$ identity on BG.) Hence we may choose an algebra basis $\{x_\alpha : \alpha \in \mathcal{A}\}$ for $\mathcal{M}(X)$, such that $\varepsilon(1 \otimes x_\alpha) = 0$ for all $\alpha \in \mathcal{A}$. It follows that $(R \otimes \mathcal{M}(X), d)$ is minimal.

Now suppose that X is simply-connected of finite $\mathbb{Q}$-type, and let PX, respectively ΩX, be the space of paths, respectively loops, in X based at x_0. Then G acts on PX and ΩX in the obvious way. From Theorem (2.5.1) applied to the fibre space $\Omega X \to (PX)_G \to X_G$, there is a model $(R \otimes \mathcal{M}(X) \otimes \mathcal{M}(\Omega X), D)$ for $(PX)_G$. If, for each $\alpha \in \mathcal{A}$,

$$dx_\alpha = \bigoplus_\beta r_{\alpha\beta} x_\beta + \xi_\alpha,$$

where $r_{\alpha\beta} \in R$, and $\xi_\alpha \in \overline{\mathcal{M}(X)}^2$, viewed as an ideal in $R \otimes \mathcal{M}(X)$, then it can be shown that there are elements

$$\overline{x}_\alpha \in R \otimes \mathcal{M}(X) \otimes \mathcal{M}(\Omega X),$$

for each $\alpha \in \mathcal{A}$, such that

$$D\overline{x}_\alpha = x_\alpha - \sum_\beta r_{\alpha\beta} \overline{x}_\beta + \eta_\alpha,$$

where $\eta_\alpha \in \overline{\mathcal{M}(X)}$, viewed as an ideal in $R \otimes \mathcal{M}(X) \otimes \mathcal{M}(\Omega X)$, and the set $\{\overline{x}_\alpha; \alpha \in \mathcal{A}\}$ maps bijectively to an algebra basis for $\mathcal{M}(\Omega X)$ under the homomorphism

$$R \otimes \mathcal{M}(X) \otimes \mathcal{M}(\Omega X) \to \mathcal{M}(\Omega X).$$

Now Theorem (2.5.6) can be applied to the pull-back Diagram 2.8

$$\begin{array}{ccc} (\Omega X)_G & \longrightarrow & (PX)_G \\ \downarrow & & \downarrow \\ BG & \xrightarrow{s} & X_G \end{array}$$

Diagram 2.8

to get a model $(R \otimes \mathcal{M}(\Omega X), \delta)$ for $(\Omega X)_G$ such that $\delta \overline{x}_\alpha = -\sum_\beta r_{\alpha\beta} \overline{x}_\beta$, for all $\alpha \in \mathcal{A}$.

2.6 Sullivan-de Rham theory for rational Alexander-Spanier cohomology.

To study the rational cohomology of general G-spaces, for G a compact connected Lie group, one needs a cohomology theory with good continuity (or tautness) properties (see Section 3.2). Typically this means using the cohomology theory of Alexander-Spanier, Čech or sheaf theory. One usually considers only paracompact G-spaces; and on such spaces the three theories coincide: indeed, with constant coefficients, the first two theories always coincide (see [Dowker, 1952]). It would be convenient, therefore, to have a theory of minimal models for such cohomology with rational coefficients. In this section we outline a Sullivan-de Rham type theory for rational Alexander-Spanier cohomology.

Given a topological space X and an open covering $\mathcal{U}$ of X, recall that the Čech nerve of $\mathcal{U}$, $\check{\mathcal{U}}$, is the simplicial complex with vertices the members of $\mathcal{U}$, and an n-simplex any set $\{U_0, U_1, \ldots, U_n\}$ of $n+1$ distinct members of $\mathcal{U}$ with $\bigcap_{i=0}^n U_i \neq \emptyset$. The Vietoris nerve of $\mathcal{U}$, $\overline{\mathcal{U}}$, is the simplicial complex with vertices the points of X, and an n-simplex any set $\{x_0, x_1, \ldots, x_n\}$ of $n+1$ distinct points of X contained in an element of $\mathcal{U}$. By [Dowker, 1952], the geometric realizations of these simplicial complexes, $|\check{\mathcal{U}}|$ and $|\overline{\mathcal{U}}|$, are canonically homotopy equivalent.

Now, given a simplicial complex K, one can form a simplicial set K_0 by letting the n-simplexes of K_0 be all ordered sets, i.e. $(n+1)$-tuples, $(v_0, \ldots, v_n)$ of vertices of K such that $\{v_0, \ldots, v_n\}$ is a simplex (of dimension $\leq n$) of K. The boundary maps, respectively degeneracy maps, of K_0 are the standard ones of omission, respectively repetition, of vertices. The normalized cochain complex of K_0 is naturally isomorphic to the ordered cochain complex of K.

(2.6.1) Definition
The Alexander-Spanier Sullivan-de Rham algebra of a space X is the CGDA
$$\overline{A}(X) = \varinjlim_{\mathcal{U}} A(\overline{\mathcal{U}}_0)$$
where the direct limit is taken over all open coverings of X directed by refinement.

Since homology commutes with direct limits, $H\overline{A}(X)$ is naturally isomorphic to
$$\varinjlim_{\mathcal{U}} H^*(\overline{\mathcal{U}}; \mathbb{Q}) = \overline{H}^*(X; \mathbb{Q}),$$
the rational Alexander-Spanier cohomology of X.

Given a space X and open covering of X, $\mathcal{U}$, let $\text{sing}(X,\mathcal{U})$ be the simplicial set of singular simplexes of X which are subordinate to $\mathcal{U}$: i.e. $\sigma: \Delta^n \to X$ is in $\text{sing}(X,\mathcal{U})$ if and only if $\sigma(\Delta^n) \subseteq U$ for some $U \in \mathcal{U}$. If $v_0, \ldots, v_n$ denote the vertices of Δ^n then there is a map $\text{sing}(X,\mathcal{U}) \to \overline{\mathcal{U}}_0$ given by $\sigma \mapsto (\sigma(v_0), \ldots, \sigma(v_n))$. Thus there are maps $A(X) = A(\text{sing}(X)) \to A(\text{sing}(X,\mathcal{U})) \leftarrow A(\overline{\mathcal{U}}_0)$, and the left hand map is a weak equivalence by [Spanier, 1966], Chapter 4, Theorem 4.14. Hence we get maps
$$A(X) \stackrel{T(X)}{\to} \varinjlim_{\mathcal{U}} A(\text{sing}(X,\mathcal{U})) \stackrel{W(X)}{\leftarrow} \overline{A}(X).$$
This gives a natural transformation
$$HT(X)^{-1}HW(X): \overline{H}^*(X; \mathbb{Q}) \to H^*(X; \mathbb{Q}).$$

(2.6.2) Definition

We shall say that X is an agreement space if $HT(X)^{-1}HW(X)$ is an isomorphism.

If X is connected, then $\overline{A}(X)$ is c-connected, and, hence, by Theorem (2.2.5), $\overline{A}(X)$ has a minimal model. We shall denote the minimal model, which is unique up to isomorphism, by $\overline{\mathcal{M}}(X)$.

(2.6.3) Definition

If X is a connected space, then let
$$\overline{\Pi}^*_\psi(X) := \pi^*(\overline{\mathcal{M}}(X)) = Q\overline{\mathcal{M}}(X).$$
This is the Alexander-Spanier (pseudo-dual) rational homotopy of X. Also, following Definition (2.3.10), we set
$$\overline{\mathcal{L}}^*(X) = \mathcal{L}^*(\overline{\mathcal{M}}(X)) = \lambda(\overline{\mathcal{M}}(X)),$$
the associated co-Lie algebra.

We shall now state some results from [Allday, Halperin, 1984], indicating proofs only where the stated result is an improvement of that in [Allday, Halperin, 1984]. The first result is a version of the Grivel-Halperin-Thomas theorem (Theorem (2.5.1)).

(2.6.4) Theorem

Let $F \stackrel{i}{\to} E \stackrel{p}{\to} B$ be a Hurewicz fibration such that:

(i) F, E and B are connected;

(ii) E is paracompact; and
(iii) B is a hausdorff space with the homotopy type of a 1-connected CW-complex with finite skeleta.

Then Diagram 2.9 is a commutative diagram of CGDAs

$$\begin{array}{ccccc}
\overline{A}(B) & \xrightarrow{\overline{A}(p)} & \overline{A}(E) & \xrightarrow{\overline{A}(i)} & \overline{A}(F) \\
\uparrow \mathrm{id} & & \uparrow \beta_1 & & \uparrow \gamma_1 \\
\overline{A}(B) & \longrightarrow & \overline{A}(B) \otimes \overline{\mathcal{M}}(F) & \longrightarrow & \overline{\mathcal{M}}(F) \\
\uparrow \alpha & & \uparrow \beta_2 & & \uparrow \gamma_2 \\
\mathcal{M}(B) & \longrightarrow & \mathcal{M}(B) \otimes \overline{\mathcal{M}}(F) & \longrightarrow & \overline{\mathcal{M}}(F)
\end{array}$$

Diagram 2.9

where the middle row is a minimal KS-extension, the bottom row is a Λ-minimal Λ-extension, all the vertical arrows are weak-equivalences, and α can be chosen to be any minimal model of $\overline{A}(B)$.

(2.6.5) Remarks

(1) There are obvious analogues of Corollaries (2.5.2) and (2.5.3) and Theorem (2.5.6).

The analogue of Theorem (2.5.6) is as follows. Let Diagram 2.10

$$\begin{array}{ccc}
E' & \longrightarrow & E \\
\downarrow p' & & \downarrow p \\
B' & \xrightarrow{f} & B
\end{array}$$

Diagram 2.10

be a pull-back diagram where p is a Hurewicz fibration with E and B and the fibre, F, being connected. Suppose that E and E' are paracompact, and that both B and B' are hausdorff spaces having the homotopy types of 1-connected CW-complexes with finite skeleta. Then Diagram 2.11 is a commutative diagram of CGDAs

$$\begin{array}{ccccc}
\overline{A}(B') & \xrightarrow{A(p')} & \overline{A}(E') & \longrightarrow & \overline{A}(F) \\
\uparrow \mathrm{id} & & \uparrow \beta' & & \uparrow \gamma' \\
\overline{A}(B') & \longrightarrow & \overline{A}(B') \otimes \overline{\mathcal{M}}(F) & \longrightarrow & \overline{\mathcal{M}}(F)
\end{array}$$

Diagram 2.11

where the bottom row is a minimal KS-extension, β' and γ' are weak equivalences, and the differential on $\overline{A}(B') \otimes \overline{\mathcal{M}}(F)$ comes from the differential on $\overline{A}(B) \otimes \overline{\mathcal{M}}(F)$ given by Theorem (2.6.4) together with the isomorphism

$$\overline{A}(B') \otimes \overline{\mathcal{M}}(F) \cong \overline{A}(B') \otimes_{\overline{A}(B)} (\overline{A}(B) \otimes \overline{\mathcal{M}}(F)).$$

Suppose further that $\overline{A}(f) : \overline{A}(B) \to \overline{A}(B')$ is surjective. Let $\alpha'' : \mathcal{M}(B') \to \overline{A}(B')$ be a minimal model. Then there is a minimal model $\alpha : \mathcal{M}(B) \to A(B)$, and a homomorphism $\phi : \mathcal{M}(B) \to \mathcal{M}(B')$ such that

$$\alpha''\phi = \overline{A}(f)\alpha;$$

and Diagram 2.12 is a commutative diagram

$$\begin{array}{ccccc}
\overline{A}(B') & \longrightarrow & \overline{A}(E') & \longrightarrow & \overline{A}(F) \\
\uparrow \alpha'' & & \uparrow \beta'' & & \uparrow \gamma'' \\
\mathcal{M}(B') & \longrightarrow & \mathcal{M}(B') \otimes \overline{\mathcal{M}}(F) & \longrightarrow & \overline{\mathcal{M}}(F)
\end{array}$$

Diagram 2.12

where the bottom row is a Λ-minimal Λ-extension, α'', β'' and γ'' are weak equivalences; and the differential on $\mathcal{M}(B') \otimes \overline{\mathcal{M}}(F)$ comes from the differential on $\mathcal{M}(B) \otimes \overline{\mathcal{M}}(F)$ given by Theorem (2.6.4) together with the isomorphism

$$\mathcal{M}(B') \otimes \overline{\mathcal{M}}(F) \cong \mathcal{M}(B') \otimes_{\mathcal{M}(B)} (\mathcal{M}(B) \otimes \overline{\mathcal{M}}(F)).$$

(2) Theorem (2.6.4) differs from Theorem 4.1 of [Allday, Halperin, 1984] by not requiring $\bar{H}^*(F; \mathbb{Q})$ to have finite type. The only time that the finite type condition was used in [Allday, Halperin, 1984] was in the proof of Lemma 2.6. Lemma 2.6 is valid for general paracompact spaces, however, by [Dold, 1980], Proposition 3.7, p. 363.

(2.6.6) Corollary
If G is a compact connected Lie group, and X is a connected paracompact G-space, then Diagram 2.13 is a commutative diagram of CGDAs

$$\begin{array}{ccccc}
\overline{A}(BG) & \longrightarrow & \overline{A}(X_G) & \longrightarrow & \overline{A}(X) \\
\uparrow & & \uparrow & & \uparrow \\
\mathcal{M}(BG) & \longrightarrow & \mathcal{M}(BG) \otimes \overline{\mathcal{M}}(X) & \longrightarrow & \overline{\mathcal{M}}(X)
\end{array}$$

Diagram 2.13

where the bottom row is a Λ-minimal Λ-extension, and the vertical maps are weak-equivalences.

(2.6.7) Remarks
(1) The corollary follows directly from the theorem applied to $X \to X_G \to BG$, since X_G is paracompact if X is. (See Section 3.2 for a proof of this.)
(2) There are obvious analogues of Examples (2.5.7)(1) and the first part of (2).
(3) It follows from the theorem that if E (resp. F) is an agreement space, then so is F (resp. E). Thus X_G is an agreement space, if X is.

We finish with a result relating $\overline{\Pi}_\psi^*(X)$ to the ordinary pseudo-dual rational homotopy $\Pi_\psi^*(|\overline{\mathcal{U}}|)$ as $\mathcal{U}$ ranges over the open coverings of X, together with a corollary.

(2.6.8) Proposition
If X is a connected space, then
$$\overline{\Pi}_\psi^*(X) \cong \varinjlim_{\mathcal{U}} \Pi_\psi^*(|\overline{\mathcal{U}}|).$$
Indeed
$$\overline{\mathcal{L}}^*(X) \cong \varinjlim_{\mathcal{U}} \mathcal{L}^*(|\overline{\mathcal{U}}|).$$

(2.6.9) Remarks
(1) Of course we can replace $\overline{\mathcal{U}}$ by $\check{\mathcal{U}}$, for all $\mathcal{U}$, in the proposition.
(2) The proposition is proved in [Allday, Halperin, 1984] only when X is compact and metrizable. We indicate below a general proof using ideas and terminology taken from [Bousfield, Gugenheim, 1976].

Let $\mathcal{A}_0$ be the category of augmented CGDAs, and let $\mathcal{C}_0$ be the full subcategory of cofibrant augmented CGDAs. Bousfield and Gugenheim construct a functor $K : \mathcal{A}_0 \to \mathcal{C}_0$. In the notation of [Bousfield, Gugenheim, 1976], Section 4.7, $K(A) = L_i$, where $i : \mathbb{Q} \to A$ is the unit map. Thus, for each object A of $\mathcal{A}_0$, there is an acyclic fibration (i.e. a fibration which is also a weak-equivalence) $\Psi(A) : K(A) \to A$, and Ψ is a natural transformation $IK \to \mathrm{id}_{\mathcal{A}_0}$ where $I : \mathcal{C}_0 \to \mathcal{A}_0$ is the inclusion functor.

Now let $\{A_j : j \in \mathcal{J}\}$ be a direct system of objects of $\mathcal{A}_0$. One can show that $\varinjlim_j K(A_j)$ is cofibrant. This can be seen by means of a detailed analysis of the Bousfield-Gugenheim construction: or one can construct a natural transformation $\Theta : K \to K^2$, where
$$K^2(A) = K(K(A)),$$

such that
$$\Psi(K(A))\Theta(A) = \mathrm{id}_{K(A)},$$
for all objects A of $\mathcal{A}_0$, and then use Θ to exhibit $\varinjlim_j K(A_j)$ as a retract of
$$K(\varinjlim_j K(A_j)).$$

The functor K applied to the maps $A_j \to \varinjlim_j A_j$ induces a weak-equivalence $\varinjlim_j K(A_j) \to K(\varinjlim_j A_j)$. By [Bousfield, Gugenheim, 1976], 6.5 and 6.13, this map induces an isomorphism
$$\pi \varinjlim_j K(A_j) \to \pi K(\varinjlim_j A_j).$$
Since direct limits are exact and commute with homology, $\varinjlim_j \pi K(A_j)$ is isomorphic to $\pi K(\varinjlim_j A_j)$. Proposition (2.6.8) now follows by applying this to the direct system $\{A(|\check{\mathcal{U}}|); \mathcal{U}$ an open covering of $X\}$.

(2.6.10) Corollary
Suppose that X is connected and finitistic (i.e. every open covering has a refinement with finite-dimensional Čech nerve: e.g. X compact or X has finite covering dimension). Suppose that $\dim_{\mathbb{Q}} \overline{\Pi}_\psi^*(X) < \infty$. Then $\dim_{\mathbb{Q}} \overline{H}^*(X;\mathbb{Q}) < \infty$.

Proof.
Let $x_1, \ldots, x_n$ be a basis for $\overline{\Pi}_\psi^*(X)$.
Then, by Propositon (2.6.8), there is an open covering, $\mathcal{U}$, of X, such that $|\check{\mathcal{U}}|$ is finite-dimensional, and $x_1, \ldots, x_n$ can all be represented in $\Pi_\psi^*(|\check{\mathcal{U}}|)$. So $\Pi_\psi^*(|\check{\mathcal{U}}|) \to \overline{\Pi}_\psi^*(X)$ is surjective. The result now follows by [Halperin, 1978], Corollary 5.13, applied with $R = A(|\check{\mathcal{U}}|)$, $L = \overline{A}(X)$, $Z = Q\overline{\mathcal{M}}(X)$, and $Z_1 = 0$. □

2.7 Formal spaces, formal maps and the Eilenberg-Moore spectral sequence

In this section we shall give a brief outline of the theory of formal CGDAs, formal topological spaces and formal morphisms. Our main references are [Halperin, Stasheff, 1979] and [Vigué-Poirrier, 1981]: other useful references are [Deligne, Morgan, Griffiths, Sullivan, 1975], [Felix, Halperin, 1982a], [Greub, Halperin, Vanstone, 1976], [Sullivan, 1977], [Thomas, 1982], [Tanré, 1983], [Felix, 1984] and [Felix, Tanré 1988]. In

2.7. Formal spaces, maps & the Eilenberg-Moore spectral sequence

addition to our discussion of formality we shall define the notion of coformality (π-formality), and we shall give a version of the Eilenberg-Moore spectral sequence coming from [Vigué-Poirrier, 1981].

(2.7.1) Definitions
(1) Let (A, d) be a c-connected CGDA. Let $(H(A), 0)$ denote the CGA $H(A, d)$ with trivial differential. Then $A = (A, d)$ is said to be formal if there is a diagram
$$(A, d) \xleftarrow{\alpha} (M, \delta) \xrightarrow{\beta} (H(A), 0),$$
where (M, δ) is a minimal CGDA, and α and β are weak-equivalences: i.e. if A and $H(A)$ have a common minimal model.
(2) A path-connected topological space X is said to be formal if $A(X)$ is formal.

(2.7.2) Examples
(1) In [Greub, Halperin, Vanstone, 1976] the following homogeneous spaces are shown to be formal:
(i) G/K, where G is any compact connected Lie group and K is any closed connected subgroup with $\text{rank}(K) = \text{rank}(G)$;
(ii) $U(n)/U(n_1) \times \ldots \times U(n_m)$, where $m \geq 1$ and $\sum_{i=1}^{m} n_i \leq n$;
(iii) $Sp(n)/SU(n)$ for $n \leq 4$;
(iv) $U(n)/SO(n)$;
(v) $SO(n)/SO(m)$, whenever $m \leq n$;
(vi) $U(2n)/Sp(n)$ and $Sp(n)/SO(n)$.
(2) In [Greub, Halperin, Vanstone, 1976] the following homogeneous spaces are shown to be not formal:
(i) $Sp(n)/SU(n)$ for $n \geq 5$;
(ii) $SU(6)/SU(3) \times SU(3)$.
(3) In [Deligne, Morgan, Griffiths, Sullivan, 1975] it is shown that any compact Kähler manifold is formal.

(2.7.3) Remark
In the examples above the quoted authors generally use the real numbers, $\mathbb{R}$, as the ground field. It can be shown, however, see, e.g., [Halperin, Stasheff, 1979], that if K is any field of characteristic 0, then a c-connected CGDA (A, d) such that $H(A, d)$ has finite type is formal if and only if $(A \otimes K, d \otimes 1)$ is formal (in the obvious sense which generalizes Definition (2.7.1) to CGDAs defined over K).

The main tool of [Halperin, Stasheff, 1979] is a bigraded model of $(H(A), 0)$ and the resulting perturbed model for (A, d). A bigraded

model for $(H(A), 0)$ is a minimal model $\rho : S(V) \to H(A)$ where the generating graded $\mathbb{Q}$-vector space V is actually bigraded. One has
$$V = \bigoplus_{m \geq 0} \bigoplus_{n \geq 1} V_m^n$$
and with the induced bigrading on $S(V)$ the differential d has degree -1 with respect to the lower degree. As usual d has degree $+1$ with respect to the upper degree. The homology $H(S(V), d)$ is then bigraded, and in the following theorem, which summarizes the main properties of a bigraded model, $H_0^*(S(V))$ indicates elements of lower degree 0, and $H_+^*(S(V))$ indicates elements of positive lower degree.

(2.7.4) Theorem
Let H be a connected CGA. Then there is a minimal CGDA $(S(V), d)$ and a weak-equivalence $\rho : (S(V), d) \to (H, 0)$ such that
(i) $V = \bigoplus_{m \geq 0} \bigoplus_{n \geq 1} V_m^n$ is bigraded and $\rho(\bigoplus_{m \geq 1} V_m^*) = 0$;
(ii) d has degree -1 with respect to the lower degree;
(iii) $\rho^* | H_0^*(S(V), d) : H_0^*(S(V), d) \to H$ is an isomorphism; and
(iv) $H_+^*(S(V), d) = 0$.

Furthermore, if $\alpha : (M, \delta) \to (H, 0)$ is any other minimal model such that $M = \bigoplus_{m \geq 0} \bigoplus_{n \geq 0} M_m^n$ is a bigraded algebra, $\alpha(\bigoplus_{m \geq 1} M_m^*) = 0$, and α and δ satisfy (ii)-(iv), then there is a bihomogeneous isomorphism
$$\phi : (S(V), d) \to (M, \delta)$$
of bidegree $(0, 0)$ such that $\alpha \phi = \rho$.

(2.7.5) Definition
$\rho : (S(V), d) \to (H, 0)$, as in Theorem (2.7.4) is called a bigraded model of H.

(2.7.6) Remark
In the bigraded model of Theorem (2.7.4),
$$V_0^* \cong QH := H^+/(H^+ \cdot H^+),$$
the space of algebra generators of H.

(2.7.7) Definition
Let H be a connected CGA, and let
$$\rho : (S(V), d) \to (H, 0)$$
be a bigraded model. For $n \geq 0$, set
$$\mathcal{F}_n(S(V)) = \bigoplus_{m=0}^{n} S(V)_m;$$
and for $n < 0$, set $\mathcal{F}_n(S(V)) = 0$.

Now another vital theorem of [Halperin, Stasheff, 1979] is the following.

(2.7.8) Theorem
Let (A, d_A) be a c-connected CGDA, and let
$$\rho : (S(V), d) \to (H(A, d_A), 0)$$
be a bigraded model of $H(A, d_A)$. Then there is a differential D on $S(V)$ such that $(S(V), D)$ is a KS-complex, and there is a weak-equivalence $\pi : (S(V), D) \to (A, d_A)$ such that:

(i)   for any $n \geq 0$ and for any $x \in V_n^*$, $D(x) - d(x) \in \mathcal{F}_{n-2}(S(V))$; and
(ii) for any $y \in S(V)_0$, $[\pi(y)] = \rho(y)$ in $H(A, d_A)$.
(Note that (i) implies that $D(y) = 0$ for any $y \in S(V)_0$. Hence the homology class $[\pi(y)]$ is defined.)

Furthermore, if $(S(V), D')$ and $\pi' : (S(V), D') \to (A, d_A)$ also satisfy the above conditions, then there is an isomorphism
$$\phi : (S(V), D) \to (S(V), D')$$
such that:
(a) for any n and for any $x \in \mathcal{F}_n(S(V))$, $\phi(x) - x \in \mathcal{F}_{n-1}(S(V))$; and
(b) $\pi'\phi \simeq \pi$ in the sense of Definition (2.2.4)(4).

(2.7.9) Corollary
Let (A, d_A) be a CGDA with
$$H(A, d_A) \cong \Lambda(y_1, \ldots, y_s) \otimes \mathbb{Q}[x_1, \ldots, x_r]/I,$$
where $y_1, \ldots, y_s$ are homogeneous of odd degrees, $x_1, \ldots, x_r$ are homogeneous of positive even degrees, and either $I = (0)$ or $I = (f_1, \ldots, f_n)$, where $f_1, \ldots, f_n$ are homogeneous elements of $\mathbb{Q}[x_1, \ldots, x_r]$ which do not involve any terms which are linear in $x_1, \ldots, x_r$, such that $f_1, \ldots, f_n$ is a regular sequence of polynomials. (See Definitions (A.5.1).) Then (A, d_A) is formal.
(The number of odd degree generators, s, may be 0.)

Proof.
In constructing
$$\rho : (S(V), d) \to (H(A, d_A), 0)$$
we can take V_0^* to be generated by $x_1, \ldots, x_r$, and $y_1, \ldots, y_s$. Then we can let V_1^* be generated by $w_1, \ldots, w_n$, and set $dw_i = f_i$ for $1 \leq i \leq n$.

Since $(f_1, \ldots, f_n)$ is a regular sequence, we can take $V_m^* = 0$ for $m \geq 2$ (See Theorem (A.5.5).)
Now $D = d$ by Theorem (2.7.8)(i). □

(2.7.10) Example
If G is a compact connected Lie group then $H^*(G; \mathbb{Q})$ is an exterior algebra and $H^*(BG; \mathbb{Q})$ is a polynomial ring. Thus both G and BG are formal by Corollary (2.7.9).

(2.7.11) Remarks
(1) Any wedge of formal path-connected spaces with non-degenerate base-points (i.e. for each space the inclusion of the base-point is a cofibration) is formal; and any finite product of formal path-connected spaces of finite $\mathbb{Q}$-type is formal, see e.g., [Halperin, Stasheff, 1979] and [Tanré, 1983].
(2) A connected CGA H is said to be intrinsically formal if any CGDA (A, d_A) with $H(A, d_A) \cong H$ is formal. A CGA of the form given in Corollary (2.7.9) is said to be hyperformal. Thus Corollary (2.7.9) asserts that hyperformal implies intrinsically formal. If $H(A, d_A)$ is 0 in positive even degrees, then (A, d_A) is formal ([Halperin, Stasheff, 1979]). Thus there are CGAs which are intrinsically formal but not hyperformal. There are also CGDAs (A, d_A) which are formal but are such that $H(A, d_A)$ is not intrinsically formal (as is seen from the existence of non-formal spaces).

Another important tool in [Halperin, Stasheff, 1979] is a theorem on the realization of homology isomorphisms.

(2.7.12) Definition
Let (A, d_A) and (B, d_B) be c-connected CGDAs. Let $f : H(A, d_A) \to H(B, d_B)$ be an isomorphism. Then f is said to be realizable by a homotopy equivalence if, given any minimal models $\alpha : (\mathcal{M}(A), \delta_A) \to (A, d_A)$ and $\beta : (\mathcal{M}(B), \delta_B) \to (B, d_B)$, there is an isomorphism $\phi : (\mathcal{M}(A), \delta_A) \to (\mathcal{M}(B), \delta_B)$ such that $\beta^* \phi^* (\alpha^*)^{-1} = f$.

(2.7.13) Theorem
Let (A, d_A) and (B, d_B) be c-connected CGDAs. Let
$$f : H(A, d_A) \to H(B, d_B)$$
be an isomorphism. Suppose that $H(A, d_A)$ has finite type. Then there is a computable sequence of obstructions
$$\mathcal{O}_1(f), \mathcal{O}_2(f), \ldots, \mathcal{O}_n(f), \ldots$$

2.7. Formal spaces, maps & the Eilenberg-Moore spectral sequence

such that f is realizable by a homotopy equivalence if and only if
$$\mathcal{O}_n(f) = 0$$
for all $n \geq 1$.

(2.7.14) Remarks
(1) See [Halperin, Stasheff, 1979] for details concerning the definition of the obstructions $\mathcal{O}_n(f)$.
(2) [Halperin, Stasheff, 1979] show that f is realizable by a homotopy equivalence if and only if $f \otimes 1 : H(A, d_A) \otimes K \to H(B, d_B) \otimes K$ is realizable by a homotopy equivalence, where K is any field of characteristic 0, and where $H(A, d_A)$ and $H(B, d_B)$ have finite type: hence Remark (2.7.3), since (A, d_A) is formal if and only if the isomorphism $H(A, d_A) \to H(H(A, d_A), 0)$ can be realized by a homotopy equivalence.

The results of [Vigué-Poirrier, 1981] generalize those of [Halperin, Stasheff, 1979]. We shall describe some of them.

(2.7.15) Theorem
Let H and H' be two connected CGAs (connected CGDAs with zero differential) and let $\beta : H \to H'$ be a homomorphism which is injective in degree 1. Let $\rho : (S(V), d) \to (H, 0)$ be a bigraded model. Then Diagram 2.14 is a commutative diagram

$$\begin{array}{ccc} (H, 0) & \xrightarrow{\beta} & (H', 0) \\ \uparrow{\rho} & & \uparrow{\rho'} \\ (S(V), d) & \longrightarrow & (S(V) \otimes S(W), d') & \longrightarrow & (S(W), d'') \end{array}$$

Diagram 2.14

where the bottom row is a Λ-minimal Λ-extension, ρ' (as well as, of course, ρ) is a weak-equivalence, and W is bigraded,
$$W = \bigoplus_{m \geq 0} \bigoplus_{n \geq 1} W_m^n$$
in such a way that:
(i) *d' has degree -1 with respect to the lower degree;*
(ii) *$\rho'^* | H_0^*(S(V) \otimes S(W), d') : H_0^*(S(V) \otimes S(W), d') \to H'$ is an isomorphism;*
(iii) *$\rho'(\bigoplus_{m \geq 1} V_m^* \oplus W_m^*) = 0$; and*
(iv) *$H_+^*(S(V) \otimes S(W), d') = 0$.*

The bigrading on $S(V) \otimes S(W) \cong S(V \oplus W)$ which is used here is that determined by the standard rule that $(V \oplus W)_m^n = V_m^n \oplus W_m^n$.

2. Some aspects of rational homotopy theory

Furthermore this construction is unique in a way analogous to the uniqueness in Theorem (2.7.4).

(2.7.16) Definition
Diagram 2.14 is called a bigraded model of β.

(2.7.17) Remark
If H is free, then
$$\rho' : (S(V) \otimes S(W), d') \to (H', 0)$$
is a bigraded model for H' if and only if
$$\beta^{-1}(H'^+ \cdot H'^+) \subseteq H^+ \cdot H^+.$$
As in Definition (2.7.7), let
$$\mathcal{F}_n(S(V) \otimes S(W)) = \bigoplus_{m=0}^{n} (S(V) \otimes S(W))_m$$
for $n \geq 0$, and let
$$\mathcal{F}_n(S(V) \otimes S(W)) = 0$$
for $n < 0$. Then the generalization of Theorem (2.7.8) is as follows.

(2.7.18) Theorem
Let (A, d_A) and $(A', d_{A'})$ be two c-connected CGDAs, and let $\alpha : (A, d_A) \to (A', d_{A'})$ be a homomorphism such that α^* is injective in degree 1. Let Diagram 2.15

$$\begin{array}{ccc} H(A, d_A) & \xrightarrow{\alpha^*} & H(A', d_{A'}) \\ \uparrow \rho & & \uparrow \rho' \\ (S(V), d) & \longrightarrow & (S(V) \otimes S(W), d') \end{array}$$

Diagram 2.15

be a bigraded model of α^*. Let $(S(V), D)$ be a KS-complex and let $\pi : (S(V), D) \to (A, d_A)$ be a weak-equivalence satisfying the conclusions of Theorem (2.7.8). Then there is a differential D' on $S(V) \otimes S(W)$ and Diagram 2.16 is a commutative diagram where the bottom row is a Λ-extension and π' (as well as π) is a weak-equivalence. In addition:

(i) for any $n \geq 0$ and for any $x \in (V \oplus W)^*_n$,
$$D'(x) - d'(x) \in \mathcal{F}_{n-2}(S(V) \otimes S(W));$$
(ii) for any $y \in (S(V) \otimes S(W))_0$, $[\pi'(y)] = \rho'(y)$ in $H(A', d_{A'})$.

2.7. Formal spaces, maps & the Eilenberg-Moore spectral sequence

Furthermore there is a uniqueness statement analogous to that of Theorem (2.7.8).

$$
\begin{array}{ccc}
(A, d_A) & \xrightarrow{\alpha} & (A', d_{A'}) \\
\uparrow \pi & & \uparrow \pi' \\
(S(V), D) & \longrightarrow & (S(V) \otimes S(W), D') & \longrightarrow & (S(W), D'')
\end{array}
$$

Diagram 2.16

(2.7.19) Definition

The second diagram in Theorem (2.7.18) is called a filtered model for α. Similarly, $\pi : (S(V), D) \to (A, d_A)$ in Theorem (2.7.8) is called a filtered model for (A, d_A). In particular, Theorem (2.7.18) says that under the stated conditions any filtered model for (A, d_A) can be extended to a filtered model for α.

Now one can define a formal homomorphism as follows.

(2.7.20) Definition

Let (A, d_A) and (B, d_B) be formal c-connected CGDAs, and let

$$\mu_A : (\mathcal{M}(A), \delta_A) \to (A, d_A)$$

and

$$\mu_B : (\mathcal{M}(B), \delta_B) \to (B, d_B)$$

be minimal models. Consequently there are weak-equivalences

$$\psi_A : (\mathcal{M}(A), \delta_A) \to (H(A, d_A), 0)$$

and

$$\psi_B : (\mathcal{M}(B), \delta_B) \to (H(B, d_B), 0)$$

such that $\psi_A^* = \mu_A^*$ and $\psi_B^* = \mu_B^*$. (This follows from the definition of formality for CGDAs together with Theorem (2.2.9).) Now let

$$f : (A, d_A) \to (B, d_B)$$

be a homomorphism. Then f is said to be formal if there is a homomorphism

$$\overline{f} : (\mathcal{M}(A), \delta_A) \to (\mathcal{M}(B), \delta_B)$$

such that

$$\mu_B \overline{f} \simeq f \mu_A$$

and

$$\psi_B \overline{f} \simeq f^* \psi_A,$$

for at least one set of choices of $(\mathcal{M}(A), \delta_A)$, $(\mathcal{M}(B), \delta_B)$, μ_A, μ_B, ψ_A, ψ_B as above.

If X and Y are formal path-connected topological spaces, then a map $f : X \to Y$ is said to be formal if $A(f) : A(Y) \to A(X)$ is a formal homomorphism of CGDAs.

From [Vigué-Poirrier, 1981] we have the following.

(2.7.21) Proposition
Let (A, d_A) and (B, d_B) be two formal c-connected CGDAs, and let
$$\alpha : (A, d_A) \to (B, d_B)$$
be a homomorphism such that α^* is injective in degree 1. Then α is formal if and only if, with the notation of Theorem (2.7.18), a filtered model of α can be chosen with $D = d$ and $D' = d'$.

From the definition of a formal homomorphism one can prove the following.

(2.7.22) Proposition
Let (A, d_A) be a hyperformal CGDA (see Remark (2.7.11)(2)) with
$$H(A, d_A) = \Lambda(y_1, \ldots, y_s) \otimes \mathbb{Q}[x_1, \ldots, x_n]/I,$$
where
$$I = (f_1, \ldots, f_m),$$
and if (B, d_B) is any formal CGDA with
$$H^{n_i}(B, d_B) = 0$$
for $1 \le i \le m$, where $n_i = \deg(f_i) - 1$, then any homomorphism
$$f : (A, d_A) \to (B, d_B)$$
is formal.

(2.7.23) Corollary
If A is any formal path-connected space with
$$H^*(X; \mathbb{Q}) = 0$$
in odd degrees, and if Y is any hyperformal path-connected space, then any map $f : X \to Y$ is formal.

(2.7.24) Remarks
(1) To prove Proposition (2.7.22) note that, by Theorem (2.2.9), there exist $\overline{f}$, $\hat{f}$: $\mathcal{M}(A) \to \mathcal{M}(B)$ such that
$$\mu_B \overline{f} \simeq f \mu_A$$

2.7. Formal spaces, maps & the Eilenberg-Moore spectral sequence

and
$$\psi_B \hat{f} \simeq f^* \psi_A.$$
So $\bar{f}^* = \hat{f}^*$. The condition $H^{n_i}(B, d_B) = 0$, for $1 \leq i \leq m$, removes the only obstructions to constructing a homotopy from $\bar{f}$ to $\hat{f}$.

(2) In [Vigué-Poirrier, 1981] there is also given an obstruction theory for the realisation of an arbitrary homomorphism between connected CGDAs which is injective in degree 1. Specifically let $(\mathcal{M}, \delta)$ and $(\mathcal{M}', \delta')$ be two minimal CGDAs, and let $\beta : H(\mathcal{M}, \delta) \to H(\mathcal{M}', \delta')$ be a homomorphism, which is injective in degree 1. Then the theory gives some obstructions to finding a homomorphism $\alpha : (\mathcal{M}, \delta) \to (\mathcal{M}', \delta')$ such that $\alpha^* = \beta$. One consequence of the theory is that α can always be found if $(\mathcal{M}, \delta)$ is hyperformal.

In contrast to the case of isomorphisms discussed in [Halperin, Stasheff, 1979] it might be possible to solve a general realization problem over an extension field of $\mathbb{Q}$ but not over $\mathbb{Q}$: cf. Remarks (2.7.14)(2).

Another part of [Vigué-Poirrier, 1981] concerns the Eilenberg-Moore spectral sequence. Let (A, d_A), $(A', d_{A'})$ and (C, d_C) be three c-connected CGDAs. Let
$$\alpha : (A, d_A) \to (A', d_{A'}) \text{ and } \gamma : (A, d_A) \to (C, d_C)$$
be homomorphisms such that α^* is injective in degree 1. Let Diagram 2.17

$$\begin{array}{ccc} (A, d_A) & \xrightarrow{\alpha} & (A', d_{A'}) \\ \uparrow \pi & & \uparrow \pi' \\ (S(V), D) & \xrightarrow{i} & (S(V) \otimes S(W), D') \end{array}$$

Diagram 2.17

be a filtered model of α. Let
$$(C \otimes S(W), \Delta') \cong (C, d_C) \otimes_{(S(V), D)} (S(V) \otimes S(W), D')$$
where (C, d_C) is a $(S(V), D)$ module via $\gamma\pi$. Filter $(C \otimes S(W), \Delta')$ by
$$\mathcal{F}^p(C \otimes S(W)) = \bigoplus_{n \leq p} C \otimes S(W)_n$$
if $p \geq 0$, and
$$\mathcal{F}^p(C \otimes S(W)) = 0$$
if $p < 0$.

(2.7.25) Theorem

In the above situation the filtration on
$$(C \otimes S(W), \Delta')$$
gives a second quadrant spectral sequence of algebras with:
(1) $E_0^{-p,q} \cong (C \otimes S(W)_p)^{q-p}$ and $d_0 = d_C \otimes 1$;
(2) $E_1^{-p,*} \cong H^*(C, d_C) \otimes S(W)_p$; and
(3) $E_2^{-p,*} \cong \operatorname{Tor}_{H^*(A, d_A)}^{-p}(H^*(C, d_C), H^*(A', d_{A'}))$.

The spectral sequence converges to $H^*(C \otimes S(W), \Delta')$ in the sense of [Cartan, Eilenberg, 1956], Chapter XV, §4.

Furthermore, from the E_2-term on, the sepctral sequence is isomorphic to the Eilenberg-Moore spectral sequence obtained from a filtration on the bar construction.

In addition, if Diagram 2.18

$$\begin{array}{ccc} H(A, d_A) & \xrightarrow{\alpha^*} & H(A', d_{A'}) \\ \uparrow \rho & & \uparrow \rho' \\ (S(V), d) & \longrightarrow & (S(V) \otimes S(W), d') \end{array}$$

Diagram 2.18

is a bigraded model of α^*, which gives rise (in the manner of Theorem (2.7.18)) to the filtered model of α used in the construction of the spectral sequence, then the spectral sequence differential d_1, is the differential on $H^*(C, d_C) \otimes S(W)$ induced by the isomorphism
$$H^*(C, d_C) \otimes S(W) \cong (H^*(C, d_C), 0) \otimes_{(S(V), d)} (S(V) \otimes S(W), d'),$$
where $H^*(C, d_C)$ is a $(S(V), d)$-module via $\gamma^* \rho$.

By a suitable alteration of Theorem (2.5.6) one obtains the following theorem.

(2.7.26) Theorem

Let Diagram 2.19

$$\begin{array}{ccc} E' & \longrightarrow & E \\ \downarrow & & \downarrow \pi \\ B' & \longrightarrow & B \end{array}$$

Diagram 2.19

be a pull-back diagram satisfying the conditions of Theorem (2.5.6). Let Diagram 2.20

$$\begin{array}{ccc} A(B) & \xrightarrow{A(\pi)} & A(E) \\ \uparrow & & \uparrow \\ S(V) & \longrightarrow & S(V) \otimes S(W) \end{array}$$

Diagram 2.20

be a filtered model of $A(\pi)$. (Note that, from the Serre spectral sequence of $F \to E \xrightarrow{\pi} B$, π^* is injective in degree 1, since F is path-connected.) Then the spectral sequence obtained from

$$A(B') \otimes S(W) \cong A(B') \otimes_{S(V)} (S(V) \otimes S(W))$$

as in Theorem (2.7.25) is, from E_2 on, isomorphic to the Eilenberg-Moore spectral sequence of the pull-back diagram, Diagram 2.19. In particular,

(i) $E_2^{-p} \cong \mathrm{Tor}_{H^*(B;\mathbb{Q})}^{-p}(H^*(B';\mathbb{Q}), H^*(E;\mathbb{Q}))$; and
(ii) the spectral sequence converges to $H^*(E';\mathbb{Q})$.

Now all of the material of this section can easily be adapted to apply to the Alexander-Spanier-Sullivan-de Rham theory of Section 2.6. In particular one gets the following variant of Theorem (2.7.26).

(2.7.27) Theorem

Let Diagram 2.21

$$\begin{array}{ccc} E' & \longrightarrow & E \\ \downarrow & & \downarrow{p} \\ B' & \longrightarrow & B \end{array}$$

Diagram 2.21

be a pull-back diagram satisfying the conditions stated in Remarks (2.6.5) (1). Let Diagram 2.22

$$\begin{array}{ccc} \overline{A}(B) & \xrightarrow{\overline{A}(p)} & \overline{A}(E) \\ \uparrow & & \uparrow \\ S(V) & \longrightarrow & S(V) \otimes S(W) \end{array}$$

Diagram 2.22

be a filtered model of $\overline{A}(p)$. Then the spectral sequence obtained from $\overline{A}(B') \otimes S(W) \cong \overline{A}(B') \otimes_{S(V)} (S(V) \otimes S(W))$ as in Theorem (2.7.25) satisfies

(i) $E_2^{-p} \cong \operatorname{Tor}_{H^*(B;\mathbf{Q})}^{-p}(H^*(B';\mathbf{Q}), \overline{H}^*(E;\mathbf{Q}))$; and
(ii) the spectral sequence converges to $\overline{H}^*(E';\mathbf{Q})$.

We finish with the following definition and examples.

(2.7.28) Definition

A path-connected topological space X is said to be π-formal (or coformal) if given any minimal model of X, $(S(V), d)$, with generating space V, then, for any $v \in V$, dv is in the image of $V \otimes V$ in $S(V)$: i.e. if the differential of any generator is purely quadratic.

(2.7.29) Examples

(1) S^n is π-formal for all $n \geq 1$. $\mathbf{C}P^n$ is not π-formal if $n \geq 2$.
(2) Any compact connected Lie group G and its classifying space BG are π-formal. More generally if X is hyperformal and if, in the notation of Corollary (2.7.9), $f_1, \ldots, f_n$ are homogeneous quadratic polynomials in $x_1, \ldots, x_r$, then X is π-formal.
(3) Any finite product of π-formal spaces of finite $\mathbf{Q}$-type is π-formal.

See [Tanré, 1983] and [Felix, 1984] for more examples.

3

Localization

In this chapter we set forth the main general machinery in the cohomology - and rational homotopy - theory of actions of torus and p-torus groups. The principal result is Theorem (3.1.6), the Localization Theorem of Borel, tom Dieck, W.-Y. Hsiang and Quillen: a greater part of the chapter, however, is devoted to developing some of the immediate consequences of this theorem. In keeping with our general policy most of the results of this chapter will be stated and proved initially for finite-dimensional G-CW-complexes so that only singular cohomology theory need be used. For the sake of reference, however, we shall always indicate for what more general G-spaces the results hold when the cohomology theory used is that of Čech or Alexander-Spanier. Since all G-spaces under consideration will be paracompact, the latter is also the cohomology theory associated with the Eilenberg-MacLane spectrum, and it is equivalent to sheaf-theoretic cohomology as defined in [Bredon, 1967a] or [Godement, 1958].

It is important to observe that all the results of this chapter are G-homotopy invariant. Thus, for example, Theorem (3.1.6) holds for G-spaces which are G-homotopy equivalent to finite-dimensional G-CW-complexes (with the given finiteness conditions on the number of orbit types).

At first, general G-spaces are allotted entire sections to themselves, i.e. Sections 3.2 and 3.4. In Sections 3.5-3.8 general G-spaces are discussed in remarks at the end of the section. In Sections 3.9-3.11 general G-spaces and G-CW-complexes are treated simultaneously. For readers interested in general G-spaces the following technical remarks are useful. Since some results in Alexander-Spanier cohomology, such as the Künneth theorem, seem to require some compactness assumptions, it is often convenient to consider $X_G^N = E_G^N \times_G X$ instead of $X_G = EG \times_G X$, where E_G^N is a N-connected compact free G-space. Since G is assumed to be a compact Lie group, $G < U(n)$, for some n; and so we may take $EG = EU(n)$, the infinite Steifel space $V_n(\mathbb{C}^\infty)$, and we may take $E_G^N = V_n(\mathbb{C}^{n+N})$ (see, e.g., [Husemoller, 1966]). By the Vietoris-Begle Mapping Theorem, however,

$$\bar{H}^i(X_G; k) = \bar{H}_G^i(X; k) \to \bar{H}^i(X_G^N; k),$$

the homomorphism induced by the inclusion $X_G^N \subseteq X_G$, is an isomorphism for all $i < N$, where $\bar{H}^*$ is Alexander-Spanier cohomology, k is a commutative ring with unit (or, indeed, any abelian group), and X is a paracompact G-space (see [Quillen, 1971a], Section 1). Note too that if X is a paracompact G-space and if G is a compact Lie group, then X_G is paracompact (see Section 3.2 below).

3.1 The Localization Theorem

Let G be a compact Lie group, let X be a G-space, and let k be a commutative ring with unit. When k is understood, we shall abbreviate $H^*(BG; k)$ by H_G^*. Thus $H_G^*(X; k) := H^*(X_G; k)$ is a H_G^*-module (in fact, a H_G^*-algebra in the graded sense) via $p^* : H_G^* \to H_G^*(X; k)$, where $p : X_G \to BG$ is the bundle map. Recall that by Theorem (1.2.8) the geometrically defined equivariant cohomology theory $H^*(X_G; k)$ coincides with the more algebraically defined theory $H(\beta_G^*(X; k))$ when G is finite.

We shall be most interested in the following special cases.

(3.1.1) Examples
(1) Case (0). Here G is a torus, T^r, the product of r copies of the circle group, S^1, for $r \geq 0$; and $k = \mathbb{Q}$. We shall abbreviate H_G^* further by R: thus, if $G = T^r$, then $R = \mathbb{Q}[t_1, \ldots, t_r]$, the polynomial ring over $\mathbb{Q}$ on r generators, $t_1, \ldots, t_r$, each of degree 2. If K is a subtorus of G, i.e. a closed connected subgroup of G, then we shall often abbreviate H_K^* by R_K.

(2) Case (2). Here G is an elementary abelian 2-group, or 2-torus, $(\mathbb{Z}_2)^r$, the product of r copies of the group of integers modulo 2, $\mathbb{Z}_2$, for $r \geq 0$; and $k = \mathbb{F}_2$, the field of integers modulo 2. Again H_G^* will be denoted by R; and so, if $G = (\mathbb{Z}_2)^r$, then $R = \mathbb{F}_2[t_1, \ldots, t_r]$, the polynomial

ring over $\mathbb{F}_2$ on r generators, $t_1, \ldots, t_r$, each of degree 1. If K is any subgroup of G, which will be called a subtorus of G, then H_K^* will also be denoted by R_K.

(3) Case (odd p). Here G is an elementary abelian p-group, or p-torus, $(\mathbb{Z}_p)^r$, where p is an odd prime number; and $k = \mathbb{F}_p$. Now, however,
$$H_G^* = \Lambda(s_1, \ldots, s_r) \otimes \mathbb{F}_p[t_1, \ldots, t_r],$$
the tensor product (over $\mathbb{F}_p$) of the exterior algebra over $\mathbb{F}_p$ on r generators, $s_1, \ldots, s_r$, each of degree 1, and the polynomial ring over $\mathbb{F}_p$ on r generators, $t_1, \ldots, t_r$, each of degree 2. For $1 \leq i \leq r$, $t_i = \beta_p(s_i)$, where β_p is the mod. p Bockstein operator. In this case R will denote the polynomial ring $\mathbb{F}_p[t_1, \ldots, t_r] \subseteq H_G^*$. Similarly, for any subgroup $K < G$, also called a subtorus of G, R_K will denote the polynomial part of H_K^*. We shall also let $H_K = H_K^*/\sqrt{(0)}$, H_K^* modulo the ideal of nilpotent elements. So $R_K \cong H_K$.

(3.1.2) Definition

If G is a compact Lie group and K is a closed subgroup of G, then let $[K] = \{gKg^{-1}; g \in G\}$, the conjugacy class of K. Note that for a fixed $g \in G$ the map $\tilde{g} \mapsto \tilde{g}g^{-1}$ from G to G induces a G-homeomorphism $G/K \to G/(gKg^{-1})$ of the orbits corresponding to K and gKg^{-1} respectively. Hence the conjugacy class of K determines a unique orbit type. If X is a G-space, let $\theta(X) = \{[G_x]; x \in X\}$, the set of isotropy types of X. Then X has finitely many orbit types (FMOT) if $\theta(X)$ is finite.

For any subgroup K of G, let K^0 denote the component of the identity in K. Let $\theta^0(X) = \{[G_x^0]; x \in X\}$, the set of connective isotropy types of X. Then X has finitely many connective orbit types (FMCOT) if $\theta^0(X)$ is finite. (In the Localization Theorem below, X is required to have FMOT; but when the ring k is a field of characteristic 0, as in case (0) of (3.1.1), then only FMCOT is needed. If $G = S^1$, then, clearly, any G-space has FMCOT.)

Before getting to the Localization Theorem for G-CW-complexes, we shall give two more definitions and some examples.

(3.1.3) Definition

Let X be a compact Lie group, let X be a G-space, and let k be a commutative ring with unit. For any $x \in X$, let $j_x : G(x)_G \to X_G$ be the map of Borel constructions induced by the inclusion of the orbit of x. Since
$$H_G^*(G(x); k) \cong H_G^*(G/G_x; k) \cong H^*(BG_x; k),$$

we get
$$j_x^* : H_G^*(X;k) \to H_{G_x}^*$$
and
$$j_x^* p^* : H_G^* \to H_{G_x}^*,$$
where $p : X_G \to BG$ is the bundle map as usual. Up to ismorphism, $j_x^* p^*$ depends only on the isotropy type $[G_x]$. If $f \in H_G^*(X;k)$, resp. $f \in H_G^*$, then let
$$X^f = \{x \in X; j_x^*(f) \neq 0\},$$
resp.
$$X^f = \{x \in X; j_x^* p^*(f) \neq 0\}.$$
If S is any subset of $H_G^*(X;k)$, resp. H_G^*, then let
$$X^S = \bigcap_{f \in S} X^f = \{x \in X; j_x^*(f) \neq 0, \text{ for all } f \in S\}, \text{ resp.}$$
$$\{x \in X; j_x^* p^*(f) \neq 0, \text{ for all } f \in S\}.$$

Note that if X is paracompact and one is using Alexander-Spanier cohomology, then, in either case, X^f, and hence also X^S, is a closed invariant subspace of X; this follows from the tautness property (see Section 3.2). Furthermore, if X is a G-CW-complex, then, in either case, X^f, and hence also X^S, is a G-CW-subcomplex of X.

(3.1.4) Definition
Let G and k be as in cases (0), (2) or (odd p) of Examples (3.1.1). Let K be a subtorus of G. Then the inclusion $K < G$ induces a homomorphism $R \to R_K$. (This is clear even in case (odd p), since the polynomial generators of H_G^* are the images of the exterior generators under the Bockstein operator.) Let $PK = \ker[R \to R_K]$, a prime ideal in R generated by homogeneous linear polynomials in the polynomial generators $t_1, \ldots, t_r$. Also let V be the k-linear subspace of R spanned by $t_1, \ldots, t_r$. For any prime ideal $\mathfrak{p} \triangleleft R$, let $\sigma\mathfrak{p}$ be the ideal in R generated by $\mathfrak{p} \cap V$. Then $\sigma\mathfrak{p} = PK$, for some subtorus $K < G$: and, if $\mathfrak{p} \supseteq PL$, for some subtorus $L < G$, then $PK \supseteq PL$; and so $K < L$. We call $\sigma\mathfrak{p}$ the support of $\mathfrak{p}$.

(3.1.5) Example
Again let G and k be as in cases (0), (2) or (odd p). Let X be a G-space. Let $\mathfrak{p}$ be any prime ideal in R, and let $\sigma\mathfrak{p} = PK$. Let V be as in Definition (3.1.4). Let $S(K)$ be the multiplicative subset of R generated by $V - PK$, the set of homogeneous linear polynomials not

in PK. Let $T(K)$, also a multiplicative subset of R, be $R - PK$, and let $S(\mathfrak{p}) = R - \mathfrak{p}$. So $S(K) \subseteq S(\mathfrak{p}) \subseteq T(K)$. Hence
$$X^{S(K)} \supseteq X^{S(\mathfrak{p})} \supseteq X^{T(K)}.$$
On the other hand, it is easy to see that
$$X^K \subseteq X^{T(K)}$$
and
$$X^{S(K)} \subseteq X^K,$$
where X^K is the fixed point set of the subtorus K. Thus
$$X^{S(K)} = X^{S(\mathfrak{p})} = X^{T(K)} = X^K.$$
We come now to the Localization Theorem for G-CW-complexes.

(3.1.6) Theorem
Let G be a compact Lie group, and let k be a commutative ring with unit. Let X be a finite-dimensional G-CW-complex. Assume that X has FMOT; or, if k is a field of characteristic 0, assume that X has FMCOT. Let S be a multiplicative subset of the centre of H_G^*: i.e. $fa = af$, for all $f \in S$ and all $a \in H_G^*$. Let $\phi^* : H_G^*(X; k) \to H_G^*(X^S; k)$ be induced by the inclusion of the G-CW-subcomplex $X^S \subseteq X$. Then the localized homomorphism
$$S^{-1}\phi^* : S^{-1}H_G^*(X; k) \to S^{-1}H_G^*(X^S; k)$$
is an isomorphism of $S^{-1}H_G^*$-algebras.
Furthermore, if Y is a G-CW subcomplex of X, then
$$S^{-1}H_G^*(X, Y; k) \to S^{-1}H_G^*(X^S, Y^S; k)$$
is an isomorphism of $S^{-1}H_G^*$-algebras also.

Proof.
Since localization is an exact functor (see, e.g., [Atiyah, MacDonald, 1969] or [Bourbaki, 1961] for the basic properties of localization) one can view $S^{-1}H_G^*(X; k)$ and $S^{-1}H_G^*(X^S; k)$ as equivariant cohomology theories in the variable X. Note though that these theories are not strongly additive since localization does not commute with products, in general. The inclusion $X^S \subseteq X$ gives a natural transformation
$$\tau(X) : S^{-1}H_G^*(X; k) \to S^{-1}H_G^*(X^S; k).$$
By the definition of X^S one has that $\tau(G/K)$ is an isomorphism for any 'G-point' G/K. ($S^{-1}H_G^*(G/G_x; k) = 0$ if $x \notin X^S$, since then there exists an element $f \in S$ which is mapped to $0 \in H_{G_x}^* = H_G^*(G/G_x; k)$.) Hence Theorem (1.1.3) gives the result for finite G-CW-complexes. The

more general case of finite-dimensional G-CW-complexes with FMOT follows by Remark (1.1.4)(2), since

$$S^{-1}H_G^*(\coprod G/K; k) = 0$$

for any disjoint union of copies of a fixed orbit type G/K, where $K = G_x$ for some $x \notin X^S$. Finally, when char$(k) = 0$, one has

$$S^{-1}H_G^*(\coprod_{i \in I} G/K_i; k) = 0$$

if all G/K_i represent the same connective orbit type occurring in $X - X^S$, i.e. there exists an $x \notin X^S$ such that K_i^0 is conjugate to G_x^0 for all $i \in I$ (see (3.1.7)(3)). Hence, by the same argument as above, the assumption FMCOT suffices in this case.

The result for pairs follows from the long exact sequence for pairs. □

(3.1.7) Remarks
(1) The assumption of finite-dimensionality is needed. For example the theorem does not hold if $G = S^1$, $k = \mathbb{Q}$, $X = EG$ with the usual free G-action, and $S = H_G^* - (0)$. Also the assumption FMOT (resp. FMCOT) is essential (see (3.1.16)).
(2) For a treatment of relative G-CW-complexes see [tom Dieck, 1987].
(3) The reason why FMCOT suffices when k is a field of characteristic 0 is the following. If K is any compact Lie group, if K^0 is the component of the identity in K, and if k is a field of characteristic 0, then the map $H^*(BK; k) \to H^*(BK^0; k)$, induced by the inclusion $K^0 < K$, is injective. This follows, for example, from the transfer homomorphism associated with the covering map $BK^0 \to BK$: cf. Theorem (3.10.15) and Lemma (3.9.2).
(4) Let B be a commutative ring with unit, let M be a B-module, and let $f \in B$. Let M_f, resp. B_f, denote M, resp. B, localized with respect to the multiplicative set generated by f. Now with G a compact Lie group, k a commutative ring with unit, and X a finite-dimensional G-CW-complex, let $B = H_G^{ev}(X; k)$, the commutative ring of even degree elements in $H_G^*(X; k)$; or, if $2 = 0$ in k, let $B = H_G^*(X; k)$. Then, for any $f \in B$, $H_G^*(X^f; k)$ is a B-module via the restriction $H_G^*(X; k) \to H_G^*(X^f; k)$. As in Theorem (3.1.6), it follows that the localized homomorphism $H_G^*(X; k)_f \to H_G^*(X^f; k)_f$ is an isomorphism of B_f-modules: and no restriction on the number of orbit types is needed for this.

It is interesting to ask when $X^f = X^{S(f)}$, where

$$S(f) = \{1, f, f^2, \ldots, f^n, \ldots\},$$

the multiplicative set generated by f. It is so in cases (0) and (2) of Examples (3.1.1), and for any compact Lie group G when k is a field of characteristic 0.

This follows since $H^*(BG;k)$ is an integral domain whenever k is a field of characteristic 0 (see Remark (3) above), and similarly in case (2). For the same reason, in case (odd p), $X^f = X^{S(f)}$ if $f \in R$. For general f in case (odd p), by studying nilpotent elements in H_G^*, it follows that $X^{S(f)} = X^{f^m}$, for some integer $m \geq 1$, if k is Noetherian and X has FMOT: this is because, for Noetherian k, H_G^* has finite type over k by the Quillen-Venkov theorem ([Quillen, 1971a], Theorem 2.1).

(5) (Cf. [Quillen, 1971a]) Let X be a G-CW-complex of finite G-dimension n. Let $f \in H_G^*(X;k)$ and suppose that $j_x^*(f) = 0$, for all $x \in X$, where j_x^* is the restriction to the orbit of x as in Definition (3.1.3): i.e. $X^f = \emptyset$. Then $f^{n+1} = 0$.

To prove this let $f_j \in H_G^*(X^j;k)$ be the restriction of f to the G-j-skeleton of X. Clearly $f_0 = 0$. By the long exact sequence for the pair (X^j, X^{j-1}), $f_{j-1}^j = 0$ implies that $f_j^{j+1} = 0$.

This result implies the Localization Theorem; and that is the approach which we shall use in Section 3.2. Conversely, of course, the version of the Localization Theorem given in Remark (4) above shows that f is nilpotent, if $X^f = \emptyset$.

(6) (See [tom Dieck, 1987].) The proof of Theorem (3.1.6) can also be done as follows.

Let X^n denote the G-n-skeleton of X. On X^0 the result is obvious from the definition of X^S and the assumption on the number of orbit types. Let $\{G/K_i \times D^n; i \in \mathcal{I}_n\}$ be the set of G-n-cells. Then

$$H_G^*(X^n, X^{n-1}; k) \cong \prod_{i \in \mathcal{I}_n} H_G^*(G/K_i \times D^n, G/K_i \times S^{n-1}; k)$$

$$\cong \prod_{i \in \mathcal{I}_n} H_{K_i}^*(D^n, S^{n-1}; k)$$

If a cell $G/K_j \times D^n$ is not contained in X^S, then there exists $s_j \in S$ such that s_j restricts to 0 under the map $H_G^* \to H_{K_j}^*$. So then $S^{-1} H_{K_j}^*(D^n, S^{n-1}; k) = 0$. By the assumption on the number of orbit types, there is only a finite number of distinct terms in the product. Hence the result follows.

The result for pairs follows from the long exact sequence for pairs.

The following corollary of Theorem (3.1.6) is the form of the Localization Theorem in which we are most interested. Recall the terminology and notation introduced in Examples (3.1.1) and (3.1.5) and Definition (3.1.4).

(3.1.8) Corollary

Assume case (0), (2) or (odd p). Let X be a finite-dimensional G-CW-complex. In case (0) assume that X has FMCOT. Let Y be a G-CW-subcomplex of X. Then restriction gives isomorphisms

$$S(K)^{-1}H_G^*(X,Y;k) \xrightarrow{\cong} S(K)^{-1}H_G^*(X^K,Y^K;k)$$

and

$$H_G^*(X,Y;k)_{PK} \xrightarrow{\cong} H_G^*(X^K,Y^K;k)_{PK},$$

for any subtorus $K < G$. Furthermore, if $\mathfrak{p} \subseteq R$ is a prime ideal with $\sigma\mathfrak{p} = PK$, then

$$H_G^*(X,Y;k)_\mathfrak{p} \to H_G^*(X^K,Y^K;k)_\mathfrak{p}$$

is an isomorphism also. □

(Recall that for any commutative ring R, prime ideal $\mathfrak{p} \subseteq R$, and R-module M, $R_\mathfrak{p}$ resp. $M_\mathfrak{p}$ denotes R, resp. M, localized with respect to the multiplicative set $R - \mathfrak{p}$: i.e. R, resp. M, localized at $\mathfrak{p}$.)

(3.1.9) Remarks

(1) In cases (0) and (odd p) the multiplicative sets with respect to which one localizes consist entirely of elements of even degree. Thus in the localized modules one can separate elements of even degree from those of odd degree: i.e. the localized modules are $\mathbb{Z}_2$-graded. More information concerning the grading of the cohomology of fixed point sets can be obtained by studying the filtrations: see, e.g., Corollary (3.1.14), and the proof (which is left as an exercise) that $m \leq n$ in Examples (3.1.10).
(2) Of course, if one localizes with respect to a multiplicative set consisting entirely of homogeneous elements, as is often the case, then the localized module is $\mathbb{Z}$-graded.

Here now are three simple examples of the Localization Theorem, in the guise of Corollary (3.1.8), in practice, together with an example, which shows how the Localization Theorem in its general form may fail to give any useful information.

(3.1.10) Examples

(1) In case (0), let $G = S^1$, and let X be a finite-dimensional G-CW-complex such that $H^*(X;\mathbb{Q}) \cong H^*(S^{2n};\mathbb{Q})$ for some $n > 0$. Since $H^{2n+1}(BG;\mathbb{Q}) = 0$, the Serre spectral sequence for

$$X \to X_G \to BG$$

collapses; so $H_G^*(X;\mathbb{Q})$ is a free R-module generated by $1 \in H_G^0(X;\mathbb{Q})$

and an element of degree $2n$. Localizing at the prime ideal $(0) \subseteq R$, one gets that $H_G^*(X^G; \mathbb{Q})_{(0)}$ is a $R_{(0)}$-vector-space on two elements of even degree. Since
$$H_G^*(X^G; \mathbb{Q}) \cong R \otimes H^*(X^G; \mathbb{Q}),$$
it follows that
$$H^*(X^G; \mathbb{Q}) \cong H^*(S^{2m}; \mathbb{Q})$$
for some $m \geq 0$. It is also clear from the Localization Theorem that $m \leq n$, since some non-zero R-multiple of the generator of $H^{2m}(X^G; \mathbb{Q})$ must be in the image of $H_G^*(X; \mathbb{Q})$.

(2) In case (odd p), let $G = \mathbb{Z}_p$, and let X be a finite-dimensional G-CW-complex such that
$$H^*(X; \mathbb{F}_p) \cong H^*(S^{2n}; \mathbb{F}_p)$$
for some $n > 0$. Let v generate $H^{2n}(X; \mathbb{F}_p)$. Since $\mathbb{F}_p$ has no non-trivial G-action, the coefficients in the Serre spectral sequence of $X \to X_G \to BG$ (with respect to cohomology with coefficients in $\mathbb{F}_p$) are constant. The spectral sequence is determined by
$$d_{2n+1}(v) \in H^{2n+1}(BG; \mathbb{F}_p).$$
Letting
$$H_G^* = \Lambda(s) \otimes \mathbb{F}_p[t],$$
one has that $d_{2n+1}(v) = \alpha s t^n$ for some $\alpha \in \mathbb{F}_p$. Since transgression anti-commutes with the Bockstein operator and since $\beta_p(v) = 0$, it follows that $\alpha = 0$; and so the spectral sequence collapses. (A more general way to see that $d_{2n+1}(v) = 0$ is thus:
$$0 = d_{2n+1}(v^2) = d_{2n+1}(v) \otimes 2v \in E_{2n+1}^{2n+1,2n} = E_2^{2n+1,2n}$$
$$= H^{2n+1}(BG; \mathbb{F}_p) \otimes H^{2n}(X; \mathbb{F}_p).)$$
As in Example (1) above, it follows that
$$H^*(X^G; \mathbb{F}_p) \cong H^*(S^{2m}; \mathbb{F}_p)$$
for some m, $0 \leq m \leq n$.

(3) In cases (0) and (odd p), let X be a finite-dimensional G-CW-complex such that $H^*(X; k) \cong H^*(S^{2n+1}; k)$ for some $n \geq 0$. Of course it is possible for X^G to be empty in this situation; but if $X^G \neq \emptyset$, then it follows, as in Examples (1) and (2) above, that $H^*(X^G; k) \cong H^*(S^{2m+1}; k)$ for some m, $0 \leq m \leq n$. Similarly in case (2), if X is a finite-dimensional G-CW-complex such that $H^*(X; k) \cong H^*(S^n; k)$ for some $n \geq 1$, and if $X^G \neq \emptyset$, then $H^*(X^G; k) \cong H^*(S^m; k)$ for some m, $0 \leq m \leq n$.

There is another very simple way to see these results about spheres. In case (0), (2) or (odd p), let X be a finite-dimensional G-CW-complex such that $H^*(X;k) \cong H^*(S^n;k)$ for some $n \geq 0$. Let CX be the cone on X; and extend the action of G to CX in the obvious way (so that G acts trivially on the unit interval I in the action on the cylinder $X \times I$). Then clearly $(CX)^G = C(X^G)$, where, if $X^G = \emptyset$, $C\emptyset$ is interpreted as a single point. Now $H_G^*(CX, X; k)$ is a free H_G^*-module on a single generator of degree $n+1$. Thus it follows immediately from the Localization Theorem that $\dim_k H^*(CX^G, X^G; k) = 1$, and $H^m(CX^G, X^G; k) \cong k$ for some m, $0 \leq m \leq n+1$. And, in cases (0) and (odd p), it follows immediately from the Localization Theorem that $m \equiv n+1 \pmod{2}$, since only elements of even degree are inverted.

(4) Let G be a compact connected Lie group and let $T < G$ be a maximal torus. Let $X = G/T$, and let G act on X in the standard transitive way by left transition of left cosets. Let the ring $k = \mathbb{Q}$. The bundle space $X_G \simeq BT$, and $p^* : H_G^* \to H_G^*(X;\mathbb{Q})$ is the morphism $H_G^* \to H_T^*$ induced by the map $BT \to BG$ associated with the inclusion of T in G. For any $x \in X$, the orbit $G(x) = X$, and so j_x^* is the identity. Thus, for any multiplicative subset $S \subseteq H_G^*$, $X^S = X$ if $0 \notin S$, and $X^S = \emptyset$ if $0 \in S$.

We conclude this section with some further immediate consequences of the Localization Theorem and Remark (3.1.7)(4). Throughout the remainder of the section we shall assume that X is a finite-dimensional G-CW-complex. As usual $p : X_G \to BG$ is the bundle map, but - and this makes a difference only in case (odd p) - by p^* we shall mean the induced map $R \to H_G^*(X;k)$.

In case (0), for any $x \in X$, $H_{G_x}^* \to H_{G_x^0}^*$ is injective, as in Remark (3.1.7)(3): in fact, in this case, it is an isomorphism. Thus

$$\ker[H_G^* \to H_{G_x}^*] = \ker[H_G^* \to H_{G_x^0}^*] = PG_x^0;$$

and so we shall often write PG_x instead of PG_x^0 in case (0).

(3.1.11) Corollary

$$\sqrt{(\ker p^*)} = \bigcap_{x \in X} PG_x.$$

Proof.

Let $f \in \sqrt{(\ker p^*)}$. So $p^*(f^m) = 0$, for some $m \geq 1$. Hence

$$j_x^* p^*(f^m) = 0,$$

for all $x \in X$. But

$$j_x^* p^* : R \to R_{G_x}$$

3.1. The Localization Theorem

has kernel PG_x. Thus $f \in PG_x$, for all $x \in X$, since PG_x is prime.

Conversely suppose that
$$f \in \bigcap_{x \in X} PG_x.$$
Then $j_x^* p^*(f) = 0$, for all $x \in X$. Hence $p^*(f)$ is nilpotent by Remark (3.1.7)(4): and so $f \in \sqrt{(\ker p^*)}$. □

(3.1.12) Corollary
If X has FMCOT, then $X^G \neq \emptyset$ if and only if p^* is injective.

Proof.
If $X^G \neq \emptyset$, and $x \in X^G$, then $PG_x = PG = (0)$. Hence $\ker p^* = (0)$ by Corollary (3.1.11).

Conversely suppose that $\ker p^* = (0)$. Then it follows from Corollary (3.1.11), and the assumption of FMCOT in case (0), that $PG_x = (0)$ for some $x \in X$. Hence $X^G \neq \emptyset$. □

(3.1.13) Corollary
Assume case (0), that X has FMCOT, and that
$$\dim_\mathbb{Q} H^*(X;\mathbb{Q}) < \infty.$$
Then
$$\dim_\mathbb{Q} H^*(X^G;\mathbb{Q}) < \infty,$$
and
$$\chi(X) = \chi(X^G).$$
In particular, $X^G \neq \emptyset$ if $\chi(X) \neq 0$.

Proof.
In the Serre spectral sequence for $p : X_G \to BG$ with rational cohomology we have $E_2^{*,q} \cong R \otimes H^q(X;\mathbb{Q})$; and the spectral sequence converges to $H_G^*(X;\mathbb{Q})$. Localize at $(0) \subseteq R$. Then letting $K = R_{(0)}$, the spectral sequence becomes a sequence of complexes over K, and we have
$$\chi(X^G) = \chi_K(R_{(0)} \otimes H^*(X^G;\mathbb{Q})) = \chi_K(H_G^*(X;\mathbb{Q})_{(0)})$$
$$= \chi_K(R_{(0)} \otimes_R E_\infty^{*,*}) = \chi_K(R_{(0)} \otimes_R E_2^{*,*})$$
$$= \chi(X).$$
(Note that $E_\infty = E_r$, for some finite r, since $\dim_\mathbb{Q} H^*(X;\mathbb{Q}) < \infty$.) □

(3.1.14) Corollary
As in (3.1.13), assume case (0), X has FMCOT, and $\dim_{\mathbb{Q}} H^*(X;\mathbb{Q}) < \infty$. Then, for any $m \geq 0$,

$$\sum_{i=0}^{\infty} \dim_{\mathbb{Q}} H^{m+2i}(X^G;\mathbb{Q}) \leq \sum_{i=0}^{\infty} \dim_{\mathbb{Q}} H^{m+2i}(X;\mathbb{Q}).$$

Proof.
Let $E_r(X)$, resp. $E_r(X^G)$, be the Serre spectral sequence in rational cohomology of $X_G \to BG$, resp. $(X^G)_G \to BG$. The inclusion $\phi : X^G \to X$ induces a map of spectral sequences

$$\phi_r^{*,*} : E_r(X)^{*,*} \to E_r(X^G)^{*,*}.$$

$E_r(X) = E_\infty(X)$ for some finite r, and $E_r(X^G) = E_\infty(X^G)$ for all $r \geq 2$.

Choose $r \geq 2$ such that $E_r(X) = E_\infty(X)$. Let $\mathcal{F}_q H_G^*(X;\mathbb{Q})$, $q \geq 0$, be the increasing filtration on $H_G^*(X;\mathbb{Q})$ such that

$$\mathcal{F}_q H_G^{p+q}(X;\mathbb{Q})/\mathcal{F}_{q-1} H_G^{p+q}(X;\mathbb{Q}) \cong E_\infty(X)^{p,q} = E_r(X)^{p,q}.$$

Define $\mathcal{F}_q H_G^*(X^G;\mathbb{Q})$ similarly. Then $\phi^* : H_G^*(X;\mathbb{Q}) \to H_G^*(X^G;\mathbb{Q})$ is filtration preserving, and $p^*(R) \subseteq \mathcal{F}_0 H_G^*(X;\mathbb{Q})$.

Let $K = R_{(0)}$. Then

$$\sum_{i=0}^{\infty} \dim_{\mathbb{Q}} H^{m+2i}(X^G;\mathbb{Q})$$
$$= \dim_K (H_G^*(X^G;\mathbb{Q})/\mathcal{F}_{m-1} H_G^*(X^G;\mathbb{Q}))_{(0)}^{(m)},$$

where the subscript (0) indicates, as usual, localization at (0), and the superscript (m) indicates elements of degrees congruent to m modulo 2. Since ϕ^* is filtration preserving and $\phi_{(0)}^*$ is an isomorphism, $\phi_{(0)}^*$ maps

$$(H_G^*(X;\mathbb{Q})/\mathcal{F}_{m-1} H_G^*(X;\mathbb{Q}))_{(0)}^{(m)}$$

onto

$$(H_G^*(X^G;\mathbb{Q})/\mathcal{F}_{m-1} H_G^*(X^G;\mathbb{Q}))_{(0)}^{(m)}.$$

Thus

$$\sum_{i=0}^{\infty} \dim_{\mathbb{Q}} H^{m+2i}(X^G;\mathbb{Q}) \leq \dim_K (H_G^*(X;\mathbb{Q})/\mathcal{F}_{m-1} H_G^*(X;\mathbb{Q}))_{(0)}^{(m)}$$
$$= \sum_{i=0}^{\infty} \dim_K E_r(X)_{(0)}^{*,m+2i} \leq \sum_{i=0}^{\infty} \dim_K E_2(X)_{(0)}^{*,m+2i}$$
$$= \sum_{i=0}^{\infty} \dim_{\mathbb{Q}} H^{m+2i}(X;\mathbb{Q}).$$

□

(3.1.15) Corollary
With the assumptions of Corollary (3.1.14) and the notation of its proof, $\dim_\mathbb{Q} H^*(X^G;\mathbb{Q}) = \dim_\mathbb{Q} H^*(X;\mathbb{Q})$ if and only if $E_2(X) = E_\infty(X)$, i.e. X is totally non-homologous to zero (TNHZ) in $X_G \to BG$ with respect to rational cohomomology.

We conclude this section with a simple example which shows that the assumption of FMCOT is needed in case (0).

(3.1.16) Example
Let $G = T^2$, and let $K_1, K_2, \ldots$ be the subcircles of G in some order. Let M_i be the mapping cylinder of the quotient map $G \to G/K_i$. For $i \geq 1$, let X_i be the union of M_i and M_{i+1} along G. Let X be the result of joining X_1 to X_2 along G/K_2, X_2 to X_3 along G/K_3, and so on. Thus X is a connected finite-dimensional G-CW-complex with isotropy groups $\{1\}, K_1, K_2, \ldots$. In particular $X^G = \emptyset$.

Suppose $p^*(a) = 0$. Then $p^*(a)$ restricts to zero on the orbit G/K_i, for all $i \geq 1$. Hence $a \in PK_i$ for all $i \geq 1$. So a is divisible by infinitely many distinct homogeneous linear polynomials. Thus $a = 0$; and so p^* is injective. (Cf. Corollary (3.1.12).) And $H_G^*(X;\mathbb{Q})_{(0)} \neq 0$.

3.2 The Localization Theorem for general G-spaces

In order to extend the Localization Theorem to more general G-spaces, it seems to be necessary to use a cohomology theory with good tautness properties: Alexander-Spanier cohomology, Čech cohomology, and sheaf cohomology are three such theories. The first two of these are naturally equivalent for all spaces, and all three theories are equivalent for paracompact spaces. All spaces considered in this section are paracompact.

Let G be a compact Lie group, k a commutative ring with unit, and let X be a paracompact G-space. Let $\overline{H}$ denote Alexander-Spanier cohomology. The following tautness property is crucial: if W is a closed invariant subspace of X, then
$$\overline{H}_G^*(W;k) \cong \varinjlim_N \overline{H}_G^*(N;k),$$
where, in the direct limit, N ranges over all closed (or all open) invariant neighbourhoods of W directed downwards by inclusion (see, e.g., [Quillen, 1971a]).

Note that if X is a paracompact G-space, then X_G is paracompact. $EG \times X$ is paracompact by [Bourbaki, 1966], Chapter IX, §4, Ex. 20(d). So $X_G = (EG \times X)/G$ is paracompact by a standard argument.

Consider the following cases:

(a) X is compact;
(b) X is paracompact and $\mathrm{cd}_k(X) < \infty$;
(c) X is paracompact and finitistic.

Here cd_k means sheaf cohomology dimension over k: see [Bredon, 1967a] or [Quillen, 1971a] for the definition and basic facts. X is said to be finitistic if every open covering of X has a finite-dimensional refinement, i.e. a refinement with finite-dimensional Čech nerve: see, e.g., [Bredon, 1972]. Any compact space is finitistic clearly; but, as we shall see below, the compact case is not only technically more simple than the finitistic case, it also requires no condition on the number of orbit types in the Localization Theorem - hence it is convenient to consider the compact case separately. Any paracompact space of finite covering dimension is also finitistic clearly. Indeed, according to [Hattori, 1985], a paracompact space X is finitistic if and only if there is a compact subspace $A \subseteq X$ such that B has finite covering dimension for all closed subspaces $B \subseteq X$ which are disjoint from A. Any paracompact space X of finite covering dimension also satisfies $\mathrm{cd}_k(X) < \infty$ for any commutative ring with unit k. ([Bredon, 1967a], Chapter IV, Section 7.) In particular, if G is a compact Lie group, then any finite-dimensional G-CW-complex belongs to both cases (b) and (c).

Clearly there are finitistic spaces which do not have finite sheaf cohomology dimension: e.g., the compact space $\prod_{n=1}^{\infty} S^n$. Owing to a result of Dranishnikov [1988], it is known that there are spaces in case (b) which are not finitistic.

We begin our approach to the Localization Theorem for general G-spaces with Quillen's proposition ([Quillen, 1971a]). Let G be a compact Lie group, k a commutative ring with unit, and let X be a G-space. Recall the homomorphism $j_x^* : \overline{H}_G^*(X; k) \to H_{G_x}^*$ as in Definition (3.1.3), but using Alexander-Spanier cohomology instead of singular cohomology.

(3.2.1) Proposition

Suppose that X belongs to case (a), (b) or (c) above. Let $f \in \overline{H}_G^*(X; k)$, and suppose that $j_x^*(f) = 0$, for all $x \in X$: i.e. $X^f = \emptyset$. Then f is nilpotent: i.e. $f^n = 0$, for some positive integer n.

Proof.

Case (a). This is the simplest, which is one reason for treating it separately from case (c). By the tautness property, for each $x \in X$,

3.2. Localization Theorem for general G-spaces

there is an open invariant neighbourhood $V(x)$ of the orbit $G(x)$, such that f restricts to zero in $\overline{H}_G^*(V(x); k)$.

If X is compact, then a finite number of $V(x)$s $V_1 = V(x_1), \ldots, V_n = V(x_n)$, say, cover X. By the long exact sequence, for each i, $1 \leq i \leq n$, there exists $u_i \in \overline{H}_G^*(X, V_i; k)$ such that u_i restricts to f. So $u = u_1 \ldots u_n$ restricts to f^n. But $u \in \overline{H}_G^*(X, \bigcup_{i=1}^n V_i; k) = 0$. Thus $f^n = 0$. (One can also, of course, use the Mayer-Vietoris sequence here, as in [Quillen, 1971a].)

Case (b). The key in this case is that $\mathrm{cd}_k(X) < \infty$ implies $\mathrm{cd}_k(X/G) < \infty$ ([Quillen, 1971a]). Consider the Leray spectral sequence for the map $X_G \to X/G$. Since f restricts to zero on every orbit, it follows that f is represented in $E_\infty^{p,q}$ for some $p \geq 1$. Since $\mathrm{cd}_k(X/G) < \infty$, $E_\infty^{i,*} = 0$ for all $i \geq n = \mathrm{cd}_k(X/G) + 1$. Hence $f^n = 0$. (See [Quillen, 1971a] for more details.)

Case (c). The crucial point here is that if X is paracompact and finitistic, then so too is X/G: this is the main theorem of [Deo, Tripathi, 1982]. As for case (a), for each $x \in X$, let $V(x)$ be an open invariant neighbourhood of $G(x)$, such that f restricts to zero on each $V(x)$. Let $q : X \to X/G$ be the orbit map, which is an open map. By the Deo-Tripathi theorem, the open covering $\{q(V(x)); x \in X\}$ of X/G has a finite-dimensional open refinement $\{A_\alpha; \alpha \in \mathcal{A}\}$, say. Let $\{\lambda_\alpha; \alpha \in \mathcal{A}\}$ be a partition of unity subordinate to $\{A_\alpha; \alpha \in \mathcal{A}\}$, and let $B_\alpha = \lambda_\alpha^{-1}(0, 1] \subseteq A_\alpha$. Then $\mathcal{B} = \{B_\alpha; \alpha \in \mathcal{A}\}$ is a finite-dimensional locally finite refinement of $\{A_\alpha; \alpha \in \mathcal{A}\}$.

Let $\mathcal{C} = \{C_\alpha; \alpha \in \mathcal{A}\}$ be a shrinking of $\mathcal{B}$: in particular $\overline{C}_\alpha \subseteq B_\alpha$, for all $\alpha \in \mathcal{A}$, and $\mathcal{C}$ is finite-dimensional and locally finite. Let $Y_\alpha = q^{-1}(B_\alpha)$ and $Z_\alpha = q^{-1}(C_\alpha)$. Then $\mathcal{Y} = \{Y_\alpha; \alpha \in \mathcal{A}\}$ and $\mathcal{Z} = \{Z_\alpha; \alpha \in \mathcal{A}\}$ are finite-dimensional locally finite coverings of X by invariant open sets; and $\mathcal{Z}$ is a shrinking of $\mathcal{Y}$.

For $\alpha \in \mathcal{A}$, let $\nu_\alpha : X/G \to [0, 1]$ be an Urysohn function such that $\nu_\alpha(\overline{C}_\alpha) = 1$ and $\nu_\alpha(X/G - B_\alpha) = 0$. Let $\mu_\alpha = \nu_\alpha q : X \to [0, 1]$. Then $\mu_\alpha(\overline{Z}_\alpha) = 1$ and $\mu_\alpha(X - Y_\alpha) = 0$. As in [Milnor, Stasheff, 1974], pp. 66-7, for any finite non-empty set $S \subseteq \mathcal{A}$, let

$$U(S) = \{x \in X; \min_{\alpha \in S} \mu_\alpha(x) > \max_{\alpha \notin S} \mu_\alpha(x)\}.$$

Thus $U(S) \subseteq \bigcap_{\alpha \in S} Y_\alpha$; and, owing to the finite dimensionality of Y, there is an integer n, such that $U(S) = \emptyset$, whenever the number of elements in S, $|S|$, exceeds n.

Since $\mathcal{Y}$ is locally finite and $\mathcal{Z}$ is a covering, $\{U(S); S \subseteq \mathcal{A} \text{ and } |S| < \infty\}$ is an invariant open covering of X. Furthermore, if $|S_1| = |S_2|$ and

$S_1 \neq S_2$, then $U(S_1) \cap U(S_2) = \emptyset$, since $x \in U(S_1) \cap U(S_2)$ implies that
$$\min_{\alpha \in S_1} \mu_\alpha(x) > \max_{\alpha \notin S_1} \mu_\alpha(x) \geq \min_{\alpha \in S_2} \mu_\alpha(x) > \max_{\alpha \notin S_2} \mu_\alpha(x) \geq \min_{\alpha \in S_1} \mu_\alpha(x).$$
Let $U_i = \bigcup \{U(S); |S| = i\}$. Thus $\{U_1, \ldots, U_n\}$ is a covering of X by invariant open sets. Since U_i is a disjoint union of $U(S)$s,
$$\overline{H}_G^*(U_i; k) \cong \prod_{|S|=i} \overline{H}_G^*(U(S); k).$$
Further, since $U(S) \subseteq \bigcap_{\alpha \in S} Y_\alpha \subseteq$ some $Y_\alpha \subseteq$ some $V(x)$, f restricts to zero on each $U(S)$; and hence f restricts to zero on each U_i. One may now proceed exactly as for case (a), and conclude that $f^n = 0$. □

(3.2.2) Remark
If $f \in \mathrm{im}[p^* : H_G^* \to H_G^*(X;k)]$, then Proposition (3.2.1) for cases (a) and (c) is valid with singular cohomology instead of Alexander-Spanier cohomology. This is because, for such f, one can use the existence of tubes and slices in the place of the tautness property. Also Proposition (3.2.1) for cases (a) and (c) and any $f \in H_G^*(X;k)$ is valid using singular cohomology if, for every $x \in X$, there is a slice at x, which is k-acyclic with respect to singular cohomology.

As usual let G be a compact Lie group, let k be a commutative ring with unit, and let X be a G-space. As in Remark (3.1.7)(4), let $B = \overline{H}_G^{\mathrm{ev}}(X;k)$, or, if $2 = 0$ in k, let $B = \overline{H}_G^*(X;k)$.

(3.2.3) Lemma
Suppose that X belongs to case (a), (b) or (c) above. Let Y be a closed invariant subspace of X. Let f be in the centre of H_G^*, resp. $f \in B$. Then the localized restriction homomorphism
$$\overline{H}_G^*(X, Y; k)_f \to \overline{H}_G^*(X^f, Y^f; k)_f$$
is an isomorphism of $(H_G^*)_f$-algebras, resp. B_f-algebras.
Proof.
By the long exact sequence and the exactness of localization, it is enough to prove the result when $Y = \emptyset$. If $X^f = \emptyset$, then, by Proposition (3.2.1), $p^*(f)$, resp. f, is nilpotent; and so $\overline{H}_G^*(X;k)_f = 0$.

If $X^f \neq \emptyset$, then let V be a closed invariant neighbourhood of X^f, and let W be the complement of the interior of V. So $W^f = (V \cap W)^f = \emptyset$. Thus from the Mayer-Vietoris sequence
$$\ldots \to \overline{H}_G^j(X;k) \to \overline{H}_G^j(V;k) \oplus \overline{H}_G^j(W;k) \to \overline{H}_G^j(V \cap W; k) \to \ldots$$
by the exactness of localization, and by the first paragraph above, one gets an isomorphism $\overline{H}_G^*(X;k)_f \stackrel{\cong}{\to} \overline{H}_G^*(V;k)_f$. The result now follows

from the tautness property, since localization commutes with direct limits. □

We can now distinguish three cases.

(3.2.4) Definition
Let G be a compact Lie group, k a commutative ring with unit, and let X be a G-space. We shall be concerned with the following three cases.

Case(A): X is compact.
Case(B): X is paracompact, $\mathrm{cd}_k(X) < \infty$, and X has FMOT or, if k is a field of characteristic 0, X has FMCOT.
Case(C): X is paracompact and finitistic, and X has FMOT or, if k is a field of characteristic 0, X has FMCOT.

(3.2.5) Remark

Note that no assumption on the number of orbit types is made in case (A). On the other hand, the following theorem of [Mann, 1962] is relevant to cases (B) and (C): if M is an orientable topological manifold, if $H_*(M; \mathbb{Z})$ is finitely generated, and if any compact Lie group is acting on M, then M has FMOT.

We come now to the general form of the Localization Theorem. As usual let G be a compact Lie group and let k be a commutative ring with unit.

(3.2.6) Theorem
Suppose that X is a G-space belonging to case (A), (B) or (C) of Definition (3.2.4). Let Y be a closed invariant subspace of X, and let $S \subseteq H_G^*$ be a multiplicative subset of the centre of H_G^*. Then the localized restriction homomorphism

$$S^{-1}\phi^* : S^{-1}\overline{H}_G^*(X, Y; k) \to S^{-1}\overline{H}_G^*(X^S, Y^S; k)$$

is an isomorphism of $S^{-1}H_G^*$-algebras.
Proof.
As before, it is enough to prove the result when $Y = \emptyset$. We begin with cases (B) and (C). Let $x_1, \ldots, x_n$ be points in X, the isotropy subgroups of which (resp. if k is a field of characteristic 0, the connective isotropy subgroups of which) form a complete set of representatives of the isotropy types (resp. connective isotropy types) not in X^S. So, for each i, $1 \leq i \leq n$, there is $s_i \in S$ such that $j_{x_i}^*(s_i) = 0$. Let $s = s_1 \ldots s_n$. Then $X^S = X^s$; and the result follows from Lemma (3.2.3).

Now suppose that X is compact (case (A)). Make S into a directed set by the partial ordering $f \leq g$ if g is a multiple of f. So $f \leq fg$ and $g \leq fg$. Since $X^S = \bigcap_{f \in S} X^f$, by the continuity property of Alexander-Spanier cohomology ([Spanier, 1966]. Chapter 6, Theorem 6.6),
$$\overline{H}^*(X^S; k) \cong \varinjlim_{f \in S} \overline{H}^*(X^f; k).$$
To extend this to equivariant cohomology, and to take care of the compactness requirement of the continuity property, let E_G^N be the first skeleton of EG which is N-connected, where N is some large integer. Let $X_G^N = E_G^N \times_G X$. By the continuity,
$$\overline{H}^*((X^S)_G^N; k) \cong \varinjlim_{f \in S} \overline{H}^*((X^f)_G^N; k).$$
But, for any paracompact G-space Y, $\overline{H}_G^i(Y; k) \cong \overline{H}^i(Y_G^N; k)$, for all $i < N$, (see [Quillen, 1971a]). Hence $\overline{H}_G^*(X^S; k) \cong \varinjlim_{f \in S} \overline{H}_G^*(X^f; k)$, since N is arbitrary: and the theorem follows since localization commutes with direct limits. □

It is clear that the argument for case (A) in Theorem $(3.2.6)$ works equally well for any multiplicative set $S \subseteq B$, where B is as in Lemma $(3.2.3)$. So we have the following corollary of the proof of Theorem $(3.2.6)$.

(3.2.7) Corollary
Let G, k, X and Y be as in case (A) of Theorem $(3.2.6)$. Let $S \subseteq B$ be any multiplicative set. Then the localized restriction homomorphism
$$S^{-1}\overline{H}_G^*(X, Y; k) \to S^{-1}\overline{H}_G^*(X^S, Y^S; k)$$
is an isomorphism of $S^{-1}B$-algebras.

(3.2.8) Remark
Let G, k, X and Y be as in Theorem $(3.2.6)$. Let T be a multiplicative set contained in $B \cap \text{im}[p^* : H_G^* \to \overline{H}_G^*(X; k)]$, and let $f \in B$. Let $S = \{tf^m ; t \in T, m \geq 0\}$ be the multiplicative subset of B generated by T and f. Then combining Lemma $(3.2.3)$ and Theorem $(3.2.6)$, we have in cases (B) and (C) that
$$S^{-1}\overline{H}_G^*(X, Y; k) \to S^{-1}\overline{H}_G^*(X^T \cap X^f, Y^T \cap Y^f; k)$$
is an isomorphism of $S^{-1}B$-algebras. Quite generally, if k is a field of characteristic 0, then $X^T \cap X^f = X^S$; and, if X has FMCOT, then $X^S = X^{tf}$, for some $t \in T$. In case (2) of Examples $(3.1.1)$ also $X^T \cap X^f = X^S$ and $X^S = X^{tf}$, for some $t \in T$. In case (odd p) $X^S = X^T \cap X^{f^m}$, for

some integer $m > 0$, and $X^S = X^{tf^m}$, for some $t \in T$, (see Remarks (3.1.7)(4)). Thus, with S as above, in cases (B) and (C),

$$S^{-1}\overline{H}_G^*(X,Y;k) \to S^{-1}\overline{H}_G^*(X^S,Y^S;k)$$

is an isomorphism of $S^{-1}B$-algebras, whenever k is a field of characteristic 0, or G, k and X are as in cases (2) or (odd p).

(3.2.9) Remark

There are obvious analogues of Corollaries (3.1.11)-(3.1.15). The assumption of FMCOT in Corollaries (3.1.12)-(3.1.15) is not needed in case (A). This is clear for Corollaries (3.1.13)-(3.1.15). For Corollary (3.1.12) one can use the Localization Theorem (p^* injective $\Rightarrow H_G^*(X;\mathbb{Q})_{(0)} \neq 0$), or one can use the fact that in case (A) the number of maximal connective isotropy subgroups (and, hence, the number of minimal PG_x) is finite. To see the latter, let $\mathfrak{p}$ be any prime ideal of R which contains $\ker p^*$. Suppose that $\mathfrak{p} \not\supseteq PG_x$, for all $x \in X$. Then there is, for all $x \in X$, $s_x \in PG_x$ such that $s_x \notin \mathfrak{p}$. By tautness, for all $x \in X$, there is an open invariant neighbourhood $V(x)$ of $G(x)$ such that $p^*(s_x)$ restricts to zero on $V(x)$. In case (A), there exist $x_1, \ldots, x_n \in X$ such that $X = \bigcup_{i=1}^n V(x_i)$. Let $s = s_{x_1} \ldots s_{x_n}$. Then $p^*(s)$ restricts to zero on every orbit of X; and, hence, $p^*(s)$ is nilpotent by Proposition (3.2.1). So some positive power of s is in $\ker p^*$, and, hence, in $\mathfrak{p}$ - a contradiction. It follows that the minimal prime ideals of R which contain $\ker p^*$ all have the form PG_x, for certain $x \in X$. And these minimal prime ideals are finite in number since R is Noetherian. Hence in cases (A), (B) or (C), combined with cases (0), (2) or (odd p), there are $x_1, \ldots, x_m \in X$ such that $\sqrt{(\ker p^*)} = \bigcap_{i=1}^m PG_{x_i}$.

An argument similar to the one just given yields the following.

(3.2.10) Lemma

Suppose that G and X belong to case (A) and to cases (0), (2) or (odd p). Let $u \in H_G^*(X;k)$. Then there exist $a_1, \ldots, a_m \in R$ such that $(u - p^*(a_1)) \ldots (u - p^*(a_m))$ is nilpotent. In particular $H_G^*(X;k)$ is an integral extension of the ring $p^*(R)$.

Proof.

For each $x \in X$, consider $j_x^*(u) \in H_{G_x}^*$, resp. in case (odd p), consider the image of $j_x^*(u)$ in $H_{G_x} = H_{G_x}^*/\sqrt{(0)} \cong R_{G_x}$. Since $R \to R_{G_x}$ is onto, there is $a_x \in R$ such that $j_x^*(u - p^*(a_x))$ is zero, resp. nilpotent. So in all cases, some power of $u - p^*(a_x)$ restricts to zero on an open invariant neighbourhood $V(x)$ of $G(x)$. Hence the result follows as above. □

We conclude this section with a simple example, which shows the importance of using Alexander-Spanier cohomology instead of singular cohomology when dealing with general G-spaces.

(3.2.11) Example
Let $X = \prod_{n=1}^{\infty} S^{2n}$, and let $G = S^1$ act on X by means of standard rotations on each factor. X^G is then the set of all sequences $(x_1, x_2, \ldots, x_n, \ldots)$, where, for each n, x_n is either the north pole or the south pole of S^{2n}. Thus X^G is compact and totally disconnected, and its cardinality is c, the power of the continuum. In particular, X^G is profinite: it is the inverse limit of a countable system of finite discrete spaces. So, by the continuity property of Alexander-Spanier cohomology, $\overline{H}^*(X^G; \mathbb{Q})$ has countably infinite dimension over $\mathbb{Q}$. On the other hand, the singular cohomology $H^*(X^G; \mathbb{Q})$ has dimension $\aleph_0^c$ - most certainly uncountable. Note too that X is an agreement space in the sense of Definition (2.6.2), even though X is not locally contractible (see Remarks (4.3.13)(1)).

If we let $G = T^2 = S^1 \times S^1$, and let $K_1, K_2, \ldots$ be the subcircles of G in some order, then we can let G act on X so that the action on the factor S^{2n} is standard rotations by the quotient circle G/K_n. Now each K_n is an isotropy subgroup; and so G is acting on the compact space X without FMCOT.

3.3 Equivariant rational homotopy

In this section we define equivariant rational homotopy and prove a localization theorem for it. Although the main application is to torus actions, we shall work in a more general setting in keeping with Theorem (3.1.6). Throughout this section G will be a compact connected Lie group, and, unless otherwise stated, X will be a connected finite-dimensional G-CW-complex with finitely many connective orbit types.

(3.3.1) Definitions
Suppose that $X^G \neq \emptyset$, and let $x_0 \in X^G$. As in Examples (2.5.7)(2), x_0 induces a section $s : BG \to X_G$, and we have a commutative diagram (Diagram 3.1)

$$\begin{array}{ccc} R \otimes \mathcal{M}(X) & \xrightarrow{\varepsilon} & R \\ \downarrow & & \downarrow \\ A(X_G) & \xrightarrow{A(s)} & A(BG) \end{array}$$

Diagram 3.1

3.3. Equivariant rational homotopy

where $R = H_G^* = H^*(BG; \mathbb{Q})$. Let

$$Q_{(G,x_0)}(X) = \ker \varepsilon / (\ker \varepsilon)^2.$$

This is a free graded R-module, isomorphic to $R \otimes \Pi_\psi^*(X)$, with a differential inherited from $R \otimes \mathcal{M}(X)$. Denote its homology R-module by $\Pi_{(G,x_0)}^*(X)$: this is the equivariant (pseudo-dual) rational homotopy of X based at x_0. Up to isomorphism, $\Pi_{(G,x_0)}^*(X)$ is independent of the particular choice of augmentation, ε, representing $A(s)$.

Let K be another compact connected Lie group, let Y be a K-CW-complex with $Y^K \neq \emptyset$, and let $y_0 \in Y^K$. Suppose that $\psi : G \to K$ is a homomorphism, and that $f : X \to Y$ is a ψ-equivariant map (i.e. $f(gx) = \psi(g)f(x)$, for all $x \in X$, $g \in G$) such that $f(x_0) = y_0$. Then f induces a H_K^*-module homomorphism

$$f^* : \Pi_{(K,y_0)}^*(Y) \to \Pi_{(G,x_0)}^*(X).$$

Furthermore, for fixed choices of minimal models $H_G^* \to A(BG)$ and $H_K^* \to A(BK)$, f^* is independent of all other choices used to define it, up to isomorphism of homomorphisms. One way to see this is to combine the discussion of functoriality of $H_G^*(X)$ in [Quillen, 1971a] with the method of [Allday, 1979], Section 4. (The assertion in the latter reference that f^* is independent of the choices of base-points, within particular components, is not proven.)

Let $S \subseteq R = H_G^*$ be a multiplicative subset with $0 \notin S$. Then $X^S \neq \emptyset$. Let $F(S)$ be the component of X^S which contains the basepoint x_0. Since G is connected, $F(S)$ is a G-CW-subcomplex of X. The inclusion $\phi : F(S) \to X$ induces a R-module homomorphism

$$\phi^* : \Pi_{(G,x_0)}^*(X) \to \Pi_{(G,x_0)}^*(F(S)).$$

The following is a general form of the Localization Theorem for equivariant rational homotopy.

(3.3.2) Theorem
With the above notation and assumptions, the localized homomorphism of $S^{-1}R$-modules

$$S^{-1}\phi^* : S^{-1}\Pi_{(G,x_0)}^*(X) \to S^{-1}\Pi_{(G,x_0)}^*(F(S))$$

is an isomorphism.

To prove this theorem, we shall begin with some lemmata. First let $\{F(S)_i; i \in I\}$ be the set of components of X^S: in particular $F(S) = F(S)_j$ for some $j \in I$. Consider the induced homomorphism

$$\psi^* : H_G^*(X;\mathbb{Q}) \to H_G^*(X^S;\mathbb{Q}) \cong \prod_{i\in I} H_G^*(F(S)_i;\mathbb{Q}).$$

Let p be the projection of the latter product onto $H_G^*(F(S);\mathbb{Q})$. Let $\psi_0^* = p\psi^*$: i.e. ψ_0^* is induced by the inclusion $F(S) \to X$. Let $f \in \Pi_{i\in I} H_G^*(F(S)_i;\mathbb{Q})$ be the element such that $p(f) = 1$ and all other projections of f are 0. Let $e = f/1 \in S^{-1}\Pi_{i\in I}H_G^*(F(S)_i;\mathbb{Q})$. By Theorem (3.1.6), there is $x \in S^{-1}H_G^*(X;\mathbb{Q})$ such that $S^{-1}\psi^*(x) = e$. Since $S^{-1}p(e)$ is a unit, $S^{-1}\psi_0^*$ induces a homomorphism

$$\theta : S^{-1}H_G^*(X;\mathbb{Q})[x^{-1}] \to S^{-1}H_G^*(F(S);\mathbb{Q}).$$

(3.3.3) Lemma
θ is an isomorphism.

Proof.
Clearly θ is onto. Suppose $\theta(u/x^n) = 0$. So $S^{-1}\psi_0^*(u) = 0$. Hence $eS^{-1}\psi^*(u) = 0$: and so, $S^{-1}\psi^*(xu) = 0$. By the Localization Theorem, $xu = 0$. Thus $u/x^n = 0$. So θ is one-to-one.

(Note that this lemma is an example of the extended localization theorem discussed in Corollary (3.2.7) and Remark (3.2.8).) □

Next consider the minimal model $R \otimes \mathcal{M}(X)$ for X_G. By the exactness of localization, $H(S^{-1}R \otimes \mathcal{M}(X)) \cong S^{-1}H_G^*(X;\mathbb{Q})$. Since the localization may have lost some information about the grading except for the distinction between even and odd degrees (see Remark (3.1.9)), we view $S^{-1}R \otimes \mathcal{M}(X)$ as a differential graded commutative algebra graded over $\mathbb{Z}_2$ which we shall call a D2GCA.

Let $A = A^0 + A^1$ be a D2GCA, where A^i is the set of elements of degree i ($i = 0,1$), and let $x \in H^0(A)$ be represented by a cycle $u \in A^0$. Let V be the free 2GCA generated by an element $v \in V^0$ and an element $w \in V^1$. Form $A \otimes V$ and extend the differential by setting $dv = 0$ and $dw = uv - 1$. Then, under the inclusion $A \to A \otimes V$, x becomes a unit in $H(A \otimes V)$. Hence there is induced a homomorphism $\alpha : H(A)[x^{-1}] \to H(A \otimes V)$, where $\alpha(x^{-1}) = \langle v \rangle$, the homology class of v.

(3.3.4) Lemma
α is an isomorphism.

3.3. Equivariant rational homotopy

Proof.
A typical element $y \in A \otimes V$ has the form
$$y = a_0 + a_1 v + \ldots + a_n v^n + (b_0 + b_1 v + \ldots + b_n v^n)w,$$
where $a_i, b_i \in A$, $0 \leq i \leq n$.
$$dy = \sum_{m=0}^{n} da_m v^m + (\sum_{m=0}^{n} db_m v^m)w + (-1)^k (\sum_{m=0}^{n} b_m v^m)(uv - 1),$$
where $k = \deg(b_0)$. So
$$dy = \sum_{m=0}^{n+1} \{da_m + (-1)^k(b_{m-1}u - b_m)\}v^m + (\sum_{m=0}^{n} db_m v^m)w$$
where $b_{-1} = b_{n+1} = a_{n+1} = 0$.

In particular, if y is a cycle, then $b_m = d(\sum_{p=0}^{m} a'_p u^{m-p})$, where $a'_p = (-1)^k a_p$ for $0 \leq m \leq n$, and $b_n u = 0$. Thus $y = \sum_{m=0}^{n} a_m v^m + dfw$, where $f \in A[v]$; and hence, $y = g + d(fw)$, where $g \in A[v]$ is a cycle of the form $\sum_{m=0}^{n+1} c_m v^m$ with each c_m, $0 \leq m \leq n+1$, also a cycle. Hence, in homology,
$$\langle y \rangle = \sum_{m=0}^{n+1} \langle c_m \rangle \langle v \rangle^m = \alpha(\sum_{m=0}^{n+1} \langle c_m \rangle x^{-m}).$$
Thus α is onto.

Now suppose $\alpha(\langle c \rangle x^{-m}) = 0$, where c is a cycle in A. Then $cv^m = dy$, for some $y \in A \otimes V$. With y as above, it follows that $c = da_m + (-1)^k(b_{m-1}u - b_m)$, and $da_j + (-1)^k(b_{j-1}u - b_j) = 0$, for $j \neq m$. This implies that b_{m-1} and $b_m u^{n-m+1}$ are boundaries; and, hence, $\langle cu^{n-m+1} \rangle = 0$: i.e. $\langle c \rangle x^{n-m+1} = 0$. So $\langle c \rangle x^{-m} = 0$: and so α is one-to-one. $\square$

The augmentation $\varepsilon : R \otimes \mathcal{M}(X) \to R$ in Definition (3.3.1) gives rise to an augmentation
$$\eta = S^{-1}\varepsilon : S^{-1}R \otimes \mathcal{M}(X) \to S^{-1}R.$$
Thus, abbreviating $S^{-1}R$ by K, $K \otimes \mathcal{M}(X)$ together with η may be viewed as a differential $\mathbb{Z}_2$-graded commutative K-algebra augmented over K: let us call the category of such objects $\mathcal{A}$. Now we may define a notion of homotopy in $\mathcal{A}$, by analogy with that for CGDAs given in Definitions (2.2.4)(4) and Definition (2.3.4).

We define a KS-complex in $\mathcal{A}$ just as in Definition (2.2.2), (i)-(iii), where now $S(E)$ means the free $\mathbb{Z}_2$-graded commutative K-algebra on some well-ordered basis E: indeed if $E = \{x_i; i \in I\} \cup \{y_j; j \in J\}$ with $\deg(x_i) = 0$, $\forall i \in I$, and $\deg(y_j) = 1$, $\forall j \in J$, then $S(E) = K[x_i; i \in I] \otimes \Lambda_K(y_j; j \in J)$, the polynomial algebra over K on the x_is tensored with the exterior algebra over K on the y_js. Without loss of generality, replacing x_i by $x_i - \eta(x_i)$, if necessary, we may assume that $E \subseteq \ker \eta$. If A is an object in $\mathcal{A}$ augmented by $\eta : A \to K$, then we set $QA = \ker \eta/(\ker \eta)^2$; QA inherits a differential from A, and we define the homotopy of A, $\pi(A)$, to be the $\mathbb{Z}_2$-graded K-module $H(QA)$.

To define the notion of homotopy between morphisms in $\mathcal{A}$, we let ∇ be the free D2GCA generated by a single element s of degree 0 and a single element t of degree 1, with differential given by $ds = t$. For an object B in $\mathcal{A}$ augmented by η_B we set $B^I = K \otimes K + \ker \eta_B \otimes \nabla$: clearly this is an object in $\mathcal{A}$. Let $d_0, d_1 : B^I \to B$ be the morphisms such that $d_0(s) = 0$, $d_1(s) = 1$, and d_0 and d_1 act like the identity on elements coming from B. In particular, $d_0 j_B$ and $d_1 j_B$ are both the identity on B, where $j_B : B \to B^I$ is the morphism given by $j_B(x) = \eta_B(x) \otimes 1 + (x - \eta_B(x)) \otimes 1$. Now suppose that A, augmented by η_A, is another object of $\mathcal{A}$, and $f, g : A \to B$ are two morphisms. Then we say that f and g are homotopic, $f \simeq g$, if there exists a morphism $H : A \to B^I$ such that $d_0 H = f$ and $d_1 H = g$.

(3.3.5) Lemma
Let A and B be KS-complexes in $\mathcal{A}$, and suppose that $f : A \to B$ is a morphism such that the induced homology map, $f_* : H(A) \to H(B)$ is an isomorphism. Then there exists a morphism $g : B \to A$ such that $fg \simeq \mathrm{id}_B$.

Proof.
Let $\{x_j; j \in \mathcal{J}\}$ be a well-ordered basis for B, with $x_j \in \ker \eta_B$, $\forall j \in \mathcal{J}$. Let $0 \in \mathcal{J}$ be the first element: and so $dx_0 = 0$. Hence there is a cycle $a \in A$ and element $b \in \ker \eta_B$ such that $f(a) = x_0 + db$. Set $g(x_0) = a$ and $H(x_0) = x_0 + db \otimes s + (-1)^\beta b \otimes t$, where $\beta = \deg(b)$.

Now suppose that g and H have been extended to B_j, the subalgebra of B generated by $\{x_i; i < j\}$. Let
$$H(dx_j) = dx_j + u_1 s + \ldots + u_n s^n + (v_0 + v_1 s + \ldots + v_n s^n)t,$$
where $u_i, v_i \in \ker \eta_B$ for $1 \leq i \leq n$. $H(dx_j)$ is defined since $dx_j \in B_j$, and the inductive assumption is that $d_1 H(dx_j) = dx_j + u_1 + \ldots + u_n = fg(dx_j)$. Since $dH(dx_j) = 0$, it follows that $dv_{m-1} = (-1)^\gamma m u_m$, for $1 \leq m \leq n$, where $\gamma = \deg(v_0) = \ldots = \deg(v_n)$, and $dv_n = 0$. Thus
$$H(dx_j) = dx_j + d(\sum_{m=0}^{n}(-1)^\gamma \frac{1}{m+1} v_m s^{m+1}).$$

Set
$$z = x_j + \sum_{m=0}^{n}(-1)^\gamma \frac{1}{m+1}v_m.$$

Then $dz = fg(dx_j)$. Since f_* is an isomorphism, and $g(dx_j)$ is a cycle in A, it follows that there exists $y_j \in A$ such that $dy_j = g(dx_j)$. So $z - f(y_j)$ is a cycle in B; and, hence, there exist a cycle $c_j \in A$ and an element $w \in \ker \eta_B$, such that $f(c_j) = z - f(y_j) + dw$.

Now set $g(x_j) = y_j + c_j$ and
$$H(x_j) = x_j + \sum_{m=0}^{n}(-1)^\gamma \frac{1}{m+1}v_m s^{m+1} + dws + (-1)^{\gamma+1}wt.$$

Then $d_0 H(x_j) = x_j$ and $d_1 H(x_j) = z + dw = fg(x_j)$. Clearly $dH(x_j) = H(dx_j)$. Hence the lemma follows by induction. □

Given a morphism $f : A \to B$, let $f_\# : \pi A \to \pi B$ be the induced homomorphism on homotopy.

(3.3.6) Lemma
If $f, g : A \to B$ are two morphisms in $\mathcal{A}$ such that $f \simeq g$, then $f_\# = g_\#$.

Proof.
Let $H : A \to B^I$ be a homotopy between f and g. Since $Q(B^I) \cong Q(B) \otimes \nabla$, it follows easily by a typical argument that $j_{B\#} : \pi(B) \to \pi(B^I)$ is an isomorphism. Thus $d_{0\#} = d_{1\#} = j_{B\#}^{-1}$. So $f_\# = d_{0\#}H_\# = d_{1\#}H_\# = g_\#$. □

(3.3.7) Proposition
Let $f : A \to B$ be a morphism of KS-complexes in $\mathcal{A}$, and suppose that $f_* : H(A) \to H(B)$ is an isomorphism. Then $f_\#$ is an isomorphism.

Proof.
By Lemma (3.3.5) there exists $g : B \to A$ such that $fg \simeq \mathrm{id}_B$. Hence $f_* g_* = \mathrm{id}_{H(B)}$; and so g_* is an isomorphism. By Lemma (3.3.5) again there exists $h : A \to B$ such that $gh \simeq \mathrm{id}_A$. So, by Lemma (3.3.6), $f_\# g_\# = \mathrm{id}_{\pi B}$ and $g_\# h_\# = \mathrm{id}_{\pi A}$. Thus $g_\#$ is an isomorphism; and so $f_\#$ is an isomorphism. □

(3.3.8) Remarks
(1) In the proof above, we used the fact that if $f, g : A \to B$ are homotopic morphisms in $\mathcal{A}$, then $f_* = g_*$. This follows as in Lemma (3.3.6), since j_{B*} is an isomorphism, by another straightforward argument.

(2) Proposition (3.3.7) can be proved by homotopical algebra [Quillen, 1967]. One can prove that $\mathcal{A}$ is a closed model category by imitating the proof for CGDAs in [Bousfield, Gugenheim, 1976], Chapter 4.

Now the morphism $f : A \to B$ in Proposition (3.3.7) is a weak-equivalence of cofibrant-fibrant objects: hence it is a homotopy equivalence by [Quillen, 1967]. In particular, $g : B \to A$ can be chosen so that $fg \simeq \mathrm{id}_B$ and $gf \simeq \mathrm{id}_A$. (This also follows from the proof of Proposition (3.3.7), since there $fg \simeq \mathrm{id}_B$ and $gh \simeq \mathrm{id}_A$, hence $f = f\,\mathrm{id}_A \simeq f(gh) = (fg)h \simeq \mathrm{id}_B h = h$. So $gf \simeq gh \simeq \mathrm{id}_A$. Here, however, the assertions that $gh \simeq \mathrm{id}_A$ implies $f(gh) \simeq f\,\mathrm{id}_A$ and $f \simeq h$ implies $gf \simeq gh$ are not trivial.)

Proof of Theorem (3.3.2.).

The inclusions of the base-point x_0 in X and $F(S)$ give rise to R-algebra augmentations $\varepsilon_1 : R \otimes \mathcal{M}(X) \to R$ and $\varepsilon_2 : R \otimes \mathcal{M}(F(S)) \to R$. Furthermore ϕ induces an augmentation preserving homomorphism $f : R \otimes \mathcal{M}(X) \to R \otimes \mathcal{M}(F(S))$. Localizing with respect to S we have $g = S^{-1}f : K \otimes \mathcal{M}(X) \to K \otimes \mathcal{M}(F(S))$ preserving the augmentations $\eta_i = S^{-1}\varepsilon_i$, $i = 1, 2$, where $K = S^{-1}R$ as above. Let $A = K \otimes \mathcal{M}(X)$, $B = K \otimes \mathcal{M}(F(S))$, both of which are KS-complexes in the category $\mathcal{A}$, and let $u \in A$ be a cycle representing the element $x \in S^{-1}H_G^*(X;\mathbb{Q})$ chosen above. Then $g(u) = 1 + dy$, for some $y \in B$.

Form $A \otimes V$ as above Lemma (3.3.4), and extend g to $\tilde{g} : A \otimes V \to B$ by $\tilde{g}(v) = 1$ and $\tilde{g}(w) = y$; $\tilde{g}$ is augmentation preserving if we extend η_1 to $\tilde{\eta}_1 : A \otimes V \to K$ by setting $\tilde{\eta}_1(v) = 1$. From the universal mapping property for the localization $j : H(A) \to H(A)[x^{-1}]$ it follows that $\tilde{g}_*\alpha = \theta$, since $\tilde{g}_*\alpha j = \theta j$ (α, θ as in Lemmata (3.3.4) and (3.3.3)). Thus, by Lemmata (3.3.3) and (3.3.4), $\tilde{g}$ is a weak-equivalence (i.e. $\tilde{g}_*$ is an isomorphism) of KS-complexes in $\mathcal{A}$; and, hence, by Proposition (3.3.7), $\tilde{g}_\#$ is an isomorphism.

Let $i : A \to A \otimes V$ be the inclusion. Let $\tilde{u} = u - 1$, $\tilde{v} = v - 1$ and $t = \tilde{u} + \tilde{v}$. Then $A \otimes V = A \otimes \mathbb{Q}[t] \otimes \Lambda(w)$ and $dw = t$, modulo $(\ker \tilde{\eta}_1)^2$. A very simple computation now shows that $i_\#$ is an isomorphism; and, hence, $g_\# = \tilde{g}_\# i_\#$ is an isomorphism. The theorem now follows by the exactness of localization. □

(3.3.9) Corollary

Let T be a torus, X a finite-dimensional T-CW-complex with finitely many connective orbit types (FMCOT), suppose that $X^T \neq \emptyset$, and let $x_0 \in X^T$. For any subtorus $K < T$, let $F(K)$ be the component of X^K which contains x_0. Let $R = H_T^* = H^*(BT;\mathbb{Q})$. Then the inclusion

3.3. Equivariant rational homotopy

$\phi : F(K) \to X$ induces isomorphisms of localized R-modules

$$S(K)^{-1}\phi^* : S(K)^{-1}\Pi^*_{(T,x_0)}(X) \xrightarrow{\cong} S(K)^{-1}\Pi^*_{(T,x_0)}(F(K)),$$

and

$$\phi^*_\mathfrak{p} : \Pi^*_{(T,x_0)}(X)_\mathfrak{p} \xrightarrow{\cong} \Pi^*_{(T,x_0)}(F(K))_\mathfrak{p}$$

where $S(K) \subseteq R$ is the multiplicative set defined in Examples (3.1.5), and $\mathfrak{p}$ is any prime ideal in R such that $\sigma\mathfrak{p} = PK$ (as in Definition (3.1.4)).

(3.3.10) Remark

Let T, X, x_0 be as in Corollary (3.3.9). Let $F = F(T)$ be the component of X^T which contains x_0. Then it is easy to see that $\Pi^*_{(T,x_0)}(X) \cong \Pi^*_{(T,x_1)}(X)$ for any $x_1 \in F$. Thus we shall usually write $\Pi^*_{(T,F)}(X)$ instead of $\Pi^*_{(T,x_0)}(X)$, even though this notation may not be fully adequate for functoriality (see the discussion following Definitions (3.3.1)). Normally, however, base-points can be chosen to take care of themselves; and so the chosen component F of X^T is all that really matters.

(3.3.11) Corollary

Let T be a torus and let X be a finite-dimensional T-CW-complex with FMCOT and with $X^T \neq \emptyset$. Let F be a component of X^T. If X is π-finite (see Definitions (2.4.1)(2)), then F is π-finite, and

$$\chi\pi(X) = \chi\pi(F).$$

Proof.
Since the differential on $R \otimes \mathcal{M}(F)$ is untwisted, it follows at once that $\Pi^*_{(T,F)}(F) \cong R \otimes \Pi^*_\psi(F)$.

Let $K = R_{(0)}$, the field of fractions of R. By Corollary (3.3.9), $K \otimes \Pi^*_\psi(F) \cong \Pi^*_{(T,F)}(X)_{(0)}$. For a finite-dimensional graded K-vector space, V^*, let $\chi_K(V) = \sum_{n=0}^\infty (-1)^n \dim_K V^n$. We have

$$\dim_\mathbb{Q} \Pi^*_\psi(F) = \dim_K \Pi^*_{(T,F)}(X)_{(0)} \leq \dim_K(Q_{(T,F)}(X) \otimes_R K)$$
$$= \dim_\mathbb{Q} \Pi^*_\psi(X),$$

where $Q_{(T,F)}(X) = Q_{(T,x_0)}(X)$ for some $x_0 \in F$. So F is π-finite.
Also $\chi\pi(F) = \chi_K(\Pi^*_{(T,F)}(X)_{(0)}) = \chi_K(Q_{(T,F)}(X) \otimes_R K) = \chi\pi(X)$. □

The inequality $\dim_\mathbb{Q} \Pi^*_\psi(F) \leq \dim_\mathbb{Q} \Pi^*_\psi(X)$, in the proof above, can be improved to the following, first observed by Yoshida ([Yoshida, 1979]).

(3.3.12) Corollary

Let T be a torus and let X be a finite-dimensional T-CW-complex with FMCOT and with $X^T \neq \emptyset$. Let F be a component of X^T. Then, for any $n \geq 1$.

$$\sum_{k=0}^{\infty} \dim_{\mathbb{Q}} \Pi_{\psi}^{n+2k}(F) \leq \sum_{k=0}^{\infty} \dim_{\mathbb{Q}} \Pi_{\psi}^{n+2k}(X).$$

Proof.

For a graded object V^* and integer n, let $V^{(n)} = \bigoplus_{m \equiv n(2)} V^m$, the set of elements having the same parity as n, and let $\mathcal{F}_n V = \bigoplus_{m=0}^{n} V^m$.

The Localization Theorem (3.3.9) implies that the homomorphism $Q_{(T,F)}(X) \to Q_{(T,F)}(F) \cong R \otimes \Pi_{\psi}^*(F)$ becomes an epimorphism after localizing at (0). Now compose the epimorphism

$$Q_{(T,F)}^{(n)}(X)_{(0)} \to K \otimes \Pi_{\psi}^{(n)}(F)$$

with the quotient map

$$K \otimes \Pi_{\psi}^{(n)}(F) \to K \otimes \Pi_{\psi}^{(n)}(F) / K \otimes \mathcal{F}_{n-1} \Pi_{\psi}^{(n)}(F).$$

Identifying $Q_{(T,F)}^{(n)}(X)_{(0)}$ with $K \otimes \Pi_{\psi}^{(n)}(X)$, the kernel of this composition contains $K \otimes \mathcal{F}_{n-1} \Pi_{\psi}^{(n)}(X)$. Thus $K \otimes \Pi_{\psi}^{(n)}(X) / K \otimes \mathcal{F}_{n-1} \Pi_{\psi}^{(n)}(X)$ maps onto $K \otimes \Pi_{\psi}^{(n)}(F) / K \otimes \mathcal{F}_{n-1} \Pi_{\psi}^{(n)}(F)$: and so the result follows.

□

(3.3.13) Definition

Let T be a torus, and let X be a T-CW-complex with $X^T \neq \emptyset$. Let F be a component of X^T. Suppose that the differential on $Q_{(T,F)}(X)$ is 0: and so $\Pi_{(T,F)}^*(X) \cong R \otimes \Pi_{\psi}^*(X)$. Then we shall say that F is Π_{ψ}^*-full (or, simply, full) in X.

(3.3.14) Remark

Fullness plays a role in equivariant rational homotopy analogous to the role in equivariant cohomology played by the condition that X be TNHZ in $X_T \to BT$. This analogy is further borne out by the next propositon. Note, however, that the conditions of fullness and TNHZ are independent: either one can occur with or without the other, or neither may occur.

Before stating the proposition we need two things. First, the identity map $\mathrm{id}_X : X \to X$, viewed as an equivariant map from X with trivial action (of the trivial group) into X as a T-space, induces a homomorphism $j^* : \Pi_{(T,F)}^*(X) \to \Pi_{\psi}^*(X)$; j^* is induced by the map $R \otimes \mathcal{M}(X) \to \mathcal{M}(X)$

associated with $j : X \to X_T$. Second, if X is simply-connected and $x_0 \in F$ is a base-point, then T acts on ΩX, the loop space of X based at x_0, in the obvious way: if $\alpha : I \to X$ is a loop at x_0 and $g \in T$, then $(g\alpha)(t) = g\alpha(t)$, $\forall t \in I$. Clearly $(\Omega X)^T = \Omega F$, the loop space of F based at x_0.

(3.3.15) Proposition
Let T be a torus and let X be a T-CW-complex with $X^T \neq \emptyset$. Let F be a component of X^T. Then the following three statements are equivalent:

(1) *F is full in X;*
(2) *$\Pi^*_{(T,F)}(X)$ is a free R-module;*
(3) *$j^* : \Pi^*_{(T,F)}(X) \to \Pi^*_\psi(X)$ is onto.*

Furthermore, if X os π-finite and finite-dimensional with FMCOT, then (1) is equivalent to

(4) $\dim_{\mathbb{Q}} \Pi^*_\psi(F) = \dim_{\mathbb{Q}} \Pi^*_\psi(X)$.

If X is simply-connected of finite $\mathbb{Q}$-type, then (1) is equivalent to

(5) *ΩX is TNHZ in $(\Omega X)_T \to BT$.*

Proof.
(1) $\Rightarrow$ (2) and (1) $\Rightarrow$ (3) clearly. Under the isomorphism $Q_{(T,F)}(X) \cong R \otimes \Pi^*_\psi(X)$, (3)$\Rightarrow$(1), by induction on degrees of generators in $\Pi^*_\psi(X)$.

To prove that (2)$\Rightarrow$(1), consider the bicomplex $C^{*,*} = Q_{(T,F)}(X) \otimes \Lambda(s_1, \ldots, s_n)$ where $n = \dim T$ and

(i) $\deg(s_i) = 1$ for $1 \leq i \leq n$,
(ii) d_1 is the given differential on $Q_{(T,F)}(X)$ extended to $C^{*,*}$ by setting $d_1(s_i) = 0$ for $1 \leq i \leq n$,
(iii) d_2 is the differential on $C^{*,*}$ defined by the formula $d_2(x \otimes s_i) = (-1)^{\deg(x)} t_i x \otimes 1$ for $1 \leq i \leq n$, where $R = \mathbb{Q}[t_1, \ldots, t_n]$, and $d_2(x \otimes 1) = 0$, and
(iv) $C^{p,q}$ is spanned by all $x \otimes y \in Q_{(T,F)}(X) \otimes \Lambda(s_1, \ldots, s_n)$, where $p = \deg(x) + 2\deg(y)$ and $q = -\deg(y)$.

So $C^{*,*}$ is a fourth quadrant bicomplex, and $d_1 : C^{p,q} \to C^{p+1,q}$, and $d_2 : C^{p,q} \to C^{p,q+1}$. Let H_i denote homology with respect to d_i, $i = 1, 2$, and H denote homology with respect to $d_1 + d_2$. Then (see, e.g., [Cartan, Eilenberg, 1956] Chapter XV, §6) we have a second quadrant spectral sequence

$$E_2^{p,q} = H_2 H_1(C^{*,*})^{q,p}$$

and a fourth quadrant spectral sequence

$$F_2^{p,q} = H_1 H_2(C^{*,*})^{p,q}$$

both converging to $H(C^{*,*})$;
$$H_1(C^{*,*})^{p,q} \cong \Pi^{p+2q}_{(T,F)}(X) \otimes \Lambda^{-q}(s_1,\ldots,s_n)$$
and
$$H_2(C^{*,*})^{p,q} \cong \operatorname{Tor}_R^{-q,p+2q}(Q_{(T,F)}(X),\mathbf{Q}).$$
Hence $F_2^{p,q} = F_\infty^{p,q} = 0$ if $q < 0$, $= \Pi_\psi^p(X)$ if $q = 0$. Also
$$E_2^{q,p} \cong \operatorname{Tor}_R^{-q,p+2q}(\Pi^*_{(T,F)}(X),\mathbf{Q}).$$

So (2)$\Rightarrow$ $E_2^{q,p} = E_\infty^{q,p} = 0$ if $q < 0$; and $E_2^{0,p} = E_\infty^{0,p}$ equal to the quotient of $\Pi^p_{(T,F)}(X)$ by elements of degree p in $\bar{R}\Pi^*_{(T,F)}(X)$, where $\bar{R}$ is the kernel of the homomorphism $R \to \mathbf{Q}$ given by $t_i \mapsto 0$ for $1 \leq i \leq n$. The edge homomorphism $E_1^{0,p} \to \Pi_\psi^p(X)$ is j^*. Thus (2)$\Rightarrow$(3).

If X is π-finite and finite-dimensional with FMCOT, then (1)$\Rightarrow$(4) by applying the Localization Theorem (3.3.9) to
$$Q_{(T,F)}(X) \cong R \otimes \Pi_\psi^*(X) \to Q_{(T,F)}(F) \cong R \otimes \Pi_\psi^*(F).$$
If (1) does not hold, then $\Pi^*_{(T,F)}(X)_{(0)} \cong H(Q_{(T,F)}(X)_{(0)})$ has dimension (over $K = R_{(0)}$) less than $\dim_\mathbf{Q} \Pi_\psi^*(X)$. So (4)$\Rightarrow$(1).

Finally, if X is simply-connected of finite $\mathbf{Q}$-type, then, by Examples (2.5.7)(2), (1) $\Leftrightarrow$ $\delta = 0$ in the model $(R \otimes \mathcal{M}(\Omega X), \delta)$ for $(\Omega X)_T$. So (1)$\Leftrightarrow$(5). □

3.4 Equivariant rational homotopy for general G-spaces

In general, even for an action of a compact connected Lie group on a topological manifold, it is not clear that the fixed point set will always be an agreement space (see Definition (2.6.2)). Nevertheless the ideas and results of Section 2.6 permit one to generalize the results of Section 3.3 to more general G-spaces. For a connected paracompact G-space, where G is a compact connected Lie group, with $X^G \neq \emptyset$ and $x_0 \in X^G$, one can use Corollary (2.6.6) and the analogue of the first part of Examples (2.5.7)(2) to define $\overline{\Pi}^*_{(G,x_0)}(X)$ in the obvious way. The analogue of Theorem (3.3.2) then follows, with essentially the same proof, if X satisfies any of the conditions (A), (B), (C) of Section 3.2 (with $k = \mathbf{Q}$), provided that, in cases (B) and (C), $F(S)$ satisfies one additional condition, which we shall describe now.

Let $\{F(S)_i : i \in \mathcal{I}\}$ be the set of components of X^S, with $F(S)$, the component at x_0, being $F(S)_{i_0}$.

(3.4.1) Definition
Let $q_i : F(S)_i \to X^S$ be the inclusion, and $q_i^* : \overline{H}_G^*(X^S; \mathbf{Q}) \to \overline{H}_G^*(F(S)_i; \mathbf{Q})$ the induced homomorphism for all $i \in \mathcal{I}$. We shall say

that $F(S)$ is cohomologically separable, or $F(S)$ satisfies (CS), if there exists $y \in S^{-1}\overline{H}_G^*(X^S; k)$ such that $S^{-1}q_i^*(y) = 0$, for all $i \neq i_0$, and $S^{-1}q_{i_0}^*(y) = 1$.

(3.4.2) Remark
If $F(S)$ is open in X^S, then
$$\overline{H}_G^*(X^S; \mathbb{Q}) \cong \overline{H}_G^*(F(S); \mathbb{Q}) \oplus \overline{H}_G^*(X^S - F(S); \mathbb{Q}).$$
Thus $F(S)$ satisfies (CS) in this case. Clearly $F(S)$ is open in X^S if X is a G-CW-complex, since then $X^S - F(S) = \bigcup_{i \neq i_0} F(S)_i$ is a subcomplex. Also, of course, $F(S)$ is open in X^S if the number of components of X^S is finite. If $X^S = X^K$ for some torus $K < G$, and if $\dim_\mathbb{Q} \overline{H}^*(X; \mathbb{Q}) < \infty$, then it follows from the Localization Theorem that X^S has a finite number of components (see Corollary (3.1.13)). If X^S has a finite number of components, $F(S)_1, \ldots, F(S)_\nu$, say, then let $e_i \in \overline{H}_G^0(X^S; \mathbb{Q})$ be the element corresponding to the identity in $\overline{H}_G^0(F(S)_i; \mathbb{Q})$ under the isomorphism
$$\overline{H}_G^*(X^S; \mathbb{Q}) \cong \bigoplus_{j=1}^{\nu} \overline{H}_G^*(F(S)_j; \mathbb{Q}).$$
Then $q_i^*(e_i) = 1$ and $q_j^*(e_i) = 0$, if $j \neq i$.

The condition (CS) is not always satisfied, however. Let Y be the subset of the plane $\mathbb{R}^2$ consisting of the lines $x = \pm 1$, and, for all integers $n \geq 2$, the rectangles with vertices $(\pm(1 - \frac{1}{n}), \pm n)$. Let $X = \Sigma^2 Y$, the unreduced double suspension of Y. Then $G = S^1$ acts on X with $X^G = Y$. The components $x = \pm 1$ of X^G do not satisfy (CS).

The Localization Theorem for equivariant Alexander-Spanier rational homotopy is as follows.

(3.4.3) Theorem
Let G be a compact connected Lie group, and let S be a multiplicative subset of $R = H^*(BG; \mathbb{Q})$, with $0 \notin S$. Suppose that X is a connected G-space with $X^G \neq \emptyset$, and let $x_0 \in X^G$. Let $F(S)$ be the component of X^S which contains x_0. Suppose further that at least one of the following conditions is satisfied:

(A) *X is compact;*
(B+) *X is paracompact, $\mathrm{cd}_\mathbb{Q}(X) < \infty$, G acts on X with FMCOT, and $F(S)$ satisfies (CS);*
(C+) *X is paracompact and finitistic, G acts on X with FMCOT, and $F(S)$ satisfies (CS).*

Then the localized homomorphism of $S^{-1}R$-modules
$$S^{-1}\phi^* : S^{-1}\overline{\Pi}^*_{(G,x_0)}(X) \to S^{-1}\overline{\Pi}^*_{(G,x_0)}(F(S))$$
is an isomorphism.

Yet again there are no additional conditions when X is compact. We shall now explain why this is so. First, however, note that the proof of (3.4.3) in cases $(B+)$ and $(C+)$ follows precisely the proof of Theorem (3.3.2): only for case (A) are some refinements needed.

(3.4.4) Lemma

With the notation of Definition (3.4.1), suppose that X^S is compact. Then for each $i \in \mathcal{I}$ with $i \neq i_0$, there exists $y_i \in \overline{H}^*_G(X^S; \mathbf{Q})$ such that $q_i^*(y_i) = 0$ and $q_{i_0}^*(y_i) = 1$.

Proof.

Since X^S is compact, there exist open subsets U_i, V_i of X^S such that $F(S) \subseteq U_i$, $F(S)_i \subseteq V_i$, $X^S = U_i \cup V_i$ and $U_i \cap V_i = \emptyset$. (See [Munkres, 1975], Chapter 5, Exercise 4). Since any orbit (being connected, since G is connected) must be contained in U_i or V_i, U_i and V_i are invariant. So
$$\overline{H}^*_G(X^S; \mathbf{Q}) \cong \overline{H}^*_G(U_i; \mathbf{Q}) \oplus \overline{H}^*_G(V_i; \mathbf{Q}).$$
Take $y_i = (1, 0)$ under this isomorphism. □

Now to prove Theorem (3.4.3) in case (A), we follow the proof of Theorem (3.3.2), except that we must invert a family of elements $\{x_i; i \in \mathcal{I}, i \neq i_0\}$ in $S^{-1}\overline{H}^*_G(X; \mathbf{Q})$ corresponding to the elements y_i of Lemma (3.4.4). The necessary version of Lemma (3.3.3), however, follows from the result concerning B-module localization given in Corollary (3.2.7). Any other alterations that are needed, such as the necessary extension of Lemma (3.3.4), can be handled by induction. □

Finally note that there are obvious analogues of Corollaries (3.3.9), (3.3.11), and (3.3.12), Remark (3.3.10), Definition (3.3.13), and Proposition (3.3.15), where for the last we must assume that X satisfies (A), $(B+)$ or $(C+)$ in order to include (4), and we must assume that X is an agreement space in order to include (5).

3.5 The Evaluation Theorem for torus actions

Let G be a torus of dimension n, $R = H^*(BG; \mathbf{Q}) = \mathbf{Q}[t_1, \ldots, t_n]$, and let X be a finite-dimensional connected G-CW-complex with FMCOT.

3.5. The Evaluation Theorem for torus actions

Let E be any field of characteristic 0, and for an n-tuple $\alpha = (\alpha_1, \ldots, \alpha_n)$ of elements of E, denote by $\tilde{\alpha}$ the homomorphism $R \to E[t]$, where t is an indeterminate of degree 2, given by $\tilde{\alpha}(t_i) = \alpha_i t$, for $1 \leq i \leq n$. For an R-module N and an $E[t]$-module M, let $M \otimes_{\tilde{\alpha}} N$ denote $M \otimes_R N$, where M is made into a R-module by means of $\tilde{\alpha}$.

Let $I(\alpha)$ be the kernel of $\tilde{\alpha}$, and let $K(\alpha)$ be the subtorus of G such that $PK(\alpha) = \sigma I(\alpha)$, as in Definition (3.1.4). Now let $R \otimes \mathcal{M}(X)$ be a model for X_G, as in Example (2.5.4). We can form

$$E[t, t^{-1}] \otimes_{\tilde{\alpha}} (R \otimes \mathcal{M}(X))$$

with the differential extended by requiring $dt = dt^{-1} = 0$: it is a D2GCA (see Section 3.3). Similarly, if $X^G \neq \emptyset$, and $x_0 \in X^G$, we can form $E[t, t^{-1}] \otimes_{\tilde{\alpha}} Q_{(G,x_0)}(X)$. Now the evaluation theorem (α-theorem) for torus actions is the following.

(3.5.1) Theorem
With the conditions and assumptions above

(1) $\quad H^*(E[t, t^{-1}] \otimes_{\tilde{\alpha}} (R \otimes \mathcal{M}(X))) \cong E[t, t^{-1}] \otimes H^*(X^{K(\alpha)}; \mathbb{Q})$,

where $\otimes = \otimes_{\mathbb{Q}}$; *and, if* $X^G \neq \emptyset$, *and* $x_0 \in X^G$,

(2) $\quad H^*(E[t, t^{-1}] \otimes_{\tilde{\alpha}} Q_{(G,x_0)}(X)) \cong E[t, t^{-1}] \otimes \Pi^*_\psi(c(\alpha))$,

where $c(\alpha)$ *is the component of* $X^{K(\alpha)}$ *which contains* x_0.

Proof.
Since $PK(\alpha) \subseteq I(\alpha)$, $\tilde{\alpha}$ factors into the composition of the quotient $i^*_{K(\alpha)} : R \to R/PK(\alpha) = R_{K(\alpha)} := H^*_{K(\alpha)}$ with a map $\tilde{\alpha} : H^*_{K(\alpha)} \to E[t, t^{-1}]$. Since $i^*_{K(\alpha)}$ is onto, it follows easily that

$$E[t, t^{-1}] \otimes_{\tilde{\alpha}} (R \otimes \mathcal{M}(X)) \cong E[t, t^{-1}] \otimes_{\tilde{\alpha}} (H^*_{K(\alpha)} \otimes \mathcal{M}(X)).$$

Now $H^*_{K(\alpha)} \otimes \mathcal{M}(X)$ is a model for $X_{K(\alpha)}$ by Examples (2.5.7)(1). Furthermore, in $H^*_{K(\alpha)}$, $\sigma I(\tilde{\alpha}) = (0)$. Thus, to prove (1), it suffices to assume that $\sigma I(\alpha) = (0)$, and so $K(\alpha) = G$.

Now let S be the multiplicative subset of R generated by the non-zero elements of $H^2(BG; \mathbb{Q})$, namely the sums $\sum_{i=1}^n r_i t_i$, $r_i \in \mathbb{Q}$, not all zero. Since $\sigma I(\tilde{\alpha}) = (0)$, $\tilde{\alpha}(s) \neq 0$, for all $s \in S$, and each $\tilde{\alpha}(s)$, $s \in S$, is invertible in $E[t, t^{-1}]$. Thus $\tilde{\alpha}$ induces a map $\hat{\alpha} : S^{-1}R \to E[t, t^{-1}]$. Generally for R-modules M and N,

$$S^{-1}(M \otimes_R N) \cong S^{-1}M \otimes_{S^{-1}R} S^{-1}N \cong S^{-1}M \otimes_R N \cong M \otimes_R S^{-1}N$$

([Bourbaki, 1961], Chapter 2, §2, no.7, Proposition 18(i)). Hence

$$E[t, t^{-1}] \otimes_{\tilde{\alpha}} (R \otimes \mathcal{M}(X)) \cong E[t, t^{-1}] \otimes_{\hat{\alpha}} (S^{-1}R \otimes \mathcal{M}(X)).$$

$S^{-1}R \otimes \mathcal{M}(X)$ may be viewed as a complex of free $S^{-1}R$-modules, and its homology

$$H^*(S^{-1}R \otimes \mathcal{M}(X)) \cong S^{-1}H^*_G(X; \mathbf{Q}) \cong S^{-1}H^*_G(X^G; \mathbf{Q})$$
$$\cong S^{-1}R \otimes H^*(X^G; \mathbf{Q})$$

is also a free $S^{-1}R$-module. (The second isomorphism in the preceding sentence is, of course, the Localization Theorem.) Thus, by Lemma (3.5.2) below,

$$H^*(E[t,t^{-1}] \otimes_{\tilde{\alpha}} (R \otimes \mathcal{M}(X))) \cong E[t,t^{-1}] \otimes_{\tilde{\alpha}} H^*(S^{-1}R \otimes \mathcal{M}(X))$$
$$\cong E[t,t^{-1}] \otimes_{\tilde{\alpha}} S^{-1}R \otimes H^*(X^G; \mathbf{Q})$$
$$\cong E[t,t^{-1}] \otimes H^*(X^G; \mathbf{Q}).$$

Part (2) follows in exactly the same way using the homotopy Localization Theorem (3.3.2). □

To complete the proof of Theorem (3.5.1) we need a universal coefficients formula. Let A be a commutative ring, and let

$$C_* = \{C_n, d_n : C_n \to C_{n+1} | n \in \mathbf{Z}\}$$

be a cochain complex of A-modules over $\mathbf{Z}$. Note that a $\mathbf{Z}_2$-graded complex, $d_0 : C_0 \to C_1$, $d_1 : C_1 \to C_0$, $d_0 d_1 = d_1 d_0 = 0$, can be converted into a cochain complex over $\mathbf{Z}$ by setting $C_{2n} = C_0$, $d_{2n} = d_0$, $C_{2n+1} = C_1$, $d_{2n+1} = d_1$, for all $n \in \mathbf{Z}$.

(3.5.2) Lemma

Let A and the cochain complex C_* be as above. Suppose that $\mathrm{gldh}(A) < \infty$ (see Definitions (A.4.1)(2)). Suppose further that the A-modules C_n and $H_n(C_*) = \ker d_n / \operatorname{im} d_{n-1}$ are projective for all $n \in \mathbf{Z}$. Let M be any A-module. Then

$$H_n(M \otimes_A C_*) \cong M \otimes_A H_n(C_*), \text{ for all } n \in \mathbf{Z}.$$

Before we prove the lemma, note that we apply it in the proof of Theorem (3.5.1) with $A = S^{-1}R$, $M = E[t,t^{-1}]$ and C_* the complex obtained from $S^{-1}R \otimes \mathcal{M}(X)$ for part (1), and from $S^{-1}R \otimes_R Q_{(G,x_0)}(X) \cong S^{-1}R \otimes \Pi^*_\psi(X)$ for part (2). We have that $\mathrm{gldh}(R) = n$, the rank of G, by Theorem (A.4.8); and $\mathrm{gldh}(S^{-1}R) \leq n$, by Remarks (A.4.13)(2).

(Note that the lemma follows directly from (B.1.14). We give an independent and different proof below.)

Proof of Lemma (3.5.2).

Let $Z_i = \ker d_i$, $B_i = \operatorname{im} d_{i-1}$ and $H_i = H_i(C_*) = Z_i/B_i$. Then we have exact sequences

$$0 \to Z_i \to C_i \to B_{i+1} \to 0$$

3.5. The Evaluation Theorem for torus actions

and
$$0 \to B_i \to Z_i \to H_i \to 0.$$

Let N be any A-module. Since C_i and H_i are projective, it follows that $\operatorname{Ext}_A^j(Z_i, N) \cong \operatorname{Ext}_A^{j+1}(B_{i+1}, N)$ for all $j \geq 1$, and $\operatorname{Ext}_A^j(Z_i, N) \cong \operatorname{Ext}_A^j(B_i, N)$ for all $j \geq 1$. Hence, for all $j \geq 1$, $\operatorname{Ext}_A^j(B_i, N) \cong \operatorname{Ext}_A^{j+n}(B_{i+n}, N) = 0$, if $n \geq \operatorname{gldh}(A)$. Hence each B_i, and each Z_i, is a projective A-module.

So $C_i \cong Z_i \oplus B_{i+1} \cong H_i \oplus B_i \oplus B_{i+1}$; and the result follows easily (cf. Appendix B and [Dold, 1960]). □

Now we can regard E as an $E[t]$-module via the homomorphism $\varepsilon_1 : E[t] \to E$ defined by $t \mapsto 1$. This map induces a homomorphism $\varepsilon_1 : E[t, t^{-1}] \to E$. Since $\varepsilon_1 \tilde{\alpha}(\sum_{i=1}^n r_i t_i) = \sum_{i=1}^n r_i \alpha_i$, $\sigma \ker(\alpha) = \sigma I(\alpha) = PK(\alpha)$, where $\alpha := \varepsilon_1 \tilde{\alpha}$.

Note that $\alpha : R \to E$, $\alpha(t_i) := \alpha_i$, does not preserve the $\mathbb{Z}$-grading, but only the $\mathbb{Z}_2$-grading given by the even and odd dimensional parts (since $|t_i| = 2$). Therefore tensor products (over R) of $\mathbb{Z}$-graded R-modules with E considered as an R-module via α will only inherit a $\mathbb{Z}_2$-grading. If M^* is a $\mathbb{Z}$-graded module then let $M^{(*)}$ denote the same module with the $\mathbb{Z}_2$-grading given by the even and odd dimensional parts, i.e.

$$M^{\mathrm{ev}} := \bigoplus_{j \text{ even}} M^j, \quad M^{\mathrm{od}} := \bigoplus_{j \text{ odd}} M^j.$$

We can apply Lemma (3.5.2) with $A = E[t, t^{-1}]$, $M = E$, and $C_* = E[t, t^{-1}] \otimes_{\tilde{\alpha}} (R \otimes \mathcal{M}(X))$ or $E[t, t^{-1}] \otimes_{\tilde{\alpha}} Q_{(G, x_0)}(X)$ to get the following corollary to Theorem (3.5.1).

(3.5.3) Corollary
With the conditions, assumptions and notation of Theorem (3.5.1),

(1) $H(E \otimes_\alpha (R \otimes \mathcal{M}(X))) \cong E \otimes H^{(*)}(X^{K(\alpha)}; \mathbb{Q})$;

and, if $X^G \neq \emptyset$, and $x_0 \in X^G$,

(2) $H(E \otimes_\alpha Q_{(G, x_0)}(X)) \cong E \otimes \Pi_\psi^{(*)}(c(\alpha))$,

where $c(\alpha)$ is the component of $X^{K(\alpha)}$, which contains x_0.
(Note that

$$E \otimes H^{(*)}(X^{K(\alpha)}; \mathbb{Q}) \cong H^{(*)}(X^{K(\alpha)}; E)$$

if $\dim_\mathbb{Q} H^*(X^{K(\alpha)}; \mathbb{Q}) < \infty$.)

(3.5.4) Remarks

If $\dim_{\mathbb{Q}} E \geq n = \dim G$, then every subtorus of G is $K(\alpha)$ for some $\tilde{\alpha} : R \to E[t]$. On the other hand for any E (of characteristic 0), including $E = \mathbb{Q}$, the set of all $\tilde{\alpha}|H^2(BG;\mathbb{Q})$ contains all linear functionals (over $\mathbb{Q}$) on $H^2(BG;\mathbb{Q})$, and hence the set of all $K(\alpha)$ contains all subcircles of G. Since the action of G on X is almost-free (i.e. all isotropy groups are finite) if and only if every subcircle has an empty fixed point set, it follows that the action is almost-free if and only if the complex $E \otimes_{\alpha} (R \otimes \mathcal{M}(X)) \cong E \otimes \mathcal{M}(X)$ is acyclic for all non-zero $\alpha : R \to E$, where E is any field of characteristic 0. Furthermore, $E \otimes \mathcal{M}(X)$ is acyclic if and only if the unit element $1 \otimes 1$ is a boundary.

Applications of Theorem (3.5.1) and Corollary (3.5.3) will appear later.

(3.5.5) General G-spaces

If we replace singular theory, H^*, $\mathcal{M}$, Π^*_ψ, etc., by Alexander-Spanier theory, $\bar{H}^*$, $\bar{\mathcal{M}}$, $\bar{\Pi}^*_\psi$, etc., then it is clear that the proofs above carry over to give Theorem (3.5.1)(1) and Corollary (3.5.3)(1) for any G-space X satisfying conditions (A), (B) or (C) of Definition (3.2.4), and Theorem (3.5.1)(2) and Corollary (3.5.3)(2) for any G-space X satisfying conditions (A), $(B+)$ or $(C+)$ of Theorem (3.4.3), where in $(B+)$ and $(C+)$ it is $c(\alpha)$ which must satisfy (CS). (Here, of course, we still assume that G is a torus and that the ground field is $\mathbb{Q}$.)

We conclude this section with a discussion of the minimal Hirsch-Brown model (see (B.2.4)).

(3.5.6) Definitions

(1) By (B.2.4), there is a differential on $R \otimes H^*(X;\mathbb{Q})$, which gives it the structure of a differential graded R-module, such that the standard maps $R \to R \otimes H^*(X;\mathbb{Q}) \to H^*(X;\mathbb{Q})$ are maps of cochain complexes where R and $H^*(X;\mathbb{Q})$ have trivial differentials, and such that there exists a cochain homotopy equivalence over R from $R \otimes \mathcal{M}(X)$ to $R \otimes H^*(X;\mathbb{Q})$. Let $\mu_G^*(X;\mathbb{Q})$ denote $R \otimes H^*(X;\mathbb{Q})$ with such a differential, and let $\mu(X) : R \otimes \mathcal{M}(X) \to \mu_G^*(X;\mathbb{Q})$ denote a resulting cochain homotopy equivalence over R. ($\mu_G^*(X;\mathbb{Q}) = R\tilde{\otimes}H^*(X;\mathbb{Q})$ in the notation of Section B.2.)

(2) Now let E be any extension field of $\mathbb{Q}$, and let $\alpha = (\alpha_1, \ldots, \alpha_n) \in E^n$, where $n = \dim G$. As above we can make $E[t]$ into a R-module via $t_i \mapsto \alpha_i t$, $1 \leq i \leq n$, and we can form

$$E[t] \otimes_{\tilde{\alpha}} (R \otimes \mathcal{M}(X)) \quad \text{and} \quad E[t] \otimes_{\tilde{\alpha}} \mu_G^*(X;\mathbb{Q});$$

3.5. The Evaluation Theorem for torus actions

similarly we can form
$$E \otimes_\alpha (R \otimes \mathcal{M}(X)) \text{ and } E \otimes_\alpha \mu_G^{(*)}(X;\mathbf{Q}).$$
Let
$$\mu_{G,\tilde\alpha}^*(X;\mathbf{Q}) := E[t] \otimes_{\tilde\alpha} \mu_G^*(X;\mathbf{Q});$$
$$\mu_{G,\alpha}^{(*)}(X;\mathbf{Q}) := E \otimes_\alpha \mu_G^{(*)}(X;\mathbf{Q}).$$
Then $\mu(X)$ extends in the standard way to a cochain homotopy equivalence over $E[t]$, which we shall denote by
$$\mu(X,\tilde\alpha) := E[t] \otimes_{\tilde\alpha} (R \otimes \mathcal{M}(X)) \to \mu_{G,\tilde\alpha}^*(X;\mathbf{Q});$$
define
$$\mu(X,\alpha) : E \otimes_\alpha (R \otimes \mathcal{M}(X)) \to \mu_{G,\alpha}^{(*)}(X;\mathbf{Q})$$
similarly.

The following corollary thus follows directly from Theorem (3.5.1)(1) and (3.5.3)(1).

(3.5.7) Corollary
If X is a finite-dimensional connected G-CW-complex with FMCOT, then, with $K(\alpha)$ as defined near the beginning of this section, there are isomorphisms

(1) $S(t)^{-1}(t)^{-1}H(\mu_{G,\tilde\alpha}^*(X;\mathbf{Q})) \cong E[t,t^{-1}] \otimes H^*(X^{K(\alpha)};\mathbf{Q})$, *where $S(t)$ is the multiplicative subset of $E[t]$ generated by t, and*

(2) $H(\mu_{G,\alpha}^{(*)}(X;\mathbf{Q})) \cong E \otimes H^{(*)}(X^{K(\alpha)};\mathbf{Q}).$

(3.5.8) Remark
Using Alexander-Spanier theory, $\bar{\mathcal{M}}$ and $\bar{H}^*$, instead of singular theory, then clearly Corollary (3.5.7) holds for any connected G-space satisfying conditions (A), (B) or (C).

(3.5.9) Remark
The algebraic models derived in this section for the Borel construction when G is a torus are to a large extent analogous to the algebraic models discussed in the Sections 1.3 and 1.4 when G is a 2-torus or p-torus (odd p) respectively. More precisely: $R \otimes \mathcal{M}(X)$ corresponds to $\beta_G^*(S^*(X;\mathbf{F}_p))$, and $E \otimes_\alpha (R \otimes \mathcal{M}(X))$ to $\beta_{G,\alpha}(S^*(X;\mathbf{F}_p))$, etc. (The notation used in Chapter 1 is not analogous to the one used in this chapter as far as the order of factors in the chain models is concerned. This comes from the fact that we wanted to emphasize the 'deformation theoretical' aspects of the constructions and arguments in Chapter 1, while now, when more machinery from algebraic topology - like spectral

sequences and minimal models for fibrations - is applied, the notation is chosen in accordance with standard use.) The role of the extension field E is played by the field k of characteristic p in the p-torus case. Corollary (3.5.7)(2) corresponds to the Theorems (1.3.5) and (1.3.24) (cf. also (1.3.27) and (1.3.28)) in case (2), and to Theorem (1.4.5) in case (odd p) if $G = \mathbb{Z}_p$. In case (odd p) the situation for p-tori of rank larger than 1 is technically much more complicated and is dealt with in detail in Sections 1.4 and 3.11. In a sense the case (0) is the simplest, due to the nice multiplicative structure which already exists on the chain model $R \otimes \mathcal{M}(X)$. Almost all the consequences and applications of (1.3.5) and (1.3.24) given in Section 1.3 for case (2) can be obtained in a completely analogous way in case (0) using (3.5.7)(2). In fact, one even gets '$\mathbb{Z}_2$-graded' versions due to the $\mathbb{Z}_2$-grading of $E \otimes_\alpha (R \otimes \mathcal{M}(X))$ and $\mu_{G,\alpha}^{(*)}(X;\mathbb{Q})$ in the case at hand, e.g. the inequalities corresponding to (1.3.8)(1) hold separately for the even and odd dimensional part. Of course, some of the results have to be 'translated' in a proper way:

(i) clearly, for G a torus ($G = S^1 \times \ldots \times S^1$) the coefficients are taken in a field of characteristic 0;

(ii) we do not get statements corresponding to the case where G is a finite 2-group, but not a 2-torus;

(iii) the examples (1.3.11)(1) and (1.3.19) have to be modified appropriately, in particular, in the analogue of (1.3.19) one cannot conclude that X^G consists of at most two components, since the argument there relied on the fact that $\mathbb{F}_2$ has only two different elements.

Moreover (3.5.1)(2) and (3.5.3)(2) make it possible to obtain certain analogous results (an analogue of (1.3.8) for example) for the rational homotopy of a G-space. We leave the rather straightforward details to the reader, since all the outcoming results (including certain improvements over cases (2) and (odd p), which are possible in the case (0)) are presented in the following sections, using somewhat different arguments which also cover the case (odd p).

3.6 The ideals $\mathfrak{p}(K,c)$

In this section we shall work with cases (0), (2) and (odd p): and we shall use the notation established for these cases in Example (3.1.1). Also we shall use the notation of Definition (3.1.4) and Example (3.1.5). By subtorus we shall mean any closed connected subgroup in case (0), and any subgroup in cases (2) and (odd p). Note that in case (odd p) the ring R, also denoted R_G, is isomorphic to the reduced ring

3.6. The ideals $\mathfrak{p}(K,c)$

$H_G = H^*(BG; \mathbb{F}_p)/\sqrt{(0)}$. Unless otherwise stated X will be a finite-dimensional G-CW-complex with FMCOT. $p^* : R \to H_G^*(X; k)$ is the homomorphism induced by the bundle map $p : X_G \to BG$; and, for any subtorus $K < G$, let $i_K^* : R \to R_K$ be induced by $i_K : BK \to BG$. (Thus, in the notation of Definition (3.1.4), $PK = \ker(i_K^*)$.)

(3.6.1) Definitions
(1) Let $H_G(X) := \bigoplus_{j=0}^\infty H_G^{2j}(X; k)$ in cases (0) and (odd p), and let $H_G(X) := H_G^*(X; k)$ in case (2). Thus, in each case, $H_G(X)$ is a strictly commutative ring, and a R-algebra via p^*.
(2) Let $\mathcal{T}(X)$ be the set of all pairs (K, c), where K is a subtorus of G such that $X^K \neq \emptyset$, and c is a component of X^K.
(3) For any $(K, c) \in \mathcal{T}(X)$, define $(K, c)^* : H_G(X) \to R_K$ to be the homomorphism induced by the composition $BK \xrightarrow{\approx} \{x_0\}_K \to X_G$, where x_0 is any point in c. Since c is connected, $(K, c)^*$ is independent of the choice of $x_0 \in c$. In case (odd p), we use here the remark above that R_K is isomorphic to the reduced ring $H_K^*/\sqrt{(0)}$, in order to make R_K the codomain of $(K, c)^*$.
(4) $\mathfrak{p}(K, c) := \ker(K, c)^*$. (It is a prime ideal in $H_G(X)$, since R_K is an integral domain.)
(5) For any $x \in X$, let G_x^0 be G_x in cases (2) and (odd p) and the identity component of G_x in case (0). Let $F(G, x)$ be the component of $X^{G_x^0}$ which contains x. We shall abbreviate $F(G, x)$ to $F(x)$ when G is understood. If $(K, c) \in \mathcal{T}(X)$ and $x \in c$, then $K < G_x^0$ and $c \supseteq F(G, x)$.

Since $(K, c)^* p^* = i_K^*$, the following lemma is immediate.

(3.6.2) Lemma
(1) $(p^*)^{-1}(\mathfrak{p}(K, c)) = PK$.
(2) There is an isomorphism $R_K = R/PK \to H_G(X)/\mathfrak{p}(K, c)$ given by $r + PK \mapsto p^*(r) + \mathfrak{p}(K, c)$.

(3.6.3) Definition
Let QK be the ideal in $H_G(X)$ generated by $p^*(PK)$. So $QK \subseteq \mathfrak{p}(K, C)$.

(3.6.4) Lemma
Let A be any k-linear subspace of $H_G(X)$ and suppose that $\mathfrak{p}(K, c) \subseteq A$. Then
$$A = \mathfrak{p}(K, c) + p^*(R) \cap A.$$
In particular, if A is an ideal in $H_G(X)$ and A' is the ideal generated by $p^*(R) \cap A$, then $A = \mathfrak{p}(K, c) + A'$ as ideals. Hence, if $\mathfrak{p}(L, d) \subseteq \mathfrak{p}(K, c)$, then $\mathfrak{p}(K, c) = \mathfrak{p}(L, d) + QK$.

Proof.
Let $a \in A$, and choose $r \in R$ such that $r + PK$ maps to $a + \mathfrak{p}(K,c)$ under the isomorphism of Lemma (3.6.2)(2). Then $a - p^*(r) \in \mathfrak{p}(K,c) \subseteq A$. Thus $a = a - p^*(r) + p^*(r)$ shows that $A \subseteq \mathfrak{p}(K,c) + p^*(R) \cap A$. The reverse inequality and the rest of the lemma are immediate. □

(3.6.5) Remark
If $K < L$ and $c \supseteq d$, and we choose $x_0 \in d$, then Diagram 3.2

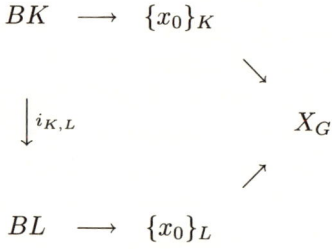

Diagram 3.2

shows that $(K,c)^* = i_{K,L}^*(L,d)^*$; and, hence, $\mathfrak{p}(K,c) \supseteq \mathfrak{p}(L,d)$.
The next lemma follows at once from Remarks (3.1.7)(5).

(3.6.6) Lemma
In the ring $H_G(X)$,
$$\sqrt{(0)} = \bigcap_{x \in X} \mathfrak{p}(G_x^0, F(G,x)).$$

(3.6.7) Remark
If $\dim_k H^*(X;k) < \infty$, then, for any subtorus $K < G$, X^K has a finite number of components. This follows from Corollary (3.1.13) in case (0), and from Corollary (3.10.2) or (1.3.8), resp. (1.4.8), in cases (2) and (odd p). Thus, if $\dim_k H^*(X;k) < \infty$, then the number of ideals of the form $\mathfrak{p}(G_x^0, F(G,x))$ is finite.

(3.6.8) Corollary
If $\dim_k H^*(X;k) < \infty$, and if $\mathfrak{p}$ is any prime ideal of $H_G(X)$, then $\mathfrak{p} \supseteq \mathfrak{p}(K,c)$ for some $(K,c) \in \mathcal{T}(X)$.

Proof.
$\mathfrak{p} \supseteq \sqrt{(0)}$, and hence $\mathfrak{p} \supseteq \mathfrak{p}(G_x^0, F(G,x))$ for some $x \in X$. □

3.6. The ideals $\mathfrak{p}(K,c)$

An essential difference between case (0) and the other two cases is that when G is a torus and $K < G$ is a subtorus with $X^K \neq \emptyset$, then each component of X^K is G-invariant. When G is a p-torus, $K < G$ and $X^K \neq \emptyset$, and c is a component of X^K, then the smallest G-invariant subspace of X which contains c, is $Gc = \bigcup \{gc; g \in G\}$, which is a union of components of X^K. Of course $Gc = c$ in case (0).

(3.6.9) Lemma
Let (K,c) and (K,c') be in $\mathcal{T}(X)$.

(1) *If $Gc = Gc'$, then $(K,c)^* = (K,c')^*$.*
(2) *If $Gc \neq Gc'$, $\mathfrak{p}(K,c) + \mathfrak{p}(K,c')$ contains an element in $p^*(S(K))$, where $S(K)$ is the multiplicative subset of R defined in Example (3.1.5).*

Proof.
(1) We may suppose that $c' = gc$. If we choose $x_0 \in c$, then we may define $(K,c)^*$ by means of $BK \to \{x_0\}_K \to X_G$ and $(K,c')^*$ by means of $BK \to \{gx_0\}_K \to X_G$. It follows that $(K,c')^* = g^*(K,c)^*$, where $g^* : R_K \to R_K$ is induced by the map $BK = EG/K \to BK$ given by $K(z) \mapsto K(zg)$, for any $z \in EG$. But then $g^* i_K^* = i_K^*$; and, hence, g^* is the identity since i_K^* is surjective.

(2) If $Gc \neq Gc'$, then, with $Y = X^K - (Gc \cup Gc')$,
$$H_G^*(X^K;k) \cong H_G^*(Gc;k) \oplus H_G^*(Gc';k) \oplus H_G^*(Y;k).$$
Let $u, v \in H_G^*(X^K;k)$ correspond to $(1,0,1)$ and $(0,1,0)$, respectively, under this isomorphism. Let $\phi^* : H_G^*(X;k) \to H_G^*(X^K;k)$ be the restriction. By the Localization Theorem, there exist $\bar{u}, \bar{v} \in H_G^*(X;k)$ and $a, b \in S(K)$ such that $\phi^*(\bar{u}) = au$, and $\phi^*(\bar{v}) = bv$. Let $u' = b\bar{u}$, $v' = a\bar{v}$, $e = ab$. Then $\phi^*(u' + v') = e(u+v) = \phi^*(p^*(e))$. So, by the Localization Theorem, there exists $f \in S(K)$ such that $f(u' + v') = p^*(fe)$. Clearly, $fu' \in \mathfrak{p}(K,c')$, $fv' \in \mathfrak{p}(K,c)$ and $fe \in S(K)$. $\square$

(3.6.10) Corollary
Suppose that $\dim_K H^*(X;k) < \infty$, and let $\mathfrak{p}$ be any prime ideal in $H_G(X)$, let $PK = \sigma(p^*)^{-1}(\mathfrak{p})$. Then

(1) $\mathfrak{p} \supseteq \mathfrak{p}(K,c)$ *for some* $(K,c) \in \mathcal{T}(X)$, *and*
(2) *if $\mathfrak{p} \supseteq \mathfrak{p}(L,d)$ for some $(L,d) \in \mathcal{T}(X)$, then $\mathfrak{p}(K,c) \supseteq \mathfrak{p}(L,d)$.*

Proof.

By Corollary (3.6.8), $\mathfrak{p} \supseteq \mathfrak{p}(L,d)$ for some $(L,d) \in \mathcal{T}(X)$. Then $(p^*)^{-1}(\mathfrak{p}) \supseteq (p^*)^{-1}(\mathfrak{p}(L,d)) = PL$, by Lemma (3.6.2)(1). Hence $PK \supseteq PL$, $K < L$, and $X^K \supseteq X^L$. Let c be the component of X^K which contains d. Then $\mathfrak{p}(K,c) \supseteq \mathfrak{p}(L,d)$ by Remark (3.6.5). Also $\mathfrak{p} \supseteq \mathfrak{p}(L,d) + QK = \mathfrak{p}(K,c)$ by Lemma (3.6.4).

To prove (2), suppose that $\mathfrak{p} \supseteq \mathfrak{p}(K,c')$. Then $\mathfrak{p} \supseteq \mathfrak{p}(K,c) + \mathfrak{p}(K,c')$. If $Gc \neq Gc'$, then, by Lemma (3.6.9)(2), $\mathfrak{p}$ contains an element of $p^*(S(K))$. This contradicts the fact that $\sigma(p^*)^{-1}(\mathfrak{p}) = PK$. So $Gc = Gc'$, and $\mathfrak{p}(K,c) = \mathfrak{p}(K,c')$ by Lemma (3.6.9)(1). □

(3.6.11) Definition

Let $\mathfrak{p} \in \mathrm{Spec}(H_G(X))$, the set of prime ideals of $H_G(X)$. Suppose that there exists $(K,c) \in \mathcal{T}(X)$ such that $\mathfrak{p} \supseteq \mathfrak{p}(K,c)$, and $\mathfrak{p}(K,c) \supseteq \mathfrak{p}(L,d)$ whenever $\mathfrak{p} \supseteq \mathfrak{p}(L,d)$ for some $(L,d) \in \mathcal{T}(X)$. Then we shall call $\mathfrak{p}(K,c)$ the support of $\mathfrak{p}$ and denote it by $\sigma\mathfrak{p}$. Note that the proof of Corollary (3.6.10) shows that $(p^*)^{-1}(\sigma\mathfrak{p}) = \sigma(p^*)^{-1}(\mathfrak{p})$ if $\sigma\mathfrak{p}$ exists; and $\sigma\mathfrak{p}$ exists if $\mathfrak{p}$ contains any ideal of the form $\mathfrak{p}(L,d)$ for some $(L,d) \in \mathcal{T}(X)$. Furthermore, Corollary (3.6.10) shows that $\sigma\mathfrak{p}$ exists for all $\mathfrak{p} \in \mathrm{Spec}(H_G(X))$ if $\dim_K H^*(X;k) < \infty$.

(3.6.12) Definition

Given $(K,c) \in \mathcal{T}(X)$, let $T(K,c)$, resp. $S(K,c)$, be the multiplicative set of all, resp. all homogeneous, elements in $H_G(X) - \mathfrak{p}(K,c)$.

(3.6.13) Proposition

$$X^{T(K,c)} = X^{S(K,c)} = Gc.$$

Furthermore, if $\mathfrak{p} \in \mathrm{Spec}(H_G(X))$ *and* $\sigma\mathfrak{p} = \mathfrak{p}(K,c)$, *then* $X^S = Gc$, *where* $S = H_G(X) - \mathfrak{p}$.

Proof.

Let $x_0 \in Gc$, $u \in T(K,c)$. Since $(K,c)^*$ factors through $H_G(G(x_0))$, $j^*_{x_0}(u) \neq 0$. Thus $Gc \subseteq X^{T(K,c)}$.

Now suppose that $x \in X^{S(K,c)}$. Then $x \in X^{S(K)} = X^K$. Suppose $x \in X^K - Gc = Z$, say.

By the Localization Theorem, as in the proof of Lemma (3.6.9)(2), there exists $w \in H_G(X)$ such that w restricts to $0 \in H_G(Z)$, and w restricts to $e \in H_G(Gc)$, for some $e \in S(K)$. Hence $w \in S(K,c)$ and $j^*_x(w) = 0$. This contradiction shows that $X^{S(K,c)} \subseteq Gc$.

Finally suppose $\mathfrak{p} \in \mathrm{Spec}(H_G(X))$, $\sigma\mathfrak{p} = \mathfrak{p}(K,c)$ and $S = H_G(X) - \mathfrak{p}$.

Then $S \subseteq T(K,c)$, and so $X^S \supseteq X^{T(K,c)} = Gc$. On the other hand, suppose $x \in X^S$. Since $p^*(S(K)) \subseteq S$, $X^S \subseteq X^{S(K)} = X^K$. Suppose $x \in Z = X^K - Gc$. Then, as above, there exists $w \in H_G(X)$ such that w restricts to $0 \in H_G(Z)$ and w restricts to $e \in H_G(Gc)$ for some $e \in S(K)$. Then $j_x^*(w) = 0$ and $w - p^*(e) \in \mathfrak{p}(K,c)$, $j_x^*(w) = 0$ and $x \in X^S$ implies that $w \in \mathfrak{p}$. Hence $p^*(e) \in \mathfrak{p}$, which is a contradiction. Thus $X^S \subseteq Gc$. □

We can now prove a Localization Theorem for $H_G(X)$-modules.

(3.6.14) Theorem
Let X be a finite-dimensional G-CW-complex with FMCOT, let $Y \subseteq X$ be a G-CW-subcomplex, and assume case (0), (2) or (odd p). Let $(K,c) \in \mathcal{T}(X)$, and let $\psi(c): Gc \cap Y \to Y$ and $\phi(c): (Gc, Gc \cap Y) \to (X, Y)$ be the inclusions. Then

$$S^{-1}\psi(c)^* : S^{-1}H_G^*(Y;k) \to S^{-1}H_G^*(Gc \cap Y;k)$$

and

$$S^{-1}\phi(c)^* : S^{-1}H_G^*(X,Y;k) \to S^{-1}H_G^*(Gc, Gc \cap Y;k)$$

are isomorphisms of $S^{-1}H_G(X)$-algebras, where $S = S(K,c)$ or $T(K,c)$, as defined in (3.6.12), or $S = H_G(X) - \mathfrak{p}$, where $\mathfrak{p} \in \operatorname{Spec} H_G(X)$ and $\mathfrak{p}$ has support $\sigma\mathfrak{p} = \mathfrak{p}(K,c)$.

(Recall that $Gc = c$ in case (0). Also, of course, $Gc = c$ in cases (2) or (odd p) if $K - G$.)

Proof.
Thanks to the long exact sequence for (X,Y), it is enough to show that $S^{-1}H_G^*(Y;k) \to S^{-1}H_G^*(Gc \cap Y;k)$ is an isomorphism. Let S' be the multiplicative subset of $H_G(X)$ generated by $p^*(S(K))$ and an element w as given by the proof of Proposition (3.6.13). Then $S' \subseteq S(K,c) \subseteq T(K,c)$ and $S' \subseteq H_G(X) - \mathfrak{p}$ if $\sigma\mathfrak{p} = \mathfrak{p}(K,c)$; and $Y^{S'} = X^{S'} \cap Y = Gc \cap Y$. Now apply Corollary (3.2.7) or Remark (3.2.8). □

Another interesting consequence of Proposition (3.6.13) is that the function $\mathfrak{p}: \mathcal{T}(X) \to \operatorname{Spec}(H_G(X))$, given by $(K,c) \mapsto \mathfrak{p}(K,c)$ is order reversing when $\mathcal{T}(X)$ is given the following geometric partial ordering.

(3.6.15) Definition
Given (K,c) and (L,d) in $\mathcal{T}(X)$, then $(K,c) \leq (L,d)$ if $K < L$ and $Gc \supseteq Gd$.

(3.6.16) Proposition
Given (K,c) and (L,d) in $T(X)$, then $(K,c) \leq (L,d)$ if and only if $\mathfrak{p}(K,c) \supseteq \mathfrak{p}(L,d)$.

Proof.
If $(K,c) \leq (L,d)$, then $\mathfrak{p}(K,c) \supseteq \mathfrak{p}(L,d)$ by Remark (3.6.5) and Lemma (3.6.9)(1).

If $\mathfrak{p}(K,c) \supseteq \mathfrak{p}(L,d)$, then $S(K,c) \subseteq S(L,d)$. Hence, by Proposition (3.6.13), $Gc = X^{S(K,c)} \supseteq X^{S(L,d)} = Gd$. Since $PK = (p^*)^{-1}(\mathfrak{p}(K,c)) \supseteq (p^*)^{-1}(\mathfrak{p}(L,d)) = PL$, $K < L$. $\square$

We shall now give an application of the ideals $\mathfrak{p}(K,c)$ to uniform actions which are defined as follows.

(3.6.17) Definition
We shall say that the action of G on X is uniform if, given any two subtori K, $L < G$ with $K < L$ and $X^L \neq \emptyset$, then every component of X^K contains a component of X^L: i.e. if $K < L$ and $X^L \neq \emptyset$, then, for any $(K,c) \in T(X)$, there exists $(L,d) \in T(X)$ such that $(K,c) \leq (L,d)$.

The following theorem characterizes uniform actions when $X^G \neq \emptyset$. Recall that we are assuming that X is a finite-dimensional G-CW-complex with FMCOT, and we are assuming case (0), (2), or (odd p).

(3.6.18) Theorem
Suppose $X^G \neq \emptyset$, and let N be the R-submodule (and ideal) of $H_G(X)$ consisting of all R-torsional elements. (Thus, by the Localization Theorem, $N = \ker[\phi^* : H_G(X) \to H_G^*(X^G; k)]$.) Then the action is uniform if and only if $N \subseteq \sqrt{(0)}$: i.e. every torsional element is nilpotent.

Proof.
Let $\{F_i; i \in I\}$ be the set of components of X^G. Clearly $N \subseteq \bigcap_{i \in I} \mathfrak{p}(G, F_i)$. On the other hand, by examining the homomorphisms
$$H_G(X) \xrightarrow{\phi^*} H_G^*(X^G; k) \cong H_G^* \otimes H^*(X^G; k) \to \prod_{i \in I} H_G^* \otimes H^*(F_i; k),$$
where $H_G^* = H^*(BG; k)$, it follows that if $u \notin \sqrt{N}$, then, for some $i \in I$, $u \notin \mathfrak{p}(G, F_i)$. Hence $\bigcap_{i \in I} \mathfrak{p}(G, F_i) \subseteq \sqrt{N}$; and so $\bigcap_{i \in I} \mathfrak{p}(G, F_i) = \sqrt{N}$.

Now $\sqrt{(0)} = \bigcap_{x \in X} \mathfrak{p}(G_x^0, F(G,x))$, by Lemma (3.6.6). Hence, if the action is uniform $\sqrt{(0)} = \bigcap_{i \in I} \mathfrak{p}(G, F_i)$; and so $N \subseteq \sqrt{(0)}$.

Now suppose $N \subseteq \sqrt{(0)}$, but there exists $(K,c) \in T(X)$ such that $c \cap X^G = \emptyset$. Then $Gc \cap X^G = (Gc)^G = \emptyset$. Let $\psi_c : Gc \to X$ be

3.6. The ideals $\mathfrak{p}(K, c)$

the inclusion, and let $p^* : R \to H_G(X)$ be as above (i.e. induced by $p : X_G \to B_G$). By the Localization Theorem, there exists $r \in R$ with $r \neq 0$, such that $\psi_c^* p^*(r) = 0$. ($\psi_c^* p^* = p_c^*$, where $p_c : (Gc)_G \to BG$.) By Theorem (3.6.14), there exists $a \in H_G(X) - \mathfrak{p}(K, c)$ such that $ap^*(r) = 0$.

Since $r \neq 0$, $a \in N$. By assumption $N \subseteq \sqrt{(0)}$, and, clearly, $\sqrt{(0)} \subseteq \mathfrak{p}(K, c)$. So $a \in \mathfrak{p}(K, c)$, a contradiction. Thus, if $N \subseteq \sqrt{(0)}$, then the action is uniform. $\square$

(3.6.19) Corollary
If X is TNHZ in $X_G \to BG$, then the action is uniform.

Proof.
$N = (0)$ in this case. $\square$

(Corollary(3.6.19) is easy to see if $\dim_k H^*(X; k) < \infty$. For then it follows from the fact that $\dim_k H^*(X^G; k) \leq \dim_k H^*(X; k)$, with equality if and only if X is TNHZ in $X_G \to BG$. See Theorem (3.10.4).)

The corollary and any contemplation of examples show that uniform actions are common. Here is an example of a non-uniform action.

(3.6.20) Example
Let $T = S^1 \times S^1$ and let $X = S^2 \times S^3$. Regard S^2 as the unreduced suspension of S^1, and let $[z, t] \in S^2$ be the suspension class of $(z, t) \in S^1 \times I$, I being the unit interval. Also view S^1 as a maximal torus of $S^3 = Sp(1) = SU(2)$. Since $\pi_1(S^3) = 0$, there exists $h : S^1 \times I \to S^3$ such that $h(z, 0) = 1$, the group identity element in S^3, and $h(z, 1) = z$, for all $z \in S^1$. Now let T act on X by the formula

$$(w_1, w_2) \cdot ([z, t], g) = ([zw_1^{-1}, t], h(zw_1^{-1}, t)w_2 h(z, t)^{-1} g w_2^{-1}).$$

where $w_1, w_2, z \in S^1$, $t \in I$, and $g \in S^3$.

It is easy to check that this is a well-defined action: and $X^T = S^1$, whereas $X^K = S^1 + S^1$ (two components), where K is the subcircle $w_1 = w_2^2$. Thus the action is non-uniform.

(3.6.21) Remark
If G acts uniformly on X with $X^G \neq \emptyset$, then, for any $(K, c) \in \mathcal{T}(X)$, $Gc = c$. (There is $x \in X^G$ with $x \in c$.)

We shall now give a very interesting theorem due to Quillen ([Quillen, 1971b]), which relates the ideals $\mathfrak{p}(K, c)$ to the action of the Steenrod algebra in cases (2) and (odd p). As usual X will be a finite-dimensional G-CW-complex. In case (2), let $P^i = Sq^i$, the Steenrod operation of degree i, and, in case (odd p), let P^i be the Steenrod operation of degree $2i(p-1)$, for $i \geq 0$. Thus, in both cases, P^i operates on $H_G(X)$. $\mathfrak{p} \in \mathrm{Spec}(H_G(X))$ is said to be stable under the Steenrod operations if $P^i \mathfrak{p} \subseteq \mathfrak{p}$, for all $i \geq 0$.

(3.6.22) Theorem
Assume case (2) or (odd p), and assume that the support $\sigma\mathfrak{p}$ exists for all $\mathfrak{p} \in \mathrm{Spec}(H_G(X))$. (E.g., assume that $\dim_k H^*(X;k) < \infty$: see Definition (3.6.11).) Let $\mathfrak{p} \in \mathrm{Spec}(H_G(X))$. Then $\mathfrak{p} = \mathfrak{p}(K,c)$ for some $(K,c) \in \mathcal{T}(X)$, i.e. $\mathfrak{p} = \sigma\mathfrak{p}$, if and only if $\mathfrak{p}$ is homogeneous and stable under the Steenrod operations.

Proof.
If t_j is a polynomial generator of R_K, $K < G$, then $P^1 t_j = t_j^p$. So P^i operates on R_K, for all $i \geq 0$, and, clearly, $P^i(K,c)^* = (K,c)^* P^i$. Hence each $\mathfrak{p}(K,c)$ is homogeneous and stable under the Steenrod operations.

Now let $\mathfrak{p} \in \mathrm{Spec}(H_G(X))$, and suppose that $\mathfrak{p}$ is homogeneous and stable under the Steenrod operations. Let $\sigma\mathfrak{p} = \mathfrak{p}(K,c)$, and let $\mathfrak{q} = (K,c)^*(\mathfrak{p}) \in \mathrm{Spec}(R_K)$. Then $P^i \mathfrak{q} = (K,c)^* P^i \mathfrak{p} \subseteq \mathfrak{q}$, for all $i \geq 0$. By [Serre, 1965b], Proposition (1), $\mathfrak{q} = \ker[R_K \to R_L]$, for some $L < K$. So $PL \subseteq (p^*)^{-1}(\mathfrak{p})$. Hence $PL \subseteq \sigma p^{*-1}(\mathfrak{p}) = PK$. $L = K$, therefore: and so $\mathfrak{q} = (0)$, and $\mathfrak{p} = \mathfrak{p}(K,c)$. □

(The theorem in [Quillen, 1971b], Section 12, is more general, since for Quillen G is any compact Lie group.)

(3.6.23) General G-spaces
In cases (B) or (C) of Section 3.2, and cases (0), (2) or (odd p), using Alexander-Spanier cohomology instead of singular cohomology, one can recover all the results of this section provided:
(1) One assumes the following condition.

(CS): given any $(K,c) \in \mathcal{T}(X)$, let $\{Gc_i; i \in I\}$ be the set of distinct Gc', where c' ranges over the components of X^K, indexed so that $Gc = Gc_{i_0}$. Let $q_i : Gc_i \to X^K$, $i \in I$, be the inclusions. Then there exists $y \in S(K)^{-1} \overline{H}_G(X^K)$, such that $S(K)^{-1} q_i^*(y) = 0$, for all $i \in I$ with $i \neq i_0$, and $S(K)^{-1} q_{i_0}^*(y) = 1$. (Cf. Definition (3.4.1).)

(2) Wherever one assumes that $\dim_k H^*(X;k) < \infty$ above (i.e. in Corollaries (3.6.8) and (3.6.10)), one assumes that $\dim_k \overline{H}^*(X;k) < \infty$.

Note that (CS) is satisfied if Gc is open in X^K for every $(K,c) \in \mathcal{T}(X)$. In particular, (CS) holds if $\dim_k \overline{H}^*(X;k) < \infty$.

Note also that (CS) is adequate for Lemma (3.6.9)(2), since
$$\ker(S(K)^{-1}(K,c)^*) = S(K)^{-1}\mathfrak{p}(K,c),$$
and
$$S(K)^{-1}\mathfrak{p}(K,c) + S(K)^{-1}\mathfrak{p}(K,c') = S(K)^{-1}\{\mathfrak{p}(K,c) + \mathfrak{p}(K,c')\}.$$
Also (CS) suffices for the proofs of (3.6.13) and Theorem (3.6.14): e.g., in the proof of Proposition (3.6.14), one can choose $w \in \overline{H}_G(X)$ such that $w/1 \in S(K)^{-1}\overline{H}_G(X)$ restricts to $0 \in S(K)^{-1}\overline{H}_G(Gc_i)$ for $i \neq i_0$, and to $s/1 \in S(K)^{-1}\overline{H}_G(Gc)$ for some $s \in S(K)$; then, for each $i \neq i_0$, there is an $s_i \in S(K)$ such that $s_i w \in \overline{H}_G(X)$ restricts to $0 \in \overline{H}_G(Gc_i)$; and, hence, $X^{S'} = Gc$, where S' is the multiplicative set generated by $p^*(S(K))$ and w in $\overline{H}_G(X)$.

In case (A), i.e. X a compact G-space, and cases (0), (2) or (odd p), then, using Alexander-Spanier cohomology, all the results of this section are valid without any further assumptions. Lemma (3.4.4) and its analogues in cases (2) and (odd p) dispense with the need for any assumption such as (CS). (Note that the analogues of Lemma (3.4.4) in cases (2) and (odd p) are valid, although the proof is a little more complicated.) Theorem (3.6.14) is a special case of Corollary (3.2.7). The assumption that $\dim_k \overline{H}^*(X;k) < \infty$ is not needed for Corollaries (3.6.8) or (3.6.10) by a typical covering argument, as follows. Let $\mathfrak{p} \in \mathrm{Spec}(\overline{H}_G(X))$, X compact. Suppose $\mathfrak{p}$ does not contain any ideal of the form $\mathfrak{p}(G_x^0, F(G,x))$. Then, for any $x \in X$, there is $u_x \in \mathfrak{p}(G_x^0, F(G,x))$ such that $u_x \notin \mathfrak{p}$. So $j_x^*(u_x)$ is nilpotent; and hence, for some $n_x > 0$, $u_x^{n_x}$ restricts to 0 or some neighbourhood V_x of $G(x)$. X is covered by $V_{x_1}, \ldots, V_{x_m}$, say, and it follows that $u = u_{x_1}^{n_{x_1}} \ldots u_{x_m}^{n_{x_m}}$ is nilpotent. Thus $u \in \mathfrak{p}$, a contradiction.

In the proof of Theorem (3.6.18) we use the fact that any element of positive degree in $H^*(X^G;k)$ is nilpotent. This is true since X^G is finite-dimensional. Conditions (A), (B) or (C) also imply that any element of positive degree in $\overline{H}^*(X^G;k)$ is nilpotent.

Finally note that Theorem (3.6.22) is valid for general G-spaces, whenever $\sigma\mathfrak{p}$ exists for all $\mathfrak{p} \in \mathrm{Spec}(\overline{H}_G(X))$, since Steenrod operations can be defined for Alexander-Spanier cohomology; see, e.g., [Bredon, 1967a] or [Steenrod, Epstein, 1962].

See [Quillen, 1971a, b] and [Allday, 1976] for more about the ideals $\mathfrak{p}(K,c)$, especially in the case of a more general compact Lie group action.

3.7 Chang-Skjelbred modules and Hsiang-Serre ideals

In this section we give some results of Chang and Skjelbred concerning the primary decomposition of certain ideals, which have geometric significance. These results first appeared in [Chang, Skjelbred, 1974], but they have reappeared in other situations: hence we shall begin with an abstract treatment suitable for all applications, and then give the main examples later in the section.

(3.7.1) Definitions
A pre-Chang-Skjelbred module is a system of modules and prime ideals consisting of the following ingredients:

(1) A field k and a strictly commutative non-negatively graded connected Noetherian k-algebra with identity $R = R^* = \bigoplus_{i=0}^{\infty} R^i$. ($R$ connected means that $R^0 = k$.) Note that any finitely generated k-algebra is Noetherian.
(2) A partially ordered set $\mathcal{I}$ and a set of homogeneous prime ideals $\text{Strat}(R) \subset \text{Spec}(R)$ indexed by $\mathcal{I}$. Thus $\text{Strat}(R) = \{P_j : j \in \mathcal{I}\}$ with each P_j a homogeneous prime ideal in R. Furthermore, the indexing is strictly order reversing: $P_i \subseteq P_j$ if and only if $i \geq j$, for $i, j \in \mathcal{I}$.
(3) A function $\sigma : \text{Spec}(R) \to \text{Strat}(R)$ with the following properties:
    (i) $\mathfrak{p} \supseteq \sigma\mathfrak{p}$, for all $\mathfrak{p} \in \text{Spec}(R)$;
    (ii) if $\mathfrak{p} \supseteq P_j$, for some $j \in \mathcal{I}$, then $\sigma\mathfrak{p} \supseteq P_j$.
    (It follows that $\sigma P_j = P_j$, for all $j \in \mathcal{I}$; and, if $\mathfrak{p} \supseteq \mathfrak{q}$, for $\mathfrak{p}, \mathfrak{q} \in \text{Spec}(R)$, then $\sigma\mathfrak{p} \supseteq \sigma\mathfrak{q}$.)
(4) A R-module M, and, for each $j \in \mathcal{I}$, a R-module M_j, and, for each $j \in \mathcal{I}$, a R-module homomorphism $\phi_j : M \to M_j$.

A pre-Chang-Skjelbred module will be called a Chang-Skjelbred module if it satisfies the following two conditions:

(5) (The localization condition.) Given $\mathfrak{p} \in \text{Spec}(R)$, let $\sigma\mathfrak{p} = P_j$. Then $(\phi_j)_\mathfrak{p} : M_\mathfrak{p} \to (M_j)_\mathfrak{p}$, i.e. ϕ_j localized at $\mathfrak{p}$, is an isomorphism.
(6) (The monomorphism condition.) For all $j \in \mathcal{I}$, the localization homomorphism $M_j \to (M_j)_{P_j}$ is a monomorphism.

(3.7.2) Example

As a preliminary example let X be a finite-dimensional G-CW-complex with FMCOT in case (0), (2) or (odd p). Let R be as usual, σ as in Definition (3.1.4), $\mathcal{I}$ the set of subtori of G, $M = H_G^*(X;k)$, and for a subtorus $K \in \mathcal{I}$, $P_K = PK$ and $M_K = H_G^*(X^K;k)$. Condition (5) is just the Localization Theorem, and condition (6) follows from the special structure conditions discussed below.

(3.7.3) Definition

A pre-Chang-Skjelbred module is said to have **special structure** if it satisfies the following conditions:

(SS1) For each $j \in \mathcal{I}$, there is a strictly commutative non-negatively graded connected k-algebra A_j, and a graded k-algebra homomorphism $\theta_j : R \to R/P_j \otimes_k A_j$ such that $(1 \otimes \varepsilon_j)\theta_j = q_j$, where $q_j : R \to R/P_j$ is the quotient map and $\varepsilon_j : A_j \to k$ is the standard augmentation (i.e. $\varepsilon_j(r) = 0$ if $\deg(r) > 0$).

(SS2) For each $j \in \mathcal{I}$, M_j is a non-negatively graded R-module, and there is a non-negatively graded A_j-module W_j, and a graded R-module isomorphism $\psi_j : M_j \to R/P_j \otimes_k W_j$, where the R-module structure on the codomain is given by θ_j.

(3.7.4) Remarks

(1) In Example (3.7.2), for $K \in \mathcal{I}$, $G = K \times L$, where $L \cong G/K$. Thus $R \cong R_K \otimes R_L = R/P_K \otimes R_L$; and so (SS1) follows by taking A_K to be $R_L \cong R_{G/K}$. For (SS2), take $W_K = H_L^*(X^K; k)$, in cases (0) and (2), and take $W_K = \Lambda(s_1, \ldots, s_m) \otimes H_L^*(X^K; k)$ in case (odd p), where $\Lambda(s_1, \ldots, s_m)$ is the exterior algebra part of $H_K^* = H^*(BK; k)$ - that is $H_K^* = \Lambda(s_1, \ldots, s_m) \otimes R_K$.

(2) We call such a complicated object a Chang-Skjelbred *module* because it can be described as a sheaf-theoretic module over $\mathrm{Spec}(R)$. The monomorphism condition can be described very nicely using two different topologies on $\mathrm{Spec}(R)$, the Zariski topology and a topology defined analogously using the ideals from $\mathrm{Strat}(R)$.

(3) When the rest of the structure is understood, we shall usually refer to M as the Chang-Skjelbred module (or pre-Chang-Skjelbred module).

(4) The modules M, M_j (for $j \in \mathcal{I}$) are not assumed to be graded, except where stated in the special structure conditions or elsewhere.

(5) Note that the monomorphism condition implies that, for any $j \in \mathcal{I}$, the localization homomorphism $M_j \to (M_j)_{\mathfrak{p}}$ is a monomorphism whenever $\sigma\mathfrak{p} = P_j$.

(6) From Diagram 3.3

$$\begin{array}{ccc} M & \xrightarrow{\phi_j} & M_j \\ \downarrow & & \downarrow \\ M_{\mathfrak{p}} & \longrightarrow & (M_j)_{\mathfrak{p}} \end{array}$$

Diagram 3.3

it follows that if M is a Chang-Skjelbred module, then, for any $j \in \mathcal{I}$ and $\mathfrak{p} \in \mathrm{Spec}(R)$ with $\sigma\mathfrak{p} = P_j$,

$$\ker(\phi_j) = \ker[M \to M_\mathfrak{p}] = \{x \in M; \text{ for some } a \in R - \mathfrak{p}, ax = 0\}.$$

(3.7.5) Lemma
A pre-Chang-Skjelbred module, which satisfies the special structure conditions (SS1) and (SS2), satisfies the monomorphism condition (6).

Proof.
Let $x \in M_j, r \in R$. Suppose $r \notin P_j$, and $x \neq 0$. By means of θ_j and ψ_j we may set $x = a_1 \otimes x_1 + \ldots + a_n \otimes x_n$ and $r = b \otimes 1 + b_1 \otimes c_1 + \ldots + b_m \otimes c_m$, where $a_i \in R/P_j$ and $a_i \neq 0$ for $1 \leq i \leq n$, b, $b_i \in R/P_j$ for $1 \leq i \leq m$, $c_i \in A_j$ and $\deg(c_i) > 0$ for $1 \leq i \leq m$, and $x_1, \ldots, x_n$ are homogeneous elements of W_j, which are linearly independent over k. Suppose

$$\deg(x_1) = \ldots = \deg(x_{i_1}) < \deg(x_{i_1+1}) \leq \ldots \leq \deg(x_n).$$

If $rx = 0$, then $ba_1 \otimes x_1 + \ldots + ba_{i_1} \otimes x_{i_1} = 0$. Since $r \notin P_j$, $b \neq 0$. But the independence of $x_1, \ldots, x_n$ implies that $ba_1 = \ldots = ba_{i_1} = 0$, a contradiction. $\square$

(3.7.6) Definitions
(1) Let M be a pre-Chang-Skjelbred module. Let $N \subseteq M$ be a R-submodule. Then we may view N as a pre-Chang-Skjelbred module by setting $N_j = \phi_j(N)$, and restricting ϕ_j to N. If M is a Chang-Skjelbred module, then it is easy to see that N, as so defined, is also a Chang-Skjelbred module: we shall call it the Chang-Skjelbred submodule defined by N.
(2) We shall say that a (pre-) Chang-Skjelbred module M is finitely generated if M is finitely generated as a R-module.

(3.7.7) Lemma
Let M be a finitely generated Chang-Skjelbred module. Then $\mathrm{ann}(M) \in \mathrm{Strat}(R)$, if $\mathrm{ann}(M) \in \mathrm{Spec}(R)$.
(Recall that $\mathrm{ann}(M) = \{r \in R; rx = 0, \forall x \in M\} \triangleleft R$.)

Proof.
Suppose $\mathrm{ann}(M) = \mathfrak{p} \in \mathrm{Spec}(R)$. Since M is finitely generated $M_\mathfrak{p} \neq 0$. Let $\sigma\mathfrak{p} = P_j$. Then $M_j \neq 0$ by the localization condition (5), and there exists $x \in M$ such that $\phi_j(x) \neq 0$. By the monomorphism condition (6), $\mathrm{ann}(\phi_j(x)) \subseteq P_j$. But $\mathfrak{p} \subseteq \mathrm{ann}(\phi_j(x))$. Hence $\mathfrak{p} = P_j$. $\square$

(3.7.8) Corollary
If M is a Chang-Skjelbred module, then
$$\mathrm{Ass}(M) \subseteq \mathrm{Strat}(R).$$

Proof.
Recall that $\mathrm{Ass}(M) = \{\mathfrak{p} \in \mathrm{Spec}(R); \mathfrak{p} = \mathrm{ann}(x), \text{ for some } x \in M\}$. Let $\mathfrak{p} = \mathrm{ann}(x) \in \mathrm{Ass}(M)$. Let N be the submodule of M generated by x. So $\mathfrak{p} = \mathrm{ann}(N)$. But N defines a finitely generated Chang-Skjelbred submodule of M; and so Lemma (3.7.7) applies. □

Suppose $\mathfrak{p} \in \mathrm{Spec}(R)$. Let $i_\mathfrak{p} : R \to R_\mathfrak{p}$ be the localization homomorphism. Let J be an ideal in $R_\mathfrak{p}$. In what follows we shall denote the ideal $i_\mathfrak{p}^{-1}(J)$ in R by $J \cap R$, even if $i_\mathfrak{p}$ is not one-to-one.

Let M be a finitely generated R-module and let $\mathfrak{p} \in \mathrm{Spec}(R)$. Then $\mathrm{ann}(M)_\mathfrak{p} = \mathrm{ann}(M_\mathfrak{p})$. (Note $\mathrm{ann}(M)_\mathfrak{p} \subseteq \mathrm{ann}(M_\mathfrak{p})$, for any R-module M; and the reverse inclusion follows if M is finitely generated.) Now suppose that M is a Chang-Skjelbred module, and that $\sigma\mathfrak{p} = P_j$. Then $\mathrm{ann}(M_\mathfrak{p}) = \mathrm{ann}((M_j)_\mathfrak{p})$; and, by the monomorphism condition,
$$\mathrm{ann}((M_j)_\mathfrak{p}) \cap R = \mathrm{ann}(M_j).$$
Thus we have the following.

(3.7.9) Lemma
If M is a finitely generated Chang-Skjelbred module, $\mathfrak{p} \in \mathrm{Spec}(R)$ and $\sigma\mathfrak{p} = P_j$, then $\mathrm{ann}(M)_\mathfrak{p} \cap R = \mathrm{ann}(M_j)$.

(3.7.10) Theorem
Let M be a non-zero finitely generated Chang-Skjelbred module. Let $\mathrm{ann}(M) = \bigcap_{i=1}^{r+s} Q_i$ be a reduced primary decomposition of $\mathrm{ann}(M)$ in the Noetherian ring R, where Q_i, for $1 \leq i \leq r$, are the isolated components, and Q_{r+i}, for $1 \leq i \leq s$, are embedded components. Then
(1) $\sqrt{Q_i} \in \mathrm{Strat}(R)$ *for* $1 \leq i \leq r+s$, *and*
(2) $Q_i = \mathrm{ann}(M_{j(i)})$ *for* $1 \leq i \leq r$, *where* $\sqrt{Q_i} = P_{j(i)}$.

Proof.
For $1 \leq i \leq r+s$, $\sqrt{Q_i} \in \mathrm{Ass}(R/\mathrm{ann}(M))$. So there exists $r_i \in R$ such that $\sqrt{Q_i} = \mathrm{ann}(r_i M)$. So $\sqrt{Q_i} \in \mathrm{Strat}(R)$ by Lemma (3.7.7).

For $1 \leq i \leq r$, $Q_i = \mathrm{ann}(M)_{P_{j(i)}} \cap R$, where $P_{j(i)} = \sqrt{Q_i}$. So $Q_i = \mathrm{ann}(M_{j(i)})$ by Lemma (3.7.9). (See Proposition (A.3.8).) □

(3.7.11) Definition
Let M be a Chang-Skjelbred module. Then we shall say that $j \in \mathcal{I}$ belongs to M if $M_j \neq 0$, but $M_i = 0$, whenever $i > j$.

By the localization and monomorphism conditions, for any $\mathfrak{p} \in \mathrm{Spec}(R)$ with $\sigma \mathfrak{p} = P_j$,
$$M_\mathfrak{p} \neq 0 \Leftrightarrow (M_j)_\mathfrak{p} \neq 0 \Leftrightarrow M_j \neq 0 \Leftrightarrow M_{P_j} \neq 0.$$
Now recall that $\mathrm{Supp}(M) = \{\mathfrak{p} \in \mathrm{Spec}(R); M_\mathfrak{p} \neq 0\}$. Thus $\mathfrak{p} \in \mathrm{Supp}(M) \Rightarrow \sigma \mathfrak{p} \in \mathrm{Supp}(M)$; and the set of minimal elements of $\mathrm{Supp}(M)$, $\min \mathrm{Supp}(M) = \{P_j; j \text{ belongs to } M\}$.

Also $\mathrm{Ass}(M) \subseteq \mathrm{Supp}(M)$, and the set of minimal elements of $\mathrm{Ass}(M)$, $\min \mathrm{Ass}(M) = \min \mathrm{Supp}(M)$. (See Proposition (A.2.3)(9).) If M is finitely generated, then
$$\mathrm{Supp}(M) = V(\mathrm{ann}(M)) := \{\mathfrak{p} \in \mathrm{Spec}(R); \mathfrak{p} \supseteq \mathrm{ann}(M)\}.$$
(See Theorem (A.2.6).)

(3.7.12) Theorem
Let M be a non-zero finitely generated Chang-Skjelbred module. Let $\mathrm{ann}(M) = \bigcap_{i=1}^{r+s} Q_i$ be a reduced primary decomposition of $\mathrm{ann}(M)$ in R, where Q_i, for $1 \leq i \leq r$, are the isolated components, and Q_{r+i}, for $1 \leq i \leq s$, are embedded components. Let $\sqrt{Q_i} = P_{j(i)} \in \mathrm{Strat}(R)$. Then

(1) *$j \in \mathcal{I}$ belongs to M if and only if $j = j(i)$ for some $i \in \{1, \ldots, r\}$; and*

(2) *if $r+1 \leq i \leq r+s$, then $\mathrm{ann}(M_{j(i)}) \subsetneqq \bigcap_{t > j(i)} \mathrm{ann}(M_t)$.*

Proof.

(1) $\{P_{j(i)}; 1 \leq i \leq r\} = \min V(\mathrm{ann}(M)) = \min \mathrm{Supp}(M) = \{P_j; j \text{ belongs to } M\}$, by the discussion above.

(2) Suppose $r+1 \leq i \leq r+s$, and $t > j(i)$. So $P_t \subset P_{j(i)}$. If $P_t \notin V(\mathrm{ann}(M))$, then $P_t \notin \mathrm{Supp}(M)$; and so $M_t = 0$.

Hence $\bigcap_{t > j(i)} \mathrm{ann}(M_t) = \bigcap \{\mathrm{ann}(M_t); \mathrm{ann}(M) \subseteq P_t \subset P_{j(i)}\}$. On the other hand, by Lemma (3.7.9),
$$\mathrm{ann}(M_t) = \mathrm{ann}(M)_{P_t} \cap R = \bigcap \{Q_m; P_{j(m)} \subseteq P_t\}$$
by Proposition (A.3.6). Thus
$$\bigcap_{t > j(i)} \mathrm{ann}(M_t) = \bigcap_{j(m) > j(i)} Q_m;$$
and
$$\bigcap_{j(m) > j(i)} Q_m \neq \bigcap_{j(m) \geq j(i)} Q_m,$$

by the irredundancy of the primary decomposition. Finally,
$$\bigcap_{j(m)\geq j(i)} Q_m = \operatorname{ann}(M)_{P_{j(i)}} \cap R = \operatorname{ann}(M_{j(i)}).$$

□

(3.7.13) Remarks
(1) With M and $\operatorname{ann}(M) = \bigcap_{i=1}^{r+s} Q_i$ as in Theorems (3.7.10) and (3.7.12), then
$$\{P_{j(i)} : 1 \leq i \leq r+s\} = \operatorname{Ass}(R/\operatorname{ann}(M)) \subseteq \operatorname{Ass}(M).$$

(2) Suppose that a Chang-Skjelbred module M is also a k-algebra with identity 1, and that the R-module structure on M is given by an identity preserving k-algebra homomorphism $R \to M$. Then $\operatorname{ann}(M) = \operatorname{ann}(1)$, and hence Theorems (3.7.10) and (3.7.12) apply to $\operatorname{ann}(M)$ even if M is not finitely generated.

(3.7.14) Definition
Let R, $\operatorname{Strat}(R)$ and σ be as in Definitions (3.7.1), (1)-(3). Let J be an ideal in R, and let $J = \bigcap_{i=1}^{r+s} Q_i$ be a reduced primary decomposition of J with Q_i, for $1 \leq i \leq r$, the isolated components, and Q_{r+i}, for $1 \leq i \leq s$, embedded components. Let $\mathfrak{p}_i = \sqrt{Q_i}$ for $1 \leq i \leq r+s$. We shall say that J is a Hsiang-Serre ideal if the minimal primes of J, $\mathfrak{p}_i$, for $1 \leq i \leq r$, are all in $\operatorname{Strat}(R)$; and we shall say that J is a strong Hsiang-Serre ideal if all the primes, $\mathfrak{p}_i$, for $1 \leq i \leq r+s$, are in $\operatorname{Strat}(R)$. Thus, if M is a finitely generated Chang-Skjelbred module, then $\operatorname{ann}(M)$ is a strong Hsiang-Serre ideal. Note that $J \triangleleft R$ is a Hsiang-Serre ideal if and only if, for any $\mathfrak{p} \in \operatorname{Spec}(R)$, $\mathfrak{p} \supseteq J \Rightarrow \sigma\mathfrak{p} \supseteq J$.

Now suppose that X is a finite-dimensional G-CW-complex with FM-COT, assume case (0), (2) or (odd p) and assume the usual notation used for these cases previously. From Example (3.7.2), $H_G^*(X;k)$ is a Chang-Skjelbred module over R (with its usual meaning). From Remark (3.7.13)(2) we get the following corollary (cf. Corollary (3.1.11)).

(3.7.15) Corollary
Let $p^* : R \to H_G^*(X;k)$ be induced from $p : X_G \to BG$. Then $\ker p^*$ is a strong Hsiang-Serre ideal, and the subtori which belong to $\ker p^*$ are precisely the maximal isotropy subgroups, resp. the identity components of the maximal isotropy subgroups, in cases (2) and (odd p), respectively case (0).

Proof.

$\ker p^* = \mathrm{ann}(1)$, where $1 \in H_G^0(X;k)$ is the identity. A subtorus K belongs to $\ker p^*$ (as in Definition (3.7.11)) if and only if PK is a minimal prime of $\ker p^*$ (by Theorem (3.7.12)), if and only if $H_G^*(X^K;k) \neq 0$ and $H_G^*(X^L;k) = 0$ whenever $L \supset K$. □

With X and cases (0), (2) or (odd p) as above, we now give three examples of Chang-Skjelbred modules. Note that Theorems (3.7.10) and (3.7.12) will then apply to any finitely generated submodule. In all the examples the monomorphism condition will follow from the special structure conditions as in Lemma (3.7.5).

(3.7.16) Examples
(1) Let Y be a G-CW-subcomplex of X. Let $M = H_G^*(X,Y;k)$ and let R, $\mathcal{I}$, $\mathrm{Strat}(R)$ and σ be as in Example (3.7.2), which is the special case where $Y = \emptyset$. For any $K \in \mathcal{I}$, let
$$M_K = H_G^*(X^K, Y^K; k).$$
For special structure, let $A_K = R_L$ and, in cases (0) and (2), let
$$W_K = H_L^*(X^K, Y^K; k),$$
where $G = K \times L$, as in Remarks (3.7.4)(1). Note that
$$\psi_K : H_G^*(X^K, Y^K; k) \to R_K \otimes H_L^*(X^K, Y^K; k)$$
is just the Künneth isomorphism. As before, in case (odd p), let
$$W_K = \Lambda(s_1, \ldots, s_m) \otimes H_L^*(X^K, Y^K; k),$$
where $\Lambda(s_1, \ldots, s_m)$ is the exterior algebra part of $H_K^* = H^*(BK; k)$.

An important advantage of this generalization of Example (3.7.2) is as follows. Let $u \in H_G^*(Y;k)$, and let
$$\phi(Y)^* : H_G^*(X;k) \to H_G^*(Y;k)$$
be induced by the inclusion $\phi(Y) : Y \to X$, and let $I_u = \{r \in R : ru \in \mathrm{Im}\,\phi(Y)^*)\}$, which is clearly an ideal in R. Then $I_u = \mathrm{ann}(\delta u)$, where
$$\delta : H_G^*(Y;k) \to H_G^*(X,Y;k)$$
is the connecting homomorphism. Thus I_u is a strong Hsiang-Serre ideal. We shall say that a subtorus $K \subseteq G$ belongs to u if it belongs to (the R-module generated by) δu.

(2) Suppose here that X is connected and $\dim_k H^*(X;k) < \infty$. With the notation of Section 3.6, $H_G(X)$ is then a finitely generated k-algebra, and hence it is Noetherian. Let $R = H_G(X)$, $\mathcal{I} = \mathcal{T}(X)$,
$$\mathrm{Strat}(R) = \{\mathfrak{p}(K,c); (K,c) \in \mathcal{I}\},$$

and let
$$\sigma : \operatorname{Spec}(R) \to \operatorname{Strat}(R)$$
be as in Definition (3.6.11). Let Y be any G-CW-subcomplex of X. Let $M = H_G^*(X, Y; k)$, and let $M_{(K,c)} = H_G^*(Gc, Gc \cap Y; k)$. The localization condition is Theorem (3.6.14). For special structure, note that $R/\mathfrak{p}(K,c) \cong R_K$ (in the usual notation), by Lemma (3.6.2)(2). In case (odd p), let Λ_K be the exterior algebra part of H_K^*. In case (0) and (2), let $\theta_{(K,c)}$ be the composition
$$H_G(X) \to H_G(Gc) \cong R_K \otimes H_L(Gc),$$
where $G = K \times L$; and, in case (odd p), let $\theta_{(K,c)}$ be the composition
$$H_G(X) \to H_G(Gc) \cong R_K \otimes [\Lambda_K \otimes H_L^*(Gc; k)]^{\operatorname{ev}}.$$
So $A_{(K,c)} = H_L(Gc)$ or $[\Lambda_K \otimes H_L^*(Gc; k)]^{\operatorname{ev}}$, as the case may be. Note that $A_{(K,c)}$ is a connected k-algebra, since $(Gc)_G = (EG \times Gc)/G$ is connected, being the image of $EG \times c$ under the orbit map. Set
$$W_{(K,c)} = H_L^*(Gc, Gc \cap Y; k),$$
resp.
$$\Lambda_K \otimes H_L^*(Gc, Gc \cap Y; k)$$
in cases (0) and (2), resp. (odd p).

Similarly $H_G^*(Y; k)$ is a Chang-Skjelbred module with special structure with the same $R(= H_G(X))$, $\mathcal{I}$, $\operatorname{Strat}(R)$ and σ.

(3) Here we shall consider only case (0); and we shall suppose that X is connected, and that $X^G \neq \emptyset$. Let $x_0 \in X^G$. Let $R = H^*(BG; \mathbb{Q})$, $\mathcal{I} = \{K; K \text{ is a subtorus of } G\}$, $\operatorname{Strat}(R) = \{PK; K \in \mathcal{I}\}$ and σ be as in Example (3.7.2). Let $M = \Pi_{(G,x_0)}^*(X)$, and for $K \in \mathcal{I}$, let
$$M_K = \Pi_{(G,x_0)}^*(F(K)),$$
where $F(K)$ is the component of X^K which contains x_0. Now the localization condition is Corollary (3.3.9). The special structure condition (SS1) is as in (1) above, i.e. $\theta_K : R \to R_K \otimes R_L$, where $G = K \times L$; and (SS2) follows with
$$W_K = \Pi_{(L,x_0)}^*(F(K)),$$
since $F(K)_G = BK \times F(K)_L$ implies that
$$Q_{(G,x_0)}(F(K)) \cong R_K \otimes Q_{(L,x_0)}(F(K)),$$
and, hence,
$$\Pi_{(G,x_0)}^*(F(K)) \cong R_K \otimes \Pi_{(L,x_0)}^*(F(K)).$$

We shall now give two useful ways to construct new Chang-Skjelbred modules from given ones. Special structure appears to be needed here. We shall assume given R, $\mathcal{I}$, Strat(R), σ, A_j, for $j \in \mathcal{I}$, and

$$\theta_j : R \to R/P_j \otimes_k A_j,$$

for $j \in \mathcal{I}$ as in Definitions (3.7.1)(1)-(3), and Definition (3.7.3)(SS1). We shall keep these data fixed throughout the following discussion and proposition. Recall that, if $\theta : A \to B$ is a homomorphism of commutative rings, and M and N are B-modules viewed also as A-modules via θ, then there is a natural homomorphism (of A-modules)

$$\alpha : M \otimes_A N \to M \otimes_B N;$$

α is always surjective; and, if θ is surjective, then α is an isomorphism. For this reason (to be made clear below), we shall assume that

$$(\theta_j)_\mathfrak{p} : R_\mathfrak{p} \to (R/P_j \otimes_k A_j)_\mathfrak{p}$$

is surjective for all $j \in \mathcal{I}$, and all $\mathfrak{p} \in \mathrm{Spec}(R)$ with $\sigma\mathfrak{p} = P_j$. Note that this is the case in the examples above. We shall abbreviate R/P_j by R_j, $R_j \otimes_k A_j$ by B_j, and we shall write $\otimes$ for $\otimes_k$.

Now let M and N be Chang-Skjelbred modules (over R), both of which satisfy the special structure condition (SS2) via graded R-module isomorphisms $\psi_j : M_j \to R_j \otimes U_j$ and $\chi_j : N_j \to R_j \otimes V_j$, for all $j \in \mathcal{I}$.

Given any R-module L and an integer $n \geq 1$, let $T_R^n(L)$, $A_R^n(L)$ and $\Lambda_R^n(L) = T_R^n(L)/A_R^n(L)$ be, respectively, the n-fold tensor product over R of L with itself (i.e. $L \otimes_R \ldots \otimes_R L$, n copies), the submodule of $T_R^n(L)$ generated by the alternating tensors (i.e tensors $x_1 \otimes \ldots \otimes x_n$, where $x_i = x_j$ for some pair i, j with $1 \leq i < j \leq n$), and the nth exterior power of L over R.

Returning to M and N, for any $j \in \mathcal{I}$, set $(M \otimes_R N)_j = M_j \otimes_{B_j} N_j$, $T_R^n(M)_j = T_{B_j}^n(M_j)$, and $\Lambda_R^n(M)_j = \Lambda_{B_j}^n(M_j)$.

(3.7.17) Proposition
With the assumptions, notation and definitions above, $M \otimes_R N$, $T_R^n(M)$ and $\Lambda_R^n(M)$ are Chang-Skjelbred modules (with special structure).

Proof.
First consider $M \otimes_R N$. Let $\mathfrak{p} \in \mathrm{Spec}(R)$, and let $\sigma\mathfrak{p} = P_j$. Then
$$(M \otimes_R N)_\mathfrak{p} \cong M_\mathfrak{p} \otimes_{R_\mathfrak{p}} N_\mathfrak{p} \cong (M_j)_\mathfrak{p} \otimes_{R_\mathfrak{p}} (N_j)_\mathfrak{p}$$
$$\cong (M_j)_\mathfrak{p} \otimes_{(B_j)_\mathfrak{p}} (N_j)_\mathfrak{p} \cong (M_j \otimes_{B_j} N_j)_\mathfrak{p}.$$
So we have the localization condition. Note that the surjectivity of $(\theta_j)_\mathfrak{p}$ is used in the third isomorphism.

3.7. Chang-Skjelbred modules

For (SS2), we use the middle four interchange (see [MacLane, 1967], Chapter VI, Section 8):
$$M_j \otimes_{B_j} N_j \cong (R_j \otimes U_j) \otimes_{R_j \otimes A_j} (R_j \otimes V_j) \cong (R_j \otimes_{R_j} R_j) \otimes (U_j \otimes_{A_j} V_j)$$
$$\cong R_j \otimes (U_j \otimes_{A_j} V_j).$$
$T_R^n(M)$ is a Chang-Skjelbred module by induction. In particular,
$$T_R^n(M)_j = T_{B_j}^n(M_j) \cong R_j \otimes T_{A_j}^n(U_j).$$
The localization condition for $\Lambda_R^n(M)$ follows by the 5-lemma from Diagram 3.4

$$\begin{array}{ccccccccc}
0 & \longrightarrow & A_R^n(M)_\mathfrak{p} & \longrightarrow & T_R^n(M)_\mathfrak{p} & \longrightarrow & \Lambda_R^n(M)_\mathfrak{p} & \longrightarrow & 0 \\
& & \downarrow & & \downarrow & & \downarrow & & \\
0 & \longrightarrow & A_{B_j}^n(M_j)_\mathfrak{p} & \longrightarrow & T_{B_j}^n(M_j)_\mathfrak{p} & \longrightarrow & \Lambda_{B_j}^n(M_j)_\mathfrak{p} & \longrightarrow & 0
\end{array}$$

Diagram 3.4

It is easy to see that the homomorphism on alternating tensors is surjective.

To show that
$$\Lambda_{B_j}^n(M_j) \cong R_j \otimes \Lambda_{A_j}^n(U_j),$$
it is enough to show that
$$A_{B_j}^n(M_j) \cong R_j \otimes A_{A_j}^n(U_j)$$
via the iterated middle four interchange. On the one hand, via the inverse middle four interchange,
$$r \otimes (u \otimes u) \mapsto (r \otimes u) \otimes (1 \otimes u) = (r \otimes 1)[(1 \otimes u) \otimes (1 \otimes u)].$$
One the other hand, suppose
$$r_1 \otimes u_1 = r_2 \otimes u_2 \in R_j \otimes U_j.$$
Let $\{\omega_\alpha\}$ be a k-vector-space basis for U_j. Then there are $r_\alpha \in R_j$, zero for all but finitely many α, such that
$$r_1 \otimes u_1 = r_2 \otimes u_2 = \sum_\alpha r_\alpha \otimes \omega_\alpha.$$
So, under the middle four interchange,
$$(r_1 \otimes u_1) \otimes (r_2 \otimes u_2) \mapsto \sum_{\alpha,\beta} r_\alpha r_\beta \otimes (\omega_\alpha \otimes \omega_\beta).$$
But
$$r_\alpha r_\beta \otimes (\omega_\alpha \otimes \omega_\beta) + r_\beta r_\alpha \otimes (\omega_\beta \otimes \omega_\alpha) =$$
$$r_\alpha r_\beta \otimes [(\omega_\alpha + \omega_\beta) \otimes (\omega_\alpha + \omega_\beta)] - r_\alpha r_\beta \otimes (\omega_\alpha \otimes \omega_\alpha)$$
$$- r_\alpha r_\beta \otimes (\omega_\beta \otimes \omega_\beta).$$
This completes the proof. □

(3.7.18) Examples
(1) Let R, $\mathcal{I}$, Strat(R), σ, X, and Y be as in Examples (3.7.16)(2). Let Z be another G-CW-subcomplex of X. Then
$$H_G^*(X,Y;k) \otimes_{H_G(X)} H_G^*(Z;k) = M,$$
say, is a Chang-Skjelbred module; and, for any $(K,c) \in \mathcal{T}(X)$,
$$M_{(K,c)} = H_G^*(Gc, Gc \cap Y; k) \otimes_{H_G(Gc)} H_G^*(Gc \cap Z; k).$$
If $H_G^*(X,Y;k)$ and $H_G^*(Z,k)$ are finitely generated $H_G(X)$-modules, e.g., if $\dim_k H^*(X,Y;k) < \infty$ and $\dim_k H^*(Z;k) < \infty$, then, by Theorem (A.2.8)(1),
$$\text{Supp}(M) = \text{Supp}(H_G^*(X,Y;k)) \cap \text{Supp}(H^*(Z;k)).$$
Thus (K,c) belongs to M if and only if (K,c) is maximal in $\mathcal{T}(X)$ with the properties that $Gc \cap Z \neq \emptyset$ and the restriction $H_G^*(Gc;k) \to H_G^*(Gc \cap Y;k)$ is not an isomorphism.

(2) Consider Example (3.7.16)(3), and suppose that X is π-finite. Let
$$\mathcal{M}(X) = \mathbb{Q}[z_1, \ldots, z_r] \otimes \Lambda(y_1, \ldots, y_s),$$
where $\deg(z_i)$ is even, for $1 \leq i \leq r$, and $\deg(y_i)$ is odd, for $1 \leq i \leq s$. Suppose that the z_i, $1 \leq i \leq r$, are chosen so that $\varepsilon(z_i) = 0$, for $1 \leq i \leq r$, where
$$\varepsilon: \mathcal{M}(X_G) = R \otimes \mathcal{M}(X) \to R$$
is induced by the chosen fixed point $x_0 \in X$. Suppose further that the differential on $Q_{(G,x_0)}(X)$ has the form $d\bar{z}_i = 0$, for $1 \leq i \leq r$, and
$$d\bar{y}_i = \sum_{j=1}^r a_{ij} \bar{z}_j,$$
for $1 \leq i \leq s$, where $a_{ij} \in R$, and $\bar{z}_i$, $\bar{y}_i$ are the images of z_i, y_i in $Q_{(G,x_0)}(X)$. Then $\Pi_{(G,x_0)}^{\text{ev}}(X)$ is the quotient of a free R-module on generators ζ_i, $1 \leq i \leq r$, corresponding to the $\bar{z}_i$, by relations
$$\sum_{i=1}^r a_{ij} \zeta_j = 0,$$
for $1 \leq i \leq s$. Let $A = (a_{ij})$ be the $s \times r$ matrix of polynomials in R with (i,j)-entry a_{ij}. For $1 \leq m \leq \min\{s,r\}$, let $\Lambda^m A$ be the $\binom{s}{m} \times \binom{r}{m}$ matrix of $m \times m$ minor determinants of A. Let
$$M = \Pi_{(G,x_0)}^{\text{ev}}(X),$$
and, for $r - \min\{s,r\} \leq m \leq r-1$, let $\mathcal{F}_m(M)$ be ideal in R generated by the entries of $\Lambda^{r-m} A$. In addition set $\mathcal{F}_r(M) = R$; and, if $s < r$, set $\mathcal{F}_m(M) = (0)$ for $0 \leq m < r-s$. The ideals $\mathcal{F}_m(M)$, for $0 \leq m \leq r$, are

called the Fitting ideals or Fitting invariants of M; see, e.g., [Kaplansky, 1970], [Kunz, 1985] or [Evans, Griffith, 1985]. Clearly,

$$\mathcal{F}_0(M) \subseteq \ldots \subseteq \mathcal{F}_m(M) \subseteq \mathcal{F}_{m+1}(M) \subseteq \ldots \subseteq \mathcal{F}_r(M).$$

Now it is a nice exercise in linear algebra to show that

$$\sqrt{(\operatorname{ann}(\Lambda_R^{m+1} M))} = \sqrt{(\mathcal{F}_m(M))},$$

i.e.

$$\operatorname{Supp}(\Lambda_R^{m+1} M) = V(\mathcal{F}_m(M)),$$

for $0 \le m \le r$; and, in fact,

$$\operatorname{ann}(\Lambda_R^r M) = \mathcal{F}_{r-1}(M) = J(A),$$

the ideal in R generated by the entries of A. Thus $\mathcal{F}_m(M)$ is a Hsiang-Serre ideal for $0 \le m \le r-1$, and $\mathcal{F}_{r-1}(M)$ is a strong Hsiang-Serre ideal.

Note that the results of this example apply to any finitely generated Chang-Skjelbred module with special structure. Note also that the Fitting invariants, $\mathcal{F}_m(M)$, do not depend on the choice of presentation of M; and, for any multiplicative set $S \subseteq R$, $S^{-1}\mathcal{F}_m(M) = \mathcal{F}_m(S^{-1}M)$ ([Kaplansky, 1970]).

With X and $M = \Pi^{\text{ev}}_{(G,x_0)}(X)$ as in Example (3.7.18)(2), there is an interesting proof of the fact that the Fitting invariants of M are Hsiang-Serre ideals using Corollary (3.5.3)(2) instead of the Chang-Skjelbred method. In Corollary (3.5.3), take $E = \mathbb{C}$, the field of complex numbers. If the torus G has dimension n, then $R = \mathbb{Q}[t_1, \ldots, t_n]$; and given an evaluation α, with values in $\mathbb{C}$, let $J(\alpha) = \ker[\alpha : R \to \mathbb{C}]$. Then $\mathcal{F}_m(M) \subseteq J(\alpha)$ if and only if $\dim_{\mathbb{Q}} \Pi^{\text{ev}}_\psi(c(\alpha)) > m$: this follows since

$$\mathcal{F}_m(M) \subseteq J(\alpha) \Leftrightarrow \Lambda^{r-m} A(\alpha) = 0 \Leftrightarrow \operatorname{rank} A(\alpha) < r - m$$

$$\Leftrightarrow \dim_{\mathbb{Q}} \Pi^{\text{ev}}_\psi(c(\alpha)) > m,$$

where $A(\alpha)$ is the $s \times r$ complex matrix obtained by evaluating each entry of A by means of α. In particular, $\mathcal{F}_{r-1}(M) \subseteq J(\alpha)$ if and only if $c(\alpha)$ is full. Equally,

$$\mathcal{F}_m(M) \subseteq PK(\alpha) \Leftrightarrow \Lambda^{r-m} \overline{A} = 0 \Leftrightarrow \operatorname{rank} \overline{A} < r - m$$

$$\Leftrightarrow \dim_{\mathbb{Q}} \Pi^{\text{ev}}_\psi(c(\alpha)) > m,$$

where $\overline{A}$ is the $s \times r$ matrix of polynomials in $R_{K(\alpha)} = R/PK(\alpha)$ obtained by reducing the entries of A modulo $PK(\alpha)$, since $Q_{(K(\alpha),x_0)}(X)$ is obtained from $Q_{(G,x_0)}(X)$ by reducing modulo $PK(\alpha)$, by Examples (2.5.7)(1). Hence $\mathcal{F}_m(M) \subseteq J(\alpha)$ if and only if $\mathcal{F}_m(M) \subseteq PK(\alpha)$.

Since $\mathbb{C}$ has infinite transcendence degree over $\mathbb{Q}$, for any $\mathfrak{p} \in \operatorname{Spec}(R)$, there exists $\alpha \in \mathbb{C}^n$ such that $\mathfrak{p} = J(\alpha)$; see, e.g., [Kunz, 1985], Chapter

I, §4. Thus $\mathfrak{p} \supseteq \mathcal{F}_m(M)$ implies that $\sigma\mathfrak{p} \supseteq \mathcal{F}_m(M)$: and so $\mathcal{F}_m(M)$ is a Hsiang-Serre ideal.

(3.7.19) Remark
[Serre, 1965b], Proposition (1), shows the following. In cases (2) or (odd p) with R the polynomial part of $H^*(BG;k)$, then an ideal $J \triangleleft R$ is a Hsiang-Serre ideal if it is homogeneous and stable under the Steenrod operations. (See Theorem (3.6.22) and the paragraph preceding it.) In particular, in cases (2) and (odd p), the fact that $\ker[p^* : R \to H^*_G(X;k)]$ is a Hsiang-Serre ideal follows without using the Localization Theorem.

We return now to the ideals I_u defined in Example (3.7.16)(1). Thus X is a finite-dimensional G-CW-complex with FMCOT, Y is a G-CW-subcomplex, $\phi(Y) : Y \to X$ is the inclusion, $u \in H^*_G(Y;k)$, $I_u = \{a \in R : au \in \operatorname{im}\phi(Y)^*\}$, and we are assuming case (0), (2) or (odd p). The following useful lemma applies, for example, when $Y = X^G$ and X is TNHZ in $X_G \to BG$, so that $H^*_G(X;k)$ is a free R-module, or, more generally, when the torsion-free quotient of $H^*_G(X;k)$ is free.

(3.7.20) Lemma
Suppose that $H^*_G(Y;k)$ is torsion-free as a R-module. Let $u \in H^*_G(Y;k)$ be homogeneous. Suppose that there is a free R-submodule $M \subseteq H^*_G(X;k)$ such that $\phi(Y)^*|M$ is injective and $\{au; a \in I_u\} \subseteq \phi(Y)^*(M)$. Then I_u is principal; and the subtori which belong to u have corank 1, if $I_u \neq R$ or (0).

Proof.
Let $I_u = (a_1, \ldots, a_m)$, where $\deg(a_1) \leq \ldots \leq \deg(a_m)$, and $a_1 \neq 0$. For $1 \leq i \leq m$, let $a_i u = \phi(Y)^*(z_i)$ where $z_i \in M$. Let $\{x_\alpha; \alpha \in \mathcal{A}\}$ be a R-module basis for M, and let $z_i = \sum_\alpha a_{i\alpha} x_\alpha$, for $1 \leq i \leq m$. Since $\phi(Y)^*|M$ is injective, $a_1 z_i = a_i z_1$; and so $a_1 a_{i\alpha} = a_i a_{1\alpha}$, for all α and $1 \leq i \leq m$. Let q be a prime factor of a_1 (in the UFD R). If $q|a_{1\alpha}$, for all α, then $a_1/q \in I_u$, contradicting the assumption that a_1 has minimal degree in I_u. Hence $a_1|a_i$; and so $I_u = (a_1)$.

If $I_u \neq R$ or (0), and if PK is a minimal prime of I_u, then $\operatorname{ht}(PK) = 1$, clearly, since R is a UFD. But $\operatorname{ht}(PK) = \operatorname{cork}(K)$. (Note that $I_u \neq (0)$, by the Localization Theorem, if, for example, $Y = X^G \neq \emptyset$.) □

(3.7.21) General G-spaces
All the results and examples of this section carry over to the general setting without alteration. In the examples, of course, one must assume

appropriate conditions which ensure that the relevant version of the Localization Theorem is valid. No further refinements are readily available in case (A): one still wants, e.g., $\overline{H}_G(X)$ to be Noetherian in Example (3.7.16)(2), and $\dim_k \overline{H}^*(X;k) < \infty$ is still the simplest geometric condition which ensures this.

3.8 Hsiang's Fundamental Fixed Point Theorem

In this section we shall work with cases (0), (2) and (odd p). Recall that in case (0), G is a torus, the ground field of coefficients $k = \mathbb{Q}$ and $R = H^*(BG;\mathbb{Q}) = \mathbb{Q}[t_1,\ldots,t_n]$, where $n = \dim G$ and each t_i has degree 2; in case (2), $G = (\mathbb{Z}_2)^n$ is a 2-torus, k is the field $\mathbb{F}_2$ and $R = H^*(BG;\mathbb{F}_2) = \mathbb{F}_2[t_1,\ldots,t_n]$, where each t_i has degree 1; and in case (odd p), $G = (\mathbb{Z}_p)^n$ is a p-torus, where p is a prime number $\neq 2$, k is the field $\mathbb{F}_p$ and $R = \mathbb{F}_p[t_1,\ldots,t_n]$ is the polynomial part of $H^*(BG;\mathbb{F}_p)$, where each t_i has degree 2. In case (odd p),

$$H^*(BG;\mathbb{F}_p) = \mathbb{F}_p[t_1,\ldots,t_n] \otimes \Lambda(s_1,\ldots,s_n),$$

where each s_i has degree 1.

Let X be a finite-dimensional G-CW-complex with FMCOT. We shall assume that X is connected and that $\dim_k H^*(X;k) < \infty$. So $H_G^*(X;k)$ is a finitely generated R-module, and X^G has at most a finite number of components. (See Proposition (3.10.1) and Corollary (3.10.2).) For simplicity we shall assume that $X^G \neq \emptyset$, and let F_i, $1 \leq i \leq \nu$, be the components of X^G. We shall abbreviate $H_G^*(X;k) = M$ and $H_G(X) = B$ (see Definition (3.6.1)(1)). Also recall the definition of $\mathcal{T}(X)$ from Definition (3.6.1)(2). Note that with these assumptions M is a finitely generated B-module, a finitely generated k-algebra and a finitely generated B-algebra; also B is a strictly commutative Noetherian ring, and M is a Chang-Skjelbred module over B as in Examples (3.7.16)(2).

By Corollary (3.7.8), $\mathrm{Ass}_B(M) \subset \mathrm{Strat}(R)$; and, since M is finitely generated, $\mathrm{Ass}_B(M)$ is finite. Clearly each $\mathfrak{p}(G,F_i)$, $1 \leq i \leq \nu$, is a minimal element of $\mathrm{Ass}_B(M)$. Let $\mathfrak{p}(K_i,c_i)$, for $1 \leq i \leq \gamma$, $\gamma \geq 0$, $(K_i,c_i) \in \mathcal{T}(X)$, be the other minimal elements of $\mathrm{Ass}_B(M)$, and let $\mathfrak{p}(L_i,d_i)$, for $1 \leq i \leq \beta$, $\beta \geq 0$, $(L_i,d_i) \in \mathcal{T}(X)$, be the non-minimal elements of $\mathrm{Ass}_B(M)$. Let $\alpha = \nu + \gamma$. For simplicity let $\mathfrak{p}_i = \mathfrak{p}(G,F_i)$ for $1 \leq i \leq \nu$, $\mathfrak{p}_i = \mathfrak{p}(K_{i-\nu},c_{i-\nu})$ for $\nu+1 \leq i \leq \nu+\gamma$, and $\mathfrak{p}_i = \mathfrak{p}(L_{i-\alpha},d_{i-\alpha})$ for $\alpha+1 \leq i \leq \alpha+\beta$. So

$$\mathrm{Ass}_B(M) = \{\mathfrak{p}_i; 1 \leq i \leq \alpha+\beta\}$$

and

$$\min \mathrm{Ass}_B(M) = \{\mathfrak{p}_i; 1 \leq i \leq \alpha\}.$$

Now, in M, the zero B-submodule, 0, has a reduced primary decomposition $0 = \bigcap_{i=1}^{\alpha+\beta} N_i$, where N_i is a $\mathfrak{p}_i$-primary B-submodule of M (Theorem (A.3.10)). From Theorem (A.3.14) it follows that N_i is a graded B-module if $1 \leq i \leq \alpha$, and N_i may be chosen to be graded for $\alpha + 1 \leq i \leq \alpha + \beta$. Furthermore, from Proposition (A.3.8)(3), for $1 \leq i \leq \alpha$, $N_i = \{x \in M; sx = 0 \text{ for some } s \in B - \mathfrak{p}_i\}$, whereas, for $\alpha + 1 \leq i \leq \alpha + \beta$, $\{x \in M; sx = 0 \text{ for some } s \in B - \mathfrak{p}_i\} = \bigcap\{N_j; 1 \leq j \leq \alpha + \beta \text{ and } \mathfrak{p}_j \subseteq \mathfrak{p}_i\}$, by Proposition (A.3.6).

On the other hand, if $\sigma\mathfrak{p} = \mathfrak{p}(K,c)$, for $(K,c) \in \mathcal{T}(X)$, then by Theorem (3.6.14) and the monomorphism condition of Section 3.7, $x \in M; sx = 0$ for some $s \in B - \mathfrak{p}\} = \ker[\phi(c)^* : M \to H_G^*(Gc; k)]$, where $\phi(c) : Gc \to X$ is the inclusion. Abbreviating $\phi(F_i)$ by ϕ_i, for $1 \leq i \leq \nu$, we have then that $N_i = \ker \phi_i^*$ for $1 \leq i \leq \nu$, and $N_{\nu+i} = \ker \phi(c_i)^*$ for $1 \leq i \leq \gamma$.

In particular, for $1 \leq i \leq \alpha$, N_i is actually a homogeneous ideal in the graded ring $M = H_G^*(X; k)$. Also, for $1 \leq i \leq \alpha$, let $\sqrt{(N_i \cap B)} = \mathfrak{p}_i$. The latter is clear for $1 \leq i \leq \nu$; for $i = \nu + j, 1 \leq j \leq \gamma$, let $G = K_j \times L_j$, where $L_j \cong G/K_j$. Then the maximality of (K_j, c_j) in $\mathcal{T}(X)$ implies that L_j acts almost-freely (freely in cases (2) and (odd p)) on Gc_j; hence, any $z \in \tilde{H}_{L_j}^*(Gc_j; k)$ is nilpotent, by the Localization Theorem. In fact N_i can be chosen to be a homogeneous ideal in M with $\sqrt{(N_i \cap B)} = \mathfrak{p}_i$, for $\alpha + 1 \leq i \leq \alpha + \beta$ also: but we shall not need this.

We summarize some of the above in the following proposition.

(3.8.1) Proposition
With the above assumptions and notations, in the $H_G(X)$-module $H_G^*(X; k)$ the zero submodule, 0, has a reduced primary decomposition $0 = \bigcap_{i=1}^{\alpha+\beta} N_i$, where the prime ideal belonging to N_i is minimal for $1 \leq i \leq \alpha = \nu + \gamma$ and embedded for $\alpha + 1 \leq i \leq \alpha + \beta$; and

(i) $N_i = \ker \phi_i^*$ *for $1 \leq i \leq \nu$, and*
(ii) $N_{\nu+i} = \ker \phi(c_i)^*$ *for $1 \leq i \leq \gamma$.*
Furthermore
(a) *the action is uniform if and only if $\gamma = 0$, and*
(b) $H_G^*(X; k)$ *is torsion-free as a R-module if and only if $\gamma = 0$ and $\beta = 0$.*

Proof.
(a) This follows easily from the definition (Definition (3.6.17)), since $X^G \neq \emptyset$.

(b) This follows since, by the Localization Theorem, the R-torsion submodule of
$$M = \{x \in M; rx = 0 \text{ for some } r \in R \text{ with } r \neq 0\}$$
$$= \ker[\phi^* : M \to H_G^*(X^G; k)] = \bigcap_{i=1}^{\nu} \ker \phi_i^*.$$

$\square$

(3.8.2) Remark
We do not know of an example of a finite-dimensional G-CW-complex X where $H_G^*(X; k)$ is torsion-free but not free as a R-module. In case (0), if $\dim G = 2$, and if $H_G^*(X; k)$ is torsion-free, then $H_G^*(X; k)$ is free. On the other hand, there is an example of a torus action on $X = S^2 \times S^3 \times S^3$ where the torsion-free quotient of $H_G^*(X; k)$ is not free. See [Allday, 1985].

(3.8.3) Remark
For $1 \leq i \leq \nu$ and $1 \leq j \leq \gamma$, ϕ_i^* and $\phi(c_j)^*$ induce monomorphisms
$$H_G^*(X; k)/N_i \to H_G^*(F_i; k)$$
and
$$H_G^*(X; k)/N_{\nu+j} \to H_G^*(Gc_j; k),$$
respectively: and, viewed as R-module homomorphisms, these monomorphisms become isomorphisms when localized at (0) and PK_j, respectively. This follows at once from the Localization Theorem.

Now $H_G^*(X; k)$ is a finitely generated R-algebra. So we can find a finitely generated free graded R-algebra $A = A^*$, and a R-algebra epimorphism $\pi : A \to H_G^*(X; k)$. In case (0), let
$$A = R[x_1, \ldots, x_r] \otimes \Lambda(y_1, \ldots, y_s),$$
where the x_i and y_j are homogeneous of positive even and odd degrees. In case (2), let $A = R[x_1, \ldots, x_r]$, where the x_i are homogeneous of positive degrees. In case (odd p), let
$$A = R[x_1, \ldots, x_r] \otimes \Lambda(s_1, \ldots, s_n, y_1, \ldots, y_s),$$
where again the x_i and y_i are homogeneous of positive even and odd degrees, respectively. In case (odd p), we suppose (by slight abuse of notation) that $\pi(s_i) = p^*(s_i)$, for $1 \leq i \leq n$, where
$$p^* : H^*(BG; \mathbb{F}_p) \to H_G^*(X; \mathbb{F}_p)$$

is, as usual, induced by $p: X_G \to BG$. In cases (0) and (odd p) let $A_0 = \bigoplus_{i=0}^{\infty} A^{2i}$ be the evenly graded part of A; in case (2) let $A_0 = A$. Let $\pi_0 = \pi | A_0 : A_0 \to H_G^*(X)$. Let $J = \ker \pi$.

For any $(K,c) \in \mathcal{T}(X)$, let $\varepsilon(c) = (K,c)^*\pi_0 : A_0 \to R_K$, and let $P(K,c) = \ker \varepsilon(c) = \pi_0^{-1}\mathfrak{p}(K,c)$. Let $\varepsilon_i = \varepsilon(F_i)$, for $1 \le i \le \nu$. In cases (0) and (odd p), let $A_1 = \bigoplus_{i=0}^{\infty} A^{2i+1}$. In case (0), let $U = (y_1, \ldots, y_s)$ be the homogeneous ideal in A generated by $y_1, \ldots, y_s$: and, in case (odd p), let $U = (s_1, \ldots, s_n, y_1, \ldots, y_s)$. In case (2), let $A_1 = 0$ and $U = (0)$. Let $U_0 = U \cap A_0$. For any $P \in \mathrm{Spec}(A_0)$, let $P^+ = P + U = P + A_1$, a homogeneous prime ideal in A. Clearly, $P = P \cap R[x_1, \ldots, x_r] + U_0$.

(3.8.4) Lemma

For any $(K,c) \in \mathcal{T}(X)$, there exist $b_1, \ldots, b_r \in R$ such that:

(1) $P(K,c) \cap R[x_1, \ldots, x_r] = PK + (x_1 - b_1, \ldots, x_r - b_r)$, *the ideal in* $R[x_1, \ldots, x_r]$ *generated by PK and $\{x_i - b_i | 1 \le i \le r\}$, and*

(2) $P(K,c)^+ = PK + (x_1 - b_1, \ldots, x_r - b_r) + U$, *the ideal in A generated by PK, $\{x_i - b_i | 1 \le i \le r\}$ and U.*

Proof.

It is enough to prove (2). Clearly $PK + U \subseteq P(K,c)^+$. Choose $b_i \in R$ such that $\varepsilon(c)(x_i) = \varepsilon(c)(b_i) \in R_K = R/PK$. So $x_i - b_i \in P(K,c)$.

Suppose $f(x_1, \ldots, x_r) \in R[x_1, \ldots, x_r] \cap P(K,c)$. Then
$$f(x_1, \ldots, x_r) - f(b_1, \ldots, b_r) = f(x_1, \ldots, x_r) - f(b_1, x_2, \ldots, x_r)$$
$$+ f(b_1, x_2, \ldots, x_r) - f(b_1, b_2, x_3, \ldots, x_r) + \ldots + f(b_1, \ldots, b_{n-1}, x_r)$$
$$- f(b_1, \ldots, b_r) \in (x_1 - b_1, \ldots, x_r - b_r).$$
Since $\varepsilon(c)(f) = 0$ and $\varepsilon(c)(x_i) = \varepsilon(c)(b_i)$, $1 \le i \le r$, $f(b_1, \ldots, b_r) \in PK$. $\square$

The next lemma is also a straightforward exercise in algebra.

(3.8.5) Lemma

(1) $\mathrm{Ass}_{A_0}(A/J) = \{\pi_0^{-1}(\mathfrak{p}); \mathfrak{p} \in \mathrm{Ass}_B(M)\}$.

(2) N *is a $\mathfrak{p}$-primary B-submodule of M if and only if $\pi^{-1}(N)$ is a $\pi_0^{-1}(\mathfrak{p})$-primary A_0-submodule of A.*

(3.8.6) Remark

From Lemma (3.8.4)(1), there exist $a_{ij} \in R$, for $1 \le i \le \nu$, $1 \le j \le r$, such that $P(G, F_i) = (x_1 - a_{i1}, \ldots, x_r - a_{ir}) + U_0$, for $1 \le i \le \nu$. Clearly, for $1 \le i, j \le \nu$,

3.8. Hsiang's Fundamental Fixed Point Theorem

$$[P(G, F_i) + P(G, F_j)] \cap R = (a_{i1} - a_{j1}, \ldots, a_{ir} - a_{jr})$$
$$= (p^*)^{-1}(\mathfrak{p}(G, F_i) + \mathfrak{p}(G, F_j)).$$

Now a sum of Hsiang-Serre ideals is a Hsiang-Serre ideal. (If $\mathfrak{p} \supseteq \mathfrak{a} + \mathfrak{b}$ then $\mathfrak{p} \supseteq \mathfrak{a}$ and $\mathfrak{p} \supseteq \mathfrak{b}$. If $\mathfrak{a}$ and $\mathfrak{b}$ are Hsiang-Serre ideals, then $\sigma\mathfrak{p} \supseteq \mathfrak{a}$ and $\sigma\mathfrak{p} \supseteq \mathfrak{b}$. $\sigma\mathfrak{p} \supseteq \mathfrak{a} + \mathfrak{b}$, therefore.) So $\mathfrak{p}(G, F_i) + \mathfrak{p}(G, F_j)$ is a Hsiang-Serre ideal in $H_G(X)$. Hence $(a_{i1} - a_{j1}, \ldots, a_{ir} - a_{jr})$ is a Hsiang-Serre ideal in R: call it $\mathfrak{a}_{ij}$. If PK is a minimal prime of $\mathfrak{a}_{ij}$, then K is maximal with the property that there is a component c of X^K such that $c \supseteq F_i \cup F_j$. (Since

$$(p^*)^{-1}\sqrt{(\mathfrak{p}(G, F_i) + \mathfrak{p}(G, F_j))} = \sqrt{((p^*)^{-1}(\mathfrak{p}(G, F_i) + \mathfrak{p}(G, F_j)))},$$

$PK = (p^*)^{-1}(\mathfrak{p}(K; c))$, where $\mathfrak{p}(K, c)$ is a minimal prime of $\mathfrak{p}(G, F_i) + \mathfrak{p}(G, F_j)$ in $H_G(X)$.) K is called a maximal connecting subtorus of F_i and F_j. By the Krull Height Theorem (see (A.1.5)),

$$\operatorname{cork}(K) = \operatorname{ht}(PK) \leq r.$$

Hence the number of even generators of positive degree in $H_G^*(X; k)$ bears on the size of connecting subtori of pairs of components of X^G. For example, in case (0) suppose

$$H^*(X; k) \cong k[\xi]/(\xi^{n+1}),$$

where $\deg(\xi) = 2$, e.g., $X = \mathbb{C}P^n$. Then X is TNHZ in $X_G \to BG$, and we may take $r = 1$. Hence all maximal connecting subtori of pairs of components of X^G have corank one.

Using Lemma (3.8.5) we can summarize the discussion above in the following version of Hsiang's Fundamental Fixed Point Theorem ([Hsiang, 1975]).

(3.8.7) Theorem
Let X be a connected finite-dimensional G-CW-complex with FMCOT in case (0), (2) or (odd p). Suppose that $X^G \neq \emptyset$ and that $\dim_k H^*(X; k) < \infty$. Let $\pi : A \to H_G^*(X; k)$ be a R-algebra epimorphism, where A is a free R-algebra of the following forms in the three cases:

(i) *$A = R[x_1, \ldots, x_r] \otimes \Lambda(y_1, \ldots, y_s)$ in case (0), where the x_i, $1 \leq i \leq r$, and y_j, $1 \leq j \leq s$, are homogeneous of positive even and odd degrees, respectively;*

(ii) *$A = R[x_1, \ldots, x_r]$ in case (2), where the x_i, $1 \leq i \leq r$, are homogeneous of positive degrees; and*

(iii) *$A = R[x_1, \ldots, x_r] \otimes \Lambda(s_1, \ldots, s_n, y_1, \ldots, y_s)$ in case (odd p), where the x_i, $1 \leq i \leq r$, and y_j, $1 \leq j \leq s$, are homogeneous of positive even and odd degrees, respectively, and, by abuse of notation, $\pi(s_i) = p^*(s_i)$, for $1 \leq i \leq n$, where $s_1, \ldots, s_n$ are the generators of $H^1(BG; \mathbb{F}_p)$ and n is the rank of G (i.e. $G = (\mathbb{Z}_p)^n$).*

Let F_i, $1 \leq i \leq \nu$, be the components of X^G; and let (K_i, c_i), $1 \leq i \leq \gamma$, be the maximal elements of $\mathcal{T}(X)$ (see Definition (3.6.1)(2) and Definition (3.6.15)), which are distinct from $(G, F_1), \ldots, (G, F_\nu)$. (γ may be 0: i.e. $(G, F_1), \ldots, (G, F_\nu)$ may be the only maximal elements of $\mathcal{T}(X)$.)

Let $H_G(X) = \bigoplus_{i=0}^{\infty} H_G^{2i}(X; k)$ in cases (0) and (odd p), and let $H_G(X) = H_G^*(X; \mathbb{F}_2)$ in case (2). Let $(L_i, d_i) \in \mathcal{T}(X)$, $1 \leq i \leq \beta$, be such that $\mathfrak{p}(G, F_i)$, $1 \leq i \leq \nu$, $\mathfrak{p}(K_i, c_i)$, $1 \leq i \leq \gamma$, and $\mathfrak{p}(L_i, d_i)$, $1 \leq i \leq \beta$, are the associated primes of $H_G^*(X; K)$ with respect to the $H_G(X)$-module structure. (See Corollary (3.7.8) and Example (3.7.16)(2).) By Proposition (3.6.16) and the discussion following Definition (3.7.11), $\mathfrak{p}(G, F_i)$, $1 \leq i \leq \nu$, and $\mathfrak{p}(K_i, c_i)$, $1 \leq i \leq \gamma$, are the minimal associated primes. Thus, by definition, $\mathfrak{p}(L_i, d_i)$, $1 \leq i \leq \beta$, are the non-minimal associated primes. Again, however, β may be 0: i.e. all the associated primes may be minimal.)

Let $A_0 = \bigoplus_{i=0}^{\infty} A^{2i}$ be the evenly graded part of A, in cases (0) and (odd p), and let $A_0 = A$ in case (2). Let $\pi_0 = \pi | A_0 : A_0 \to H_G(X)$. For any $(K, c) \in \mathcal{T}(X)$, let $P(K, c) = \ker((K, c)^* \pi_0) = \pi_0^{-1}(\mathfrak{p}(K, c))$. So $P(K, c)$ is a prime ideal in A_0, for any $(K, c) \in \mathcal{T}(X)$.

Let $J = \ker \pi$. Then J, viewed as a A_0-submodule of A, has a reduced primary decomposition

$$J = \bigcap_{i=1}^{\nu} J_i \cap \bigcap_{i=1}^{\gamma} Q_i \cap \bigcap_{i=1}^{\beta} V_i,$$

where J_i is $P(G, F_i)$-primary, for $1 \leq i \leq \nu$, Q_i is $P(K_i, c_i)$-primary, for $1 \leq i \leq \gamma$, and V_i is $P(L_i, d_i)$-primary for $1 \leq i \leq \beta$.

Furthermore:

(1) J_i, $1 \leq i \leq \nu$, and Q_i, $1 \leq i \leq \gamma$, are homogeneous ideals in A, $J_i \cap A_0$ is a $P(G, F_i)$-primary ideal in A_0, for $1 \leq i \leq \nu$, and $Q_i \cap A_0$ is a $P(K_i, c_i)$-primary ideal in A_0, for $1 \leq i \leq \gamma$.

(2) There exist $a_{ij} \in R$, $1 \leq i \leq \nu$, $1 \leq j \leq r$, $b_{ij} \in R$, $1 \leq i \leq \gamma$, $1 \leq j \leq r$, and $c_{ij} \in R$, $1 \leq i \leq \beta$, $1 \leq j \leq r$, such that

$$P(G, F_i) \cap R[x_1, \ldots, x_r] = (x_1 - a_{i1}, \ldots, x_r - a_{ir}),$$

for $1 \leq i \leq \nu$,

$$P(K_i, c_i) \cap R[x_1, \ldots, x_r] = PK_i + (x_1 - b_{i1}, \ldots, x_r - b_{ir}),$$

for $1 \leq i \leq \gamma$, and

$$P(L_i, d_i) \cap R[x_1, \ldots, x_r] = PL_i + (x_1 - c_{i1}, \ldots, x_r - c_{ir}),$$

for $1 \leq i \leq \beta$.

(3) $J_i = \ker(\phi_i^* \pi)$, for $1 \leq i \leq \nu$, where $\phi_i^* : H_G^*(X; k) \to H_G^*(F_i; k)$ is induced by the inclusion $F_i \subset X$, $1 \leq i \leq \nu$; and $Q_i = \ker(\phi(c_i)^* \pi)$,

3.8. Hsiang's Fundamental Fixed Point Theorem

for $1 \leq i \leq \gamma$, where $\phi(c_i)^* : H_G^*(X;k) \to H_G^*(Gc_i;k)$ is induced by the inclusion $Gc_i \subset X$, $1 \leq i \leq \gamma$.

(4) The action is uniform if and only if $\gamma = 0$. $H_G^*(X;k)$ is R-torsion-free if and only if $\beta = \gamma = 0$. Indeed the R-torsion submodule of $H_G^*(X;k)$ is isomorphic to $\overline{J}/J$, where $\overline{J} = \bigcap_{i=1}^{\nu} J_i$.

(5) π induces monomorphisms $\pi_i : A/J_i \to H_G^*(F_i;k)$, for $1 \leq i \leq \nu$, and π_i, $1 \leq i \leq \nu$, becomes an isomorphism when localized at $(0) \triangleleft R$, at any $\mathfrak{p} \in \mathrm{Spec}(R)$ with $\sigma\mathfrak{p} = (0)$, or with respect to the multiplicative set S generated by the non-zero homogeneous linear polynomials in R.

π also induces monomorphisms $\rho_i : A/Q_i \to H_G^*(Gc_i;k)$, for $1 \leq i \leq \gamma$, and ρ_i, $1 \leq i \leq \gamma$, becomes an isomorphism when localized at PK_i, at any $\mathfrak{p} \in \mathrm{Spec}(R)$ with $\sigma\mathfrak{p} = PK_i$, or with respect to the multiplicative set $S(K_i)$ generated by the homogeneous linear polynomials in $R - PK_i$.

The original version of Theorem (3.8.7), as it may be found in [Hsiang, 1975], is simplified by localizing at (0) in R. In fact, if one localizes with respect to any multiplicative subset $T \subseteq R$, such that $0 \notin T$ and T contains S, the multiplicative subset of R generated by the non-zero homogeneous linear polynomials in R, then, with notation used above, one gets $0 = \bigcap_{i=1}^{\nu} T^{-1} N_i$ in $T^{-1} H_G^*(X;k)$ and $T^{-1} J = \bigcap_{i=1}^{\nu} T^{-1} J_i$ in $T^{-1} A$. This is clear directly, since

$$T^{-1} H_G^*(X;k) \cong T^{-1} H_G^*(X^G;k) \cong \bigoplus_{i=1}^{\nu} T^{-1} H_G^*(F_i;k).$$

It also follows since $T^{-1} N_{\nu+j} = T^{-1} H_G^*(X;k)$, for $1 \leq j \leq \gamma$, and $T^{-1} N_{\alpha+j} \supseteq \bigcap_{i=1}^{\nu} T^{-1} N_i$, for $1 \leq j \leq \beta$, (and where $\alpha = \nu + \gamma$). The first statement follows since

$$T^{-1} N_{\nu+j} = T^{-1} \ker \phi(c_j)^* = \ker T^{-1} \phi(c_j)^* = T^{-1} H_G^*(X;k)$$

by the Localization Theorem, because $(Gc_j)^G = \emptyset$, by assumption, for $1 \leq j \leq \gamma$. For the second statement we have that $N_{\alpha+j} \supseteq \ker \phi(d_j)^*$, as remarked above Proposition (3.8.1). If $(Gd_j)^G = \emptyset$, then

$$T^{-1} N_{\alpha+j} = T^{-1} H_G^*(X;k),$$

as before. If

$$(Gd_j)^G = F_{i_1} \cup \ldots \cup F_{i_{m_j}},$$

where $1 \leq i_1, \ldots, i_{m_j} \leq \nu$, then

$$T^{-1} N_{\alpha+j} \supseteq T^{-1} \ker \phi(d_j)^* = T^{-1} N_{i_1} \cap \ldots \cap T^{-1} N_{i_{m_j}}.$$

Now $T^{-1} J_i$ is a $T^{-1} P(G, F_i)$-primary $T^{-1} A_0$-submodule of $T^{-1} A$, and it is a $T^{-1} P(G, F_i)^+$-primary ideal in $T^{-1} A$. (See Proposition (A.3.6).) Furthermore,

$$T^{-1}P(G, F_i)^+ = (x_1 - a_{i1}, \ldots, x_r - a_{ir}) + U,$$

where $a_{ij} \in R$ for $1 \leq i \leq \nu$, $1 \leq j \leq r$, and $U = (y_1, \ldots, y_s)$ in case (0), $U = (0)$ in case (2), and $U = (s_1, \ldots, s_n, y_1, \ldots, y_s)$ in case (odd p), as above. Note that

$$T^{-1}A/T^{-1}P(G, F_j)^+ = T^{-1}A_0/T^{-1}P(G, F_j) \cong T^{-1}R:$$

and so, if $T = R - (0)$, then $T^{-1}P(G, F_j)$ is a maximal ideal in $T^{-1}A_0$, and $T^{-1}P(G, F_j)^+$ is maximal in $T^{-1}A$, for $1 \leq j \leq \nu$.

Here now is the localized version of Hsiang's Fundamental Fixed Point Theorem.

(3.8.8) Theorem

Let X, G, π and A be as in Theorem (3.8.7). Let U be the ideal in A which is generated by $y_1, \ldots, y_s$ in case (0), and which is generated by $s_1, \ldots, s_n, y_1, \ldots, y_s$ in case (odd p). Let $U = 0$ in case (2). Let T be any multiplicative subset of R such that $0 \notin T$ and T contains S, the multiplicative subset generated by all non-zero homogeneous linear polynomials in R.

Let $I = \ker[T^{-1}\pi : T^{-1}A \to T^{-1}H_G^*(X; k)]$. Then I has a reduced primary decomposition in $T^{-1}A$, $I = \bigcap_{j=1}^{\nu} I_j$, where

(1) $\sqrt{I_j} = (x_1 - a_{j1}, \ldots, x_r - a_{jr}) + U$, *where* $a_{ij} \in R$, *for* $1 \leq i \leq \nu$, $1 \leq j \leq r$;
(2) $T^{-1}A/I_j \cong T^{-1}H_G^*(F_j; k)$, *for* $1 \leq j \leq \nu$;
(3) $I_j = I_{P_j} \cap T^{-1}A$, *for* $1 \leq j \leq \nu$, *where* $P_j = \sqrt{(I_j)}$; *and*
(4) $I = I_1 \ldots I_\nu$, *the product ideal of* $I_1, \ldots, I_\nu$.

Proof.

Only (4) is new. Let $e_j \in T^{-1}H_G^*(X; k) \cong \bigoplus_{i=1}^{\nu} T^{-1}H_G^*(F_i; k)$ be the element which is 1 in $T^{-1}H_G^*(F_j; k)$ and 0 in $T^{-1}H_G^*(F_i; k)$ for $i \neq j$. Choose $a_j \in T^{-1}A$ such that $T^{-1}\pi(a_j) = e_j$. Let $a'_j = 1 - a_j$. Then $a'_j \in I_j$ and $a_j \in \bigcap_{\substack{i=1 \\ i \neq j}}^{\nu} I_i$. So $1 \in I_j + \bigcap_{\substack{i=1 \\ i \neq j}}^{\nu} I_i$.

Since this holds for all j, $1 \leq j \leq \nu$, (4) follows. □

(3.8.9) Remarks

(1) The proof of Theorem (3.8.8) in [Hsiang, 1975] is direct, using only the Localization Theorem and properties of primary decomposition. The additional information in Theorem (3.8.7), which comes from the results of Sections 3.6 and 3.7, is localized away in Theorem (3.8.8).

(2) Consider Theorem (3.8.8) in case (odd p). Let $\Lambda = \Lambda(s_1,\ldots,s_n)$. By (2) there is an exact sequence
$$0 \to I_j \to T^{-1}A \to T^{-1}H_G^*(F_j;k) \to 0$$
for $1 \leq i \leq \nu$. Regard k as a Λ-module via the homomorphism $\Lambda \to k$ given by $s_i \mapsto 0$ for $1 \leq i \leq n$. Since $T^{-1}H_G^*(F_j;k)$ is isomorphic to $T^{-1}R \otimes \Lambda \otimes H^*(F_j;k)$, it is a free Λ-module. Hence we have an exact sequence
$$0 \to I_j \otimes_\Lambda k \to T^{-1}A \otimes_\Lambda k \to T^{-1}H_G^*(F_j;k) \otimes_\Lambda k \to 0.$$
Thus
$$T^{-1}A \otimes_\Lambda k / I_j \otimes_\Lambda k \cong T^{-1}R \otimes H^*(F_j;k)$$
for $1 \leq j \leq \nu$. Note that
$$T^{-1}A \otimes_\Lambda k \cong T^{-1}R[x_1,\ldots,x_r] \otimes \Lambda(y_1,\ldots,y_s).$$
Similarly,
$$T^{-1}A \otimes_\Lambda k / I \otimes_\Lambda k \cong T^{-1}R \otimes H^*(X^G;k).$$
Also, since
$$T^{-1}A \otimes_\Lambda k / I \otimes_\Lambda k \cong \bigoplus_{j=1}^{\nu} T^{-1}A \otimes_\Lambda k / I_j \otimes_\Lambda k,$$
it follows that
$$I \otimes_\Lambda k = \bigcap_{j=1}^{\nu} I_j \otimes_\Lambda k = (I_1 \otimes_\Lambda k)\ldots(I_\nu \otimes_\Lambda k),$$
where the latter is the product ideal as in Theorem (3.8.8)(4).

(3.8.10) Remark
With the notation of Theorem (3.8.7), let
$$J \cap A_0 = (g_1,\ldots,g_m),$$
and put $g_i = f_i + \bar{g}_i$, where $f_i \in R[x_1,\ldots,x_r]$ and $\bar{g}_i \in U_0$. Let
$$J_0 = (f_1,\ldots,f_m) \triangleleft R[x_1,\ldots,x_r],$$
and let
$$J_{00} = J \cap R[x_1,\ldots,x_r].$$
It is easy to see that
$$J_{00} \subseteq J_0 \subseteq \sqrt{J_{00}},$$
the radical of J_{00} in $R[x_1,\ldots,x_r]$. Of course,
$$\sqrt{J_{00}} = \sqrt{J} \cap R[x_1,\ldots,x_r].$$
Since
$$\mathfrak{p}(G,F_1),\ldots,\mathfrak{p}(G,F_\nu),\mathfrak{p}(K_1,c_1),\ldots,\mathfrak{p}(K_\gamma,c_\gamma)$$

are the minimal prime ideals in $H_G(X)$, it follows that
$$\sqrt{J} \cap R[x_1,\ldots,x_r] = P(G,F_1) \cap \ldots \cap P(G,F_\nu) \cap P(K_1,c_1) \cap$$
$$\ldots \cap P(K_\gamma,c_\gamma) \cap R[x_1,\ldots,x_r].$$

Now let E be any field which contains R, such as the quotient field $R_{(0)}$ or its algebraic closure, or let E be R itself. For any ideal
$$\mathfrak{a} \triangleleft R[x_1,\ldots,x_r],$$
let $V_E(\mathfrak{a})$ be the variety of $\mathfrak{a}$ in E^r: i.e. $(\alpha_1,\ldots,\alpha_r) \in E^r$ is in $V_E(\mathfrak{a})$ if and only if $f(\alpha_1,\ldots,\alpha_r) = 0$, for all $f(x_1,\ldots,x_r) \in \mathfrak{a}$. Since $PK_j \subset P(K_j,c_j)$, for $1 \leq j \leq \gamma$, clearly
$$V_E(P(K_j,c_j) \cap R[x_1,\ldots,x_r]) = \emptyset.$$
So
$$V_E(J_0) = V_E(J_{00}) = \bigcup_{j=1}^{\nu} V_E(P(G,F_j) \cap R[x_1,\ldots,x_r])$$
$$= \{(a_{j1},\ldots,a_{jr}) : 1 \leq j \leq \nu\}$$
$$= V_R(J_0) = V_R(J_{00}) \subset R^r.$$

Furthermore, this variety is in one-to-one correspondence with the set of components of X^G. It is also in one-to-one correspondence with $\mathrm{Hom}_{R-\mathrm{Alg}}(H_G(X),E)$, the set of R-algebra homomorphisms from $H_G(X)$ to E.

Note also that, for $1 \leq i \leq r$,
$$(x_j - a_{1j})\ldots(x_j - a_{\nu j})(x_j - b_{1j})\ldots(x_j - b_{\gamma j}) \in \sqrt{J_{00}}.$$
Likewise, for any $u \in H_G(X)$, there exist $a_1,\ldots,a_\nu, b_1,\ldots,b_\gamma \in R$ such that
$$(u - a_1)\ldots(u - a_\nu)(u - b_1)\ldots(u - b_\gamma) \in \sqrt{(0)}.$$
To see this take $a_i = (G,F_i)^*(u)$, for $1 \leq i \leq \nu$; and, for $1 \leq i \leq \gamma$, choose $b_i \in R$, such that
$$b_i + PK_i = (K_i,c_i)^*(u) \in R/PK_i.$$

(3.8.11) Example

A simple but interesting example occurs when X is a space of $\mathbb{C}P$-type: i.e. $H^*(X;k) \cong k[\xi]/(\xi^{n+1})$, where $\deg(\xi) = 2$. In case (0), X is TNHZ in $X_G \to BG$ and $X^G \neq \emptyset$. In cases (2) and (odd p), the coefficients in the E_2-term of the Serre spectral sequence of $X_G \to BG$ are constant, since $\mathbb{F}_p$ has no non-trivial automorphism of order p. Thus X is TNHZ in $X_G \to BG$ if $X^G \neq \emptyset$. So, assuming $X^G \neq \emptyset$, we may take $A = R[x]$ in cases (0) and (2), and $A = R[x] \otimes \Lambda$, where $\Lambda = \Lambda(s_1,\ldots,s_n)$ in case (odd p), where, in each case, $\deg(x) = 2$.

3.8. Hsiang's Fundamental Fixed Point Theorem

In cases (0) and (2), $J = (f(x))$, and in case (odd p), $J = (f(x)+h(x))$, where, in all cases, $f(x) \in R[x]$ is a monic polynomial of degree $n+1$ in x (since $\xi^{n+1} = 0$), and, in case (odd p), $h(x)$ is in the ideal of A generated by $s_1, \ldots, s_n$, and $h(x)$ has degree at most n in x. By the Fundamental Fixed Point Theorem, or Remark (3.8.10), $f(x)$ must factorize:

$$f(x) = \prod_{i=1}^{\nu}(x - a_i)^{n_i+1},$$

where $a_1, \ldots, a_\nu$ are distinct elements of R, ν is the number of components of $X^G = F_1 + \ldots + F_\nu$; and $\sum_{i=1}^{\nu}(n_i + 1) = n + 1$.

From Theorem (3.8.8), and Remarks (3.8.9)(2) in case (odd p), it follows that $H^*(F_i; k) \cong k[\xi_i]/(\xi_i^{n_i+1})$ for $1 \leq i \leq \nu$. If $n_i = 0$, then F_i is k-acyclic: i.e. $H^*(F_i; k) \cong k$. If $n_i > 0$, then $\deg(\xi_i) = 2$ in cases (0) and (odd p); and $\deg(\xi_i) = 1$ or 2 in case (2).

In [Hsiang, 1974] it is shown that in case (2) either $\deg(\xi_i) = 1$ for all non-k-acyclic components F_i, or $\deg(\xi_i) = 2$ for all non-k-acyclic components F_i. To see this choose $y \in H^2_G(X; k)$ so that $i^*(u) = \xi$, where $i : X \to X_G$ is the inclusion of the fibre. Replacing u by $u - (G, F_1)^*(u)$ if necessary, we may suppose that $(G, F_1)^*(u) = 0$. Since $\text{Sq}^1(\xi) = 0$, $i^*(\text{Sq}^1(u)) = 0$; and, since $(G, F_1)^*(\text{Sq}^1(u)) = 0$, it follows that $\text{Sq}^1(u) = bu$ for some $b \in R$. Choose x so that $\pi(x) = u$. Now suppose that F_i is not k-acyclic, and that $\deg(\xi_i) = 2$. Then

$$\phi_i^*(u) = a_i + \xi_i;$$

and applying Sq^1, we get $ba_i + b\xi_i = \text{Sq}^1(a_i)$, since $H^3(F_i; k) = 0$. So $b = 0$ (and $\text{Sq}^1(a_i) = 0$). If, however, $\deg(\zeta_i) = 1$, then

$$\phi_i^*(u) = a_i + c_i\xi_i + \alpha_i\xi_i^2,$$

where $c_i \in R$ and $\alpha_i \in k$. By the Localization Theorem, we must have $c_i \neq 0$. Now applying Sq^1 gives

$$ba_i + bc_i\xi_i + \alpha_i b\xi_i^2 = \text{Sq}^1(a_i) + c_i^2\xi_i + c_i\xi_i^2.$$

So $b = c_i$ (and $\alpha_i = 1$ and $\text{Sq}^1(a_i) = ba_i$). Thus, if $b = 0$, all components have $\mathbb{C}P$-type: $n_i = 0$ or $\deg(\xi_i) = 2$. But, if $b \neq 0$, then all components have $\mathbb{R}P$-type: $n_i = 0$ or $\deg(\xi_i) = 1$. Furthermore, if $b = 0$, then $\text{Sq}^1(a_i) = 0$ implies that there are at most 2^r components, where $G = (\mathbb{Z}_2)^r$; whereas, if $b \neq 0$, then $\text{Sq}^1(a_i) = ba_i$ implies that there are at most 2^{r-1} components (where $G = (\mathbb{Z}_2)^r$). Likewise in case (odd p), if $r = 1$, i.e. $G = \mathbb{Z}_p$, then X^G has at most p components, since $a_i = \alpha_i t$ for $\alpha_i \in k = \mathbb{F}_p$; cf. (1.3.19).

For more on this example, especially in cases (2) and (odd p), and for the situation when $\deg(\xi) = 1$, 4, or 8, see [Borel, et al., 1960], Chapter XII, §4, [Bredon, 1972], Chapter VIII, Section 3, and [Hsiang, 1974].

For case (0), when $\deg(\xi) = 4$ or 8, respectively, see [Chang, Skjelbred, 1974] or [Chang, Skjelbred, 1976], respectively.

We shall now use Hsiang's Fundamental Fixed Point Theorem, Theorem (3.8.8), to give a proof of the generators and relations theorem, cf. Theorem (1.3.17) and Exercise (1.15), which will be stated below as Theorem (3.8.12). Our basic assumptions are

(i) G, k, R and $H_G^* = H^*(BG; k)$ are as usual for cases (0), (2) or (odd p);
(ii) X is a connected finite-dimensional G-CW-complex, with FMCOT in case (0);
(iii) X is TNHZ in $X_G \to BG$; and
(iv) $\dim_k H^*(X; k) < \infty$.

Since X is finite-dimensional, (iv) is equivalent to the assumption that $H^*(X; k)$ is a finitely generated k-algebra.

(3.8.12) Theorem

With assumptions (i)-(iv) above, let F be a component of X^G (which is non-empty by (iii)). In cases (0) and (odd p), suppose that $H^*(X; k)$ has a graded k-algebra presentation with $\mathfrak{g}_0$ generators of even degree, $\mathfrak{g}_1$ generators of odd degree, $\mathfrak{r}_0$ relations of even degree, and $\mathfrak{r}_1$ relations of odd degree. Suppose that $H^*(F; k)$ has a minimal graded k-algebra presentation with $\mathfrak{g}_0'$ generators of even degree, $\mathfrak{g}_1'$ generators of odd degree, $\mathfrak{r}_0'$ relations of even degree, and $\mathfrak{r}_1'$ relations of odd degree.

Then:

(1) $\mathfrak{g}_0' \leq \mathfrak{g}_0$ and $\mathfrak{g}_1' \leq \mathfrak{g}_1$;
(2) $\mathfrak{r}_0' \leq \mathfrak{r}_0$ and $\mathfrak{r}_1' \leq \mathfrak{r}_1$; and indeed,
(3) $\mathfrak{r}_0' - \mathfrak{g}_0' \leq \mathfrak{r}_0 - \mathfrak{g}_0$ and $\mathfrak{r}_1' - \mathfrak{g}_1' \leq \mathfrak{r}_1 - \mathfrak{g}_1$.

In case (2), suppose that $H^*(X; \mathbb{F}_2)$ has a graded $\mathbb{F}_2$-algebra presentation with g generators and r relations. Suppose that $H^*(F; \mathbb{F}_2)$ has a minimal graded $\mathbb{F}_2$-algebra presentation with $\mathfrak{g}'$ generators and $\mathfrak{r}'$ relations. Then

(1)' $\mathfrak{g}' \leq \mathfrak{g}$;
(2)' $\mathfrak{r}' \leq \mathfrak{r}$; and indeed
(3)' $\mathfrak{r}' - \mathfrak{g}' \leq \mathfrak{r} - \mathfrak{g}$;

cf. (A.7.6)–(A.7.11).

(3.8.13) Remarks

(1) For different proofs of the various parts of this theorem see [Hsiang, 1975], [Chang, 1976], [Puppe, V. 1974], [Puppe, V. 1978] and [Puppe, V. 1984].

(2) The quantities $\mathfrak{r}_0 - \mathfrak{g}_0$, $\mathfrak{r}_1 - \mathfrak{g}_1$, $\mathfrak{r}'_0 - \mathfrak{g}'_0$, $\mathfrak{r}'_1 - \mathfrak{g}'_1$ are called complete intersection defects (cids) by analogy with the notion of cid in local algebra. Here, however, the odd degree cids can be negative: e.g. if $H^*(X;k)$ is an exterior algebra on generators of odd degrees in cases (0) or (odd p), then $\mathfrak{r}_1 = 0$.

(3) Of course, conclusion (3), resp. (3)', together with (1), resp. (1)', imply (2), resp. (2)'.

(4) The theorem is not true in general if X is not TNHZ in $X_G \to BG$. See Example (3.8.15) below.

(5) If G is any finite p-group, where p is prime, and if $k = \mathbb{F}_p$, or any field of characteristic p, then the theorem remains valid if assumption (iii) is replaced by the following:

(iii)' $\dim_k H^*(X^G;k) = \dim_k H^*(X;k)$.

It is easy to see that the induction of Step 1 below remains valid. Of course, by Theorem (3.10.4), (iii)' is equivalent to (iii) if G is an elementary abelian p-group.

We shall prove the theorem in case (odd p). The proof for case (0) is similar but easier owing to the absence of the exterior part in H_G^*. The proof in case (2) is similar to that in case (0). The proof will be divided into a number of short steps and lemmata.

Step 1. Without loss of generality $G = \mathbb{Z}_p$; and so $H_G^* = \mathbb{F}_p[t] \otimes \Lambda(s)$, and $R = \mathbb{F}_p[t]$. This follows since we can use induction to go from the case where $G = \mathbb{Z}_p$ to the general case where $G = (\mathbb{Z}_p)^n$. To see this let $(\mathbb{Z}_p)^n = K \times L$, where $K = (\mathbb{Z}_p)^{n-1}$ and $L = \mathbb{Z}_p$. Let $F(K)$ be the component of X^K which contains F. Then F is a component of $F(K)^L$. The induction works provided X is TNHZ in $X_K \to BK$, and $F(K)$ is TNHZ in $F(K)_L \to BL$: both these facts follow from Theorem (3.10.4), since

$$\dim_k H^*(X;k) = \dim_k H^*(X^G;k) \leq \dim_k H^*(X^K;k)$$
$$\leq \dim_k H^*(X;k),$$

by assumption (iii), above.

Step 2. Let

$$0 \to \mathcal{I} \to k[\overline{u}_1, \ldots, \overline{u}_{g_0}] \otimes \Lambda(\overline{v}_1, \ldots, \overline{v}_{g_1}) \overset{\overline{\pi}}{\to} H^*(X;k) \to 0$$

be a minimal presentation of $H^*(X;k)$ as a graded k-algebra (see Definition (A.7.8)(2)). Let $\mathcal{I} = \ker \overline{\pi}$ be generated by

$$\overline{f}_i = \overline{f}_i(\overline{u}_i, \ldots, \overline{u}_{g_0}, \overline{v}_1, \ldots, \overline{v}_{g_1}), \quad \text{for } 1 \leq i \leq \mathfrak{r}_0,$$

$$\overline{h}_i = \overline{h}_i(\overline{u}_1, \ldots, \overline{u}_{g_0}, \overline{v}_1, \ldots, \overline{v}_{g_1}), \quad \text{for } 1 \leq i \leq \mathfrak{r}_1,$$

where each $\overline{f}_i$ has even degree and each $\overline{h}_i$ has odd degree. Choose $u_j \in H_G^*(X;k)$, for $1 \leq j \leq \mathfrak{g}_0$, and $v_j \in H_G^*(X;k)$, for $1 \leq i \leq \mathfrak{g}_1$, such that $i^*(u_j) = \overline{\pi}(\overline{u}_j)$ and $i^*(v_j) = \overline{\pi}(\overline{v}_j)$, where $i : X \to X_G$ is the inclusion. By assumption (iii), $H_G^*(X;k)$ is generated by $u_1, \ldots, u_{\mathfrak{g}_0}$, $v_1, \ldots, v_{\mathfrak{g}_1}$ and $s \in H_G^*$ as a R-algebra. Also, of course, $H_G^*(X;k)$ is a free H_G^*-module with a basis consisting of a finite number of homogeneous elements in the k-algebra generated by $u_1, \ldots, u_{\mathfrak{g}_0}$, and $v_1, \ldots, v_{\mathfrak{g}_1}$.

Replacing each u_j by $u_j - (G,F)^*(u_j)$, if necessary, we may assume that $(G,F)^*(u_j) = 0$ for $1 \leq j \leq \mathfrak{g}_0$. Now let

$$0 \to J \to R[x_1, \ldots, x_{\mathfrak{g}_0}] \otimes \Lambda(s, y_1, \ldots, y_{\mathfrak{g}_1}) \xrightarrow{\pi} H_G^*(X;k) \to 0$$

be a R-algebra presentation of $H_G^*(X;k)$ as in Theorem (3.8.7), where $\pi(x_j) = u_j$ for $1 \leq j \leq \mathfrak{g}_0$, and $\pi(y_i) = v_j$ for $1 \leq j \leq \mathfrak{g}_1$. Let

$$A = R[x_1, \ldots, x_{\mathfrak{g}_0}] \otimes \Lambda(s, y_1, \ldots, y_{\mathfrak{g}_1}).$$

Note that

$$P(G,F)^+ = \ker((G,F)^*\pi) + (s, y_1, \ldots, y_{\mathfrak{g}_1})$$
$$= (x_1, \ldots, x_{\mathfrak{g}_0}, s, y_1, \ldots, y_{\mathfrak{g}_1})$$
$$= \overline{M}_1, \text{ say}.$$

Let $\phi_1 : F \to X$ be the inclusion, and let $J_1 = \ker(\phi_1^*\pi)$. Then $\sqrt{J_1} = \overline{M}_1$, and J_1 is a $\overline{M}_1$-primary ideal in A by Theorem (3.8.7). Let μ denote both the ideal in A and the ideal in $H_G^*(X;k)$ generated by t, $s \in H_G^*$.

We have a k-algebra epimorphism

$$\theta : A \to k[\overline{u}_1, \ldots, \overline{u}_{\mathfrak{g}_0}] \otimes \Lambda(\overline{v}_1, \ldots, \overline{v}_{\mathfrak{g}_1})$$

given by $\theta(t) = 0$, $\theta(s) = 0$, $\theta(x_j) = \overline{u}_j$ for $1 \leq j \leq \mathfrak{g}_0$, and $\theta(y_j) = \overline{v}_j$ for $1 \leq j \leq \mathfrak{g}_1$. So $\overline{\pi}\theta = i^*\pi$. For $1 \leq j \leq \mathfrak{r}_0$,

$$\theta(\overline{f}_j(x_1, \ldots, x_{\mathfrak{g}_0}, y_1, \ldots, y_{\mathfrak{g}_1})) = \overline{f}_j(\overline{u}_1, \ldots, \overline{u}_{\mathfrak{g}_0}, \overline{v}_1, \ldots, \overline{v}_{\mathfrak{g}_1}) \in \mathcal{I}.$$

So $\pi(\overline{f}_j) \in \ker i^* = \mu : \pi(\overline{f}_j) = t\overline{a}_j + s\overline{b}_j$, say. Let $a_j, b_j \in A$ be such that $\pi(a_j) = \overline{a}_j$, $\pi(b_j) = \overline{b}_j$. Then $\theta(\overline{f}_j - ta_j - sb_j) = \theta(\overline{f}_j) \in \mathcal{I}$, and $f_j = \overline{f}_j - ta_j - sb_j \in J$. Similarly, for $1 \leq j \leq \mathfrak{r}_1$, there exists $h_j = h_j(x_1, \ldots, x_{\mathfrak{g}_0}, y_1, \ldots, y_{\mathfrak{g}_1}) \in J$, such that $\theta(h_j) = \overline{h}_j \in \mathcal{I}$.

(3.8.14) Lemma ([Chang, 1976])

J is generated by

$$f_1, \ldots, f_{\mathfrak{r}_0}, h_1, \ldots, h_{\mathfrak{r}_1}.$$

3.8. Hsiang's Fundamental Fixed Point Theorem

Proof.
Let $f \in J$. Then $\theta(f) \in \mathcal{I}$. Since $\overline{f}_1, \ldots, \overline{f}_{r_0}, \overline{h}_1, \ldots, \overline{h}_{r_1}$ generate $\mathcal{I}$ and θ is surjective, there exist $a_j, b_j \in A$ such that

$$f - \sum_{j=1}^{r_0} a_j f_j - \sum_{j=1}^{r_1} b_j h_j$$

is in $\ker \theta = \mu$:

$$f = \sum_{j=1}^{r_0} a_j f_j + \sum_{j=1}^{r_1} b_j h_j + tf' + sh', \text{ say.}$$

Now $t\pi(f') + s\pi(h') = 0$. Hence $ts\pi(f') = 0$; and so $s\pi(f') = 0$. For some $f'' \in A$, therefore, $\pi(f') = s\pi(f'')$. Thus $f' - sf'' = f'''$, say, is in J. Now $s\pi(tf'' + h') = 0$. So, for some $h'' \in A$, $\pi(tf'' + h') = s\pi(h'')$: and $tf'' + h' - sh'' = h'''$, say, is in J. Hence $tf' + sh' = tf''' + sh'''$. We can now use induction on $\deg(f)$, since $\deg(f''') = \deg(f) - 2$ and $\deg(h''') = \deg(f) - 1$, to complete the proof. $\square$

This simple lemma is the crucial point. The rest of the proof is straightforward, although we must do a little bit more work with the machinery.

Let

$$0 \to \mathcal{J} \to k[\xi_1, \ldots, \xi_{g'_0}] \otimes \Lambda(\eta_1, \ldots, \eta_{g'_1}) \xrightarrow{\pi_F} H^*(F; k) \to 0$$

be a minimal graded k-algebra presentation of $H^*(F; k)$. Let T be the multiplicative set $R \quad (0)$, and let $E = R_{(0)}$. Let $B = T^{-1}A$, $I = T^{-1}J$ and $I_1 = T^{-1}J_1$. So ϕ_1^* induces an isomorphism

$$\psi_1 : B/I_1 \to T^{-1}H_G^*(F; k) \cong E \otimes \Lambda(s) \otimes H^*(F; k)$$

by Theorem (3.8.8). Let

$$C = E \otimes \Lambda(s) \otimes k[\xi_1, \ldots, \xi_{g'_0}] \otimes \Lambda(\eta_1, \ldots, \eta_{g'_1})$$

and let

$$\mathcal{J}_1 = E \otimes \Lambda(s) \otimes \mathcal{J}.$$

So we have two presentations of

$$E \otimes \Lambda(s) \otimes H^*(F; k),$$

viewed as a $\mathbb{Z}_2$-graded E-algebra, namely

$$P_1 : 0 \to I_1 \to B \to B/I_1 \to 0$$

and

$$P_2 : 0 \to \mathcal{J}_1 \to C \to T^{-1}H_G^*(F; k) \to 0.$$

And P_2 is minimal.

Proof of Theorem (3.8.12).

By Proposition (A.7.14)(1), $\dim_E Q(C) \leq \dim_E Q(B)$, which, together with Remark (A.7.17)(1), proves part (1).

By Corollary (A.7.16), $\text{cid}(P_2) \leq \text{cid}(P_1)$. Now

$$\text{cid}(P_1) = \dim_E(I_1/M_1 I_1) - \dim_E Q(B),$$

where $M_1 = T^{-1}\overline{M}_1$. But the inclusion of I into I_1 induces a homomorphism $\gamma : I/M_1 I \to I_1/M_1 I_1$; and γ is surjective as follows. Let $\zeta \in I_1$. Since $I_1 = I_{M_1} \cap B$ by Theorem (3.8.8), $a\xi \in I$ for some $a \notin M_1$. Let $a = \alpha + b$, where $\alpha \in E$, $\alpha \neq 0$, and $b \in M_1$. Then $\xi + M_1 I_1 = \gamma(\alpha^{-1} a\xi + M_1 I)$. Hence γ is surjective.

So $\text{cid}(P_2) \leq \text{cid}(P_1) \leq \dim_E(I/M_1 I) - \dim_E Q(B)$. This inequality in even and odd degrees separately proves part (3) of the theorem, thanks to Lemma (3.8.14). □

We now give an example to show that Theorem (3.8.12) can fail if X is not TNHZ in $X_G \to BG$.

(3.8.15) Example ([Bredon, 1972], Chapter VIII, Section 10)
Let $G = S^1$. Let τ be the tangent bundle of S^8 with trivial G-action. Let ε be a trivial $\mathbb{R}^2$-bundle over S^8 with G acting on $\mathbb{R}^2$ by rotations in the standard way. Let $f : S^3 \times S^5 \to S^8$ be a map of degree 1; and let $\eta = f^*(\tau \oplus \varepsilon)$ with induced G-action. Let $X = S(\eta)$ be the total space of the unit sphere bundle of η. Then $X \simeq S^3 \times S^5 \times S^9$, and G acts on X with $X^G = S(f^*\tau)$. From the Serre spectral sequence, it follows that $H^*(X^G; \mathbb{Q})$ is generated by $\eta_1, \eta_2, \xi_1, \xi_2$, where $\deg(\eta_1) = 3$, $\deg(\eta_2) = 5$, $\deg(\xi_1) = 10$, $\deg(\xi_2) = 12$, with relations $\eta_1 \eta_2 = 0$, $\eta_1 \xi_1 = 0$, $\eta_2 \xi_2 = 0$, $\xi_1^2 = 0$, $\xi_1 \xi_2 = 0$, $\xi_2^2 = 0$ and $\eta_1 \xi_2 = \eta_2 \xi_1$. Thus $g_0 = 0$, $g_1 = 3$, $g'_0 = 2$, $g'_1 = 2$, $r_0 = 0$, $r_1 - 0$, $r'_0 = 4$, $r'_1 = 3$, $r_0 - g_0 = 0$, $r_1 - g_1 = -3$, $r'_0 - g'_0 = 2$, and $r'_1 - g'_1 = 1$.

This example is one of the simplest known examples of a $G = S^1$ action on a manifold X with X not TNHZ in $X_G \to BG$.

(3.8.16) Remark General G-spaces.
Using Alexander-Spanier (or Čech) cohomology all the results of this section hold as stated, clearly, for general G-spaces satisfying conditions (A), (B) or (C) of Section 3.2, and belonging to cases (0), (2) or (odd p). Even in case (A) the condition that $\dim_k \overline{H}^*(X;k) < \infty$ should still be assumed. Note that for connected spaces (or any spaces in case (A) or cases (2) or (odd p)) this is again equivalent to the assumption that $\overline{H}^*(X;k)$ is a finitely generated k-algebra: this follows in case (B) as before; it follows in cases (A) and (C) since finite-dimensional Čech

nerves are cofinal, and hence any class of positive degree in $\overline{H}^*(X;k)$ is nilpotent.

3.9 Remarks on the Weyl group

Although it is not, in general, our purpose to dwell on actions of groups other than tori and p-tori we include here a few basic results which relate actions of general compact Lie groups to the actions of their maximal tori or p-tori. In fact we shall discuss only compact connected Lie groups and their maximal tori: for maximal p-tori in compact Lie groups see, for example, [Quillen, 1971a,b].

Throughout the remainder of this section let G be a compact connected Lie group, let T be a maximal torus of G, let NT be the normalizer of T in G, and let $W = NT/T$ be the Weyl group of G. Let w be the order of W.

From [Borel, 1967], we have that $H^i(G/T;\mathbb{Q}) = 0$ for all odd i, and $\chi(G/T) = \dim_\mathbb{Q} H^*(G/T;\mathbb{Q}) = w$ ([Borel, 1967], Theorem 20.5). Also W acts on $H_T^* = H^*(BT;\mathbb{Q})$ and $H_G^* = H^*(BG;\mathbb{Q}) = (H_T^*)^W$, the fixed subalgebra of this action ([Borel, 1967], Proposition 20.4).

The following results, through Theorem (3.9.3), may be found in [Hsiang, 1975].

(3.9.1) Lemma
(1) $H_T^* \cong H_G^* \otimes H^*(G/T;\mathbb{Q})$ as graded H_G^*-modules.
(2) $H^i(G/NT;\mathbb{Q}) = 0$, for $i > 0$, i.e. G/NT is $\mathbb{Q}$-acyclic.

Proof.
(1) Since $H^i(G/T;\mathbb{Q}) = 0$ for i odd, the Serre spectral sequence for $G/T \to BT \to BG$ in rational cohomology collapses. So
$$H_T^* \cong H_G^* \otimes H^*(G/T;\mathbb{Q})$$
as graded H_G^*-modules.

(2) From the covering $W \to G/T \to G/NT$ it follows that
$$H^*(G/NT;\mathbb{Q}) = H^*(G/T;\mathbb{Q})^W,$$
and
$$\chi(G/NT) = \left(\frac{1}{w}\right)\chi(G/T) = 1.$$
Since $H^i(G/T;\mathbb{Q}) = 0$ for odd i,
$$\chi(G/NT) = \dim_\mathbb{Q} H^*(G/NT;\mathbb{Q}),$$
and the result follows. □

For the rest of this section, we shall consider a space X with various groups K acting on it. We shall treat general spaces simultaneously with CW-complexes. So, if X is a K-space, we shall assume either

(i) X is a finite-dimensional K-CW-complex with FMCOT, and all cohomology and all minimal models are singular; or
(ii) X satisfies conditions (A), (B) or (C) of Section 3.2, and all cohomology and all minimal models are Alexander-Spanier. We shall write H^* and $\mathcal{M}$, rather that $\overline{H}^*$ and $\overline{\mathcal{M}}$, in this case, in order to unify the presentation.

(3.9.2) Lemma

Let K be a compact Lie group and let G be the identity component of K. Let $\Gamma = K/G$. Then Γ acts on X_G, and $H_K^*(X;\mathbb{Q}) \cong H_G^*(X;\mathbb{Q})^\Gamma$.

Proof.
Since Γ is finite, this follows from the fibration $X_G \to X_K \to B\Gamma$. (Γ acts freely on X_G, and $X_K \simeq X_G/\Gamma$.) □

(3.9.3) Theorem
If G is a compact connected Lie group and T is a maximal torus of G, and if X is a G-space, then:
(1) $H_G^*(X;\mathbb{Q}) \cong H_T^*(X;\mathbb{Q})^W$ *as graded H_G^*-algebras; and*
(2) $H_T^*(X;\mathbb{Q}) \cong H_G^*(X;\mathbb{Q}) \otimes_{H_G^*} H_T^*$ *as graded H_G^*-algebras.*

Proof.
(1) From Lemma (3.9.2), $H_{NT}^*(X;\mathbb{Q}) \cong H_T^*(X;\mathbb{Q})^W$. From Lemma (3.9.1)(2) and the fibration $G/NT \to X_{NT} \to X_G$,
$$H_G^*(X;\mathbb{Q}) \cong H_{NT}^*(X;\mathbb{Q}).$$

(2) Consider the pull-back of fibrations shown in Diagram 3.5.

$$\begin{array}{ccc} G/T & = & G/T \\ \downarrow & & \downarrow \\ X_T & \xrightarrow{p} & BT \\ \downarrow \psi & & \downarrow \phi \\ X_G & \xrightarrow{q} & BG \end{array}$$

Diagram 3.5

As in Lemma (3.9.1)(1), the Leray-Serre spectral sequence of ϕ collapses yielding a H_G^*-module isomorphism $\phi^* \otimes \lambda$: $H_G^* \otimes H^*(G/T; \mathbb{Q}) \to H_T^*$, where $\lambda : H^*(G/T; \mathbb{Q}) \to H_T^*$ is a cohomology extension of the fibre for the fibration ϕ. Now $p^*\lambda$ is a cohomology extension of the fibre for the fibration ψ: and so $\psi^* \otimes p^*\lambda : H_G^*(X; \mathbb{Q}) \otimes H^*(G/T; \mathbb{Q}) \to H_T^*(X; \mathbb{Q})$ is an isomorphism.

The H_G^*-algebra homomorphisms ψ^* and p^* yield, via the direct sum in the category of H_G^*-algebras, a H_G^*-algebra homomorphism

$$\psi^* \otimes p^* : H_G^*(X; \mathbb{Q}) \otimes_{H_G^*} H_T^* \to H_T^*(X; \mathbb{Q}).$$

The composition of H_G^*-module homomorphisms

$$H_G^*(X; \mathbb{Q}) \otimes H^*(G/T; \mathbb{Q}) \to H_G^*(X; \mathbb{Q}) \otimes_{H_G^*} (H_G^* \otimes H^*(G/T; \mathbb{Q}))$$

$$\xrightarrow{1 \otimes (\phi^* \otimes \lambda)} H_G^*(X; \mathbb{Q}) \otimes_{H_G^*} H_T^* \xrightarrow{\psi^* \otimes p^*} H_T^*(X; \mathbb{Q})$$

is $\psi^* \otimes p^*\lambda$. So $\psi^* \otimes p^*$ is an isomorphism. $\square$

Now from Theorem (2.5.1) and $X \to X_T \to BT$, if X is connected, we have a model of the form $H_T^* \otimes \mathcal{M}(X)$ for X_T. In fact, with the notation of Definition (2.2.2), we have a well-ordered set $E = \{x_\alpha; \alpha \in \mathcal{A}\}$ and a differential d on $H_T^* \otimes S(E)$ such that:

(i) $d|H_T^* = 0$,
(ii) for any $\alpha \in \mathcal{A}$, $d(1 \otimes x_\alpha) \in H_T^* \otimes S(E_\alpha)$,
(iii) $(H_T^* \otimes S(E), d)$ is a model for X_T, and
(iv) if $\varepsilon : H_T^* \to \mathbb{Q}$ is the standard augmentation,

and the differential δ on $S(E)$ is defined by

$$\delta x_\alpha = (\varepsilon \otimes 1)d(1 \otimes x_\alpha) \in \mathbb{Q} \otimes S(E_\alpha) = S(E_\alpha),$$

then $(S(E), \delta)$ is a minimal model for X. (See Definition (2.2.2).) Clearly one has some options concerning how one chooses the generating set E: see [Halperin, 1977(1983)], Chapter I. The point of the following proposition is that E can be chosen so that the action of the Weyl group on X_T is represented very nicely on $H_T^* \otimes \mathcal{M}(X)$. For $\omega \in W$, let $\omega_B : BT \to BT$ be the action of ω on BT, and let $\omega_X : X_T \to X_T$ be the action of ω on X_T.

(3.9.4) Proposition

With the notation above, in the model

$$H_T^* \otimes \mathcal{M}(X) = H_T^* \otimes S(E)$$

for X_T, the generating set $E = \{x_\alpha; \alpha \in \mathcal{A}\}$ can be chosen so that

(1) ω_X *is represented by* $\omega_B^* \otimes 1 : H_T^* \otimes \mathcal{M}(X) \to H_T^* \otimes \mathcal{M}(X)$; *and*

(2) $\psi : X_T \to X_G$ is represented by
$$\phi^* \otimes 1 : H_G^* \otimes \mathcal{M}(X) \to H_T^* \otimes \mathcal{M}(X)$$
where $\phi : BT \to BG$ is induced by the inclusion $T \to G$. In particular:
(i) $d(1 \otimes x_\alpha) \in (H_T^*)^W \otimes S(E_\alpha)$, for any $\alpha \in \mathcal{A}$, and
(ii) $\omega_X^*(1 \otimes x_\alpha) = 1 \otimes x_\alpha$, for any $\alpha \in \mathcal{A}$, $\omega \in W$.
We shall not give the proof here: it is an exercise in the use of minimal models and KS-extensions. The proof would be easy if Theorem (2.5.6) applied, but it is not clear that it does.

3.10 The cohomology inequalities and other basic results

In this section we shall state, for reference, some of the fundamental results relating the size of $H^*(X^G; k)$ to that of $H^*(X; k)$ together with some other basic results. Many of the theorems given are proven in [Bredon, 1972]; and sometimes we do not give the proofs here when that is the case. We shall work mainly with cases (0), (2) and (odd p): thus, unless stated otherwise, G is a torus or p-torus, and $k = \mathbb{Q}$ or $\mathbb{F}_p$, respectively.

As in Section 3.9 we shall combine the treatments of G-CW-complexes and general G-spaces. A few of the results hold for any paracompact G-space; but otherwise we shall have to assume that our G-spaces satisfy one of the conditions which are used in the Localization Theorem. We shall say that a G-space X satisfies condition (LT) if

(i) X is a finite-dimensional G-CW-complex with FMCOT, or
(ii) X is a general G-space in case (A), (B) or (C) of Definition (3.2.4).

We shall say that a pair of G-spaces (X, A) satisfies condition (LT) if X satisfies condition (LT), and A is a G-CW-subcomplex in case (i), or any closed invariant subspace in case (ii). To unify the presentation we shall use $H^*(X; k)$ to denote singular cohomology in case (i) and Alexander-Spanier cohomology in case (ii). In other words, the cohomology theory used will always be the one associated with the Eilenberg-MacLane spectrum. We shall use the usual notation for Alexander-Spanier cohomology, $\overline{H}^*(X; k)$, only when emphasis is desirable.

Note that in some of the results induction is used to get from the case where $G = \mathbb{Z}_p$ to the case where G is a more general p-torus. Some of these inductive arguments could be applied to treat the case where G is any finite p-group. Also, when results for torus actions can be obtained by induction from the corresponding results for circle actions, then the FMCOT condition in cases (B) and (C) can usually be omitted.

3.10. Cohomology inequalities and other basic results

Before stating the first theorem, we give the following very useful proposition. We give a slightly simplified version of the Quillen-Venkov proof, which may be found in [Quillen, 1971a], Section 2, where a more general result is proved.

(3.10.1) Proposition

If $\dim_k H^*(X;k) < \infty$, *then* $H_G^*(X;k)$ *is a finitely generated R-module.*

Proof.

We begin with case (0). In the Leray-Serre spectral sequence for $X_G \to BG$,
$$E_2 \cong R \otimes H^*(X;k).$$
So each of the finitely many rows in E_2 is a finitely generated R-module. Hence each of the finitely many rows in E_∞ is a finitely generated R-module, since R is Noetherian. From the increasing filtration $\mathcal{F}_q H_G^*(X;k)$, $q \geq 0$, with
$$E_\infty^{*,q} \cong \mathcal{F}_q H_G^*(X;k)/\mathcal{F}_{q-1} H_G^*(X;k),$$
it follows that $H_G^*(X;k)$ is a finitely generated R-module.

In case (2) or (odd p) with $G = (\mathbb{Z}_p)^n$, let $T = T^n$, the n-dimensional torus, and view G as a subgroup of T. Let $Z = T \times_G X$, where G acts on T by right translation. In the Leray-Serre spectral sequence of
$$X \to Z \to T/G \approx T$$
with $\mathbb{F}_p$-coefficients, we have
$$E_2 \cong H^*(T; H^*(X; \mathbb{F}_p)),$$
where $\pi_1(T) = \mathbb{Z}^n$ may act non-trivially on $H^*(X; \mathbb{F}_p)$. Since
$$\dim_{\mathbb{F}_p} H^*(X; \mathbb{F}_p) < \infty,$$
it follows that $\dim_{\mathbb{F}_p} E_2 < \infty$; and hence
$$\dim_{\mathbb{F}_p} H^*(Z; \mathbb{F}_p) < \infty.$$

Now T acts on Z by left translation on the factor T; and, as for case (0), it follows that $H_T^*(Z; \mathbb{F}_p)$ is a finitely generated module over $H^*(BT; \mathbb{F}_p) = \mathbb{F}_p[t_1, \ldots, t_n]$, where $\deg(t_i) = 2$ for $1 \leq i \leq n$. But the map $x \mapsto G$-orbit of $(1,x)$ induces a homeomorphism $X_G \to Z_T$, which lies over the map $i : BG \to BT$ induced by the inclusion of G in T. In case (odd p), i^* is an isomorphism of $H^*(BT; \mathbb{F}_p)$ onto R, the polynomial part of $H^*(BG; \mathbb{F}_p)$. Thus in either case (2) or (odd p) we get that $H_G^*(X; \mathbb{F}_p)$ is a finitely-generated R-module. $\square$

(3.10.2) Corollary
If $\dim_k H^*(X;k) < \infty$, *and if* X *satisfies condition (LT), then*
$$\dim_k H^*(X^G;k) < \infty.$$
In particular X^G *has a finite number of components.*

Proof.

By the Localization Theorem,
$$R_{(0)} \otimes_R H_G^*(X;k) \cong R_{(0)} \otimes_R H_G^*(X^G;k) = (R_{(0)} \otimes_R H_G^*) \otimes H^*(X^G;k).$$
Since $H_G^*(X;k)$ is a finitely generated R-module, $R_{(0)} \otimes_R H_G^*(X;k)$ is a finite-dimensional $R_{(0)}$-vector-space. So $\dim_k H^*(X^G;k) < \infty$.

In the G-CW-complex case it is clear that $\dim_k H^0(X^G;k) < \infty$ implies that X^G has a finite number of components. For general G-spaces it follows from the fact that $\overline{H}^0(X^G;k)$ is the ring of locally constant k-valued functions on X^G. □

(3.10.3) Remarks
(1) Clearly Proposition (3.10.1) and the first part of Corollary (3.10.2) hold for pairs (X,Y), where X is a G-space and Y is a closed invariant subspace: i.e. if $\dim_k H^*(X,Y;k) < \infty$, then $H_G^*(X,Y;k)$ is a finitely generated R-module; and
$$\dim_k H^*(X^G, Y^G;k) < \infty,$$
if (X,Y) satisfies condition (LT).

(2) If $G = S^1$ or $\mathbb{Z}_p$, X satisfies condition (LT), and $\dim_k H^*(X;k) < \infty$, then $\dim_k H_G^*(X, X^G;k) < \infty$. This follows since $\dim_k H^*(X^G;k) < \infty$, hence
$$\dim_k H^*(X, X^G;k) < \infty;$$
and so $H_G^*(X, X^G;k)$ is a finitely generated torsional R-module.

(3.10.4) Theorem
Suppose that $\dim_k H^*(X;k) < \infty$, *and that* X *satisfies condition (LT). Then*
$$\dim_k H^*(X^G;k) \leq \dim_k H^*(X;k).$$
Furthermore, $\dim_k H^*(X^G;k) = \dim_k H^*(X;k)$ *if and only if*

(1) $i^* : H_G^*(X;k) \to H^*(X;k)$ *is surjective, where* $i : X \to X_G$ *is the inclusion; or, equivalently,*

(2) G *acts trivially on* $H^*(X;k)$ *and* $E_2 = E_\infty$ *in the Leray-Serre spectral sequence of* $X_G \to BG$ *for cohomology with coefficients in* k.

(Of course, (1), or (2), is what is meant by saying that X is TNHZ in $X_G \to BG$.)

(3.10.5) Remarks

(1) This is essentially Theorem 1.6 of [Bredon, 1972], Chapter VII. The result here, as there, extends to the pair (X, A) where A is a closed invariant subspace: i.e.
$$\dim_k H^*(X^G, A^G; k) \leq \dim_k H^*(X, A; k),$$
with equality if and only if (X, A) is TNHZ in (X_G, A_G).

(2) The proof in [Bredon, 1972] is for the case where $G = S^1$ or $\mathbb{Z}_p$. We shall indicate in (3) below how to use induction to get the result for larger tori and p-tori. Note that Bredon is working with general G-spaces in case (C). He must assume that X/G is finitistic, but this is now known by [Deo, Tripathi, 1982]. Also his assumption, in the case where $G = S^1$, that G acts with finitely many orbit types is unnecessary since one needs only FMCOT for finite-dimensional G-CW-complexes or general G-spaces in cases (B) and (C): and one needs no assumption on the number of orbit types in case (A). There is no problem in adapting Bredon's proof to the other cases under consideration here.

(3) We shall indicate here how to proceed by induction from the case where $G = S^1$ or $\mathbb{Z}_p$ to the result for general tori and p-tori. We shall give the details for the hardest case, namely case (odd p).

Let $G = (\mathbb{Z}_p)^n$, where $n \geq 1$, and suppose that the result has been proved for $(\mathbb{Z}_p)^m$, $m < n$. The case $m = 1$ is, of course, in [Bredon, 1972]. Let $G = K \times L$ where K and L are proper subtori. Then, by induction,
$$\dim_k H^*(X^G; k) = \dim_k H^*((X^K)^L; k) \leq \dim_k H^*(X^K; k)$$
$$\leq \dim_k H^*(X; k).$$
Equality implies that K acts trivially on $H^*(X; k)$. Since K is any proper subgroup of G, it follows that G acts trivially on $H^*(X; k)$. Now, in the spectral sequence, $E_2 = H^*(BG; k) \otimes H^*(X; k)$. So, with R the polynomial part of $H^*(BG; k)$ as usual, we get
$$2^n \dim_k H^*(X; k) = \dim_{R_{(0)}}(R_{(0)} \otimes_R E_2)$$
$$\geq \dim_{R_{(0)}}(R_{(0)} \otimes_R E_\infty) = \dim_{R_{(0)}}(R_{(0)} \otimes_R H_G^*(X; k))$$
$$= \dim_{R_{(0)}}(R_{(0)} \otimes_R H_G^*(X^G; k)) = 2^n \dim_k H^*(X^G; k),$$
where we have used the Localization Theorem. Thus equality implies that $E_2 = E_\infty$.

(4) Theorem (3.10.4), and more, in case (0) has already appeared in Corollaries (3.1.14) and (3.1.15). The proofs given there, however, do not

easily carry over to cases (2) and (odd p): the possibility of G acting non-trivially on $H^*(X;k)$ and the exterior part of $H^*(BG;k)$ in case (odd p) get in the way. Of course, one cannot separate the even and odd parts of the cohomology in case (2), in general. In [Bredon, 1972], Chapter VII, Section 2, inequalities analogous to those of Corollary (3.1.14) for cases (2) and (odd p) are proven under certain conditions, however. (Cf. also (1.3.8), (1.4.8) and Exercise (1.13).)

Consider the weak and strong forms of these inequalities:

(W) $\sum_{i=0}^{\infty} \dim_k H^{m+i}(X^G;k) \leq \sum_{i=0}^{\infty} \dim_k H^{m+i}(X;k)$, and
(S) $\sum_{i=0}^{\infty} \dim_k H^{m+2i}(X^G;k) \leq \sum_{i=0}^{\infty} \dim_k H^{m+2i}(X;k)$,
where in each case m is any non-negative integer.

We shall now state as a theorem some conditions under which (W) or (S) hold, referring to [Bredon, 1972], Chapter VII, Section 2 for the proof. We shall then give a theorem on the inequalities (S), which is essentially the theorem of Heller and Swan given as Theorem 2.2 in [Bredon, 1972], but we shall give a slight variation of the proof, using the Localization Theorem, which is, therefore, applicable to all kinds of spaces under consideration here. Recall that throughout this section H^* denotes singular cohomology for finite-dimensional G-CW-complexes and Alexander-Spanier cohomology for general G-spaces satisfying (A), (B) or (C). Finally note that the following results again extend to pairs (X, A), where A is a closed invariant subspace, as in [Bredon, 1972].

(3.10.6) Theorem

Suppose that $\dim_k H^*(X;k) < \infty$, and that X satisfies condition (LT). Then the inequalities (W) hold in cases (2) and (odd p).

If X is TNHZ in $X_G \to BG$ with respect to cohomology with coefficients in $k = \mathbb{F}_p$, then the inequalities (S) hold in case (odd p).

(3.10.7) Theorem

Suppose that $\dim_k H^*(X;k) < \infty$, that X satisfies condition (LT) and that, for any subgroups K and L of G such that $G = K \times L$, L acts trivially on $H^*(X^K; \mathbb{Z})$. Then the inequalities (S) hold in cases (2) and (odd p). (For X a general G-space in case (B), we assume here that $\mathrm{cd}_{\mathbb{Z}}(X) < \infty$.)

Proof.

It is enough to prove the result when $G = \mathbb{Z}_p$, for then we can use induction to get the result for larger p-tori. Let

$$B = H^*(BG; \mathbb{Z}) = \mathbb{Z}[u]/(pu),$$

3.10. Cohomology inequalities and other basic results

where $\deg(u) = 2$. Let $S \subseteq B$ be the multiplicative subset generated by u. Then $X^S = X^G$; and so, by the Localization Theorem, we have an isomorphism of $S^{-1}B$-modules:

$$S^{-1}\phi^* : S^{-1}H_G^*(X;\mathbb{Z}) \to S^{-1}H_G^*(X^G;\mathbb{Z}).$$

Now $S^{-1}B \cong \mathbb{F}_p[u, u^{-1}]$. Let E be the field of fractions (quotient field) of $S^{-1}B$. Then, reasoning as in Corollary (3.1.14), we get

$$\sum_{i=0}^{\infty} \dim_k H^{m+2i}(X^G; k)$$

$$= \dim_E (E \otimes_{S^{-1}B} S^{-1}\{H_G^*(X^G;\mathbb{Z})/\mathcal{F}_{m-1}H_G^*(X^G;\mathbb{Z})\}^{(m)})$$

$$\leq \dim_E (E \otimes_{S^{-1}B} S^{-1}\{H_G^*(X;\mathbb{Z})/\mathcal{F}_{m-1}H_G^*(X;\mathbb{Z})\}^{(m)})$$

$$= \sum_{q=m}^{\infty} \dim_E (E \otimes_{S^{-1}B} S^{-1} E_\infty^{*,q})^{(m)}$$

$$\leq \sum_{q=m}^{\infty} \dim_E (E \otimes_{S^{-1}B} S^{-1} E_2^{*,q})^{(m)}$$

$$= \sum_{i=0}^{\infty} \dim_k H^{m+2i}(X;k).$$

Here the superscript (m) indicates elements of degrees congruent to m modulo 2, $E_r^{*,*}$ is the Leray-Serre spectral sequence of $X \to X_G \to BG$ in integral cohomology, and, as usual, $k = \mathbb{F}_p$. The final equality follows since, for $n \geq 1$,

$$F_2^{2n-1,q} = H^{2n-1}(BG; H^q(X;\mathbb{Z})) \cong H^q(X;\mathbb{Z}) * \mathbb{F}_p,$$

and

$$E_2^{2n,q} = H^{2n}(BG; H^q(X;\mathbb{Z})) \cong H^q(X;\mathbb{Z}) \otimes \mathbb{F}_p.$$

The B-module structure of $E_2^{*,*}$ is given as follows: for $n \geq 1$, multiplication by u is an isomorphism from $E_2^{n,q}$ to $E_2^{n+2,q}$; and multiplication by u from $E_2^{0,q}$ to $E_2^{2,q}$ is reduction modulo p. (See [Cartan, Eilenberg, 1956], Chapter XII, §§7 and 11.)

Finally note that the Leray-Serre spectral sequence in integral cohomology for $X^G \xrightarrow{i} (X^G)_G \to BG$ collapses (i.e. $E_2 = E_\infty$): this justifies the first equality in the sequence above; and it can be seen as follows. From the naturality of the Künneth sequences comparing $(X^G)_G = BG \times X^G$ with $X^G \cong \{x_0\} \times X^G$, for $x_0 \in BG$, it follows that

$$\ker[i^* : H_G^q(X^G;\mathbb{Z}) \to H^q(X^G;\mathbb{Z})] \cong$$

$$\bigoplus_{n=1}^{\infty} \mathbb{F}_p \otimes H^{q-2n}(X^G;\mathbb{Z}) \oplus \bigoplus_{n=1}^{\infty} \mathbb{F}_p * H^{q-2n+1}(X^G;\mathbb{Z}).$$

The latter is finite-dimensional over $\mathbb{F}_p$, since $\dim_{\mathbb{F}_p} H^*(X^G; \mathbb{F}_p) < \infty$, by Theorem (3.10.4). So

$$\dim_{\mathbb{F}_p} \ker[i^* : H^q_G(X^G; \mathbb{Z}) \to H^q(X^G; \mathbb{Z})] = \sum_{n=1}^{\infty} \dim_{\mathbb{F}_p} E_2^{n,q-n}.$$

But

$$\ker[i^* : H^q_G(X^G; \mathbb{Z}) \to H^q(X^G; \mathbb{Z})] = \mathcal{F}_{q-1} H^q_G(X^G; \mathbb{Z}) \cong \bigoplus_{n=1}^{\infty} E_\infty^{n,q-n}.$$

$\square$

(3.10.8) Remark

The shorter proof given in [Bredon, 1972] works well when X is a finite-dimensional G-CW-complex, or when X is a general G-space which is paracompact and finite-dimensional or which satisfies case (B) with $\mathrm{cd}_{\mathbb{Z}}(X) < \infty$. The proof here takes care of case (A) and the general (finitistic) form of case (C). The point in the latter case is to use the Localization Theorem to avoid the question of whether or not $H^i(X/G, X^G; \mathbb{Z}) = 0$ for all large i.

We continue this section by listing several more useful results, indicating where proofs may be found in the references. Some of these results hold in greater generality than most of those above, as in the following.

(3.10.9) Proposition
Let G be any compact Lie group, and let X be any paracompact G-space. Let A be a closed invariant subspace, and suppose that G acts freely on $X - A$.

Then the projection $\psi : (X_G, A_G) \to (X/G, A/G)$ induces an isomorphism $\psi^* : \overline{H}^*(X/G, A/G; \Lambda) \overset{\cong}{\to} \overline{H}^*_G(X, A; \Lambda)$, for any abelian group of coefficients Λ.

More generally, ψ^* is an isomorphism if $H^j(BG_x; \Lambda) = 0$ for all $j > 0$ and all $x \in X - A$. In particular, this is the case if Λ is a field of characteristic 0 and G acts almost-freely on $X - A$ (i.e. G_x is finite, for all $x \in X - A$).

'If X is a G-CW-complex and A is a G-CW-subcomplex, then, of course, the result holds in singular cohomology.

Proof.
The general proof of this result follows by replacing X_G, A_G by $X_G^N = E_G^N \times_G X$ and $A_G^N = E_G^N \times_G A$, for E_G^N a compact N-connected free G-space, using [Quillen, 1971a], (1.4), and the Vietoris-Begle Mapping Theorem [Bredon, 1967a] Chapter II, 11.2, applied to the map $X_G^N - A_G^N \to (X - A)/G$.

3.10. Cohomology inequalities and other basic results 215

In the case where $A = \emptyset$ and G is acting freely on X, then $\psi : X_G \to X/G$ is a fibre bundle with fibre EG; and hence ψ is a homotopy equivalence, by [Dold, 1963]. (See also [tom Dieck, 1987], Chapter 1, Proposition (8.18)(iii).) $\square$

(3.10.10) Theorem
Let $G = S^1$, the circle group, let X be a paracompact G-space, and let A be a closed invariant subspace. Let Λ be any abelian group. Suppose either that G acts semi-freely (i.e. G acts freely on $X - X^G$) or that Λ is a field of characteristic 0. Then there is the following Smith-Gysin long exact sequence:

$$\ldots \to \overline{H}^i(X/G, A/G \cup X^G; \Lambda) \to \overline{H}^i(X, A; \Lambda)$$
$$\to \overline{H}^{i-1}(X/G, A/G \cup X^G; \Lambda) \oplus \overline{H}^i(X^G, A^G; \Lambda)$$
$$\to \overline{H}^{i+1}(X/G, A/G \cup X^G; \Lambda) \to \ldots$$

For the proof see [Bredon, 1972], Chapter III, Section 10, or [Bredon, 1967a], Chapter IV, Section 12.3.

The homomorphism

$$\overline{H}^{i-1}(X/G, A/G \cup X^G; \Lambda) \to \overline{H}^{i+1}(X/G, A/G \cup X^G; \Lambda),$$

which appears in the exact sequence, is multiplication by the generator $t \in H^2(BG; \Lambda)$ via $p^* : H^*(BG; \Lambda) \to H_G^*(X; \Lambda)$ and the isomorphism

$$\psi^* : H^*(X/G, A/G \cup X^G; \Lambda) \to H_G^*(X, A \cup X^G; \Lambda)$$

of Proposition (3.10.9).

The next lemma, taken from [Chang, Skjelbred; 1974], has many uses.

(3.10.11) Lemma
Let k be a field and let $k[t]$ be the polynomial ring generated by an element t of degree 2. Let $E_r^{p,q}$ be a first quadrant spectral sequence such that each row $E_r^{*,q}$ is a graded $k[t]$-module, and each differential d_r is a $k[t]$-module homomorphism (for $r \geq 2$). Suppose further that, for each q, $E_2^{*,q}$ is a free $k[t]$-module generated by elements in $E_2^{0,q}$.

Then

(1) *for each q and all $r \geq 2$, $E_r^{*,q}$ is generated as a $k[t]$-module by elements in $E_r^{0,q}$; and*

(2) *for each q, any $r \geq 2$ and any $p \geq r - 1$, multiplication by t is a monomorphism of $E_r^{p,q}$ into $E_r^{p+2,q}$.*

Proof.
Clearly (1) and (2) are true for $r = 2$. So assume that (1) and (2) are true for $2 \leq r \leq m$. Let $u \in E_{m+1}^{p,q}$, where $p > 0$. Then $u = [u_1]$, where

$u_1 \in E_m^{p,q}$ and $d_m(u_1) = 0 \in E_m^{p+m,q-m+1}$. By inductive assumption, $\exists v \in E_m^{0,q}$ and $\mu > 0$ such that $u_1 = t^\mu v$. So $t^\mu d_m(v) = 0$; and, by (2), since $p + m - 2 \geq m - 1$, $d_m(t^{\mu-1}v) = 0$. Hence $u = t[t^{\mu-1}v]$. Repeating this argument, (1) follows for $r = m + 1$.

If $u \in E_{m+1}^{p,q}$ with $p \geq m$, and if $tu = 0$, then, putting $u = [u_1]$ with $u_1 \in E_m^{p,q}$, $\exists w \in E_m^{p-m+2,q+m-1}$ such that $d_m(w) = tu_1$. Since $p - m + 2 \geq 2$, by (1), $\exists v \in E_m^{0,q-m+1}$ and $\nu > 0$ such that $w = t^\nu v$. So $t(u_1 - d_m(t^{\nu-1}v)) = 0$. By (2), $u_1 = d_m(t^{\nu-1}v)$: hence $u = 0$, and so (2) follows for $r = m + 1$. □

(3.10.12) Corollary

Let G be a torus, and suppose that the pair (X, A) satisfies condition (LT). Suppose that $H^i(X, A; \mathbb{Q}) = 0$ for all $i > n$. Then

(1) $H^i(X^G, A^G; \mathbb{Q}) = 0$, *for all* $i > n$, *and*
(2) $H^i(X/G, A/G; \mathbb{Q}) = 0$, *for all* $i > n$.

Proof.

By means of induction, it is enough to assume that $G = S^1$. The Leray-Serre spectral sequence for $(X_G, A_G) \to BG$ with rational coefficients has $E_2^{*,*} \cong H^*(BG; \mathbb{Q}) \otimes H^*(X, A; \mathbb{Q})$. Hence by Lemma (3.10.11), $H_G^*(X, A; \mathbb{Q})$ is generated as a $R = H^*(BG; \mathbb{Q})$-module by elements of degree at most n. It now follows that

$$H^i(X^G, A^G; \mathbb{Q}) = 0,$$

for all $i > n$, by the Localization Theorem.

From the Smith-Gysin sequence of Theorem (3.10.10),

$$H^i(X/G, A/G \cup X^G; \mathbb{Q}) \cong H^{i+2}(X/G, A/G \cup X^G; \mathbb{Q}),$$

for all $i \geq n$. By Proposition (3.10.9),

$$H^*(X/G, A/G \cup X^G; \mathbb{Q}) \cong H_G^*(X, A \cup X^G; \mathbb{Q}).$$

By the Localization Theorem and Lemma (3.10.11), or the fact that the isomorphism

$$H_G^i(X, A \cup X^G; \mathbb{Q}) \to H_G^{i+2}(X, A \cup X^G; \mathbb{Q}),$$

for $i \geq n$, is given by the multiplication by $t \in H^2(BG; \mathbb{Q})$, it follows that

$$H_G^i(X, A \cup X^G; \mathbb{Q}) = 0,$$

for all large i. Hence

$$H^i(X/G, A/G \cup X^G; \mathbb{Q}) = 0,$$

for all $i \geq n$. Now (2) follows from the long exact sequence of the triple

$$(X/G, A/G \cup X^G, A/G),$$

3.10. Cohomology inequalities and other basic results

the excision isomorphism
$$H^*(A/G \cup X^G, A/G; \mathbb{Q}) \cong H^*(X^G, A^G; \mathbb{Q}),$$
and (1). □

(3.10.13) Theorem
Let $G = S^1$, and let (X, A) be as in Corollary (3.10.12). Then, for any $r \geq 0$,
$$\dim_{\mathbb{Q}} H^{r-1}(X/G, A/G \cup X^G; \mathbb{Q}) + \sum_{i=0}^{\infty} \dim_{\mathbb{Q}} H^{r+2i}(X^G, A^G; \mathbb{Q})$$
$$\leq \sum_{i=0}^{\infty} \dim_{\mathbb{Q}} H^{r+2i}(X, A; \mathbb{Q}).$$

See [Bredon, 1972], Chapter III, Section 10, for the proof.

(3.10.14) Remark
Theorem (3.10.13) is false for higher dimensional torus actions. Consider $G = S^1 \times S^1$ acting freely on $X = S^5 \times S^5$ so that $X/G = \mathbb{C}P^2 \times \mathbb{C}P^2$. Let $A = \emptyset$. Then $\dim_{\mathbb{Q}} H^6(X/G; \mathbb{Q}) = 2$, but
$$\sum_{i=0}^{\infty} \dim_{\mathbb{Q}} H^{7+2i}(X; \mathbb{Q}) = 0.$$
We shall now give the corresponding results for p-tori.

(3.10.15) Theorem
Let G be a finite group of order γ. Let X be any paracompact G-space, let A be a closed invariant subspace, and let $\pi : (X, A) \to (X/G, A/G)$ be the orbit map. Then, for any abelian group Λ of coefficients, there exists a natural transfer homomorphism $\mu : \overline{H}^*(X, A; \Lambda) \to \overline{H}^*(X/G, A/G; \Lambda)$, such that, for any $u \in \overline{H}^*(X, A; \Lambda)$ and $v \in \overline{H}^*(X/G, A/G; \Lambda)$, $\mu\pi^*(v) = \gamma v$ and $\pi^*\mu(u) = \sum_{g \in G} g^*(u)$.

For the proof see [Bredon, 1972], Chapter III, Section 7 (or [Bredon, 1967a], Chapter II, Section 19).

(3.10.16) Lemma
Let $G = (\mathbb{Z}_p)^m$ be a p-torus, for p any prime number, and let (X, A) satisfy condition (LT). Suppose that
$$H^i(X, A; \mathbb{F}_p) = 0, \quad \text{for all } i > n.$$
Then
(1) $H^i(X^G, A^G; \mathbb{F}_p) = 0$, for all $i > n$, and
(2) $H^i(X/G, A/G; \mathbb{F}_p) = 0$, for all $i > n$.

Proof.

As in Corollary (3.10.12), we can use induction, and assume that $G = \mathbb{Z}_p$. The proof now appears in [Bredon, 1972], proof of Theorem 7.9, Chapter III. Note that general G-spaces in case (B) are covered by Bredon's proof, since his triangle (7.5) is exact for any paracompact G-space, and $\text{cd}_{\mathbb{F}_p}(X) < \infty$ implies that $\text{cd}_{\mathbb{F}_p}(X/G) < \infty$ by [Quillen, 1971a], Appendix A. □

(3.10.17) Example ([Heller, 1959])
Let X satisfy condition (LT) and suppose that

$$H^*(X; \mathbb{F}_p) \cong H^*(S^m \times S^n; \mathbb{F}_p),$$

where $0 < m < n$. Then $G = (\mathbb{Z}_p)^3$ cannot act freely on X.

To see this, let $E_r^{p,q}$ be the Leray-Serre spectral sequence for

$$X \to X_G \to BG$$

with coefficients in $\mathbb{F}_p$. By Proposition (3.10.9), if G is acting freely,

$$H^*(X_G; \mathbb{F}_p) \cong H^*(X/G; \mathbb{F}_p);$$

and so, by Lemma (3.10.16)(2), $H^{m+n+1}(X_G; \mathbb{F}_p) = 0$. So $E_\infty^{m+n+1,0} = 0$. But $E_2^{m+n+1,0}$ can be killed only by elements coming from $E_2^{n,m}$, $E_2^{m,n}$ and $E_2^{0,m+n}$.

Now

$$\dim_{\mathbb{F}_p}(E_2^{n,m} + E_2^{m,n} + E_2^{0,m+n})$$
$$= \dim_{\mathbb{F}_p}(H^n(BG; \mathbb{F}_p) + H^m(BG; \mathbb{F}_p) + H^0(BG; \mathbb{F}_p))$$
$$= \frac{1}{2}(n+1)(n+2) + \frac{1}{2}(m+1)(m+2) + 1$$
$$= \frac{1}{2}(m^2 + n^2 + 3m + 3n + 6),$$

whereas

$$\dim_{\mathbb{F}_p} E_2^{m+n+1,0} = \dim_{\mathbb{F}_p} H^{m+n+1}(BG; \mathbb{F}_p)$$
$$= \frac{1}{2}(m+n+2)(m+n+3)$$
$$= \frac{1}{2}(m^2 + n^2 + 2mn + 5m + 5n + 6).$$

Thus $E_2^{m+n+1,0}$ is too big to be killed. So G cannot act freely on X.

(3.10.18) Theorem
Let $G = \mathbb{Z}_p$, where p is any prime number, and suppose that (X, A)

3.10. Cohomology inequalities and other basic results

satisfies condition (LT). Then, for any $m \geq 0$,

$$\dim_{\mathbb{F}_p} H^m(X/G, A/G \cup X^G; \mathbb{F}_p) + \sum_{i=0}^{\infty} \dim_{\mathbb{F}_p} H^{m+i}(X^G, A^G; \mathbb{F}_p)$$

$$\leq \sum_{i=0}^{\infty} \dim_{\mathbb{F}_p} H^{m+i}(X, A; \mathbb{F}_p).$$

For the proof, see [Bredon, 1972], Chapter III, Theorem 7.9.

(3.10.19) Remarks
(1) In Theorem (3.10.18) we do not require that
$$\dim_{\mathbb{F}_p} H^*(X, A; \mathbb{F}_p) < \infty.$$
On the other hand, for any $m \geq 0$ such that
$$\sum_{i=0}^{\infty} \dim_{\mathbb{F}_p} H^{m+i}(X, A; \mathbb{F}_p) < \infty,$$
then
$$\dim_{\mathbb{F}_p} H^m(X/G, A/G \cup X^G; \mathbb{F}_p) + \sum_{i=0}^{\infty} \dim_{\mathbb{F}_p} H^{m+i}(X^G, A^G; \mathbb{F}_p)$$

$$\leq \sum_{i=0}^{\infty} \dim_{\mathbb{F}_p} H^{m+i}(X, A; \mathbb{F}_p)^G.$$

This follows from Bredon's proof, since, using his notation, the map
$$H^*(X, A; \mathbb{F}_p) \to H^*_\sigma(X, A; \mathbb{F}_p) \oplus H^*(X^G, A^G; \mathbb{F}_p)$$
factors through the coinvariants
$$H^*(X, A; \mathbb{F}_p)_G = H^*(X, A; \mathbb{F}_p)/\tau H^*(X, A; \mathbb{F}_p).$$
(Here $\tau = 1 - g$, where g generates G, and $\sigma = 1 + g + \ldots + g^{p-1}$.) If, however, V is a $\mathbb{F}_p[G]$-module which is finite-dimensional over $\mathbb{F}_p$, then, clearly $\dim_{\mathbb{F}_p} V_G = \dim_{\mathbb{F}_p} V^G$. See also [tom Dieck, 1987], Chapter III, Section 6, for an alternative proof of this extension of Theorem (3.10.18) using the Borel construction.

(2) Theorem (3.10.18) is invalid for general p-tori. Consider
$$G = \mathbb{Z}_2 \times \mathbb{Z}_2$$
acting freely on $X = S^2 \times S^2$ with $X/G = \mathbb{R}P^2 \times \mathbb{R}P^2$. Then
$$\dim_{\mathbb{F}_2} H^3(X/G; \mathbb{F}_2) = 2,$$
but
$$\sum_{i=0}^{\infty} \dim_{\mathbb{F}_2} H^{3+i}(X; \mathbb{F}_2) = 1.$$

On the other hand, by induction, Theorem (3.10.18) implies the first part of Theorem (3.10.6).

(3) If $G = \mathbb{Z}_p$, and if (X, A) satisfies condition (LT), and if
$$\dim_{\mathbb{F}_p} H^*(X, A; \mathbb{F}_p) < \infty,$$
then we have from [Bredon, 1972], Chapter III, Theorem 7.10, that
$$\chi(X, A) + (p-1)\chi(X^G, A^G) = p\chi(X/G, A/G),$$
where the Euler characteristics are defined with respect to cohomology with coefficients in $\mathbb{F}_p$. (Again note that general G-spaces in case (B) are included as in Lemma (3.10.16).)

(3.10.20) Lemma
Let G be any finite group, and let (X, A) satisfy condition (LT). (In case (B) suppose that $\text{cd}_{\mathbb{F}_p}(X) < \infty$, for all p dividing the order of G: e.g., $\text{cd}_{\mathbb{Z}}(X) < \infty$.) Suppose that $H^i(X, A; \mathbb{Z}) = 0$, for all $i > n$. Then $H^i(X/G, A/G; \mathbb{Z}) = 0$, for all $i > n$.

Proof.

If G is abelian, or, more generally, solvable, then by induction we are reduced to the case where $G = \mathbb{Z}_p$ with p prime. Let
$$u \in H^i(X/G, A/G; \mathbb{Z})$$
for some $i > n$. From the transfer of Theorem (3.10.15), we get that $pu = 0$. From Lemma (3.10.16),
$$H^i(X/G, A/G; \mathbb{F}_p) = 0.$$
Thus
$$H^i(X/G, A/G; \mathbb{Z}) = 0.$$

For non-solvable finite G, one uses the generalized transfer (as in [Bredon, 1972], Chapter III, Section 2) in the manner explained by Floyd in [Borel *et al.*, 1960], Chapter III, Theorem 5.2. □

(3.10.21) Corollary
Let $G = (\mathbb{Z}_p)^r$ for p any prime number, and suppose that (X, A) satisfies condition (LT). Suppose too that $H^i(X, A; \mathbb{Z}) = 0$, for all $i > n$, and that $H^n(X, A; \mathbb{Z}) \cong \mathbb{Z}$. If $p = 2$, assume that G acts trivially on $H^n(X, A; \mathbb{Z})$. Then $H^i(X/G, A/G; \mathbb{Z}) = 0$, for all $i > n$, and $H^n(X/G, A/G; \mathbb{Z}) \cong \mathbb{Z}$.

Proof.

By induction it is enough to assume that $r = 1$. By Theorem (3.10.17), $\dim_{\mathbb{F}_p} H^n(X/G, A/G; \mathbb{F}_p) \leq 1$. Hence $\dim_{\mathbb{F}_p} H^n(X/G, A/G; \mathbb{Z}) \otimes \mathbb{F}_p \leq 1$, by universal coefficients.

Let
$$\pi : (X, A) \to (X/G, A/G)$$

be the orbit map, and let μ be the transfer of Theorem (3.10.15) with integer coefficients. Abbreviate $H^n(X/G, A/G; \mathbb{Z})$ by H. Since $\pi^*\mu$ has non-zero image, $\pi^*(H) \cong \mathbb{Z}$. So $H \cong K \oplus \mathbb{Z}$, where $K = \ker \pi^*$.

Now $\mu\pi^*(K) = pK$: and so $pK = 0$. Thus $H \otimes \mathbb{F}_p \cong K \oplus \mathbb{F}_p$. By above, $\dim_{\mathbb{F}_p} H \otimes \mathbb{F}_p \leq 1$. Thus $K = 0$, and $H \cong \mathbb{Z}$. □

(3.10.22) Remarks
(1) In Corollary (3.10.21) for general G-spaces in case (B) one needs only to assume that $\mathrm{cd}_{\mathbb{F}_p}(X) < \infty$ rather than $\mathrm{cd}_{\mathbb{Z}}(X) < \infty$.
(2) Clearly Lemma (3.10.20) and Corollary (3.10.21) remain valid if $\mathbb{Z}$ is replaced by $\mathbb{Z}_{(p)}$, the integers localized at p, throughout. Indeed using $\mathbb{Z}_{(p)}$ in Lemma (3.10.20) the result holds for any paracompact X if $p \nmid |G|$. If $p \mid |G|$ and one is considering case (B), then only $\mathrm{cd}_{\mathbb{F}_p} < \infty$ need be assumed.
(3) To see that the assumption that G acts trivially on $H^n(X, A; \mathbb{Z})$ when $p = 2$ works well in the induction, suppose that $G = (\mathbb{Z}_2)^r = K \times L$, where $K \cong (\mathbb{Z}_2)^{r-1}$ and $L \cong \mathbb{Z}_2$. Let

$$X \xrightarrow{\pi_1} X/K \xrightarrow{\pi_2} X/G = (X/K)/L$$

be the orbit maps. By induction $H^n(X/K, A/K; \mathbb{Z}) \cong \mathbb{Z}$: let u be a generator. Let g generate L, and suppose that $g^*(u) = -u$. Then

$$\pi_1^* g^*(u) = -\pi_1^*(u) = g^* \pi_1^*(u) = \pi_1^*(u).$$

So $\pi_1^*(u) = 0$. But, letting μ be the transfer of Theorem (3.10.15), $\mu \pi_1^*(u) = 2^{r-1} u \neq 0$.

(3.10.23) Corollary
In addition to the conditions of Corollary (3.10.21), suppose that G acts freely on $X - A$. Let $\pi : (X, A) \to (X/G, A/G)$ be the orbit map, and let $u \in H^n(X, A; \mathbb{Z})$ and $v \in H^n(X/G, A/G; \mathbb{Z})$ be generators. Then $\pi^*(v) = \pm p^r u$.

Proof.
Using induction, it is enough to prove the result when $r = 1$. In that case, owing to the transfer, $\pi^*(v) = \pm pu$ or $\pm u$. Suppose that $\pi^*(v) = \pm u$. Let $\psi : (X_G, A_G) \to (X/G, A/G)$ be the map of Proposition (3.10.9). Then $i^*\psi^*(v) = \pm u$, where $i : (X, A) \to (X_G, A_G)$ is the inclusion.

Now let E_r be the Leray-Serre spectral sequence of $(X_G, A_G) \to BG$ with integer coefficients. Then $u \in E_2^{0,n} \cong H^n(X, A; \mathbb{Z})$ survives to ∞; and so $E_\infty^{*,n} \cong H^*(BG; \mathbb{Z})$. From the increasing filtration, which gives

the exact sequence
$$0 \to \mathcal{F}_{n-1}H_G^*(X,A;\mathbb{Z}) \to H_G^*(X,A;\mathbb{Z}) \to E_\infty^{*,n} \to 0,$$
we conclude that $S^{-1}H_G^*(X,A;\mathbb{Z}) \neq 0$, where S is the multiplicative set generated by the generator $t \in H^2(BG;\mathbb{Z})$. By the Localization Theorem, this contradicts the assumption that G is acting freely on $X - A$. □

(3.10.24) Remark
If we replace $\mathbb{Z}$ by $\mathbb{Z}_{(p)}$ throughout in Corollary (3.10.23), then the conclusion is this: if $v \in H^*(X/G, A/G; \mathbb{Z}_{(p)})$ is a generator, then $\pi^*(v) = p^r u$, where u is a generator of $H^*(X,A;\mathbb{Z}_{(p)})$.

We conclude this section with an exact sequence due to Skjelbred [Skjelbred, 1978a].

(3.10.25) Proposition
Let G be a compact Lie group, let X be a paracompact G-space, and let A be a closed invariant subspace. Let Λ be an abelian group. Suppose that G acts freely on $X - (A \cup X^G)$. (If Λ is a field of characteristic 0, then it is enough to assume that G acts almost-freely on $X - (A \cup X^G)$.) Then there is the following functorial long exact sequence:
$$\ldots \to \overline{H}^i(X/G, A/G; \Lambda) \xrightarrow{\alpha} \overline{H}^i(X^G, A^G; \Lambda) \oplus \overline{H}_G^i(X,A;\Lambda)$$
$$\xrightarrow{\beta} \overline{H}_G^i(X^G, A^G; \Lambda) \xrightarrow{\delta} \overline{H}^{i+1}(X/G, A/G; \Lambda) \to \ldots$$

Proof.
Let $\psi : (X_G, (A \cup X^G)_G) \to (X/G, A/G \cup X^G)$ be the projection: and let $\psi_1; (X_G, A_G) \to (X/G, A/G)$ and $\psi_2 : ((X^G)_G, (A^G)_G) \to (X^G, A^G)$ be the restrictions of ψ. Then we have the map of long exact sequences shown in Diagram 3.6.

$$\begin{array}{ccccccccc}
\ldots & \to & L^i & \xrightarrow{i^*} & M^i & \xrightarrow{j^*} & N^i & \to & L^{i+1} \\
& & \downarrow \psi^* & & \downarrow \psi_1^* & & \downarrow \psi_2^* & & \downarrow \psi^* \\
\ldots & \to & P^i & \to & Q^i & \xrightarrow{\phi^*} & R^i & \xrightarrow{\delta_G} & P^{i+1} & \to & \ldots
\end{array}$$

Diagram 3.6

where $L^i = \overline{H}^i(X/G, A/G \cup X^G; \Lambda)$, $M^i = \overline{H}^i(X/G, A/G; \Lambda)$, $N^i = \overline{H}(X^G, A^G; \Lambda)$, $P^i = \overline{H}_G^i(X, A \cup X^G; \Lambda)$, $Q^i = \overline{H}_G^i(X, A; \Lambda)$ and $R^i = \overline{H}_G^i(X^G, A^G; \Lambda)$. (See proof of Corollary (3.10.12).) By Proposition

(3.10.9), ψ^* is an isomorphism. Hence the result follows from the Barratt-Whitehead Lemma ([Greenberg, Harper, 1981], Section 17, or [Whitehead, 1978], Chapter V, Exercise 1). In particular, with the notation as indicated above $\delta = i^*\psi^{*-1}\delta_G$, $\alpha(u) = (j^*(u), \psi_1^*(u))$, for any $u \in \overline{H}^*(X/G, A/G; \Lambda)$ and $\beta(v,w) = \psi_2^*(v) - \phi^*(w)$, for any
$$(v,w) \in \overline{H}^*(X^G, A^G; \Lambda) \oplus \overline{H}_G^*(X, A; \Lambda).$$

$\square$

3.11 The Hirsch-Brown model and the Evaluation Theorem for p-torus actions on general spaces

Throughout this section we shall let $G = (\mathbb{Z}_p)^r$, where $r \geq 0$ and p is a prime number. We begin by recalling some of the constructions from Chapter 1. Let k be any commutative ring with identity.

(3.11.1) Recollections

(1) If $r = 1$, then k has a free resolution as a trivial $k[G]$-module of the form
$$\ldots \to \mathcal{E}_n(G) \xrightarrow{\partial_n} \mathcal{E}_{n-1}(G) \to \ldots \to \mathcal{E}_1(G) \xrightarrow{\partial_1} \mathcal{E}_0(G) \xrightarrow{\varepsilon} k \to 0.$$
where, for $n \geq 0$, $\mathcal{E}_n(G)$ is the free $k[G]$-module on a single generator w_n, ∂_n is given by $\partial_n(w_n) = \nu w_{n-1}$, if n is even, and $\partial_n(w_n) = \tau w_{n-1}$ if n is odd, where, for a given generator $g \in G$, $\nu = 1 + g + \ldots + g^{p-1}$ and $\tau = 1 - g$; and, for any $a \in k[G]$, $\varepsilon(aw_0) = N(a)$, where $N: k[G] \to k$ is the norm (so that $\nu a = N(a)\nu$).

Furthermore, there is a diagonal map $\Delta : \mathcal{E}_*(G) \to \mathcal{E}_*(G) \otimes \mathcal{E}_*(G)$ given by $\Delta(w_n) = \sum_{a+b=n} \Delta_{a,b}$ where

$$\Delta_{a,b} = \begin{cases} w_a \otimes w_b & \text{if } a \text{ is even} \\ w_a \otimes gw_b & \text{if } a \text{ is odd and } b \text{ is even} \\ \bigoplus_{0 \leq i < j \leq p-1} g^i w_a \otimes g^j w_b & \text{if } a \text{ is odd } b \text{ is odd} \end{cases}$$

(See [Brown, K.S. 1982], Chapter V, Section 1. The different sign used here is because we are taking $\tau = 1 - g$ instead of $g - 1$.)

(2) If $G = (\mathbb{Z}_p)^r$ for $r > 1$, set $G = G_1 \times \ldots \times G_r$, where $G_i \cong \mathbb{Z}_p$, for $1 \leq i \leq r$, and set
$$\mathcal{E}_*(G) = \mathcal{E}_*(G_1) \otimes \ldots \otimes \mathcal{E}_*(G_r).$$

Then the norm again gives an augmentation $\varepsilon : \mathcal{E}_0(G) \to k$; and $\mathcal{E}_*(G)$ is again a free $k[G]$-resolution of the trivial $k[G]$-module k. Furthermore the diagonal extends to a diagonal $\Delta : \mathcal{E}_*(G) \to \mathcal{E}_*(G) \otimes \mathcal{E}_*(G)$ in the obvious way.

If $\mathcal{E}_n(G_i)$ is generated by w_{in}, for $1 \leq i \leq r$, then $\mathcal{E}_n(G)$ is generated by
$$\{w_{1n_1} \otimes \ldots \otimes w_{rn_r} : n_1 + \ldots + n_r = n\}.$$
Note that $k[G]$ is identified with $k[G_1] \otimes \ldots \otimes k[G_r]$ in the obvious way.

(3) Now let X be a G-space. Let $S_*(X;k)$,
$$S^*(X;k) = \mathrm{Hom}_k(S_*(X;k), k),$$
denote the singular chain complex, resp. singular cochain complex, of X with coefficients in k. Then we define (cf. Section 1.2)
$$\beta_G^*(X;k) = \mathrm{Hom}_{k[G]}(\mathcal{E}_*(G), S^*(X;k)).$$
There are isomorphisms (which are natural with respect to X)
$$\beta_G^*(X;k) \cong \mathrm{Hom}_{k[G]}(\mathcal{E}_*(G) \otimes S_*(X;k), k)$$
$$\cong \mathrm{Hom}_k(\mathcal{E}_*(G) \otimes_{k[G]} S_*(X;k), k),$$
and
$$\beta_G^*(X;k) \cong \mathrm{Hom}_{k[G]}(\mathcal{E}_*(G), k[G]) \otimes_{k[G]} S^*(X;k).$$

Indeed, for any non-negative cochain complex of $k[G]$-modules, C^*, we put $\beta_G^*(C^*) = \mathrm{Hom}_{k[G]}(\mathcal{E}_*(G), C^*)$. So $\beta_G^*(C^*) \cong \mathcal{E}^*(G) \otimes_{k[G]} C^*$, where $\mathcal{E}^*(G) = \mathrm{Hom}_{k[G]}(\mathcal{E}_*(G), k[G])$.

In Chapter 1 we saw that $\beta_G^*(C^*)$ has a useful structure as a right module over the polynomial ring $k[t_1, \ldots, t_r]$, where $\deg(t_i) = 2$ for $1 \leq i \leq r$. We can make this into a left module structure by using the diagonal $T\Delta : \mathcal{E}_*(G) \to \mathcal{E}_*(G) \otimes \mathcal{E}_*(G)$, where Δ is the diagonal defined above and T is the standard degree twist - i.e. $T(x \otimes y) = (-1)^{mn} y \otimes x$, where $\deg(x) = m$ and $\deg(y) = n$. The module structure on $\mathcal{E}^*(G)$ is induced from the homomorphisms
$$\mathrm{Hom}_{k[G]}(\mathcal{E}_*(G), k) \otimes \mathcal{E}^*(G) \longrightarrow \mathrm{Hom}_{k[G]}(\mathcal{E}_*(G) \otimes \mathcal{E}_*(G), k \otimes k[G])$$
$$\xrightarrow{T\Delta^*} \mathrm{Hom}_{k[G]}(\mathcal{E}_*(G), k[G]) = \mathcal{E}^*(G).$$
As a left module over $R = k[t_1, \ldots, t_r]$, $\deg(t_i) = 2$ for $1 \leq i \leq r$, one has then $\beta_G^*(C^*) \cong R \otimes \Lambda(s;r) \otimes C^*$, where $\Lambda(s;r)$ is the free k-module with basis
$$\{s_{i_1 \ldots i_q} : 1 \leq i_1 < \ldots < i_q \leq r, 0 \leq q \leq r\},$$
where $\deg(s_{i_1 \ldots i_q}) = q$. Thus, as a k-module, $\Lambda(s;r)$ is isomorphic to the exterior algebra $\Lambda_k(s_1, \ldots, s_r)$. The basis element s_i corresponds under duality to the basis element
$$w_{10} \otimes \ldots \otimes w_{i1} \otimes \ldots \otimes w_{r0}$$

of $\mathcal{E}_*(G)$, and t_i corresponds to
$$w_{10} \otimes \ldots \otimes w_{i2} \otimes \ldots \otimes w_{r0}.$$
In the particular case where $p = 2$ and k is a field of characteristic 2 (or, more generally, whenever $2 = 0$ in k), then, as a left R-module, $\beta_G^*(C^*) \cong R \otimes C^*$, where $R = k[t_1, \ldots, t_r]$ and now $\deg(t_i) = 1$ for $1 \leq i \leq r$. (In the notation used for the general situation above R here is actually $k[s_1, \ldots, s_r]$.)

(4) Now if X is any G-space, then
$$S^*(X_G; k) \cong \mathrm{Hom}_{k[G]}(S_*(EG \times X; k), k).$$
Thus, by the Eilenberg-Zilber theorem $S^*(X_G; k)$ is chain homotopy equivalent to $\beta_G^*(X; k)$. So the Hirsch-Brown model $\beta_G^*(X; k)$ is a model for the singular equivariant cohomology of X; and hence it is useful for studying G-CW-complexes or other G-spaces for which singular theory suffices (see Theorem (1.2.8) and its proof).

Before defining $\beta_G^*(X; k)$ with respect to Alexander-Spanier cohomology, so as to apply to the study of general G-spaces, we need to comment on some facts concerning coverings. From now on (for the rest of this section) we shall assume that X is a paracompact G-space.

(3.11.2) Definitions and comments

Let $\mathcal{U}$ be an open covering of X. Recall from Section 2.6 the definitions of the Čech nerve of $\mathcal{U}$, $\check{\mathcal{U}}$, and the Vietoris nerve of $\mathcal{U}$, $\overline{\mathcal{U}}$. Since we shall be interested in cofinal systems of coverings, by [Bredon, 1972], Chapter III, Section 6, we may restrict our attention to locally finite invariant coverings. So if $U \in \mathcal{U}$ and $g \in G$, then $gU \in \mathcal{U}$. For any locally finite invariant covering $\mathcal{U}$, we have an equivariant canonical map $f_\mathcal{U} : X \to |\check{\mathcal{U}}|$, from X to the geometric realization of $\check{\mathcal{U}}$ ([Bredon, 1972], p. 134).

An invariant covering $\mathcal{U}$ is said to be Čech-G-covering if, for any $U \in \mathcal{U}$ and $g \in G$, $U \cap gU \neq \emptyset$ implies that $U = gU$. In [Bredon, 1972] a Čech-G-covering is simply called a G-covering; but the notion of a Vietoris-G-covering is also interesting: U is said to be a Vietoris-G-covering if, for any $x \in X$ and $g \in G$, if there is a U in $\mathcal{U}$ such that U contains both x and gx, then $x = gx$ - i.e. if each $U \in \mathcal{U}$ meets each orbit in at most one point. By [Bredon, 1972], Chapter III, Theorem 6.1, locally finite Čech-G-coverings are always cofinal. But Vietoris-G-coverings exist in general only if X is a free G-space. This latter fact forces us to subdivide.

Let $\mathcal{U}$ be a Čech-G-covering of X, and let $\mathrm{Sd}\,\overline{\mathcal{U}}$ be the first barycentric subdivision of $\overline{\mathcal{U}}$. Then we may define an equivariant Dowker map $\psi_\mathcal{U} : \mathrm{Sd}\,\overline{\mathcal{U}} \to \check{\mathcal{U}}$ as follows; cf. [Dowker, 1952]. G acts on $\mathrm{Sd}\,\overline{\mathcal{U}}$ in the obvious way. Choose a complete set of representatives of the orbits of the vertices of $\mathrm{Sd}\,\overline{\mathcal{U}}$. Suppose that v is a chosen representative. v is a simplex of $\overline{\mathcal{U}}$: and so v is a finite set of points of X lying in some element of $\mathcal{U}$. Choose $U \in \mathcal{U}$ such that $v \subseteq U$, and set $\psi_\mathcal{U}(v) = \check{U}$. Now extend $\psi_\mathcal{U}$ equivariantly: i.e. set $\psi_\mathcal{U}(gv) = gU$. Since $\mathcal{U}$ is a Čech-G-covering $\psi_\mathcal{U} : \mathrm{Sd}\,\overline{\mathcal{U}} \to \check{\mathcal{U}}$ is well-defined. Clearly $\psi_\mathcal{U}$ is equivariant.

By [Dowker, 1952], Theorem 1,
$$\psi_\mathcal{U}^* : H^*(|\check{\mathcal{U}}|; k) \to H^*(|\mathrm{Sd}\,\overline{\mathcal{U}}|; k)$$
is an isomorphism: indeed $|\psi_\mathcal{U}|$ is a homotopy equivalence.

(3.11.3) Definitions

(1) For any $k[G]$-cochain complex C^*, let
$$\beta_G^*(C^*) = \mathrm{Hom}_{k[G]}(\mathcal{E}_*(G), C^*)$$
(see (1.2.1)).

(2) Let X be a paracompact G-space. For any open covering $\mathcal{U}$ of X, let $C^*(\overline{\mathcal{U}}; k)$ be the ordered cochain complex of $\overline{\mathcal{U}}$ with coefficients in k. Let
$$\overline{C}^*(X; k) = \varinjlim_{\mathcal{U}} C^*(\overline{\mathcal{U}}; k)$$
where $\mathcal{U}$ ranges over the locally finite Čech-G-coverings of X. Then $\overline{C}^*(X; k)$ is naturally isomorphic to the Alexander-Spanier cochains on X as defined for example in [Spanier, 1966], Chapter 6, Section 4.

(3) Let X be a paracompact G-space. Set
$$\overline{\beta}_G^*(X; k) = \beta_G^*(\overline{C}^*(X; k)).$$

(4) if X is a paracompact G-space and $A \subseteq X$ is a closed invariant subspace, then let $\overline{C}^*(X, A; k)$ be the kernel of the epimorphism of restriction $\overline{C}^*(X; k) \to \overline{C}^*(A; k)$. Set
$$\overline{\beta}_G^*(X, A; k) := \beta_G^*(\overline{C}^*(X, A; k)).$$

(3.11.4) Remarks

(1) Let X be a paracompact G-space, and let $\mathcal{U}$ and $\mathcal{V}$ be locally finite Čech-G-coverings of X such that $\mathcal{V}$ is a refinement of $\mathcal{U}$. Just as we chose the Dowker map $\psi_\mathcal{U}$ to be equivariant, we can choose an equivariant refinement map $\check{\rho}(\mathcal{U}, \mathcal{V}) : \check{\mathcal{V}} \to \check{\mathcal{U}}$; and any two maps induce G-homotopic maps $|\check{\mathcal{V}}| \to |\check{\mathcal{U}}|$. ([Bredon, 1972], p. 135.) Furthermore, if $\overline{\rho}(\mathcal{U}, \mathcal{V}) : \overline{\mathcal{V}} \to \overline{\mathcal{U}}$ is the canonical map, then it is easy to see that $|\check{\rho}(\mathcal{U}, \mathcal{V})\psi_\mathcal{V}|$ and $|\psi_\mathcal{U}\overline{\rho}(\mathcal{U}, \mathcal{V})|$ are G-homotopic maps from $|\mathrm{Sd}\,\overline{\mathcal{V}}|$ to $|\check{\mathcal{U}}|$, where $\psi_\mathcal{U}$ and $\psi_\mathcal{V}$ are equivariant Dowker maps.

Thus, letting C_* denote ordered chains with coefficients in k, we have Diagram 3.7

3.11. Hirsch-Brown model and the Evaluation Theorem

$$\begin{array}{ccccccc}
C_*(\overline{\mathcal{U}}) & \longrightarrow & S_*(|\overline{\mathcal{U}}|) & \longrightarrow & S_*(|\operatorname{Sd}\overline{\mathcal{U}}|) & \longrightarrow & S_*(|\check{\mathcal{U}}|) \\
\uparrow & & \uparrow & & \uparrow & & \uparrow \\
C_*(\overline{\mathcal{V}}) & \longrightarrow & S_*(|\overline{\mathcal{V}}|) & \longrightarrow & S_*(|\operatorname{Sd}\overline{\mathcal{V}}|) & \longrightarrow & S_*(|\check{\mathcal{V}}|)
\end{array}$$

Diagram 3.7

where the left hand horizontal maps are the standard ones, the middle horizontal maps are induced by the canonical homeomorphisms, the right hand horizontal maps are induced by equivariant Dowker maps, the first three vertical maps (from the left) are the canonical ones, the right hand vertical map is induced by an equivariant refinement map, the left hand and middle squares commute, and the right hand square commutes up to $k[G]$-chain-homotopy.

(2) For any non-negative cochain complex, C^*, of $k[G]$-modules, the filtration

$$\mathcal{F}^p \beta_G^*(C^*)^n = \bigoplus_{i \geq p} \operatorname{Hom}_{k[G]}(\mathcal{E}_i(G), C^{n-i})$$

gives rise to a spectral sequence converging to $H(\beta_G^*(C^*))$. We shall denote this spectral sequence by $E(G, C^*)_r^{*,*}$, or just $E_r^{*,*}$, when G and C^* are clear from the context. It is easy to see that

$$E(G, C^*)_2^{p,q} \cong H^p(BG; H^q(C^*)),$$

where, in general, the coefficients $H^q(C^*)$ are twisted by the action of G: i.e. $E(G, C^*)_2^{p,q}$ is isomorphic to the group cohomology $H^p(G; H^q(C^*))$ having coefficients in the $k[G]$-module $H^q(C^*)$. (See, e.g., [Brown, K.S. 1982], Chapter VII, Section 5.)

(3) Let $\{M_\gamma : \gamma \in \Gamma\}$ be a direct system of $k[G]$-modules. Since $\mathcal{E}_*(G)$ has finite type,

$$\varinjlim_\gamma \operatorname{Hom}_{k[G]}(\mathcal{E}_*(G), M_\gamma) \cong \operatorname{Hom}_{k[G]}(\mathcal{E}_*(G), \varinjlim_\gamma M_\gamma).$$

Since homology commutes with direct limits, it follows that

$$\varinjlim_\gamma H^*(BG; M_\gamma) \cong H^*(BG; \varinjlim_\gamma M_\gamma).$$

(4) Combining (1) and (3) above, it follows that

$$H(\overline{\beta}_G^*(X; k)) \cong \varinjlim_\mathcal{U} H(\beta_G^*(C^*(\overline{\mathcal{U}}; k))) \cong \varinjlim_\mathcal{U} H(\beta_G^*(S^*(|\check{\mathcal{U}}|; k))),$$

where X is any paracompact G-space, and $\mathcal{U}$ ranges over the locally finite Čech-G-coverings of X. Thus, from Recollection (3.11.1)(4),

$$H(\overline{\beta}_G^*(X; k)) \cong \varinjlim_\mathcal{U} H_G^*(|\check{\mathcal{U}}|; k).$$

(3.11.5) Lemma
Let X be any paracompact G-space and let $A \subset X$ be a closed invariant subspace. Then
$$H(\overline{\beta}_G^*(X, A; k)) \cong \overline{H}_G^*(X, A; k).$$

Proof.

We shall give the proof when $A = \emptyset$. The proof for the general case follows similarly.

An equivariant canonical map $f_{\mathcal{U}} : X \to |\check{\mathcal{U}}|$ induces a map from the bundle $X_G \to BG$ to the bundle $|\check{\mathcal{U}}|_G \to BG$; and hence $f_{\mathcal{U}}$ induces a map of the Leray spectral sequences, which at the E_2-level is
$$H^*(BG; H^*(|\check{\mathcal{U}}|; k)) \to H^*(BG; \overline{H}^*(X; k)),$$
where the coefficients are twisted in general. If we take direct limits over the locally finite Čech-G-coverings of X, then the map at the E_2-level becomes an isomorphism by [Dold, 1980], Appendix, Proposition 3.7.

Thus
$$\overline{H}_G^*(X; k) \cong \varinjlim_{\mathcal{U}} H_G^*(|\check{\mathcal{U}}|; k)$$
and the result follows from Remarks (3.11.4)(4). □

(3.11.6) Remark
$\overline{\beta}_G^*(X; k)$ is functorial with respect to maps of G-spaces for fixed G. When G varies one must choose a map of resolutions; and so $\overline{\beta}_G^*(X; k)$ is only functorial up to chain homotopy in the variable G.

For the remainder of this section we shall take $k = \mathbb{F}_p$. Let E be an extension field of $\mathbb{F}_p$. As before $G = (\mathbb{Z}_p)^r$ for some $r \geq 0$, and p is a prime number. Recall the evaluation procedure of Chapter 1.

(3.11.7) Recollections
(1) Let $\alpha = (\alpha_1, \ldots, \alpha_r) \in E^r$. Let $R = \mathbb{F}_p[t_1, \ldots, t_r]$ be the polynomial part of $H^*(BG; \mathbb{F}_p)$. Then α defines a map $\tilde{\alpha} : R \to E[t]$ given by $t_i \mapsto \alpha_i t$, for $1 \leq i \leq r$. As in Definition (3.1.4), there is a subgroup $K(\alpha) \subset G$ such that $PK(\alpha) = \sigma \ker(\tilde{\alpha})$.

Let V be the $\mathbb{F}_p$-linear subspace of $H^*(BG; \mathbb{F}_p)$ spanned by $t_1, \ldots, t_r$; and denote by $\alpha : V \to E$ the linear map $t_i \mapsto \alpha_i$, $1 \leq i \leq r$. Now choose $\lambda_{ij} \in \mathbb{F}_p$, $1 \leq i, j \leq r$, so that:

(i) $\{\sum_{j=1}^r \lambda_{ij} t_j : 1 \leq i \leq r\}$ is a basis for V;
(ii) $\{\sum_{j=1}^r \lambda_{ij} t_j : 1 \leq i \leq r(\alpha)\}$ is a basis for $\ker[\alpha : V \to E]$, where $r(\alpha) = \dim \ker(\alpha)$; and so

3.11. Hirsch-Brown model and the Evaluation Theorem

(iii) $\{\sum_{j=1}^{r} \lambda_{ij}\alpha_j : r(\alpha)+1 \leq i \leq r\}$ is linearly independent over $\mathbb{F}_p$ in E.

Note that the p-rank of $K(\alpha)$ is $r - r(\alpha)$.

Set $\sum_{j=1}^{r} \lambda_{ij}\alpha_j = \beta_i$, for $1 \leq i \leq r$. So $\beta_i = 0$ for $1 \leq i \leq r(\alpha)$, and $\beta_{r(\alpha)+1}, \ldots, \beta_r$ are linearly independent over $\mathbb{F}_p$. Let $M = (\mu_{ij})$ be inverse to the matrix $\Lambda = (\lambda_{ij})$. So $\alpha_i = \sum_{j=1}^{r} \mu_{ij}\beta_j$, $1 \leq i \leq r$.

(2) Suppose that G is generated by $g_1, \ldots, g_r$, and that we have chosen a resolution $\mathcal{E}_*(G)$ for the trivial $\mathbb{F}_p[G]$-module $\mathbb{F}_p$, as in Recollections (3.11.1)(2) above, with G_i generated by g_i for $1 \leq i \leq r$.

Now $H^*(BG; \mathbb{F}_p)$ is the homology of
$$\mathrm{Hom}_{\mathbb{F}_p[G]}(\mathcal{E}_*(G), \mathbb{F}_p) \cong \mathbb{F}_p[t_1, \ldots, t_r] \otimes \Lambda(s; r) \otimes \mathbb{F}_p.$$

Over $\mathbb{F}_p$ the differential here is trivial; and so
$$H^*(BG; \mathbb{F}_p) = \mathbb{F}_p[t_1, \ldots, t_r] \otimes \Lambda(s; r)$$
$$= \mathbb{F}_p[t_1, \ldots, t_r] \otimes \Lambda(s_1, \ldots, s_r).$$

We shall now assume that the polynomial generators, $t_1, \ldots, t_r$, (and the exterior generators, $s_1, \ldots, s_r$) of $H^*(BG; \mathbb{F}_p)$ as used in (1) above, correspond in this way to dual basis elements for $\mathcal{E}^*(G)$ as in (3.11.1)(3).

(3) Let $L(\alpha) \subseteq G$ be generated by $g_1^{\mu_{1j}} \ldots g_r^{\mu_{rj}}$ for $r(\alpha)+1 \leq j \leq r$. We shall see below that $L(\alpha) = K(\alpha)$.

In $\mathbb{F}_p[G]$, let $\tau_i = 1 - g_i$, $1 \leq i \leq r$; and let
$$\tau'_j = 1 - g_1^{\mu_{1j}} \ldots g_r^{\mu_{rj}}, \qquad r(\alpha)+1 \leq j \leq r.$$

Thus there exist $b_{jmn} \in \mathbb{F}_p[G]$, such that, for $r(\alpha)+1 \leq j \leq r$,
$$\tau'_j = \bigoplus_{i=1}^{r} \mu_{ij}\tau_i + \bigoplus_{m,n=1}^{r} b_{jmn}\tau_m\tau_n.$$

Now in $E[G]$ let
$$\gamma = 1 - \bigoplus_{i=1}^{r} \alpha_i \tau_i - \bigoplus_{j,m,n=1}^{r} \beta_j b_{jmn}\tau_m\tau_n.$$

Thus $\gamma = 1 - \sum_{j=r(\alpha)+1}^{r} \beta_j \tau'_j$ in $E[L(\alpha)]$.

Since we are in characteristic p, $\gamma^p = 1$. Let Γ be the subgroup of $E[L(\alpha)] \subset E[G]$ generated by γ.

(If $\alpha = (\alpha_1, \ldots, \alpha_r) \neq 0$, then Γ is called a shifted subgroup of $E[G]$, or, more loosely, when E is understood, a shifted subgroup of G, of rank one. For more about shifted subgroups see Section 4.6, especially Definition (4.6.20) and Theorem (4.6.21), and some of the references given in Section 4.6.)

(4) Now consider resolutions $\mathcal{E}_*(\Gamma; E)$, $\mathcal{E}_*(L(\alpha); E)$ and $\mathcal{E}_*(G; E)$ of E as a trivial Γ-, $L(\alpha)$-, and G-module respectively, where
$$\mathcal{E}_*(G; E) = \mathcal{E}_*(G) \otimes_{\mathbb{F}_p} E.$$

Suppose also that
$$\mathcal{E}_*(L(\alpha); E) = \mathcal{E}_*(L(\alpha)) \otimes_{\mathbb{F}_p} E,$$
where $\mathcal{E}_*(L(\alpha))$ is chosen in accord with the generators $g_1^{\mu_{1j}} \ldots g_r^{\mu_{rj}}$, $r(\alpha) + 1 \leq j \leq r$, of $L(\alpha)$.

For $1 \leq i \leq r$, let $u_i = w_{10} \otimes \ldots \otimes w_{i1} \otimes \ldots \otimes w_{r0} \in \mathcal{E}_1(G)$; and, for $r(\alpha) + 1 \leq j \leq r$, define $u'_j \in \mathcal{E}_1(L(\alpha))$ similarly. Let w_1 be the generator of $\mathcal{E}_1(\Gamma; E)$. So, for $1 \leq i \leq r$, $s_i \in H^1(BG; \mathbb{F}_p)$ is dual to u_i. Let $s'_j \in H^1(BL(\alpha); \mathbb{F}_p)$, for $r(\alpha)+1 \leq j \leq r$, correspond to u'_j similarly. And let $s \in H^1(B\Gamma; E)$ correspond to w_1. Define $t'_j \in H^2(BL(\alpha); \mathbb{F}_p)$, $r(\alpha) + 1 \leq i \leq r$, and $t \in H^2(B\Gamma; E)$, similarly.

The inclusions $E[\Gamma] \to E[L(\alpha)] \to E[G]$ give rise to maps of resolutions $\mathcal{E}_*(\Gamma; E) \to \mathcal{E}_*(L(\alpha); E) \to \mathcal{E}_*(G; E)$; and, clearly, these maps of resolutions may be chosen so that

$$w_1 \mapsto \sum_{j=r(\alpha)+1}^{r} \beta_j u'_j \quad \text{and} \quad u'_j \mapsto \sum_{i=1}^{r} \mu_{ij} u_i + \sum_{m,n=1}^{r} b_{jmn} \tau_n u_m.$$

So, in cohomology, $s_i \mapsto \sum_{j=r(\alpha)+1}^{r} \mu_{ij} s'_j$ and $s'_j \mapsto \beta_j s$. Composing, $s_i \mapsto \alpha_i s$. Furthermore, examining the maps of resolutions in degree two shows that $t_i \mapsto \sum_{j=r(\alpha)+1}^{r} \mu_{ij} t'_j$ (as follows from the Bockstein), and $t'_j \mapsto \beta_j^p t$. So, composing, since $\mu_{ij} \in \mathbb{F}_p$, thus $\mu_{ij}^p = \mu_{ij}$,

$$t_i \mapsto \sum_{j=1}^{r} \mu_{ij} \beta_j^p t = \left(\sum_{j=1}^{r} \mu_{ij} \beta_j \right)^p t = \alpha_i^p t;$$

cf. [Carlson, 1983], Section 2, Part III.

Since the homomorphism $i_\Gamma^* : H^*(BG; E) \to H^*(B\Gamma; E)$ which is induced by the inclusion $E[\Gamma] \to E[G]$ is not, in general, induced by a map of classifying spaces (unless Γ is an actual subgroup of G), it is not obvious that i_Γ^* is multiplicative. In Recollection (3.11.9)(2), however, we shall show that i_Γ^* is multiplicative. So i_Γ^* is determined by the results above that $i_\Gamma^*(s_i) = \alpha_i s$ and $i_\Gamma^*(t_i) = \alpha_i^p t$, for $1 \leq i \leq r$.

In particular this shows that $L(\alpha) = K(\alpha)$.

(3.11.8) Definitions

(1) Let C^* be a cochain complex of $E[\Gamma]$-modules. With the notation of (3.11.7), let
$$\beta_\Gamma^*(C^*) := \mathrm{Hom}_{E[\Gamma]}(\mathcal{E}_*(\Gamma; E), C^*).$$
(2) if X is a paracompact G-space and if $A \subseteq X$ is a closed invariant subspace, then $\overline{C}^*(X, A; E)$ can be viewed as a cochain complex of $E[\Gamma]$-modules via the inclusions $E[\Gamma] \subseteq E[K(\alpha)] \subseteq E[G]$. Set
$$\overline{\beta}_\Gamma^*(X, A; E) = \beta_\Gamma^*(\overline{C}^*(X, A; E));$$

3.11. Hirsch-Brown model and the Evaluation Theorem

and set
$$\overline{H}^*_\Gamma(X, A; E) = H(\overline{\beta}^*_\Gamma(X, A; E)).$$

Before proceeding further we need to recall some technicalities concerning cup-products.

(3.11.9) Recollections

(1) Until now we have always assumed that, for any $E[G]$-modules M and N, the tensor product $M \otimes N = M \otimes_E N$ is viewed as a $E[G]$-module via the standard diagonal

$$\triangle_0 : E[G] \to E[G] \otimes E[G],$$

which is defined by $\triangle_0(g) = g \otimes g$ for all $g \in G$. It is useful, however, to have some other diagonals available.

Let $\gamma \in E[G]$ be as in Recollections (3.11.7)(3) with non-zero

$$\alpha = (\alpha_1, \ldots, \alpha_r) \in E^r.$$

Now choose vectors $u_i \in E^r$, $1 \leq i \leq r$, such that $u_1 = \alpha$ and $\{u_1, \ldots, u_r\}$ is linearly independent over E in E^r. Let $u_i = (u_{i1}, \ldots, u_{ir})$; and for $2 \leq i \leq r$, let $\gamma_i = 1 - \sum_{j=1}^r u_{ij}\tau_j$, and let $\gamma_1 = \gamma$. Let Γ' be the subgroup of $E[G]$ generated by $\gamma_1, \ldots, \gamma_r$. By [Carlson, 1983], Lemma 2.9, for example, Γ' is an elementary abelian p-group of rank r, and $E[\Gamma'] = E[G]$. So we may define a new diagonal

$$\triangle_s : E[G] \to E[G] \otimes E[G]$$

such that $\triangle_s(\gamma') = \gamma' \otimes \gamma'$ for all $\gamma' \in \Gamma'$. We shall refer to $\triangle_s$ as a shifted diagonal.

There is also a third type of diagonal. Let $\{h_1, \ldots, h_r\}$ be a generating set for G. Let $X_i = 1 - h_i$ for $1 \leq i \leq r$. So

$$E[G] \cong E[X_1, \ldots, X_r]/(X_1^p, \ldots, X_r^p).$$

The third type of diagonal

$$\triangle_c : E[G] \to E[G] \otimes E[G],$$

which we shall call a compatible diagonal, is defined by

$$\triangle_c(X_i) = X_i \otimes 1 + 1 \otimes X_i$$

for $1 \leq i \leq r$. (See [Carlson, 1985].)

For example, with the notation of Recollection (3.11.7)(3), let

$$h_j = g_1^{\mu_{1j}} \ldots g_r^{\mu_{rj}}$$

for $r(\alpha) + 1 \leq j \leq r$, and choose

$$h_1, \ldots, h_{r(\alpha)}$$

so that $\{h_1, \ldots, h_r\}$ is a generating set for G. Let $X = 1 - \gamma$. Then
$$X = \sum_{j=1}^{r} \beta_j X_j.$$
So $\triangle_c(X) = X \otimes 1 + 1 \otimes X$.

Let M and N be $E[G]$-modules. Let $M \otimes_i N$, for $i = 0$, s or c, denote $M \otimes N$ made into a $E[G]$-module via $\triangle_i$ for $i = 0$, s or c. respectively.

The reason for considering different diagonals comes from the following problem. Suppose that M and N are $E[G]$-modules and that M' and N' are $E[\Gamma]$-modules, where $\Gamma = \langle \gamma \rangle \subseteq E[G]$ is as above. Suppose that $\phi : M' \to M$ and $\psi : N' \to N$ are $E[\Gamma]$-module homomorphisms where M and N are considered as $E[\Gamma]$-modules via the inclusion $E[\Gamma] \subseteq E[G]$. Now, in general, $\triangle_0(\gamma) \neq \gamma \otimes \gamma$ in $E[G] \otimes E[G]$; and so $\phi \otimes_0 \psi : M' \otimes_0 N' \to M \otimes_0 N$ is not a $E[\Gamma]$-module homomorphism, in general, where the diagonal $\triangle_0$ in $E[\Gamma]$ used to define $M' \otimes_0 N'$ is the standard one for Γ, i.e. $\triangle_0(\gamma) = \gamma \otimes \gamma$. But if we use a shifted generating set $\gamma = \gamma_1, \ldots, \gamma_r$, as above, then $\phi \otimes_s \psi : M' \otimes_s N' = M' \otimes_0 N' \to M \otimes_s N$ is a $E[\Gamma]$-module homomorphism. Similarly, if $h_1, \ldots, h_r$ are as above, then $\phi \otimes_c \psi : M' \otimes_c N' \to M \otimes_c N$ is a $E[\Gamma]$-module homomorphism, where the diagonal $\triangle_c$ in $E[\Gamma]$ used to define $M' \otimes_c N'$ is given by $\triangle_c(X) = X \otimes 1 + 1 \otimes X$, where $X = 1 - \gamma$.

Note that if N is any $E[G]$-module, then $E[G] \otimes_i N$ is a free $E[G]$-module for $i = 0$, s or c. In each case, if $\{y_j; j \in J\}$ is a basis for N as a E-vector space, then $\{1 \otimes y_j; j \in J\}$ is a basis for $E[G] \otimes_i N$ as a $E[G]$-module.

(2) For each choice of diagonal $\triangle_i$ one has a map of resolutions
$$\mathcal{E}_*(G; E) \to \mathcal{E}_*(G; E) \otimes_i \mathcal{E}_*(G; E);$$
and hence, with $\mathcal{E}_* = \mathcal{E}_*(G; E)$, a cohomology product defined by
$$\mathrm{Hom}_{E[G]}(\mathcal{E}_*, E) \otimes \mathrm{Hom}_{E[G]}(\mathcal{E}_*, E) \to \mathrm{Hom}_{E[G]}(\mathcal{E}_* \otimes_i \mathcal{E}_*, E)$$
$$\to \mathrm{Hom}_{E[G]}(\mathcal{E}_*, E).$$

By [MacLane, 1967], Chapter VIII, Proposition 4.5 and Theorem 4.1, and Chapter III, Theorem 6.4, all these products on $H^*(BG; E)$ are one and the same, coinciding with the Yoneda product on $\mathrm{Ext}^*_{E[G]}(E, E)$. So all the diagonals $\triangle_i$ define the standard cup-product on $H^*(BG; E)$. This cup-product is also the same as the composition product: see [Brown, K.S. 1982], Chapter V, Theorem (4.6).

It follows that the restriction homomorphism
$$H^*(BG; E) \to H^*(B\Gamma; E)$$
is a ring homomorphism (since one may use $\triangle_s$ or $\triangle_c$).

3.11. Hirsch-Brown model and the Evaluation Theorem

(3) Let C^* be a cochain complex of $E[G]$-modules. Then $H(\beta_G^*(C^*))$ with coefficients in E is a $H^*(BG; E)$-module via

$$\mathrm{Hom}_{E[G]}(\mathcal{E}_*, E) \otimes \mathrm{Hom}_{E[G]}(\mathcal{E}_*, C^*) \to \mathrm{Hom}_{E[G]}(\mathcal{E}_* \otimes_i \mathcal{E}_*, C^*)$$
$$\to \mathrm{Hom}_{E[G]}(\mathcal{E}_*, C^*)$$

for $i = 0$, s or c. Now by [Brown, K.S. 1982], Chapter V, Theorem (4.6), or, more precisely, the proof of Theorem (4.6), all these module structures are one and the same, coinciding with the composition product. (Note that Brown's homomorphism α does not depend on the choice of diagonal, but its weak inverse α' does depend on the diagonal: but then, in cohomology, $(\alpha')^* = (\alpha^*)^{-1}$ is again independent of the choice of diagonal.)

Since $\mathcal{E}_* = \mathcal{E}_*(G; E)$ is a minimal resolution it is clear that

$$\mathrm{Hom}_{E[G]}(\mathcal{E}_*, E) = H^*(BG; E).$$

Furthermore, if X is a paracompact G-space, then E is contained in $\overline{C}^0(X; E)$ as the set of constant functions. Thus there is a homomorphism

$$\mathrm{Hom}_{E[G]}(\mathcal{E}_*, E) \to \mathrm{Hom}_{E[G]}(\mathcal{E}_*, C^*),$$

where $C^* = \overline{C}^*(X; E)$; and this homomorphism induces

$$p^* : H^*(BG; E) \to \overline{H}_G^*(X; E),$$

where $p : X_G \to BG$ is the bundle map corresponding to $X \to P$, where P is any one-point space.

The cup-product in $\overline{C}^*(X; E)$ is compatible with $\triangle_0$; and the cup-product in $\overline{H}_G^*(X; E)$ is induced in the obvious way by

$$\mathrm{Hom}_{E[G]}(\mathcal{E}_*, C^*) \otimes \mathrm{Hom}_{E[G]}(\mathcal{E}_*, C^*) \to \mathrm{Hom}_{E[G]}(\mathcal{E}_* \otimes_0 \mathcal{E}_*, C^*)$$
$$\to \mathrm{Hom}_{E[G]}(\mathcal{E}_*, C^*),$$

where $C^* = \overline{C}^*(X; E)$. Hence, in particular, the $H^*(BG; E)$-module structure on $\overline{H}_G^*(X; E)$ given above is the same as that given by p^* and the cup-product in $\overline{H}_G^*(X; E)$.

Since $\triangle_s$ and $\triangle_c$ on $E[\Gamma]$ are not compatible with the cup-product in $\overline{C}^*(X; E)$, in general, it is not possible to define a cup-product in $\overline{H}_\Gamma^*(X; E)$ by the method above.

Finally note that, in general, for any i and diagonal

$$\triangle_i : \mathcal{E}_* \to \mathcal{E}_* \otimes_i \mathcal{E}_*,$$

the sequence of homomorphisms at the beginning of this recollection gives a homomorphism

$$H^*(BG; E) \otimes \mathrm{Hom}_{E[G]}(\mathcal{E}_*, C^*) \to \mathrm{Hom}_{E[G]}(\mathcal{E}_*, C^*).$$

This is not a module structure at the cochain level, however, since
$$\triangle_i : \mathcal{E}_* \to \mathcal{E}_* \otimes_i \mathcal{E}_*$$
will not be coassociative in general, although it is, of course, chain-homotopy coassociative. But in the standard case, $i = 0$, with $\triangle_0$ as in Recollection (3.11.1)(1), then one gets the structure described in Recollection (3.11.1)(3).

(4) Given a non-negative cochain complex, C^*, of $E[G]$-modules, and a map of resolutions
$$\mathcal{E}_*(\Gamma; E) \to \mathcal{E}_*(G; E),$$
there is clearly a restriction homomorphism
$$\beta_G^*(C^*) \to \beta_\Gamma^*(C^*).$$
By (3), the induced homomorphism
$$H(\beta_G^*(C^*)) \to H(\beta_\Gamma^*(C^*))$$
is a homomorphism of $H^*(BG; E)$-modules where $H(\beta_\Gamma^*(C^*))$ is made into a $H^*(BG; E)$-module via the restriction
$$H^*(BG; E) \to H^*(B\Gamma; E).$$
This follows by using the diagonal $\triangle_i$ for $i = s$ or c, or, more directly, from the composition product.

Furthermore, with the notation of Remark (3.11.4)(2), there is induced a homomorphism of spectral sequences
$$E(G, C^*)_r \to E(\Gamma, C^*)_r.$$
Now, as follows, for example, from [Brown, K.S. 1982], Chapter VII, section 5, $E(G, C^*)_r$, resp. $E(\Gamma, C^*)_r$, is a spectral sequence of $H^*(BG; E)$-modules, resp. $H^*(B\Gamma; E)$-modules. Since, by (3) above, the module structure may be defined using $\triangle_s$ or $\triangle_c$, it follows that the homomorphism of spectral sequences is compatible with the module structures.

(5) Suppose that X is a paracompact G-space, and let A be a closed invariant subspace. In (4) above let $C^* = \overline{C}^*(X, A; E)$. Then the cup-product structure on C^* induces a cup-product structure in $E(G, C^*)_r$, again by [Brown, K.S. 1982], Chapter VII, Section 5. This is not true in general for $E(\Gamma, C^*)_r$, however, since only $\triangle_0$ is compatible with the cup-product in $\overline{C}^*(X, A; E)$.

In Example (3.11.18) below we shall give an explicit example where $E(\Gamma, C^*)_r$ is not multiplicative.

Next we establish some familiar properties of the cohomology
$$\overline{H}_\Gamma^*(X, A; E).$$

(3.11.10) Proposition
Let $G = (\mathbb{Z}_p)^r$ and let X be any paracompact G-space. Let Γ and E be as in Recollections (3.11.7).
(1) For any closed invariant subspace $A \subseteq X$, there is a long exact sequence of $H^*(B\Gamma; E)$-modules
$$\ldots \to \overline{H}^n_\Gamma(X, A; E) \to \overline{H}^n_\Gamma(X; E) \to \overline{H}^n_\Gamma(A; E) \to \overline{H}^{n+1}_\Gamma(X, A; E) \to \ldots$$
(2) For any closed invariant subspace $A \subset X$, restriction induces an isomorphism of $H^*(B\Gamma; E)$-modules
$$\overline{H}^*_\Gamma(A; E) \cong \varinjlim_V \overline{H}^*_\Gamma(V; E),$$
where V ranges over the closed the invariant neighbourhoods of A.
(3) For any closed invariant subspaces A and $B \subset X$ with $X = A \cup B$, there is a long exact Mayer-Vietoris sequence of $H^*(B\Gamma; E)$-modules
$$\ldots \to \overline{H}^n_\Gamma(X; E) \to \overline{H}^n_\Gamma(A; E) \oplus \overline{H}^n_\Gamma(B; E)$$
$$\to \overline{H}^n_\Gamma(A \cap B; E) \to \overline{H}^{n+1}_\Gamma(X; E) \to \ldots .$$

Proof.
The proof is essentially the same as the proof of Proposition (4.6.10). The only additional point here is that the connecting homomorphisms in the long exact sequences are homomorphisms of $H^*(B\Gamma; E)$-modules. This, however, follows easily by using the composition product, for example. □

Before stating the Evaluation Theorem we shall prove two lemmata, the second being a preliminary version of the theorem.

(3.11.11) Lemma
Let G, E, α, γ, Γ and $K(\alpha)$ be as in Recollections (3.11.7). Let $H \subset G$ be a subgroup and suppose that $K(\alpha) \not\subseteq H$. View $E[G/H]$ as a $E[\Gamma]$-module via the homomorphisms
$$E[\Gamma] \to E[G] \to E[G/H].$$
Then $E[G/H]$ is a free $E[\Gamma]$-module.

Proof.
For any $a \in E[G]$, let $\bar{a}$ denote the image of a in $E[G/H]$; and let $\mathfrak{m}(G/H)$ be the augmentation ideal in $E[G/H]$. By Theorem (4.6.21) (1), it is enough to show that
$$1 - \bar{\gamma} \notin \mathfrak{m}(G/H)^2.$$

Suppose $1 - \bar{\gamma} \in \mathfrak{m}(G/H)^2$. Then it is easy to see that one can construct a map of resolutions

$$\phi : \mathcal{E}_*(\Gamma; E) \to \mathcal{E}_*(G/H; E)$$

such that

$$\phi(w_i) \in \mathfrak{m}(G/H)\mathcal{E}_i(G/H; E)$$

for $i = 1, 2$, where w_i is the generator of $\mathcal{E}_i(\Gamma; E)$. It follows that the induced homomorphism

$$H^*(B(G/H); E) \to H^*(B\Gamma; E)$$

is zero in positive degrees. Now this homomorphism factors through

$$p(G/H; E)^* : H^*(B(G/H); E) \to H^*(BG; E);$$

and

$$p(G/H; E)^* = p(G/H)^* \otimes 1_E,$$

where $p(G/H)^*$ is the corresponding homomorphism with coefficients in $\mathbb{F}_p$.

Since $G \cong G/H \times H$, $p(G/H)^*$ maps the homogeneous linear polynomials in $H^*(B(G/H); \mathbb{F}_p)$ onto $PH \subseteq H^*(BG; \mathbb{F}_p)$. Since $K(\alpha) \not\subseteq H$, $PH \not\subseteq PK(\alpha)$; and so there is a homogeneous linear polynomial $u \in H^*(B(G/H); \mathbb{F}_p)$ which has non-zero image in $H^*(BK(\alpha); \mathbb{F}_p)$. But the homomorphism

$$H^*(BK(\alpha); \mathbb{F}_p) \subseteq H^*(BK(\alpha); E) \to H^*(B\Gamma; E)$$

is injective on the homogeneous linear polynomials. Thus

$$u \otimes 1 \in H^*(B(G/H); E)$$

has non-zero image in $H^*(B\Gamma; E)$. This contradiction shows that $1 - \bar{\gamma} \notin \mathfrak{m}(G/H)^2$. □

In the next lemma let G, E, α, γ, Γ and $K(\alpha)$ be as in Recollections (3.11.7). Let t be a polynomial generator for $H^*(B\Gamma; E)$: so $H^*(B\Gamma; E) = E[t]$ in case (2) and

$$H^*(B\Gamma; E) = \Lambda(s) \otimes E[t]$$

in case (odd p). Let $S(t)$ be the multiplicative subset of $H^*(B\Gamma; E)$ generated by t.

(3.11.12) Lemma
Let X be a paracompact finitistic G-space. Suppose that $X^{K(\alpha)} = \emptyset$. Then $S(t)^{-1}\overline{H}_\Gamma^*(X; E) = 0$.

Proof.
We shall make use of faithful coverings: see Definition (4.6.27) and Lemma (4.6.28).

Let $\mathcal{U}$ be a faithful finite-dimensional Čech-G-covering of X. Let $C_*(\check{\mathcal{U}}; E)$ be the oriented chain complex of the Čech nerve $\check{\mathcal{U}}$. Let $\{\sigma_i; i \in I\}$ be a complete set of representatives for the G-orbits of the n-simplexes of $\check{\mathcal{U}}$. Then
$$C_n(\check{\mathcal{U}}; E) \cong \bigoplus_{i \in I} E[G/G_{\sigma_i}].$$
Since, for each $i \in I$, there is $x_i \in X$ such that
$$G_{\sigma_i} \subseteq G_{x_i},$$
and as $X^{K(\alpha)} = \emptyset$, it follows from Lemma (3.11.11) that each $E[G/G_{\sigma_i}]$ is a free $E[\Gamma]$-module. Thus, letting $C^*(\check{\mathcal{U}}; E)$ denote the oriented cochain complex of $\check{\mathcal{U}}$, each $C^n(\check{\mathcal{U}}; E)$ is a product of free $E[\Gamma]$-modules.

Now for any cochain complex C^* of $E[\Gamma]$-modules there is a second spectral sequence with
$$E_1^{r,s} \cong H^s(B\Gamma; C^r)$$
converging to $H(\beta_\Gamma^*(C^*))$. (See [Brown, K.S. 1982], Chapter VII, Section 5; and cf. Corollary (4.6.7)(2).)

With $C^* = C^*(\check{\mathcal{U}}; E)$ as above, it follows that $E_1^{r,s} = 0$ if $s > 0$, and $E_1^{r,0} = C^r(\check{\mathcal{U}}; E)^\Gamma$. Thus, by Remark (3.11.4)(1), $H_\Gamma^*(|\check{\mathcal{U}}|; E)$ is isomorphic to the homology of the cochain complex $C^*(\check{\mathcal{U}}; E)^\Gamma$. Hence $H_\Gamma^*(|\check{\mathcal{U}}|; E)$ is zero in high degrees, since $\check{\mathcal{U}}$ is finite-dimensional. So
$$S(t)^{-1} H_\Gamma^*(|\check{\mathcal{U}}|; E) = 0.$$
$\square$

Now we are ready to state and prove the general version of the Evaluation Theorem. In the theorem $G = (\mathbb{Z}_p)^r$, E is an extension field of $\mathbb{F}_p$, $\alpha \in E^r$, $K(\alpha) \subset G$ is determined by α as in Recollection (3.11.7)(1), and γ and $\Gamma = \langle \gamma \rangle \subseteq E[G]$ are determined by α as in Recollection (3.11.7)(3). As above $S(t) \subseteq H^*(B\Gamma; E)$ is the multiplicative subset generated by a polynomial generator $t \in H^*(B\Gamma; E)$. We shall abbreviate $H^*(B\Gamma; E)$ by H_Γ^*.

(3.11.13) Theorem
With the notation above, let X be a paracompact finitistic G-space, and let $A \subset X$ be a closed invariant subspace. Then restriction induces an isomorphism
$$S(t)^{-1} \overline{H}_\Gamma^*(X, A; E) \xrightarrow{\cong} S(t)^{-1} H_\Gamma^* \otimes_E \overline{H}^*(X^{K(\alpha)}, A^{K(\alpha)}; E).$$

3. Localization

Proof.

The proof is just like the proof of the Localization Theorem. Because of the long exact sequence of Proposition (3.11.10)(1), it is enough to prove the result when $A = \emptyset$. Let V be a closed invariant neighbourhood of $X^{K(\alpha)}$, and let W be the complement of the interior of V. By Lemma (3.11.12),

$$S(t)^{-1}\overline{H}_\Gamma^*(W;E) = 0 \quad \text{and} \quad S(t)^{-1}\overline{H}_\Gamma^*(V \cap W; E) = 0.$$

So by Proposition (3.11.10)(3),

$$S(t)^{-1}\overline{H}_\Gamma^*(X;E) \cong S(t)^{-1}\overline{H}_\Gamma^*(V;E).$$

By Proposition (3.11.10)(2),

$$S(t)^{-1}\overline{H}_\Gamma^*(X;E) \cong S(t)^{-1}\overline{H}_\Gamma^*(X^{K(\alpha)};E).$$

Finally, Γ acts trivially on $\overline{C}^*(X^{K(\alpha)}, A^{K(\alpha)}; E)$ since $\Gamma \subseteq E[K(\alpha)]$. Hence

$$\overline{H}_\Gamma^*(X^{K(\alpha)}, A^{K(\alpha)}; E) \cong H_\Gamma^* \otimes_E \overline{H}^*(X^{K(\alpha)}, A^{K(\alpha)}; E),$$

by the Künneth Theorem. □

(3.11.14) Remarks

(1) In the theorem one could evaluate further by putting $t = 1$. If M is any $E[t]$-module, let $M_1 = E_1 \otimes_{E[t]} M$, where E_1 is E made into a $E[t]$-module via the homomorphism $E[t] \to E$, $t \mapsto 1$. Then the theorem gives isomorphisms

$$\overline{H}_\Gamma^*(X, A; E)_1 \xrightarrow{\cong} \overline{H}^*(X^{K(\alpha)}, A^{K(\alpha)}; E)$$

in case (2), and

$$\overline{H}_\Gamma^*(X, A; E)_1 \xrightarrow{\cong} \Lambda(s) \otimes_E \overline{H}^*(X^{K(\alpha)}, A^{K(\alpha)}; E)$$

in case (odd p).

In case (odd p) one could also put $s = 0$. Let E_0 be E made into a $\Lambda(s)$-module via the homomorphism $\Lambda(s) \to E$, $s \mapsto 0$. Then, in case (odd p), the theorem gives an isomorphism

$$E_0 \otimes_{\Lambda(s)} \overline{H}_\Gamma^*(X, A; E)_1 \xrightarrow{\cong} \overline{H}^*(X^{K(\alpha)}, A^{K(\alpha)}; E).$$

(2) For a topological space Y one might ask when

$$\overline{H}^*(Y;E) \cong \overline{H}^*(Y;\mathbb{F}_p) \otimes_{\mathbb{F}_p} E.$$

By theorems on universal coefficients this is true if:

(i) Y is compact; or
(ii) $\dim_{\mathbb{F}_p}(E) < \infty$; or
(iii) $\overline{H}^*$ can be replaced by singular cohomology, H^*, i.e. if Y is an agreement space, and $H^*(Y;\mathbb{F}_p)$ has finite type; or

(iv) Y is paracompact, locally compact and $clc^\infty_{\mathbb{F}_p}$ (e.g. Y is a paracompact $\mathbb{F}_p$-cohomology manifold) and $\overline{H}^*(Y;\mathbb{F}_p)$ has finite type. (See, e.g., [Bredon, 1967a], Chapter V, Exercise (25).)

So in Theorem (3.11.13), if $X^{K(\alpha)}$ or E satisfies (i), (ii), (iii) or (iv), then there is an isomorphism
$$S(t)^{-1}\overline{H}^*_\Gamma(X;E) \xrightarrow{\cong} S(t)^{-1} H^*_\Gamma \otimes_{\mathbb{F}_p} \overline{H}^*(X^{K(\alpha)};\mathbb{F}_p).$$

(3) For odd p, the isomorphism of the theorem is an isomorphism of $\mathbb{Z}_2$-graded modules as in the Localization Theorem. For $p = 2$ the isomorphism is ungraded.

(4) Note that G-spaces in case (B) are not included in Theorem (3.11.13).

The following lemma is useful in applying the Evaluation Theorem.

(3.11.15) Lemma

Let $G = (\mathbb{Z}_p)^r$ and let $R = \mathbb{F}_p[t_1,\ldots,t_r]$ be the polynomial part of $H^*(BG;\mathbb{F}_p)$. Let $f_1,\ldots,f_n$ be polynomials in R, such that $R \neq (f_1,\ldots,f_n)$, the ideal in R generated by $f_1,\ldots,f_n$. And let E be an algebraically closed extension of $\mathbb{F}_p$. Then $\exists \alpha \in E^r$ such that $f_1(\alpha) = \ldots = f_n(\alpha) = 0$, and $\mathrm{rank}_p(K(\alpha)) \geq r - n$.

Proof.

Let $J = (f_1,\ldots,f_n)$, and let
$$V(J) = \{\alpha \in E^r; f_1(\alpha) = \ldots = f_n(\alpha) = 0\};$$
and for any $\alpha \in E^r$, let $I(\alpha) = \{f \in R : f(\alpha) = 0\}$. So (see Recollection (3.11.7)(1)) $PK(\alpha) = \sigma I(\alpha)$.

By Hilbert's *Nullstellensatz*, $\sqrt{J} = \bigcap_{\alpha \in V(J)} I(\alpha)$. Hence, if P is a minimal prime ideal of J, $P \supseteq PK(\alpha)$, for some $\alpha \in V(J)$, since $I(\alpha) \supseteq PK(\alpha)$, and there is at most a finite number of distinct $PK(\alpha)$s. (I.e.
$$P \supseteq \sqrt{J} = \bigcap_{\alpha \in V(J)} I(\alpha) \supseteq \bigcap_{\alpha \in V(J)} PK(\alpha),$$
the latter being a finite intersection.) Thus
$$r - \mathrm{rank}_p(K(\alpha)) = \mathrm{ht}(PK(\alpha)) \leq \mathrm{ht}(P);$$
and $\mathrm{ht}(P) \leq n$ by the Krull Height Theorem. □

The following corollary of the Evaluation Theorem partially generalizes Theorem 1 of [Heller, 1959]. In the statement of the corollary $(S^n)^m$ means the product of m copies of S^n; and so
$$\overline{H}^*((S^n)^m;\mathbb{F}_p) = \Lambda(u_1,\ldots,u_m),$$
an exterior algebra over $\mathbb{F}_p$ on m generators of degree n.

(3.11.16) Corollary
Let $G = (\mathbb{Z}_p)^r$, where p is any odd prime number, and let X be a paracompact finitistic G-space. Suppose that $\overline{H}^*(X; \mathbb{F}_p)$ is isomorphic to $\overline{H}^*((S^n)^m; \mathbb{F}_p)$ as a graded $\mathbb{F}_p$-algebra, where $n > 0$ and $1 \leq m \leq 17$. Assume that G acts trivially on $\overline{H}^*(X; \mathbb{F}_p)$. Then there is an isotropy subgroup isomorphic to $(\mathbb{Z}_p)^\rho$ where $\rho \geq r - m$.

In particular, G cannot act freely on X if $r > m$.

Proof.

If n is even, then $\chi(X) = 2^m$; and so $X^G \neq \emptyset$, by Remark (3.10.19)(3). So we shall assume that n is odd.

Let $u_1, \ldots, u_m$ span $\overline{H}^n(X; \mathbb{F}_p)$. And in the Leray-Serre spectral sequence of $X_G \to BG$, let $f_1, \ldots, f_m \in R = \mathbb{F}_p[t_1, \ldots, t_r]$ be the polynomial parts of the transgressions of $u_1, \ldots, u_m$. By Lemma (3.11.15), there is a finite extension field E of $\mathbb{F}_p$ and $\alpha \in E^r$ such that

$$\operatorname{rank}_p(K(\alpha)) \geq r - m, \quad \text{and} \quad f_1(\alpha) = \ldots = f_m(\alpha) = 0.$$

We shall show that $X^{K(\alpha)} \neq \emptyset$.

Let γ and $\Gamma = \langle \gamma \rangle \subseteq E[G]$ be determined by α as in Recollection (3.11.7)(3); and consider the spectral sequence $E(\Gamma, C^*)_r$ of Remarks (3.11.4)(2), where $C^* = \overline{C}^*(X; E)$. We shall suppose that $X^{K(\alpha)} = \emptyset$ and obtain a contradiction.

Let $S \subseteq E[t] \subseteq H^*(B\Gamma; E)$ be the multiplicative subset of all non-zero polynomials. Let $K = S^{-1}E[t]$, the field of fractions of $E[t]$. Since $X^{K(\alpha)} = \emptyset$, the localized spectral sequence $S^{-1}E(\Gamma, C^*)_r$ converges to zero, by Lemma (3.11.12).

Let $E(\Gamma, C^*)_r = E_r$. Since G acts trivially on $\overline{H}^*(X; \mathbb{F}_p)$, Γ acts trivially on $\overline{H}^*(X; E)$; and so $E_2^{i,j} \cong H^i(B\Gamma; E) \otimes \overline{H}^j(X; E)$. We can view $S^{-1}E_r$ as consisting of two columns with $S^{-1}E_2^{0,j} \cong K \otimes \overline{H}^j(X; E)$ and $S^{-1}E_2^{1,j} \cong K \otimes \Lambda^1(s) \otimes \overline{H}^j(X; E)$. The only possible non-zero differentials are d_{qn+1} for $1 \leq q \leq m$. But from the homomorphism of spectral sequences discussed in Recollection (3.11.9)(4), and from the choice of α, it follows that $d_{n+1} = 0$. Furthermore, since $s^2 = 0$, it follows that $d_{2n+1} : S^{-1}E_2^{1,j} \to S^{-1}E_2^{0,j-2n}$ is zero also.

If m is odd, $m = 2j + 1$, say, then consider $S^{-1}E_2^{0,j}$ and $S^{-1}E_2^{1,j+1}$. These both have dimension $\binom{2j+1}{j}$ over K. And the only possible non-zero differentials having domain or codomain $S^{-1}E_r^{0,j}$ or $S^{-1}E_r^{1,j+1}$ must come from $S^{-1}E_r^{0,j+3}, S^{-1}E_r^{1,j+4}, S^{-1}E_r^{0,j+5}, S^{-1}E_r^{1,j+6}, \ldots$, or go to $S^{-1}E_r^{1,j-2}, S^{-1}E_r^{0,j-3}, S^{-1}E_r^{1,j-4}, \ldots$, for the appropriate level

3.11. Hirsch-Brown model and the Evaluation Theorem

r. Since the spectral sequence converges to zero, looking at dimensions over K, it follows that

$$2\binom{2j+1}{j} \leq 2\left\{\binom{2j+1}{0} + \binom{2j+1}{1} + \ldots + \binom{2j+1}{j-2}\right\}.$$

Hence

$$2^{2j} \geq 2\binom{2j+1}{j} + \binom{2j+1}{j-1}.$$

If m is even, $m = 2j$, say, then, considering $S^{-1}E_2^{0,j-1}$ and $S^{-1}E_2^{1,j}$ similarly, it follows that

$$2^{2j} \geq 2\binom{2j}{j} + 3\binom{2j}{j-1} + \binom{2j}{j-2} = 2\binom{2j+1}{j} + \binom{2j+1}{j-1},$$

again.

A quick check, helped by downward induction, shows that these inequalities are invalid for $m \leq 17$. E.g., $2^{16} = 65,536$, but

$$2\binom{17}{8} + \binom{17}{7} = 2(24,310) + 19,448 = 68,068.$$

$\square$

(3.11.17) Remark
In fact Corollary (3.11.16) holds without any restriction on m; and one only needs $\overline{H}^*(X; \mathbb{F}_p)$ to be isomorphic to $\overline{H}^*((S^n)^m; \mathbb{F}_p)$ as a graded $\mathbb{F}_p$-vector space. See Theorem (4.6.42) and Discussion (4.6.43).

Next we give an example where the spectral sequence $E(\Gamma, C^*)_r$ is not multiplicative.

(3.11.18) Example
Let $X = \mathbb{R}P^{4n+3}$. Then $G = (\mathbb{Z}_2)^2$ can act freely on X as follows. View S^{4n+3} as

$$\{(q_0, \ldots, q_n) \in \mathbb{H}^{n+1} : |q_0|^2 + \ldots + |q_n|^2 = 1\},$$

and view $\mathbb{R}P^{4n+3}$ as the quotient of S^{4n+3} with respect to the involution $(q_0, \ldots q_n) \mapsto (-q_0, \ldots, -q_n)$. Now the quaternionic 8-group

$$Q = \{\pm 1, \pm i, \pm j, \pm k\}$$

acts on S^{4n+3} by scalar multiplication: e.g.,

$$i(q_0, \ldots, q_n) = (iq_0, \ldots, iq_n).$$

This action induces a free action of $G = Q/\{\pm 1\}$ on $\mathbb{R}P^{4n+3}$.

Now let $H^*(BG; \mathbb{F}_2) = \mathbb{F}_2[t_1, t_2]$ and let $x \in H^1(X; \mathbb{F}_2)$ be a generator: i.e. $H^*(X; \mathbb{F}_2) = \mathbb{F}_2[x]/(x^{4n+4})$. In the spectral sequence for $X_G \to BG$ it is easy to see that x transgresses to $t_1^2 + t_1 t_2 + t_2^2$, the only irreducible homogeneous polynomial of degree 2. Thus $x^2 = \text{Sq}^1(x)$ transgresses to

$$\text{Sq}^1(t_1^2 + t_1 t_2 + t_2^2) = t_1 t_2 (t_1 + t_2).$$

Choose $\alpha = (\alpha_1, \alpha_2) \in \bar{\mathbb{F}}_2^2$ such that $\alpha_1^2 + \alpha_1 \alpha_2 + \alpha_2^2 = 0$, and $\alpha \neq 0$. Let Γ be determined by α as in Recollection (3.11.7)(3). Then the homomorphism of spectral sequences, $E(G, C^*)_r \to E(\Gamma, C^*)_r$, as in Recollection (3.11.9)(4), with $C^* = \overline{C}^*(X; \bar{\mathbb{F}}_2)$, shows that, in $E(\Gamma, C^*)_r$, x transgresses to zero, but x^2 transgresses to

$$\alpha_1 \alpha_2 (\alpha_1 + \alpha_2) t^3 = (\alpha_1 + \alpha_2)^3 t^3 \neq 0.$$

(3.11.19) Remark

In Example (3.11.18), as elsewhere, we have used the familiar fact that Steenrod operations commute with transgression in the Serre spectral sequence. The proof of this fact usually requires some special features of the Serre spectral sequence itself, and it is not directly applicable to the Leray spectral sequence or the spectral sequence of Remark (3.11.4)(2); see [McCleary, 1985], §6.1, or [Whitehead, 1978], Chapter XIII, Theorem (7.9*). An easy way out is to use the direct limit of the Serre spectral sequences for $|\tilde{\mathcal{U}}|_G \to BG$, which gives the same information about

$$\text{im}[p^* : H^*(BG; k) \to H^*_G(X; k)]$$

as do the other spectral sequences for $X_G \to BG$. And Steenrod operations commute with transgression in the direct limit spectral sequence. A more general solution to this technical difficulty might be found in [Singer, 1973] or [Barnes, 1985], but we have not verified this.

We shall now consider the minimal $R(E)$-free models (see (B.2.4)) associated with the models $\overline{\beta}^*_G(X; E)$, where $G = (\mathbb{Z}_p)^r$, E is an extension of $\mathbb{F}_p$, and $R(E) = E[t_1, \ldots, t_r]$. With the notation of Recollection (3.11.1)(3), then, by (1.3.23) and (B.2.4), if $p = 2$, there is a differential on $R(E) \otimes \overline{H}^*(X; E)$, making the latter into a differential graded $R(E)$-module, and there is a cochain homotopy equivalence over $R(E)$,

$$\mu(X) : \overline{\beta}^*_G(X; E) \to R(E) \tilde{\otimes} \overline{H}^*(X; E).$$

If p is an odd prime, we obtain similarly a cochain homotopy equivalence over $R(E)$,

$$\mu(X) : \overline{\beta}^*_G(X; E) \to R(E) \tilde{\otimes} H(\Lambda(s_1, \ldots, s_r) \tilde{\otimes} \overline{H}^*(X; E)).$$

In the case (odd p), the process has two stages: first there is a differential on $R(E) \tilde{\otimes} H(\Lambda(s; r) \tilde{\otimes} \overline{C}^*(X; E))$ and a cochain homotopy equivalence over $R(E)$ from $\overline{\beta}^*_G(X; E)$ to

$$R(E) \tilde{\otimes} H(\Lambda(s; r) \tilde{\otimes} \overline{C}^*(X; E));$$

3.11. Hirsch-Brown model and the Evaluation Theorem

r. Since the spectral sequence converges to zero, looking at dimensions over K, it follows that

$$2\binom{2j+1}{j} \leq 2\left\{\binom{2j+1}{0} + \binom{2j+1}{1} + \ldots + \binom{2j+1}{j-2}\right\}.$$

Hence

$$2^{2j} \geq 2\binom{2j+1}{j} + \binom{2j+1}{j-1}.$$

If m is even, $m = 2j$, say, then, considering $S^{-1}E_2^{0,j-1}$ and $S^{-1}E_2^{1,j}$ similarly, it follows that

$$2^{2j} \geq 2\binom{2j}{j} + 3\binom{2j}{j-1} + \binom{2j}{j-2} = 2\binom{2j+1}{j} + \binom{2j+1}{j-1},$$

again.

A quick check, helped by downward induction, shows that these inequalities are invalid for $m \leq 17$. E.g., $2^{16} = 65,536$, but

$$2\binom{17}{8} + \binom{17}{7} = 2(24,310) + 19,448 = 68,068.$$

$\square$

(3.11.17) Remark
In fact Corollary (3.11.16) holds without any restriction on m; and one only needs $\overline{H}^*(X; \mathbb{F}_p)$ to be isomorphic to $\overline{H}^*((S^n)^m; \mathbb{F}_p)$ as a graded $\mathbb{F}_p$-vector space. See Theorem (4.6.42) and Discussion (4.6.43).

Next we give an example where the spectral sequence $E(\Gamma, C^*)_r$ is not multiplicative.

(3.11.18) Example
Let $X = \mathbb{R}P^{4n+3}$. Then $G = (\mathbb{Z}_2)^2$ can act freely on X as follows. View S^{4n+3} as

$$\{(q_0, \ldots, q_n) \in \mathbb{H}^{n+1} : |q_0|^2 + \ldots + |q_n|^2 = 1\},$$

and view $\mathbb{R}P^{4n+3}$ as the quotient of S^{4n+3} with respect to the involution $(q_0, \ldots q_n) \mapsto (-q_0, \ldots, -q_n)$. Now the quaternionic 8-group

$$Q = \{\pm 1, \pm i, \pm j, \pm k\}$$

acts on S^{4n+3} by scalar multiplication: e.g.,

$$i(q_0, \ldots, q_n) = (iq_0, \ldots, iq_n).$$

This action induces a free action of $G = Q/\{\pm 1\}$ on $\mathbb{R}P^{4n+3}$.

Now let $H^*(BG; \mathbb{F}_2) = \mathbb{F}_2[t_1, t_2]$ and let $x \in H^1(X; \mathbb{F}_2)$ be a generator: i.e. $H^*(X; \mathbb{F}_2) = \mathbb{F}_2[x]/(x^{4n+4})$. In the spectral sequence for $X_G \to BG$ it is easy to see that x transgresses to $t_1^2 + t_1 t_2 + t_2^2$, the only

irreducible homogeneous polynomial of degree 2. Thus $x^2 = \mathrm{Sq}^1(x)$ transgresses to
$$\mathrm{Sq}^1(t_1^2 + t_1 t_2 + t_2^2) = t_1 t_2 (t_1 + t_2).$$

Choose $\alpha = (\alpha_1, \alpha_2) \in \bar{\mathbb{F}}_2^2$ such that $\alpha_1^2 + \alpha_1 \alpha_2 + \alpha_2^2 = 0$, and $\alpha \neq 0$. Let Γ be determined by α as in Recollection (3.11.7)(3). Then the homomorphism of spectral sequences, $E(G, C^*)_r \to E(\Gamma, C^*)_r$, as in Recollection (3.11.9)(4), with $C^* = \bar{C}^*(X; \bar{\mathbb{F}}_2)$, shows that, in $E(\Gamma, C^*)_r$, x transgresses to zero, but x^2 transgresses to
$$\alpha_1 \alpha_2 (\alpha_1 + \alpha_2) t^3 = (\alpha_1 + \alpha_2)^3 t^3 \neq 0.$$

(3.11.19) Remark

In Example (3.11.18), as elsewhere, we have used the familiar fact that Steenrod operations commute with transgression in the Serre spectral sequence. The proof of this fact usually requires some special features of the Serre spectral sequence itself, and it is not directly applicable to the Leray spectral sequence or the spectral sequence of Remark (3.11.4)(2); see [McCleary, 1985], §6.1, or [Whitehead, 1978], Chapter XIII, Theorem (7.9*). An easy way out is to use the direct limit of the Serre spectral sequences for $|\check{\mathcal{U}}|_G \to BG$, which gives the same information about
$$\mathrm{im}[p^* : H^*(BG; k) \to H_G^*(X; k)]$$
as do the other spectral sequences for $X_G \to BG$. And Steenrod operations commute with transgression in the direct limit spectral sequence. A more general solution to this technical difficulty might be found in [Singer, 1973] or [Barnes, 1985], but we have not verified this.

We shall now consider the minimal $R(E)$-free models (see (B.2.4)) associated with the models $\bar{\beta}_G^*(X; E)$, where $G = (\mathbb{Z}_p)^r$, E is an extension of $\mathbb{F}_p$, and $R(E) = E[t_1, \ldots, t_r]$. With the notation of Recollection (3.11.1)(3), then, by (1.3.23) and (B.2.4), if $p = 2$, there is a differential on $R(E) \otimes \bar{H}^*(X; E)$, making the latter into a differential graded $R(E)$-module, and there is a cochain homotopy equivalence over $R(E)$,
$$\mu(X) : \bar{\beta}_G^*(X; E) \to R(E) \tilde{\otimes} \bar{H}^*(X; E).$$
If p is an odd prime, we obtain similarly a cochain homotopy equivalence over $R(E)$,
$$\mu(X) : \bar{\beta}_G^*(X; E) \to R(E) \tilde{\otimes} H(\Lambda(s_1, \ldots, s_r) \tilde{\otimes} \bar{H}^*(X; E)).$$
In the case (odd p), the process has two stages: first there is a differential on $R(E) \tilde{\otimes} H(\Lambda(s; r) \tilde{\otimes} \bar{C}^*(X; E))$ and a cochain homotopy equivalence over $R(E)$ from $\bar{\beta}_G^*(X; E)$ to
$$R(E) \tilde{\otimes} H(\Lambda(s; r) \tilde{\otimes} \bar{C}^*(X; E));$$

3.11. Hirsch-Brown model and the Evaluation Theorem   243

then, by studying the internal differential on $\Lambda(s;r)\tilde{\otimes}\overline{C}^*(X;E)$ coming from $\overline{\beta}_G^*(X;E)$, and applying Corollary (B.2.4) again, it follows that there is a differential on $\Lambda(s_1,\ldots,s_r)\tilde{\otimes}\overline{H}^*(X;E)$ such that

$$H(\Lambda(s;r)\tilde{\otimes}\overline{C}^*(X;E)) \cong H(\Lambda(s_1,\ldots,s_r)\tilde{\otimes}\overline{H}^*(X;E)).$$

(3.11.20) Definition
In both cases (2) and (odd p) we shall denote the minimal $R(E)$-free model $R(E)\tilde{\otimes}\overline{H}^*(X;E)$, resp.

$$R(E)\tilde{\otimes}H(\Lambda(s_1,\ldots,s_r)\tilde{\otimes}\overline{H}^*(X;E)),$$

by $\overline{\mu}_G^*(X;E)$. Thus in both cases there is a cochain homotopy equivalence over $R(E)$,

$$\mu(X) : \overline{\beta}_G^*(X;E) \to \overline{\mu}_G^*(X;E).$$

(Note that in case (2) the minimal $R(E)$-free model coincides with the minimal Hirsch-Brown model, whereas in case (odd p) the two models differ in general; cf. (1.4.10).

We shall now study case (2) in more detail with the aim of extending Theorem (3.11.13) to an induced complex $\overline{\mu}_{G,\tilde{\alpha}}^*(X;E)$.

(3.11.21) Definition
Let $G = (\mathbb{Z}_2)^r$, E be an extension field of $\mathbb{F}_2$, and $\alpha = (\alpha_1,\ldots,\alpha_r) \in E^r$ with $\alpha \neq 0$. Suppose that G is generated by $g_1,\ldots,g_r$, and put $\gamma = 1-\sum_{i=1}^r \alpha_i(1-g_i) \in E[G]$. Let Γ be the subgroup of $E[G]$ generated by γ. Let t generate $H_\Gamma^* \cong E[t]$. So we have the homomorphism $\tilde{\alpha}$: $R(E) \to H_\Gamma^*$ given by $t_i \mapsto \alpha_i t$ for $1 \leq i \leq r$. With this $R(E)$-module structure on $E[t]$, put

$$\overline{\mu}_{G,\tilde{\alpha}}^*(X;E) = E[t] \otimes_{R(E)} \overline{\mu}_G^*(X;E).$$

(3.11.22) Remark
Let $\overline{\beta}_{G,\tilde{\alpha}}^*(X;E)$ denote

$$\mathrm{Hom}_{E[\Gamma]}(\mathcal{E}_*(\Gamma;E),\overline{C}^*(X;E))$$

where $\overline{C}^*(X;E)$ is made into a $E[\Gamma]$-module via the above inclusion $E[\Gamma] \subseteq E[G]$ which sends γ to $1-\sum_{i=1}^r \alpha_i(1-g_i)$. Thus $\overline{\beta}_{G,\tilde{\alpha}}^*(X;E)$ differs from $\overline{\beta}_\Gamma^*(X;E)$ as defined in (3.11.8) in that the higher order terms have been dropped from the expression for γ. Thus, in general, the inclusion $E[\Gamma] \subseteq E[G]$ used here does not factor through $E[K(\alpha)]$. On the other hand it is clear that $\overline{\beta}_{G,\tilde{\alpha}}^*(X;E) \cong E[t] \otimes_{R(E)} \overline{\beta}_G^*(X;E)$. Thus there is a cochain homotopy equivalence over $E[t]$,

$$\mu(X,\alpha) : \overline{\beta}_{G,\tilde{\alpha}}^*(X;E) \to \overline{\mu}_{G,\tilde{\alpha}}^*(X;E).$$

(3.11.23) Definition
Let
$$\overline{H}^*_{G,\tilde{\alpha}}(X;E) := H(\overline{\beta}^*_{G,\tilde{\alpha}}(X;E)) \cong H(\overline{\mu}^*_{G,\tilde{\alpha}}(X;E)).$$
Note that $\overline{H}^*_{G,\tilde{\alpha}}(X;E)$ is a $E[t]$-module.

In the following proposition $S(t)$ is the multiplicative subset of $E[t]$ generated by t. Recall the definition of $K(\alpha)$ from (3.11.7)(1).

(3.11.24) Proposition
Let $G = (\mathbb{Z}_2)^r$, let X be a paracompact finitistic G-space, let E be a field of characteristic 2, and let $\alpha \in E^r$ be non-zero. Then restriction induces an isomorphism
$$S(t)^{-1}\overline{H}^*_{G,\tilde{\alpha}}(X;E) \overset{\cong}{\to} S(t)^{-1}\overline{H}^*_{G,\tilde{\alpha}}(X^{K(\alpha)};E).$$

Proof.
By imitating the proof of Theorem (3.11.13), it is enough to show that $S(t)^{-1}\overline{H}^*_{G,\tilde{\alpha}}(X;E) = 0$ if $X^{K(\alpha)} = \emptyset$.

Let $\{g_1, \ldots, g_r\}$ generate G, let $\gamma_1 = 1 - \sum_{i=1}^r \alpha_i(1-g_i)$, and let γ be chosen as in Recollection (3.11.7)(3) for the given $\alpha \in E^r$. So $\gamma \in E[K(\alpha)]$; and, using the notation of the proof of Lemma (3.11.11), $\gamma - \gamma_1 \in \mathfrak{m}(G)^2$. We must show that $E[G/H]$ is a free $E[\Gamma_1]$-module if $K(\alpha) \not\subseteq H$, where $\Gamma_1 = \langle \gamma_1 \rangle \subseteq E[G]$.

If $K(\alpha) \not\subseteq H$, then $1 - \overline{\gamma} \notin \mathfrak{m}(G/H)^2$, by the proof of Lemma (3.11.11). Hence $1 - \overline{\gamma}_1 \notin \mathfrak{m}(G/H)^2$, since $\overline{\gamma} - \overline{\gamma}_1 \in \mathfrak{m}(G/H)^2$. So $E[G/H]$ is a free $E[\Gamma_1]$-module by [Carlson, 1983], Lemma 2.9, (Theorem (4.6.21)(1)), as before. □

Exercises

In exercises involving G-spaces, unless stated otherwise, assume that the G-spaces satisfy condition (LT) of the introduction to Section 3.10.

(3.1) Let G be a p-torus, resp. a torus, and let $H_1, \ldots, H_n$ be sub-p-tori, resp. subtori. Let $K = H_1 \cap \ldots \cap H_n$, resp. $K = (H_1 \cap \ldots \cap H_n)^0$, the identity component of $H_1 \cap \ldots \cap H_n$. Show that $PK = PH_1 + \ldots + PH_n$.

(3.2) Let G be a torus and let X be a compact G-space. Let M be the set of maximal connective isotropy subgroups. Show that M is finite by considering $X = \bigcup_{H \in M} X(H)$, where
$$X(H) = \{x \in X : G_x^0 \subseteq H\}.$$

Show similarly that there is a finite number of maximal isotropy subgroups.

If G is any compact connected Lie group, then show that

$$\max\{[G_x^0] : x \in X\}$$

is finite, where $[G_x^0]$ is the conjugacy class of the identity component of G_x.

(3.3) Let X be a G-space in case (0), (2) or (odd p). Suppose that $H^n(X;k) \cong k$, and $H^j(X;k) = 0$ for all $j > n$. Suppose that

$$i^* : H_G^n(X;k) \to H^n(X;k)$$

is surjective. Show that $X^G \neq \emptyset$.

(3.4) Let G be any compact connected Lie group, and let X be any paracompact G-space. Let $\pi : X \to X/G$ be the orbit map.

(1) Show that $\pi^* : H^1(X/G;k) \to H^1(X;k)$ is a monomorphism for any constant coefficients k.

(2) If X is connected and $X^G \neq \emptyset$, then show that

$$\pi^* : H^1(X/G;k) \to H^1(X;k)$$

is an isomorphism where k is any field of characteristic 0.

(3.5) ([Yau, 1977]) Let M be a connected topological n-manifold (or, more generally, a connected paracompact n-cohomology manifold over $\mathbb{Z}$). Let T be a torus acting almost - effectively on M such that $M^T \neq \emptyset$. Show that

$$\dim T \leq n - \alpha_1(M),$$

where $\alpha_1(M)$ is the maximal number of algebraically independent elements in $H^1(M;\mathbb{Q})$: i.e. $\alpha_1(M)$ is the greatest integer m such that there exist

$$\omega_1, \ldots, \omega_m \in H^1(M;\mathbb{Q})$$

with $\omega_1 \ldots \omega_m \neq 0$ in $H^m(M;\mathbb{Q})$. (By [Borel et al., 1960], Chapter IX, Corollary of Theorem 2.2, $\operatorname{cd}_{\mathbb{Z}}(M/T) \leq n - \dim T$.)

(3.6) Let G be any compact Lie group, and let X and Y be any G-spaces. Suppose that $\varphi : X \to Y$ and $\psi : Y \to X$ are G-homotopy equivalences, which are G-homotopy inverse to one another. Let

$$S \subseteq H_G^*(X;k)$$

be any multiplicative set, where k is any commutative ring. Show that $\varphi(X^S) \subseteq Y^{\psi^*(S)}$; and that by restriction φ and ψ give G-homotopy equivalences between X^S and $Y^{\psi^*(S)}$.

(3.7) Let G be a torus, and let X be any paracompact G-space with $\dim_{\mathbb{Q}} H^*(X;\mathbb{Q}) < \infty$. Let $R = H^*(BG;\mathbb{Q})$, as usual, and let $R_K = H^*(BK;\mathbb{Q})$ for any subtorus $K \subseteq G$. Suppose that $\mathrm{hd}_R H_G^*(X;\mathbb{Q}) = h$. (See Section A.4.) Show that there is a subtorus $K \subseteq G$ such that:
(i) $\mathrm{rank}(K) = h$;
(ii) $0 \to PK.H_G^*(X;\mathbb{Q}) \to H_G^*(X;\mathbb{Q}) \to H_K^*(X;\mathbb{Q}) \to 0$ is exact; and
(iii) $\mathrm{hd}_{R_K} H_K^*(X;\mathbb{Q}) = h$.

(3.8) Let X be any paracompact G-space in case (0), (2) or (odd p). Show that $H_G^*(X;k)$ is a finitely generated R-module if and only if $\dim_k H^*(X;k) < \infty$. Prove the corresponding result for pairs (X, A), where $A \subseteq X$ is closed and invariant.

(3.9) Let X be a compact G-space in case (0), (2), or (odd p). Show that $H_G^*(X;k)$ is a finitely generated k-algebra if and only if
$$\dim_k H^*(X;k) < \infty.$$

(3.10)
(1) Let X be a G-space in case (2) or (odd p). Let
$$u \in H_G^{2n}(X;k),$$
and let $R_n = R \cap H^{2n}(BG;k)$. Show that $\prod_{a \in R_n}(u-a)$ is nilpotent. Deduce that the result of Exercise (3.9) holds for any G-space (satisfying condition (LT)) in cases (2) and (odd p).
(2) Show that the result of Exercise (3.9) holds for connected G-spaces in case (0) which satisfy conditions (B) or (C).

(3.11) Let X be a G-space in case (0). Let $\langle u_1, \ldots, u_n \rangle$ be a Massey product in $H_G^*(X;\mathbb{Q})$ where $n \geq 3$. Show that $\langle u_1, \ldots, u_n \rangle$ is nilpotent.

(3.12) ([Assadi, 1988]) Let $G = (\mathbb{Z}_p)^r$ and $k = \mathbb{F}_p$. Let $f : X \to Y$ be an equivariant map of connected G-spaces. Suppose that $H^i(X;k) = 0$ for $0 < i < n$, and $H^i(Y;k) = 0$ for $i > n$. Suppose that Y satisfies condition (LT).
(1) Show that $Y^G \neq \emptyset$ if either
  (i) $H^n(X;k) = 0$, or
  (ii) n is even, p is odd, and G acts trivially on $H^n(X;k)$.
(2) If G acts trivially on $H^n(X;k)$, then show that
$$\mathrm{rank}(\Phi) \leq \dim_k H^n(X;k),$$
where $\Phi : G \times Y \to Y$ is the action on Y. (See Definition (4.1.5) for the definition of $\mathrm{rank}(\Phi)$.)
(3) Show that these results may be false if X is not connected.

(3.13) Let X be a G-space in case (0). Suppose that $X^G \neq \emptyset$, and let F be a component of X^G. Give examples where:

(1) X is TNHZ (in $X_G \to BG$) and F is full;
(2) X is TNHZ and F is not full;
(3) X is not TNHZ and F is full; and
(4) X is not TNHZ and F is not full.

(3.14) Let X be a G-space in case (0). Suppose that X has FDRH. (See Definition (2.4.1)(2).) Suppose that $X^G \neq \emptyset$, and that there is a component F of X^G such that $H^{od}(F;\mathbf{Q}) = 0$. Show that

$$H^{od}(X;\mathbf{Q}) = 0.$$

Give an example to show that the assumption that X has FDRH is needed for this result to hold in general.

(3.15) Prove Lemma (3.4.4) in cases (2) and (odd p). That is let G be a p-torus (or any finite abelian group), let X be a compact G-space, let $K \subseteq G$ be a subgroup, and suppose that c and c' are two components of X^K such that $Gc \neq Gc'$. Show that there are G-invariant open subsets of X^K, U and V, such that $U \cap V = \emptyset$, $U \cup V = X^K$, $Gc \subseteq U$ and $Gc' \subseteq V$.

(3.16) Let X be any paracompact G-space in case (0), (2) or (odd p). Suppose that $X^G \neq \emptyset$, and let F_1, F_2 be two components of X^G. Let $x_i \in F_i$ for $i = 1, 2$, and let $Y = \{x_1, x_2\}$. Let $\xi \in H_G^*(Y;k)$ correspond to $(0, 1)$ under the isomorphism

$$H_G^*(Y;k) \cong \bigoplus_{i=1}^{2} H_G^*(\{x_i\};k) \cong H^*(BG;k) \oplus H^*(BG;k).$$

Let $\psi^* : H_G^*(X;k) \to H_G^*(Y;k)$ be induced by the inclusion $\psi : Y \to X$; and let $I_\xi = \{a \in R | a\xi \in \operatorname{im} \psi^*\} = \operatorname{ann}_R(\delta\xi)$, where δ is the connecting homomorphism for the long exact sequence of the pair (X, Y). Show that $I_\xi = {p^*}^{-1}(\mathfrak{p}(G, F_1) + \mathfrak{q}(G, F_2))$, where, as usual, $p : X_G \to BG$ is th bundle map.

(3.17) Let X be a G-space in case (0). Let $K \subseteq G$ be a subtorus, and let $\beta_K : X_K \to X_G$ be the obvious map. Suppose that

$$\dim_k H^*(X;k) < \infty,$$

and let $H_G(X) = H_G^{\text{even}}(X;k)$. Let QK be the ideal in $H_G(X)$ generated by $p^*(PK)$. Show that in $H_G(X)$

$$\sqrt{\ker \beta_K^*} = \sqrt{QK} = \bigcap_{x \in X} \mathfrak{p}(K_x^0, F(K, x)).$$

(See Definition (3.6.1) for the notation.)

(3.18) Let X be a G-space in case (0), (2) or (odd p). Suppose that X is TNHZ in $X_G \to BG$. Let $\varphi : X^G \to X$ and $\psi_K : X^G \to X^K$, for any subtorus $K \subseteq G$, be the inclusions. Show that, in $H_G^*(X^G; k)$,
$$\operatorname{im} \varphi^* = \bigcap_{\operatorname{cork}(K)=1} \operatorname{im} \psi_K^*.$$

(3.19) In Example (3.6.20) let $L \subseteq T$ be the subcircle $\{(1, \omega) | \omega \in S^1\}$. Show that $X^L \approx S^3$.

(3.20) Show that any sum of Hsiang-Serre ideals is a Hsiang-Serre ideal. Show that any finite product and any finite intersection of Hsiang-Serre ideals are Hsiang-Serre ideals.

(3.21) Let X be a G-space in case (0). Suppose that X has FDRH (see Definition (2.4.1)(2))), and that $X^G \neq \emptyset$. Suppose that the conditions of Sections 3.3 or 3.4 are satisfied. Let $\xi_0 \in X^G$, and let $\epsilon : R \otimes \mathcal{M}(X) \to R$ be defined w.r.t. ξ_0. Let $\mathcal{M}(X) = \mathbb{Q}[x_1, \dots, x_r] \otimes \Lambda(y_1, \dots, y_s)$, where in $R \otimes \mathcal{M}(X)$, $\epsilon(x_i) = 0$ for $1 \leq i \leq r$. In $Q_{(G, \xi_0)}(X)$, let $dy_i = \sum_{j=1}^r a_{ij} x_j$ for $1 \leq i \leq s$, and $dx_i = \sum_{j=1}^s b_{ij} y_j$ for $1 \leq i \leq r$. Let J be the ideal in R generated by all the a_{ij} and b_{ij}. Show that J is a Hsiang-Serre ideal.

(3.22) Let X be a G-space in case (0). Suppose that $X^G \neq \emptyset$, and let $N \subseteq H^*(X^G; k)$ be a graded k-vector subspace. Let
$$\delta(R \otimes N) \subset H_G^*(X, X^G; k)$$
be the image of $R \otimes N \subseteq H_G^*(X^G; k)$ under the connecting homomorphism for the long exact sequence of the pair (X, X^G). Let
$$M = H_G^*(X, X^G; k) / \delta(R \otimes N).$$
Show that M is a Chang-Skjelbred module.

(3.23) Let M be a Chang-Skjelbred module. Let
$$L_j = \ker[\varphi_j : M \to M_j].$$
Show that $\operatorname{Ass}(L_j) = \{P \in \operatorname{Ass}(M) | P \not\subseteq P_j\}$.

(3.24) Let X be a G-space in case (0) or (odd p). Suppose that $H^*(X; k)$ is generated by a finite number of elements of odd degree, and that X is TNHZ in $X_G \to BG$. Deduce that X^G is connected.

(3.25) Let X be a G-space in case (2) such that
$$H^*(X, \mathbb{Z}) \cong H^*(\mathbb{C}P^n; \mathbb{Z})$$
as rings. (If X is a general G-space in case (B), then assume that $\operatorname{cd}_{\mathbb{Z}}(X) < \infty$.) Suppose that G acts trivially on $H^*(X; \mathbb{Z})$. If $r = \operatorname{rank}(G) = 1$, then show that $X^G \neq \emptyset$; and if $r > 1$, then assume that

$X^G \neq \emptyset$. Show that every component of X^G which is not acyclic over $\mathbb{F}_2$ has $\mathbb{C}P$-type over $\mathbb{F}_2$; (cf. Example (3.8.11).)

(3.26) ([Skjelbred, 1975]) Let X be a G-space in case (odd p) with rank$(G) = 1$ and $H^*(X;k) \cong H^*(\mathbb{H}P^n; k)$ as k-algebras. Suppose that X^G has more that $\frac{1}{2}(p+1)$ components. Show that:
(1) each component of X^G is acyclic (over k);
(2) the Steenrod operations $\mathcal{P}^1$ and $\mathcal{P}^2$ are trivial on $H^*(X;k)$; and
(3) $n < p$.

(3.27) Let (X, A) be a pair of G-spaces in case (0). Let $\mathcal{F}_n H^*_G(X, A; k)$ denote the increasing filtration on $H^*_G(X, A; k)$ coming from the Leray-Serre spectral sequence for $(X_G, A_G) \to BG$. For any $n \geq 0$ prove the inequalities of Corollary (3.1.14) for the pair (X, A) by:
(1) showing that without loss of generality $G = S^1$, and
(2) comparing the short exact sequence
$$0 \to \mathcal{F}_{n-1} H^*_G(X, A; k) \to H^*_G(X, A; k)$$
$$\to H^*_G(X, A; k)/\mathcal{F}_{n-1} H^*_G(X, A; k) \to 0$$
with the corresponding sequence for the pair (X^G, A^G), applying $- \otimes_R k_1$, and using Lemmas (3.10.11) and (A.7.4).

(3.28) Let G be a finite group and let (X, A) be a pair of G-spaces satisfying condition (LT). (In case (B) suppose that $\text{cd}_{\mathbb{F}_p}(X) < \infty$ for all $p \mid |G|$.) Suppose that G is acting freely on $X - A$ and that $H^*(X, A; \mathbb{Z})$ is finitely generated.
(1) If G is solvable, then show that $\chi(X, A) = |G| \chi(X/G, A/G)$.
(2) For any finite G show that $|G| \mid \chi(X, A)$.

(3.29) The purpose of this exercise is to give some versions of the Lefschetz Fixed Point Theorem as it applies to finite group actions (and, along the way, to improve upon Exercise (3.28)). The main point is that for maps of prime period the Lefschetz Fixed Point Theorem is a theorem of Smith Theory; and hence it applies to the more general spaces covered by the latter theory. Note that, unlike some of the references given below, we are using cohomology with closed supports instead of compact supports.

Let G be a finite group and let (X, A) be a pair of G-spaces satisfying condition (LT). (In case (B) suppose that $\text{cd}_{\mathbb{F}_p}(X) < \infty$ for all $p \mid |G|$.) Suppose that $H^*(X, A; \mathbb{Z})$ is finitely generated; and for $g \in G$ let $\Lambda(g)$ be the Lefschetz number of g defined as the alternating sum of the traces of g on $H^*(X, A; \mathbb{Z}) \otimes \mathbb{Q}$.

(1) if $G = \mathbb{Z}_p$ with generator g, then show that
$$\Lambda(g) = \chi_p(X^G, A^G) := \sum_{m=0}^{\infty} (-1)^m \dim_{\mathbb{F}_p} H^m(X^G, A^G; \mathbb{F}_p).$$
(Any rational representation of G is a direct sum of trivial representations and reduced regular representations. $H^*(X/G, A/G; \mathbb{Z})$ is finitely generated, since $H^*(X, A; \mathbb{Z})$ is finitely generated. And
$$(H^*(X, A; \mathbb{Z}) \otimes \mathbb{Q})^G \cong H^*(X, A; \mathbb{Z})^G \otimes \mathbb{Q}$$
$$\cong H_G^*(X, A; \mathbb{Z}) \otimes \mathbb{Q}$$
$$\cong H^*(X/G, A/G; \mathbb{Z}) \otimes \mathbb{Q}.$$
See too Remark (3.10.19)(3).)

(2) If G (any finite group) is acting freely on $X - A$, then show that $\chi(X, A) = |G|\chi(X/G, A/G)$. (See [tom Dieck, 1987], Chapter III, Exercise (6.17)(2) or [Brown, K.S., 1982], Chapter IX, Corollary (5.5).)

(3) If G is cyclic with generator g, and if $H^*(X^K, A^K; \mathbb{Z})$ is finitely generated for all subgroups $K \subseteq G$, then show that
$$\Lambda(g) = \chi(X^G, A^G).$$
(See [tom Dieck, 1987], Chapter III, Exercise (6.17)(3) or [Brown, K.S., 1982], Chapter IX, Exercise for Section 10, or [Petrie, Randall, 1986], Theorem 6.2(due to Verdier).)

(4) Suppose that $G = H \times K$ where $H = \mathbb{Z}_p$ and $K = \mathbb{Z}_q$ with p and q distinct primes and suppose that G is acting on $X = \mathbb{R}^n$ without fixed points. Deduce that X^H or X^K has non-finitely generated integral cohomology (whereas, of course, X^H, resp. X^K, is acyclic over $\mathbb{F}_p$, resp. $\mathbb{F}_q$).

(See the above references for further results along the lines of this exercise. See also [Ku, H.-T., Ku, M.-C., 1968].)

(3.30) Let G and (X, A) be as in Exercise (3.29). If G is a finite p-group, if $\dim_{\mathbb{F}_p} H^*(X, A; \mathbb{F}_p) < \infty$, and if n is an integer which divides the order of every orbit of G on $X - A$, then show that n divides $\chi_p(X, A)$. If G is any finite group, if $H^*(X, A; \mathbb{Z})$ is finitely generated, and if n is an integer which divides the order of every orbit of G on $X - A$, then show that n divides $\chi(X, A)$. (See [Brown, K.S., 1982], Chapter IX, Theorem (10.1).)

(3.31) Let M be a connected closed oriented topological N-manifold (or, more generally, a compact connected oriented rational cohomology N-manifold), and let G be a finite cyclic group with generator g acting on M. Let $t_j(g)$ be the trace of g on $H^j(M; \mathbb{Q})$.

Show the following:
(1) if $N = 2n+1$ and if g is orientation preserving, then $\Lambda(g) = 0$;
(2) if $N = 2n+1$ and if g is orientation reversing, then

$$\Lambda(g) = 2\sum_{j=0}^{n}(-1)^j t_j(g);$$

(3) if $N = 2n$ and if g is orientation preserving, then

$$\Lambda(g) = 2\sum_{j=0}^{n-1}(-1)^j t_j(g) + (-1)^n t_n(g);$$

(4) if $N = 2n$ and if g is orientation reversing, then $\Lambda(g) = 0$.

In particular, if $N = 3$ (and M is an integral cohomology manifold in the general case) and if $G = \mathbb{Z}_2$, generated by g, is acting in an orientation reversing way on M, then deduce that

$$\dim_{\mathbb{F}_2} H^1(M^G; \mathbb{F}_2) \leq \dim_{\mathbb{F}_2} H^1(M; \mathbb{F}_2) + t_1(g);$$

(cf. [Kobayashi, 1988].)

(3.32) ([Cusick, 1987], [Hoffman, 1987]) Let G be a finite group, and let X be a G-space satisfying condition (LT). (In case (B) assume that $\mathrm{cd}_{\mathbb{Z}}(X) < \infty$.) Suppose that $H^*(X; \mathbb{Z}) \cong H^*(\prod_{j=1}^m S^{2n_j}; \mathbb{Z})$ as rings where each $n_j > 0$. Suppose that G is acting freely on X. Show:
(1) if G acts trivially on $H^*(X; \mathbb{F}_2)$, then $G = (\mathbb{Z}_2)^r$ for some $r \leq m$; and
(2) if $n_1 = \ldots = n_m$ and if G is cyclic of order 2^r, then $2^{r-1} \leq m$.
[For (1) show that there are no elements of order 4 in

$$\ker[GL(n, \mathbb{Z}) \to GL(n, \mathbb{F}_2)].$$

For (2) show that if $A \in GL(m, \mathbb{Z})$ has order 2^r, then $2^{r-1} \leq m$.]

(3.33) Assume the conditions of Corollary (3.10.21) and that G is acting freely on $X - A$. Let $\mathcal{F}_i H_G^*(X, A; \mathbb{Z})$, $i \geq 0$ denote the increasing filtration on $H_G^*(X, A; \mathbb{Z})$ coming from the Leray-Serre spectral sequence of $(X_G, A_G) \to BG$ with integer coefficients. Show $\mathcal{F}_{n-1} H_G^n(X, A; \mathbb{Z}) = 0$.

(3.34) Suppose that $H^j(X; \mathbb{Z}) = \mathbb{Z}$ for $j = 0$, 1 and 3, $H^j(X; \mathbb{Z}) = 0$ for $j > 3$, and $H^2(X; \mathbb{Z}) = \mathbb{Z} \oplus A$, where A is a finite abelian group containing a direct summand of the form $\mathbb{Z}_{p^n} \oplus \mathbb{Z}_{p^m}$ for some prime p and $m, n > 0$. Show that S^1 cannot act freely on X.

(3.35) Give an example of a compact space on which S^1 can act almost-freely but not freely. (Seifert homology 3-spheres are examples: see [Seifert, 1932] and [Raymond, 1968].)

(3.36) (1) If G is any finite p-group, then show that Lemma (3.10.20) remains valid if one uses $\mathbb{Z}_{(p)}$, the integers localized at p, instead of $\mathbb{Z}$ throughout.

(2) Verify that Corollary (3.10.21) remains valid if one uses $\mathbb{Z}_{(p)}$ instead of $\mathbb{Z}$ throughout.

4
General results on torus and p-torus actions

In this chapter we continue in the vein of Chapter 3, giving various consequences of the Localization and Evaluation Theorems. Unless stated otherwise, we shall assume cases (0), (2) or (odd p) of Examples (3.1.1). Thus G will denote a torus or p-torus (elementary abelian p-group, p a prime number), $k = \mathbb{Q}$ or $\mathbb{F}_p$, respectively; and $R = H^*(BG; k)$ in cases (0) or (2), and R is the polynomial part of $H^*(BG; k)$ in case (odd p). (See Examples (3.1.1).)

Furthermore, as in Sections 3.9 and 3.10, we shall treat G-CW-complexes and general G-spaces simultaneously. Thus, unless stated otherwise, if X is a G-space, then we shall assume either that:

(I) X is a finite-dimensional G-CW-complex with FMCOT; or that

(II) X satisfies conditions (A), (B) or (C) of Definition (3.2.4), or conditions (A), $(B+)$ or $(C+)$ of Theorem (3.4.3) for results involving rational homotopy.

Cohomology, minimal models, pseudo-dual rational homotopy, the pseudo-dual rational homotopy co-Lie algebra, etc., will be assumed to be defined with respect to singular theory in case (I) and Alexander-Spanier theory in case (II): to unify the presentation the same notation (H^*, $\mathcal{M}$, Π_ψ^*, $\mathcal{L}^*$, respectively, for those listed above) will be used in both cases. The usual notation for Alexander-Spanier theory ($\overline{H}^*$, $\overline{\mathcal{M}}$, $\overline{\Pi}_\psi^*$, $\overline{\mathcal{L}}^*$, respectively) will be used only when emphasis is desirable.

If $\Phi : G \times X \to X$ is an action of G on X, then we shall let

$$\theta(\Phi) = \{G_x; x \in X\},$$

the set of isotropy subgroups, in cases (2) and (odd p), and we shall let $\theta(\Phi) = \{G_x^0; x \in X\}$, the set of identity components of isotropy subgroups (i.e. the set of connective isotropy subgroups) in case (0). Thus, under our blanket assumptions above, $\theta(\Phi)$ will be a finite set, except for compact G-spaces in case (0), for which no assumption on the size of $\theta(\Phi)$ is made unless stated.

4.1 Rank and Poincaré series

Recall that the rank of G, rank(G), is r if $G = T^r$ or $(\mathbb{Z}_p)^r$. If G is a torus or p-torus and $K \subseteq G$ is a subtorus or sub-p-torus, respectively, then the corank of K, cork(K), is rank(G) $-$ rank(K). From commutative algebra, rank(G) = dim(R), the Krull dimension of R, cork(K) = ht(PK), and rank(K) = coht(PK), where $PK = \ker[R \to R_K]$ is the prime ideal of R defined in Definition (3.1.4), ht denotes height and coht denotes coheight. (See Definitions (A.1.1).) Since R is a polynomial ring, coht$(\mathfrak{p})$ = dim(R) $-$ ht$(\mathfrak{p})$, for any $\mathfrak{p} \in$ Spec(R) (Theorem (A.1.6)).

If $\mathfrak{a} \triangleleft R$ is an ideal, then

$$\mathrm{ht}(\mathfrak{a}) = \min\{\mathrm{ht}(\mathfrak{p}); \mathfrak{p} \in \mathrm{Spec}(R) \text{ and } \mathfrak{p} \supseteq \mathfrak{a}\}.$$

If $\mathfrak{a} = (a_1, \ldots, a_m)$, i.e. $\mathfrak{a}$ can be generated by m elements, then ht$(\mathfrak{a}) \leq m$, by the Krull Height Theorem (Theorem (A.1.5)). If M is a R-module, then dim(M) = dim$_R(M)$ is defined to be the Krull dimension of $R/\mathrm{ann}(M)$: i.e. dim$_R(M)$ = dim$(R/\mathrm{ann}(M))$.

If M is a finitely generated R-module, then the sets Ass(M), Supp(M) and $V(\mathrm{ann}(M)) = \{\mathfrak{p} \in \mathrm{Spec}(R); \mathfrak{p} \supseteq \mathrm{ann}(M)\}$ have the same minimal elements: i.e. min Ass(M) = min Supp(M) = min $V(\mathrm{ann}(M))$ (Theorem (A.2.6)). Thus, in this case, dim$_R(M)$ = max$\{\mathrm{coht}(\mathfrak{p}); M_\mathfrak{p} \neq 0\}$. Furthermore, if $M \neq 0$, there exist submodules

$$M = M_0 \supseteq M_1 \supseteq \ldots M_n \supseteq M_{n+1} = 0$$

and $\mathfrak{p}_0, \ldots, \mathfrak{p}_n$ in Spec(R), such that $M_i/M_{i+1} \cong R/\mathfrak{p}_i$ for $0 \leq i \leq n$. Ass$(M) \subseteq \{\mathfrak{p}_0, \ldots, \mathfrak{p}_n\} \subseteq$ Supp(M), and the set $\{\mathfrak{p}_0, \ldots, \mathfrak{p}_n\}$, therefore, has the same minimal elements as Ass(M) and Supp(M) (Theorem (A.2.6)). If M is a non-zero graded R-module, then the submodules M_i, $0 \leq i \leq n$, may be chosen to be graded, the prime ideals $\mathfrak{p}_i$, $0 \leq i \leq n$, are homogeneous, and the isomorphisms $R/\mathfrak{p}_i \to M_i/M_{i+1}$, $0 \leq i \leq n$, are graded isomorphisms (of non-negative degree, if M is graded over the non-negative integers) (Theorem (A.2.11)).

Let $M = \bigoplus_{i=0}^{\infty} M^i$ be a non-negatively graded R-module, where

$$R = k[t_1, \ldots, t_r].$$

4.1. Rank and Poincaré series

If $\dim_k(M^i) < \infty$, for all $i \geq 0$, then the Poincaré series of M in the indeterminate z is $P(M,z) := \sum_{i=0}^{\infty} \dim_k(M^i) z^i$. If M is finitely generated (and non-zero), and if M_i, $\mathfrak{p}_i$, for $0 \leq i \leq n$, are as in the preceding paragraph, then

$$P(M,z) = \sum_{i=0}^{n} P(M_i/M_{i+1}, z) = \sum_{i=0}^{n} z^{d(i)} P(R/\mathfrak{p}_i, z)$$

where $d(i) \geq 0$ is the degree of the isomorphism $R/\mathfrak{p}_i \xrightarrow{\cong} M_i/M_{i+1}$. This is clear from the exact sequences

$$0 \to M_{i+1} \to M_i \to M_i/M_{i+1} \to 0.$$

Since $P(M \otimes_k N, z) = P(M,z) P(N,z)$, it follows that

$$P(R,z) = \prod_{i=1}^{r} (1 - z^{\tau_i})^{-1},$$

where $\tau_i = \deg(t_i)$. In particular, $P(R,z) = (1-z^2)^{-r}$ in cases (0) and (odd p), and $P(R,z) = (1-z)^{-r}$ in case (2), where $r = \mathrm{rank}(G)$.

(4.1.1) Lemma
With the notation above suppose that M is a finitely generated non-negatively graded R-module. Then $P(M,z)/P(R,z)$ is a polynomial in $\mathbb{Z}[z]$.

Proof.
By Hilbert's theorem on Syzygies (see Theorems (A.4.4) and (A.4.8)) there is a finite resolution,

$$0 \to P_r \to \ldots \to P_0 \to M \to 0,$$

of M by finitely generated free R-modules P_i. So

$$P(M,z) = \sum_{j=0}^{r} (-1)^j P(P_j, z).$$

(The lemma also follows more elementarily by induction: see [Quillen, 1971a] or [Matsumura, 1986], Theorem 13.2.) $\square$

(4.1.2) Definition
For a finitely generated non-negatively graded non-zero R-module M, where $R = k[t_1, \ldots, t_r]$, let $\zeta(M)$ be the multiplicity of the root $z = 1$ in the polynomial $P(M,z)/P(R,z)$. (Of course $\zeta(M) = 0$ if $z = 1$ is not a root.) Let $\rho(M) = r - \zeta(M)$. So $\rho(M)$ is the order of pole of $P(M,z)$ at $z = 1$.

Thus $P(M, z)$ has a Laurent expansion
$$P(M, z) = \sum_{m=-\infty}^{\infty} a_m(1-z)^m,$$
where $a_m = 0$ for $m < \rho(M)$.

Indeed, let $P(M, z) = h(z)P(R, z)$, where $h(z) \in \mathbb{Z}[z]$. Then
$$h(z) = (1-z)^\zeta f(z),$$
where $f(z) \in \mathbb{Z}[z]$ and $\zeta = \zeta(M)$. So, in cases (0) and (odd p), where $\deg(t_i) = 2$ for $1 \le i \le r$, we have
$$P(M, z) = \frac{f(z)}{(1-z)^\rho(1+z)^r},$$
where $\rho = \rho(M)$; and in case (2), where $\deg(t_i) = 1$ for $1 \le i \le r$,
$$P(M, z) = \frac{f(z)}{(1-z)^\rho}.$$
Since the coefficients of $P(M, z)$ are non-negative integers and $M \ne 0$, it follows that $f(1) > 0$. Hence, in the Laurent expansion, $a_{-\rho} = f(1)/2^r$ in cases (0) and (odd p), and $a_{-\rho} = f(1)$ in case (2): thus $a_{-\rho} > 0$ in all cases. Note that, in case (2), $a_m = 0$ for all $m > \deg(f) - \rho$. Also $a_m \in \mathbb{Z}$, for all m, in case (2). In cases (0) and (odd p), $a_m \in \mathbb{Q}$, for all m: indeed, $2^{m+r+\rho} a_m \in \mathbb{Z}$, for all m.

(4.1.3) Lemma
If M is a finitely generated non-negatively graded R-module, then $\rho(M) = \dim_R(M)$.

Proof.

Say that $\sum_{n=0}^{\infty} a_n z^n \le \sum_{n=0}^{\infty} b_n z^n$ if $a_n \le b_n$ for all $n \ge 0$. Then, for finitely generated non-negatively graded R-modules A and B, it is clear that $P(A, z) \le P(B, z)$ implies $\rho(A) \le \rho(B)$. Hence, writing
$$P(M, z) = \sum_{i=0}^{n} z^{d(i)} P(R/\mathfrak{p}_i, z),$$
as above, it follows that
$$\rho(M) = \max\{\rho(R/\mathfrak{p}_i); \quad \mathfrak{p}_i \text{ minimal in } \{\mathfrak{p}_0, \ldots, \mathfrak{p}_n\}\}$$
$$= \max\{\rho(R/\mathfrak{p}); \mathfrak{p} \in \min V(\operatorname{ann}(M))\}.$$
Thus it remains to show that $\rho(R/\mathfrak{p}) = \dim(R/\mathfrak{p})$. This follows from [Matsumura, 1986], Theorem 13.8.(ii). □

(4.1.4) Remarks
(1) Typically we shall apply Lemma (4.1.3) for $M = H_G^*(X; k)$. But then $\min V(\operatorname{ann}(M)) = \min \operatorname{Supp}(M)$ consists of ideals of the form PK

4.1. Rank and Poincaré series

where $K \subseteq G$ is a subtorus or sub-p-torus as the case may be. And $\rho(R/PK) = \rho(R_K) = \mathrm{rank}(K) = \mathrm{coht}(PK) = \dim(R/PK)$, clearly.

(2) Instead of Theorem 13.8 of [Matsumura, 1986] one could use the following graded version of Noether's Normalization Theorem: if A is a finitely generated non-negatively graded k-algebra of Krull dimension n, then there exist algebraically independent homogeneous $x_1, \ldots, x_n$ in A, such that A is an integral extension of $k[x_1, \ldots, x_n]$. In fact [Zariski, Samuel, 1960], Chapter VII, Theorem 25, is sufficient to prove the lemma.

Now we shall return to group actions.

(4.1.5) Definition
Let $\Phi : G \times X \to X$ be an action of a torus or p-torus on a space X. The rank of Φ, $\mathrm{rank}(\Phi)$, is defined by

$$\mathrm{rank}(\Phi) = \min\{\mathrm{rank}(G/K); K \in \theta(\Phi)\}$$
$$= \min\{\mathrm{cork}(K); K \in \theta(\Phi)\}$$
$$= \min\{\mathrm{ht}(PK); K \in \theta(\Phi)\}.$$

By the Localization Theorem, then,

$$\mathrm{rank}(\Phi) = \min\{\mathrm{ht}(\mathfrak{p}); \mathfrak{p} \in \mathrm{Supp}(H_G^*(X;k))\}.$$

If $\dim_k H^*(X;k) < \infty$, then it follows from the above that

$$\mathrm{rank}(\Phi) = \min\{\mathrm{ht}(\mathfrak{p}); \mathfrak{p} \in \mathrm{Ass}(H_G^*(X;k))\}$$
$$= \min\{\mathrm{ht}(\mathfrak{p}); \mathfrak{p} \in V(\mathrm{ann}(H_G^*(X;k)))\}$$
$$= r - \rho(H_G^*(X;k)) = \zeta(H_G^*(X;k))$$
$$= r - \dim_R H_G^*(X;k),$$

where $r = \mathrm{rank}(G)$.

(4.1.6) Remarks
(1) $\mathrm{rank}(\Phi) = 0$ if and only if $X^G \neq \emptyset$.

$$\mathrm{rank}(\Phi) = r = \mathrm{rank}(G)$$

if and only if G is acting freely in cases (2) or (odd p), or G is acting almost-freely in case (0).

(2) From Proposition (3.10.1) and Exercise (3.8) $H_G^*(X;k)$ is a finitely generated R-module if and only if $\dim_k H^*(X;k) < \infty$.

(4.1.7) Proposition
Given an action $\Phi : G \times X \to X$ with

$$\dim_k H^*(X;k) < \infty,$$

then

(1) $\operatorname{ann}(H_G^*(X;k)) = (a_1,\ldots,a_h) \triangleleft R \Rightarrow \operatorname{rank}(\Phi) \leq h$, and
(2) $\operatorname{rank}(\Phi) = r = \operatorname{rank}(G) \Leftrightarrow \dim_k H_G^*(X;k) < \infty$.

Proof.
(1) $\operatorname{ann}(H_G^*(X;k)) = (a_1,\ldots,a_h)$ and $\mathfrak{p} \in \min V(\operatorname{ann}(H_G^*(X;k)))$ implies $\operatorname{ht}(\mathfrak{p}) \leq h$, by the Krull Height Theorem (Theorem (A.1.5)).

(2) Let $\mathfrak{m}$ be the unique homogeneous maximal ideal of R. Then $\operatorname{rank}(\Phi) = r \Leftrightarrow V(\operatorname{ann}(H_G^*(X;k))) = \{\mathfrak{m}\}$. If $R = k[t_1,\ldots,t_r]$, then

$$V(\operatorname{ann}(H_G^*(X;k))) = \{\Leftrightarrow\} = \sqrt{\operatorname{ann}(H_G^*(X;k))} \Leftrightarrow$$

for some positive integer n, $t_i^n \in \operatorname{ann}(H_G^*(X;k))$ for $1 \leq i \leq r$. Since $H_G^*(X;k)$ is a finitely generated R-module the latter can occur if and only if

$$\dim_k H_G^*(X;k) < \infty.$$

$\square$

(4.1.8) Proposition
Let X be a G-space, and suppose that $\dim_k H^*(X;k) < \infty$. Make k into a R-module by the map $R = k[t_1,\ldots,t_r] \to k$, $t_i \mapsto 0$, $1 \leq i \leq r = \operatorname{rank}(G)$. Then

$$\dim_k H^*(X^G;k) = \begin{cases} \sum_{j=0}^r (-1)^j \dim_k \operatorname{Tor}_j^R(H_G^*(X;k),k) \\ \quad \text{in cases (0) or (2)} \\ 2^{-r} \sum_{j=0}^r (-1)^j \dim_k \operatorname{Tor}_j^R(H_G^*(X;k),k) \\ \quad \text{in case (odd p).} \end{cases}$$

Proof.
Let $H_G^*(X;k) = M$, and let

$$0 \to P_r \to \ldots \to P_0 \to M \to 0$$

be a resolution of M by finitely generated free R-modules, P_j. Let $K = R_{(0)}$, R localized at $(0) \triangleleft R$. From the exactness of localization,

$$\dim_K M_{(0)} = \sum_{j=0}^r (-1)^j \dim_K P_{j(0)} = \sum_{j=0}^r (-1)^j \tau_j,$$

where τ_j is the rank of P_j as a free R-module.

From the Localization Theorem,

$$\dim_K M_{(0)} = \begin{cases} \dim_k H^*(X^G;k) \\ \quad \text{in cases (0) or (2)} \\ 2^r \dim_k H^*(X^G;k) \\ \quad \text{in case (odd p).} \end{cases}$$

From the complex
$$0 \to P_r \otimes_R k \to \ldots \to P_0 \otimes_R k \to 0,$$
$$\sum_{j=0}^{r}(-1)^j \dim_k \operatorname{Tor}_j^R(M,k) = \sum_{j=0}^{r}(-1)^j \tau_j.$$
□

(4.1.9) Remarks
(1) In the proof of Proposition (4.1.8), suppose that $\{\xi_{j1},\ldots,\xi_{j\tau_j}\}$ is a basis for P_j, with $\deg(\xi_{ji}) = d_{ji} \geq 0$. Then, as before,
$$P(M,z)/P(R,z) = \sum_{j=0}^{r}(-1)^j \sum_{i=1}^{\tau_j} z^{d_{ji}}$$
$$= (1-z)^\zeta f(z).$$
Hence, if $X^G \neq \emptyset$, and so $\zeta = 0$, $f(1) = \sum_{j=0}^{r}(-1)^j \tau_j$.
Thus, in the Laurent series for $H_G^*(X;k)$,
$$a_{-r} = \begin{cases} 2^{-r} \dim_k H^*(X^G; k) \\ \quad \text{in case (0)} \\ \\ \dim_k H^*(X^G; k) \\ \quad \text{in cases (2) or (odd p).} \end{cases}$$

(2) $\sum_{j=0}^{r}(-1)^j \dim_k \operatorname{Tor}_j^R(H_G^*(X;k),k) = \sum_{j=0}^{r}(-1)^j \tau_j$, when k is made into a R-module by means of any evaluation $t_i \mapsto \alpha_i \in k$, $1 \leq i \leq r$.
(3) [Skjelbred, 1972]. By induction, one can choose the resolution
$$0 \to P_r \to \ldots \to P_j \xrightarrow{d_j} P_{j-1} \to \ldots \to P_0 \xrightarrow{\varepsilon} M \to 0$$
to be minimal: i.e. $\varepsilon(P_0) \subseteq \mathfrak{m}M$ and $d_j(P_j) \subseteq \mathfrak{m}P_{j-1}$, for $1 \leq j \leq r$, where $\mathfrak{m} = (t_1,\ldots,t_r)$ is the maximal homogeneous ideal in R. (See Theorem (A.4.12).) Then $\tau_j = \operatorname{rank}(P_j) = \dim_k \operatorname{Tor}_j^R(M,k)$, where k is a R-module via $t_i \mapsto 0$, $1 \leq i \leq r$. With the notation above, let τ_j^i be the number of $d_{js} = 2j+i$, resp. $j+i$, in cases (0) or (odd p), resp. case (2). Note that $d_{js} \geq 2j$, resp. j, by minimality. So
$$P(M,z)/P(R,z) = \begin{cases} \sum_{n=0}^{\infty}\left(\sum_{2j+i=n}(-1)^j \tau_j^i\right) z^n \\ \quad \text{in cases (0) or (odd p)} \\ \\ \sum_{n=0}^{\infty}\left(\sum_{j+i=n}(-1)^j \tau_j^i\right) z^n \\ \quad \text{in case (2)} \end{cases}$$

Since $(1-z^2)^r = \sum_{j=0}^{r}(-1)^j\binom{r}{j}z^{2j}$ and $(1-z)^r = \sum_{j=0}^{r}(-1)^j\binom{r}{j}z^j$,

$$P(M,z)/P(R,z) = \begin{cases} \sum_{n=0}^{\infty}\left(\sum_{2j+i=n}(-1)^j\beta_i(M)\binom{r}{j}\right)z^n \\ \quad\text{in cases (0) or (odd }p) \\ \sum_{n=0}^{\infty}\left(\sum_{j+i=n}(-1)^j\beta_i(M)\binom{r}{j}\right)z^n \\ \quad\text{in case (2),} \end{cases}$$

where $\beta_i(M) = \dim_k(M^i) = \dim_k(H^i_G(X;k))$.

Furthermore we can compute $\operatorname{Tor}^R_j(M,k)$ from the Koszul resolution

$$0 \to R\otimes\Lambda^r \to \ldots \to R\otimes\Lambda^j \xrightarrow{\delta_j} R\otimes\Lambda^{j-1} \to \ldots \to R\otimes\Lambda^0 \to k \to 0$$

where $\Lambda^j = \Lambda^j(s_1,\ldots,s_r)$ is the jth exterior power of the k-vector space with basis $\{s_1,\ldots,s_r\}$, δ_j is the derivation with $\delta_j(s_i) = t_i$ for $1 \le i \le r$, and $\deg(s_i) = 1$, resp. 0, in cases (0) or (odd p), resp. case (2). Consider the complex

$$\ldots \to M^{i-2}\otimes\Lambda^{j+1} \to M^i\otimes\Lambda^j \to M^{i+2}\otimes\Lambda^{j-1} \to \ldots \to M^{i+2j}\otimes\Lambda^0 \to 0,$$

in cases (0) or (odd p); and consider

$$\ldots \to M^{i-1}\otimes\Lambda^{j+1} \to M^i\otimes\Lambda^j \to M^{i+1}\otimes\Lambda^{j-1} \to \ldots \to M^{i+j}\otimes\Lambda^0 \to 0$$

in case (2). Let σ^i_j be the dimension of the homology subquotient of $M^i\otimes\Lambda^j$. Then, in cases (0) or (odd p), respectively case (2),

$$\sum_{2j+i=n}(-1)^j\beta_i(M)\binom{r}{j} = \sum_{2j+i=n}(-1)^j\sigma^i_j,$$

resp.

$$\sum_{j+i=n}(-1)^j\beta_i(M)\binom{r}{j} = \sum_{j+i=n}(-1)^j\sigma^i_j,$$

since Euler characteristic commutes with homology.

Thus, letting $P(M,z)/P(R,z) = h(z) \in \mathbb{Z}[z]$, as before, and letting $h(z) = \sum_{n=0}^{\infty}c_n z^n$, we have

$$c_n = \sum_{2j+i=n}(-1)^j\sigma^i_j = \sum_{2j+i=n}(-1)^j\tau^i_j = \sum_{2j+i=n}(-1)^j\beta_i(M)\binom{r}{j},$$

in cases (0) and (odd p); and

$$c_n = \sum_{j+i=n}(-1)^j\sigma^i_j = \sum_{j+i=n}(-1)^j\tau^i_j = \sum_{j+i=n}(-1)^j\beta_i(M)\binom{r}{j}$$

in case (2).

(4.1.10) Definition

Let X be a G-space. Let

$$X_i = \{x \in X; \operatorname{rank}(G_x) \ge i\}.$$

Thus
$$X_i = \bigcup \{X^H; H \subseteq G \text{ and } \text{rank}(H) = i\}.$$
We have, then, a filtration of X
$$X = X_0 \supseteq X_1 \supseteq \ldots \supseteq X_i \supseteq X_{i+1} \supseteq \ldots \supseteq X_r = X^G \supseteq X_{r+1} = \emptyset,$$
where $r = \text{rank}(G)$ and $\text{rank}(\Phi) = r - \max\{i; X_i \neq \emptyset\}$, where $\Phi : G \times X \to X$ is the action. Clearly each X_i is an invariant subspace of X; and, indeed, each X_i is a closed invariant subspace of X, by the following general consequence of the existence of tubes and slices (see [Bredon, 1972], Chapter II, Corollary 5.5). (If X is a G-CW-complex, then ,clearly, each X_i is a G-CW-subcomplex.)

(4.1.11) Proposition
If G is any compact Lie group and X is any completely regular (e.g. paracompact) G-space, then, for any $x \in X$, there is an invariant neighbourhood U of x in X, such that, for any $y \in U$, G_y is conjugate to a subgroup of G_x.

(4.1.12) Lemma
If H and K are distinct subgroups of G of rank i, then $X^H \cap X^K \subseteq X_{i+1}$, and $X^H - X_{i+1}$ is open in $X_i - X_{i+1}$.

Proof.
The first statement is obvious: the second follows from Proposition (4.1.11). □

(4.1.13) Corollary
$X_i - X_{i+1} = \bigcup\{X^H - X_{i+1}; H \in \theta(\Phi) \text{ and } \text{rank}(H) = i\}$ *expresses $X_i - X_{i+1}$ as a disjoint union of open subspaces.*

(4.1.14) Proposition
If $\dim_k H^*(X;k) < \infty$, then
$$\dim_k H^*(X_i;k) < \infty,$$
for all $i \geq 0$.

Proof.
First assume that $\theta(\Phi)$ is finite.
$$X_i = \bigcup\{X^H; H \in \theta(\Phi) \text{ and } \text{rank}(H) \geq i\},$$
and $\dim_k H^*(X^H;k) < \infty$, by Corollary (3.10.2). The result now follows

by induction (on the number of $H \in \theta(\Phi)$ with $\text{rank}(H) \geq i$) and the Mayer-Vietoris sequence.

$\theta(\Phi)$ may be infinite only for certain compact G-spaces in case (0). In this situation we are, of course, using Alexander-Spanier cohomology (with coefficients in $k = \mathbb{Q}$), and this is equivalent to sheaf cohomology. Let $\tilde{H}^*$ denote sheaf cohomology, and let $\tilde{H}_c^*$ denote sheaf cohomology with compact supports. Then $H^*(X_i, X_{i+1}; k) \cong \tilde{H}_c^*(X_i, X_{i+1}; k)$, since X_i is compact; $\tilde{H}_c^*(X_i, X_{i+1}; k) \cong \tilde{H}_c^*(X_i - X_{i+1}; k)$, by [Bredon, 1967a], Chapter II, Theorem 12.5, since X_{i+1} is closed; $\tilde{H}_c^*(X_i - X_{i+1}; k) \cong \oplus_K \tilde{H}_c^*(X^K - X_{i+1}; k)$, where the direct sum ranges over all $K \in \theta(\Phi)$ with $\text{rank}(K) = i$, by Corollary (4.1.13), and the definition of sheaf cohomology with compact supports;

$$\bigoplus_K \tilde{H}_c^*(X^K - X_{i+1}; k) \cong \bigoplus_K \tilde{H}_c^*(X^K, X_{i+1}^K; k) \cong \bigoplus_K H^*(X^K; X_{i+1}^K; k)$$

by the previous reasoning in reverse. Since

$$\dim_k H^*(X_r; k) = \dim_k H^*(X^G; k) < \infty,$$

where $r = \text{rank}(G)$, by Corollary (3.10.2), we can do downward induction on i. Hence, assuming the result for X_{i+1}, the result will follow for X_i, if we can show that $H^*(X^K, X_{i+1}^K; k) \neq 0$ for only a finite number of K. But

$$H^*(X^K, X_{i+1}^K; k) \neq 0 \Leftrightarrow H_G^*(X^K, X_{i+1}^K; k) \neq 0,$$

clearly, by the Leray-Serre spectral sequence.

$$H_G^*(X^K, X_{i+1}^K; k) \neq 0 \Leftrightarrow H_G^*(X^K, X_{i+1}^K; k)_{PK} \neq 0,$$

by Example (3.7.16)(1) and the Lemma (3.7.5). $H_G^*(X^K, X_{i+1}^K; k)_{PK} \cong H_G^*(X, X_{i+1}; k)_{PK}$, by the Localization Theorem. Hence

$$H^*(X^K, X_{i+1}^K; k) \neq 0 \Leftrightarrow PK \in \text{Supp}(H_G^*(X, X_{i+1}; k)).$$

Since $\text{rank}(K) = i$,

$$PK \in \text{Supp}(H_G^*(X, X_{i+1}; k)) \Leftrightarrow PK \in \min \text{Supp}(H_G^*(X, X_{i+1}; k)).$$

By inductive assumption, $\dim_k H^*(X, X_{i+1}; k) < \infty$, so $H_G^*(X, X_{i+1}; k)$ is a finitely generated R-module. Hence

$$\min \text{Supp}(H_G^*(X, X_{i+1}; k))$$

is finite. $\square$

(4.1.15) Proposition
Let X be a G-space with $\dim_k H^*(X; k) < \infty$. Let $M = H_G^*(X; k)$ and $M_i = H_G^*(X_i; k)$, for $0 \leq i \leq r = \text{rank}(G)$. Let $P(M, z) = \sum_{m=-\infty}^{\infty} a_m (1-z)^m$ and $P(M_i, z) = \sum_{m=-\infty}^{\infty} b_{im}(1-z)^m$ be the Laurent expansions as above. Then $b_{im} = a_m$, for all $m \leq -i$.

Proof.
Let $\psi_i^* : M \to M_i$ be induced by the inclusion $\psi_i : X_i \to X$. Let $K_i = \ker(\psi_i^*)$ and $C_i = \mathrm{coker}(\psi_i^*)$. Then, from the exact sequence
$$0 \to K_i \to M \to M_i \to C_i \to 0,$$
we have
$$P(M, z) = P(M_i, z) + P(K_i, z) - P(C_i, z).$$
Now C_i is a submodule of, and K_i is a quotient of, $H_G^*(X, X_i; k)$. So both $\rho(C_i)$ and $\rho(K_i)$ are less than or equal to
$$\begin{aligned}\rho(H_G^*(X, X_i; k))\} &= \dim_R(H_G^*(X, X_i; k)) \\ &= \max\{\mathrm{rank}(H); PH \in \mathrm{Supp}(H_G^*(X, X_i; k))\} \\ &\leq i - 1,\end{aligned}$$
by the Localization Theorem. The result follows. $\square$

The following theorem (due to W.-Y. Hsiang) generalizes Remarks (4.1.9) (1).

(4.1.16) Theorem
Let X be a G-space with $\dim_k H^*(X; k) < \infty$. Let $r = \mathrm{rank}(G)$, and let $s = \max\{\mathrm{rank}(K); K \in \theta(\Phi)\} \leq r$. Let $K_1, \ldots, K_\sigma$ be the members of $\theta(\Phi)$ of rank s. (Note that the number of maximal connective isotropy subgroups is always finite in cases (A) and (0), even without assuming $\dim_k H^*(X; k) < \infty$; see Remark (3.2.9) or Exercise (3.2).)
Let $P(H_G^*(X; k), z) = \sum_{m=-\infty}^{\infty} a_m (1 - z)^m$ be the Laurent expansion. Then $a_m = 0$ for $m < -s$, and
$$a_{-s} = \begin{cases} 2^{-s} \sum_{j=1}^\sigma \dim_k H^*(X^{K_j}/G; k) & \text{in case (0)} \\ \sum_{j=1}^\sigma \dim_k H^*(X^{K_j}/G; k) & \text{in cases (2) or (odd p)} \end{cases}$$

Proof.
$\mathrm{rank}(\Phi) = r - s$; and so $a_m = 0$ for $m < -s$, as above. Let
$$P(H_G^*(X_s; k), z) = \sum_{m=-\infty}^{\infty} b_m (1 - z)^m.$$
Then $a_{-s} = b_{-s}$, by Proposition (4.1.15). Since $X_{s+1} = \emptyset$, $X^{K_1}, \ldots, X^{K_\sigma}$ are pairwise disjoint. Hence
$$H_G^*(X_s; k) = \bigoplus_{j=1}^\sigma H_G^*(X^{K_j}; k) = \bigoplus_{j=1}^\sigma H^*(BK_j; k) \otimes H_{G/K_j}^*(X^{K_j}; k).$$

Now
$$P(H^*(BK_j;k),z) = \begin{cases} (1-z^2)^{-s} & \text{in case (0)} \\ (1-z)^{-s} & \text{in case (2)} \\ (1+z)^s(1-z^2)^{-s} = (1-z)^{-s} & \text{in case (odd } p). \end{cases}$$

Also
$$H^*_{G/K_j}(X^{K_j};k) \cong H^*(X^{K_j}/(G/K_j);k), \text{ by Proposition (3.10.9)},$$
$$= H^*(X^{K_j}/G;k).$$

Let $f(z) = \sum_{j=1}^{\sigma} P(H^*(X^{K_j}/G;k), z)$. Then $f(z) \in \mathbb{Z}[z]$, by Proposition (4.1.7)(2). Thus

$$P(H^*_G(X_s;k),z) = \begin{cases} (1-z^2)^{-s}f(z) & \text{in case (0)} \\ (1-z)^{-s}f(z) & \text{in cases (2) or (odd } p). \end{cases}$$

Hence $b_{-s} = 2^{-s}f(1)$ in case (0), respectively $f(1)$ in cases (2) or (odd p). □

(4.1.17) Remark
Theorem (4.1.16) studies X_s where
$$s = \max\{i; X_i \neq \emptyset\}.$$
At the other extreme, one could study $X_t - X_{t+1}$, where
$$t = \min\{i; X_i - X_{i+1} \neq \emptyset\}.$$
This is the study of principal orbit types: see [Bredon, 1972], Chapter IV, Section 3, and Montgomery in [Borel et al., 1960], Chapter IX.

4.2 Generalities on torus actions

In this section we consider torus actions only, i.e. case (0). Thus $G = T^r$, $k = \mathbb{Q}$ and $R = k[t_1, \ldots, t_r]$. Recall the blanket assumptions made in the introduction to this chapter.

We begin with the following useful lemma.

(4.2.1) Lemma
Let X be a G-space.
(1) If X has FMCOT, then there is a subcircle $L = S^1 \subseteq G$ such that $X^L = X^G$.
(2) If $\dim_k H^*(X;k) < \infty$, then there is a subcircle $L = S^1 \subseteq G$ such that the inclusion $j_L : X^G \to X^L$ induces an isomorphism $j_L^* : H^*(X^L;k) \xrightarrow{\cong} H^*(X^G;k)$.

4.2. Generalities on torus actions

Proof.

(1) Let $K_1, \ldots, K_n$ be the connective isotropy subgroups other than G itself. Then there is a subcircle $L \subseteq G$ such that $L \cap K_i$ is finite for $1 \leq i \leq n$. Clearly $X^L = X^G$.

(2) If $\dim_k H^*(X; k) < \infty$, then $\dim_k H^*(X^G; k) < \infty$, by Corollary (3.10.2). Hence $\dim_k H^*(X, X^G; k) < \infty$: and so $H_G^*(X, X^G; k)$ is a finitely generated R-module. Let
$$\{PK_1, \ldots, PK_n\} = \min \operatorname{Supp}(H_G^*(X, X^G; k));$$
and, as before, choose a subcircle $L \subseteq G$ such that $L \cap K_i$ is finite for $1 \leq i \leq n$. So $PL \notin \operatorname{Supp}(H_G^*(X, X^G; k))$: i.e.
$$H_G^*(X, X^G; k)_{PL} = H_G^*(X^L, X^G; k)_{PL} = 0.$$
Hence $H_G^*(X^L, X^G; k) = 0$; and so $H^*(X^L, X^G; k) = 0$. (See the proof of Proposition (4.1.14).) □

Now consider the principal G-bundle $G \to EG \times X \to X_G$, which is the pull-back of $G \to EG \to BG$ along $p \colon X_G \to BG$. We have a Leray-Serre spectral sequence
$$E_2^{p,q} \cong H_G^p(X; k) \otimes H^q(G; k) \text{ converging to } H^*(X; k).$$

The differential d_2 is that of the Koszul complex described in Remarks (4.1.9)(3). Thus $E_3^{*,q} \cong \operatorname{Tor}_q^R(H_G^*(X; k), k)$, where, as usual, unless otherwise stated, k is viewed as a R-module via the evaluation $t_i \mapsto 0$, $1 \leq i \leq r$. Indeed, with the notation of Remarks (4.1.9)(3), $\dim_k(E_3^{p,q}) = \sigma_q^p$.

(4.2.2) Theorem
Let X be a G-space and suppose that
$$\dim_k H^*(X; k) < \infty.$$
Then
$$\dim_k H^*(X; k) - \dim_k H^*(X^G; k) \leq 2 \sum_{j=0}^{\left[\frac{r-1}{2}\right]} \dim_k \operatorname{Tor}_{2j+1}^R(H_G^*(X; k), k).$$
Furthermore, if $G = S^1$, then
$$\dim_k H^*(X; k) - \dim_k H^*(X^G; k) = 2 \dim_k \operatorname{Tor}_1^R(H_G^*(X; k), k).$$

Proof.
By Proposition (4.1.8), letting $H_G^*(X; k) = M$,
$$\dim_k H^*(X^G; k) = \sum_{j=0}^{r} (-1)^j \dim_k \operatorname{Tor}_j^R(M, k).$$

266   4. General results on torus and p-torus actions

From the Leray-Serre spectral sequence $E_r^{p,q}$ for $G \to EG \times X \to X_G$, described above,

$$\dim_k H^*(X;k) = \dim_k E_\infty \leq \dim_k E_3 = \sum_{j=0}^{r} \dim_k \operatorname{Tor}_j^R(M,k).$$

If $G = S^1$, then $E_3 = E_\infty$. □

(4.2.3) Corollary
If $\dim_k H^*(X;k) < \infty$, then $H_G^*(X;k)$ is a free R-module if and only if X is TNHZ in $X_G \to BG$.

There is an analogue of Theorem (4.2.2) in rational homotopy.

(4.2.4) Theorem
Let X be a connected G-space and suppose that $\dim_k \Pi_\psi^*(X) < \infty$. Let F be a component of X^G. Then

$$\dim_k \Pi_\psi^*(X) - \dim_k \Pi_\psi^*(F) \leq 2 \sum_{j=0}^{[\frac{r-1}{2}]} \dim_k \operatorname{Tor}_{2j+1}^R(\Pi_{(G,F)}^*(X),k).$$

Furthermore, if $G = S^1$, then

$$\dim_k \Pi_\psi^*(X) - \dim_k \Pi_\psi^*(F) = 2 \dim_k \operatorname{Tor}_1^R(\Pi_{(G,F)}^*(X),k).$$

Proof.

$$\dim_k \Pi_\psi^*(F) = \sum_{j=0}^{r}(-1)^j \dim_k \operatorname{Tor}_j^R(\Pi_{(G,F)}^*(X),k)$$

by resolving $\Pi_{(G,F)}^*(X)$ just as for $H_G^*(X;k)$ in Proposition (4.1.8).

$$\dim_k \Pi_\psi^*(X) \leq \sum_{j=0}^{r} \dim_k \operatorname{Tor}_j^R(\Pi_{(G,F)}^*(X),k)$$

by considering the spectral sequences of the bicomplex $C^{*,*}$, which was needed in the proof of Proposition (3.3.15). To compare the notation used here with that used in the proof of Proposition (3.3.15), note that $\operatorname{Tor}_j^R(M,k) = \operatorname{Tor}_R^{j,*}(M,k)$, where $M = \Pi_{(G,F)}^*(X)$. (In the spectral sequence for $G \to EG \times X \to X_G$, above, $E_3^{p,q} = \operatorname{Tor}_R^{q,p}(H_G^*(X;k),k)$.) □

(4.2.5) Remark
According to the blanket assumptions of the introduction to this chapter, in Theorem (4.2.4), for general G-spaces in cases (B) or (C), we assume that F satisfies condition (CS) of Definition (3.4.1): i.e. we assume

$(B+)$ or $(C+)$ of Theorem (3.4.3). Condition (CS), however, is fulfilled automatically if $\dim_k \overline{\Pi}_\psi^*(X) < \infty$; for then $\dim_k \overline{H}^*(X;k) < \infty$, if X belongs to case (A), (B) or (C). In case (B) this follows immediately from the fact that $\dim_k \overline{\Pi}_\psi^*(X) < \infty$ implies that $\overline{H}^*(X;k)$ has finite type. In cases (A) and (C), we have Corollary (2.6.10). Similar to case (B), it follows immediately that $\dim_k H^*(X;k) < \infty$, if X is a connected finite-dimensional CW-complex with $\dim_k \Pi_\psi^*(X) < \infty$. So under the conditions of Theorem (4.2.4), X^G always has a finite number of components.

The following simple lemma will be useful in the proofs of Theorems (4.2.7) and (4.2.8).

(4.2.6) Lemma
Let X be a G-space and suppose that $X^G \neq \emptyset$. For any $x \in X^G$, let $\sigma(x) : BG \to X_G$ be the section defined by
$$G(z) \mapsto G(z,x),$$
for all $z \in EG$. If X^G is not connected, then there exist $x_1, x_2 \in X^G$ such that $\sigma(x_1)$ and $\sigma(x_2)$ induce different maps $H_G^*(X;k) \to R$.

Proof.
Let $X^G = V \cup W$ where V and W are open (in X^G) and non-empty, and $V \cap W = \emptyset$. Choose $x_1 \in V$ and $x_2 \in W$. Clearly one gets different induced maps
$$H_G^*(X;k) \to H_G^*(X^G;k) \cong H_G^*(V;k) \oplus H_G^*(W;k) \to R.$$
(Note that this is valid also in cases (2) and (odd p).) □

We shall state the next two theorems first for G-CW-complexes; and then we shall list the additional assumptions, which are needed for general G-spaces, in Remark (4.2.9).

(4.2.7) Theorem
Let X be a connected finite-dimensional G-CW-complex with FMCOT. Suppose that $\Pi_\psi^{2m}(X) = 0$, for all $m \geq 1$. Then X^G is connected.

Proof.
There is at most one possible map $R \otimes \mathcal{M}(X) \to R$. Hence there is at most one possible induced map $H_G^*(X;k) \to R$. So the result follows from Lemma (4.2.6). □

The next theorem is a version of Theorem (4.2.7), which uses only the ordinary rational homotopy groups $\pi_{2m}(X) \otimes \mathbb{Q}$. It is due mainly to Bredon ([Bredon, 1969]) and Skjelbred ([Skjelbred, 1972]). Our proof is modelled on the latter.

(4.2.8) Theorem
Let X be a connected finite-dimensional G-CW-complex with FMCOT. Suppose either that
(1) $\pi_{2m}(X) \otimes \mathbb{Q} = 0$, for all $m \geq 1$; or that
(2) there is a non-negative integer N such that $H^{2j}(X;k) = 0$, for all $j > N$, and $\pi_{2m}(X) \otimes \mathbb{Q} = 0$, for all m such that $1 \leq m \leq N$.
Then X^G is connected.

Proof.
We shall indicate briefly the main steps in the proof (and phrase things so as to apply to general G-spaces too).

Step 1. Suppose that X is simply-connected. Let x_1, x_2 be any two points in X^G. Then we get sections $\sigma(x_1), \sigma(x_2): BG \to X_G$, given by $G(z) \mapsto G(z, x_i)$, for $i = 1, 2$. Suppose $\pi_{2n}(X) \otimes \mathbb{Q} = 0$ for all $n \leq N$; and let B_N be the $2N$-skeleton of BG. Consider the diagram

$$B_N \xrightarrow{i_N} BG \underset{\sigma(x_2)}{\overset{\sigma(x_1)}{\rightrightarrows}} X_G \xrightarrow{\lambda} \mathbb{Q}_\infty(X_G)$$

where i_N is the inclusion, and $(\mathbb{Q}_\infty(X_G), \lambda)$ is the rational localization of X_G in the sense of [Bousfield, Kan, 1972]. By classical obstruction theory (see, e.g. [Hu, 1959]),

$$\lambda \sigma(x_1) i_N \simeq \lambda \sigma(x_2) i_N.$$

Hence

$$\sigma(x_1)^* = \sigma(x_2)^* : H_G^j(X;k) \to H^j(BG;k),$$

for all $j \leq 2N$. (Note that X_G is simply-connected, if X is, by [Bredon, 1972], Chapter II, Corollary 6.3.)

Step 2. If X is not simply-connected, but $\pi_{2n}(X) \otimes \mathbb{Q} = 0$, for all $n \leq N$, and if F_1, F_2 are components of X^G, then there are $x_i \in F_i$, $i = 1, 2$, such that

$$\sigma(x_1)^* = \sigma(x_2)^* : H_G^j(X;k) \to H^j(BG;k),$$

for all $j \leq 2N$. This follows from Step 1 by taking the universal covering space of X, and lifting the action according to [Bredon, 1972], Chapter I, Theorem 9.1. If X is a general G-space, which is nasty enough not to have a universal covering space, then we take a G-CW-approximation to X.

Step 3. Under assumption (1) the theorem now follows, by Lemma (4.2.6). Under assumption (2), we may assume, without loss of generality, that $G = S^1$, thanks to Lemma (4.2.1). The theorem now follows by Lemma (4.2.6), and Lemma (3.10.11) applied to the Leray-Serre spectral sequence of $X \to X_G \to BG$. □

(4.2.9) Remark
Theorem (4.2.7) is valid as it stands for general G-spaces in cases (A), (B) or (C). (Of course, Π^*_ψ is replaced by $\overline{\Pi}^*_\psi$.)

Theorem (4.2.8)(1) is valid as it stands for path-connected general G-spaces in cases (A), (B) or (C). The assumption of path-connectedness appears to be necessary here, whereas mere connectedness suffices for Theorem (4.2.7). The conclusion, however, is still that X^G is connected.

Theorem (4.2.8)(2) is valid as it stands for path-connected G-spaces in cases (B) or (C). Owing to the use of Lemma (4.2.1), however, if X is a path-connected compact G-space without FMCOT, then one must assume either that $\dim_k H^*(X;k) < \infty$, or that $G = S^1$.

4.3 Almost-free torus actions and the rank of a space

In this section again we consider only torus actions: i.e. case (0). The blanket assumptions of this chapter apply. Thus a G-space is assumed to be either a finite-dimensional G-CW-complex with FMCOT or a general G-space satisfying conditions (A), (B) or (C) of Definition (3.2.4). In the first definition below, however, which is due to W.-Y. Hsiang, we are interested in all possible reasonable torus actions on a space X. In the CW case this might mean all actions of a torus G on X so that X is a finite-dimensional G-CW-complex with FMCOT: but this is unnecessarily restrictive for the purpose of this section. In Remark (4.3.2) below we shall indicate a natural geometric setting for the results here.

(4.3.1) Definition
Let X be a space. Then the rank, or torus rank, of X, denoted $\mathrm{rank}_0(X)$, is defined by $\mathrm{rank}_0(X) = \sup \mathrm{rank}(\Phi)$, where Φ ranges over all continuous torus actions on X with FMCOT if X is not compact. (See Definition (4.1.5) for $\mathrm{rank}(\Phi)$.)

(4.3.2) Remark
Let X be a (paracompact) separable topological manifold. Let G be any compact Lie group acting on X. So G has FMOT on any compact

subset of X ([Bredon, 1972] or [Borel et al., 1960], Chapter VIII). Thus from [Oliver, 1976] (or [Conner, 1960] when G is a torus) it follows that X/G is locally contractible. Thus, as can be seen below, the methods of this section work using singular theory (as opposed to Alexander-Spanier theory) whenever X is a separable topological manifold. In fact the next lemma shows that we can confine our attention to almost-free torus actions, which necessarily have FMOT on any compact space (see Exercise (3.2)). And if G is a torus acting almost-freely on a locally compact locally contractible finite-dimensional separable metric space X, then X/G is locally contractible ([Conner, 1960]). Such metric spaces are then a more general class for which singular theory suffices.

(4.3.3) Lemma
If $\operatorname{rank}_0(X)$ is finite, then $\operatorname{rank}_0(X) = r$ if and only if T^r can act almost-freely on X, and T^n cannot act almost-freely on X whenever $n > r$.

Proof.
If $\operatorname{rank}_0(X) = r$, then, clearly, T^n cannot act almost-freely on X if $n > r$. Let $\Phi : T^n \times X \to X$ be an action of rank r, where $n \geq r$. Let $K_1, \ldots, K_m$ be the maximal connective isotropy subgroups. (There are finitely many K_i by the assumption of FMCOT in most cases, and by Remark (3.2.9) or Corollary (3.7.15) or Exercise (3.2) in case (A).) Then $\operatorname{rank}(K_i) \leq n - r$, for all i. So one can find a subtorus L of rank r, such that $L \cap K_i$ is finite for all i. Thus L acts almost-freely. □

(4.3.4) Remark
The proof of Lemma (4.3.3) shows that:

(i) $\operatorname{rank}_0(X) \geq r$ if and only if T^r can act almost-freely on X; and
(ii) $\operatorname{rank}_0(X) \leq r$ if and only if T^n cannot act almost-freely on X, for all $n > r$.

(4.3.5) Examples
(1) If $\chi(X) \neq 0$, then $\operatorname{rank}_0(X) = 0$; since
$$\chi(X) \neq 0 \Rightarrow \chi(X^G) \neq 0 \Rightarrow X^G \neq \emptyset,$$
for any torus action on X.
(2) Suppose that $H^j(X; k) = 0$, for all $j > n$. Then $\operatorname{rank}_0(X) \leq n$.
To see this, suppose that $G = T^r$ is acting almost-freely on X. Then by Proposition (3.10.9) and Corollary (3.10.12), $\nu \leq n$ where
$$\nu := \max\{i;\, H_G^i(X; k) \neq 0\}.$$

From the Leray-Serre spectral sequence for
$$G \to EG \times X \to X_G,$$
one gets $H^{\nu+r}(X;k) \neq 0$.

(3) $\operatorname{rank}_0(T^r) = r$ Clearly $\operatorname{rank}_0(T^r) \geq r$, and $\operatorname{rank}_0(T^r) \leq r$ by (2) above.

(4) $\operatorname{rank}_0(S^{2n+1}) = 1$. Clearly $\operatorname{rank}_0(S^{2n+1}) \geq 1$; and $\operatorname{rank}_0(S^{2n+1}) \leq 1$, by Proposition (4.1.7)(2) and the Leray-Serre spectral sequence for
$$X \to X_G \to BG$$
(with $X = S^{2n+1}$).

(5) Suppose that X is in case (B). (Note that finite-dimensional CW-complexes belong in case (B): see [Bredon, 1967a], p. 144.) Then, by (2) above, $\operatorname{rank}_0(X) \leq \operatorname{cd}_k(X)$: but note that this follows also by dimension theory (see [Bredon, 1967a], Chapter II, Proposition 15.8), since X contains closed subspaces (the orbits) homeomorphic to T^r, which has dimension r.

On the other hand, if X is an infinite product of copies of S^1, then X is compact and $\operatorname{rank}_0(X) = \infty$.

For the next result, recall that if X is π-finite, i.e. $\dim_k \Pi_\psi^*(X) < \infty$, then $\chi\pi(X) = \sum_{j=1}^{\infty}(-1)^j \dim_k \Pi_\psi^j(X)$.

(4.3.6) Theorem
Suppose that X is connected and π-finite, and that $\dim_k H^*(X;k) < \infty$. Then $\operatorname{rank}_0(X) \leq -\chi\pi(X)$.

Proof.
Suppose that $G = T^r$ is acting almost-freely on X. By Proposition (4.1.7)(2), $\dim_k H_G^*(X;k) < \infty$. By Corollary (2.5.3), and Remarks (2.6.5)(1) for general G-spaces, applied to $X \to X_G \to BG$,
$$\chi\pi(X_G) = \chi\pi(X) + \chi\pi(BG) = \chi\pi(X) + r,$$
since BG is a $K(\mathbb{Z}^r, 2)$. (See Example (2.5.4).)

So, in the terminology of Theorem (2.4.4), X_G is π-finite and c-finite. Hence $\chi\pi(X_G) \leq 0$: i.e. $r \leq -\chi\pi(X)$. □

(Note that Theorem (2.4.4) is actually a theorem about minimal models, as remarked in Section 2.4: hence it applies to connected spaces in the Alexander-Spanier theory of Section 2.6.)

(4.3.7) Remarks

(1) See [Friedlander, Halperin, 1979] for a more complicated but stronger result.

(2) For general G-spaces in Theorem (4.3.6), $\chi\pi(X)$ is defined with respect to the Alexander-Spanier theory described in Section 2.6. If X is a CW-complex, however, or any agreement space (as defined in Section 2.6), then $\chi\pi(X)$ is the same, whether defined by singular theory or Alexander-Spanier theory.

(3) It follows immediately from [Bredon, 1972], Chapter I, Theorem 9.1, that if $q : \tilde{X} \to X$ is a finite one-to-one covering projection with X connected and locally path-connected, then $\operatorname{rank}_0(X) \leq \operatorname{rank}_0(\tilde{X})$.

(4) If X is connected and π-finite, then it follows that

$$\dim_k H^*(X;k) < \infty,$$

for all types of space under consideration here: see Remark (4.2.5).

Also, in [Halperin, preprint], Proposition 1.1, Halperin, shows that if Y is simply-connected, if $H^j(Y;\mathbb{Q}) = 0$, for all $j >$ some n_0, and if

$$\pi_i(Y) \otimes \mathbb{Q} = 0,$$

for all $i >$ some n_1, then $\dim_\mathbb{Q} H^*(Y;\mathbb{Q}) < \infty$, and

$$\dim_\mathbb{Q} \pi_*(Y) \otimes \mathbb{Q} = \dim_\mathbb{Q} \Pi^*_\psi(Y) < \infty.$$

Similarly, if Y is path-connected, if $H^j(Y;\mathbb{Q}) = 0$, for all $j >$ some n_0, and if $\Pi^i_\psi(Y) = 0$, for all $i >$ some n_1, then $\dim_\mathbb{Q} \Pi^i_\psi(Y) < \infty$, for all $i \geq 2$. ([Halperin, preprint], Lemma 2.1). Since this latter result comes from an analysis of the minimal model, $\mathcal{M}(Y)$, it is valid with $\overline{H}^*$, $\overline{\Pi}^*_\psi$, instead of H^*, Π^*_ψ, when Y is connected.

(4.3.8) Corollary
Let G be any compact connected Lie group, and let H be any closed subgroup of G. Let $X = G/H$ be the associated homogeneous space of left cosets. Then

$$\operatorname{rank}_0(X) = \operatorname{rank}(G) - \operatorname{rank}(H).$$

(Recall that $\operatorname{rank}(G) = \operatorname{rank}(T)$, where T is a maximal torus of G.)

Proof.

Let T be a maximal torus of G, and let $\Phi : T \times X \to X$ be the action by left translation. Then $\operatorname{rank}(\Phi) = \operatorname{rank}(G) - \operatorname{rank}(H)$. So

$$\operatorname{rank}_0(X) \geq \operatorname{rank}(G) - \operatorname{rank}(H).$$

Since $X_G = EG \times_G G/H \sim EG/H = BH$, there is a fibre bundle $X \to BH \to BG$. So $\chi\pi(X) = \chi\pi(BH) - \chi\pi(BG)$, if defined, by

4.3. Almost-free torus actions and the rank of a space

Corollary (2.5.3). But $\chi\pi(BG) = \operatorname{rank}(G)$; and $\chi\pi(BH) = \operatorname{rank}(H)$, if H is connected (see Examples (2.4.2)). Thus, if H is connected,

$$\chi\pi(X) = \operatorname{rank}(H) - \operatorname{rank}(G);$$

and the result follows from Theorem (4.3.6).

If H is not connected, then let H_0 be the identity component of H. So $G/H_0 \to G/H$ is a finite one-to-one covering projection; and $\operatorname{rank}_0(G/H) \leq \operatorname{rank}_0(G/H_0)$, by Remark (4.3.7)(3). Hence

$$\operatorname{rank}_0(G/H) \leq \operatorname{rank}(G) - \operatorname{rank}(H_0) = \operatorname{rank}(G) - \operatorname{rank}(H).$$

$\square$

The quantity $\chi\pi(X)$ is readily computable in many cases, even if X is not simply-connected or not nilpotent, by Corollary (2.5.3), for example, or perhaps by Theorem (2.7.8). Halperin, however, in [Halperin, preprint] and [Halperin, 1985], has given three interesting variations on $\chi\pi(X)$, which are useful for certain non simply-connected spaces, and which give upper bounds for $\operatorname{rank}_0(X)$, if X is nice enough. We shall now define these alternative forms of $\chi\pi(X)$, and state Halperin's theorems without proof. We shall state the theorems first under the assumption that X is a connected CW-complex; and, then in a remark, we shall list some slightly more general conditions under which the theorems hold. None of the theorems have been established for general G-spaces without the additional assumptions listed in the remark below.

(4.3.9) Definitions
(1) For a space, X, let

$$n(X) = \sup\{j;\, H^j(X;\mathbb{Q}) \neq 0\}.$$

Similarly, for a CGDA, A, let $n(A) = \sup\{j;\, H^j(A) \neq 0\}$.

(2) If X is connected, locally path-connected, and semi-locally 1-connected, let $n_T(X) = \sup\{n(X')\}$, where X' ranges over the path-connected covering spaces of X.

Let $d_{\text{top}}(X) = n_T(X) - n_T(\tilde{X}) = n_T(X) - n(\tilde{X})$, where $\tilde{X}$ is the universal covering space of X.

(3) If X is as in (2), if $d_{\text{top}}(X) < \infty$ and if

$$\dim_{\mathbb{Q}}\left(\bigoplus_{i=2}^{\infty} \pi_i(X) \otimes \mathbb{Q}\right) < \infty,$$

then set

$$\chi\pi_{\text{top}}(X) = -d_{\text{top}} + \sum_{i=2}^{\infty}(-1)^i \dim_{\mathbb{Q}}(\pi_i(X) \otimes \mathbb{Q})$$

(4) For a path-connected space, X, $\pi_1(X)$ acts on $\pi_i(X)$ for $i \geq 1$. The action on $\pi_1(X)$ is conjugation. Thus there is a subgroup of the centre of $\pi_1(X)$ consisting of all elements of $\pi_1(X)$ which act trivially on $\pi_i(X)$, for all $i \geq 2$: denote this subgroup by $Z_1(X)$.

(5) Let X be as in (4). If $\dim_{\mathbb{Q}}(Z_1(X) \otimes \mathbb{Q}) < \infty$, and
$$\dim_{\mathbb{Q}} \left(\bigoplus_{i=2}^{\infty} \pi_i(X) \otimes \mathbb{Q} \right) < \infty,$$
then set
$$\chi'\pi(X) = \dim_{\mathbb{Q}}(Z_1(X) \otimes \mathbb{Q}) + \sum_{i=2}^{\infty}(-1)^i \dim_{\mathbb{Q}}(\pi_i(X) \otimes \mathbb{Q}).$$

(6) Let X be a path-connected space, and let $\mathcal{M}(X)$ be its minimal model. Suppose that $\mathcal{M}(X) = S(E)$, where $E = \{x_\alpha; \alpha \in \mathcal{A}\}$ satifies (i)-(v) of Definition (2.2.2). Let $E^1 = \{x_\alpha; \alpha \in \mathcal{A}$ and $\deg(x_\alpha) = 1\}$. Then $S(E^1)$ is a sub-CGDA of $\mathcal{M}(X)$. Let $d(X) = n(S(E^1))$, as defined in (1) above. Of course, if E^1 is finite, then $n(S(E^1)) =$ the number of elements in $E^1 = \dim_{\mathbb{Q}} \Pi^1_\psi(X)$. On the other hand, $d(X)$ may be finite even though $\Pi^1_\psi(X)$ is infinite-dimensional: see Examples (4.3.14), below.

(7) Let X be as in (6). If $d(X) < \infty$, and $\dim_{\mathbb{Q}} \left(\bigoplus_{i=2}^{\infty} \Pi^i_\psi(X) \right) < \infty$, then set
$$\chi''\pi(X) = -d(X) + \sum_{i=2}^{\infty}(-1)^i \dim_{\mathbb{Q}}(\Pi^i_\psi(X)).$$

Now for Halperin's theorems.

(4.3.10) Theorem
Let X be a connected finite-dimensional CW-complex. Suppose that there exists $n_1 \in \mathbb{Z}$ such that $\pi_i(X) \otimes \mathbb{Q} = 0$, for all $i > n_1$. Then $d_{\text{top}}(X) < \infty$, $\dim_{\mathbb{Q}} \left(\bigoplus_{i=2}^{\infty} \pi_i(X) \otimes \mathbb{Q} \right) < \infty$, and
$$\text{rank}_0(X) \leq -\chi\pi_{\text{top}}(X).$$

(Halperin [Halperin, preprint] shows that, under the conditions of this theorem, $\tilde{X}$, the universal covering space of X, is π-finite and c-finite. Thus $-\chi\pi_{\text{top}}(X) = d_{\text{top}}(X) - \chi\pi(\tilde{X})$. So $-\chi\pi_{\text{top}}(X) \geq d_{\text{top}}(X)$, by Theorem (2.4.4).)

(4.3.11) Theorem
Let X be a connected finite-dimensional CW-complex. Suppose that there exists $n_1 \in \mathbb{Z}$ such that $\pi_i(X) \otimes \mathbb{Q} = 0$, for all $i > n_1$. Then $\dim_{\mathbb{Q}}(Z_1(X) \otimes \mathbb{Q}) < \infty$, $\dim_{\mathbb{Q}} \left(\bigoplus_{i=2}^{\infty} \pi_i(X) \otimes \mathbb{Q} \right) < \infty$, and
$$\text{rank}_0(X) \leq -\chi'\pi(X).$$

4.3. Almost-free torus actions and the rank of a space   275

(As before,
$$-\chi'\pi(X) = \dim_{\mathbb{Q}}(Z_1(X) \otimes \mathbb{Q}) - \chi\pi(\tilde{X})$$
$$\geq \dim_{\mathbb{Q}}(Z_1(X) \otimes \mathbb{Q}).)$$

(4.3.12) Theorem
Let X be a connected finite-dimensional CW-complex. Suppose that there exists $n_1 \in \mathbb{Z}$ such that $\Pi^i_\psi(X) = 0$, for all $i > n_1$. Then $d(X) < \infty$, $\dim_{\mathbb{Q}}\left(\bigoplus_{i=2}^\infty \Pi^i_\psi(X)\right) < \infty$, and
$$\mathrm{rank}_0(X) \leq -\chi''\pi(X).$$

(4.3.13) Remarks
(1) Theorems (4.3.10) and (4.3.11) are valid for any path-connected space, X, satisfying the following four conditions:

(i) X is paracompact and $\mathrm{cd}_{\mathbb{Q}}(X) < \infty$ (i.e. case (B));
(ii) X is locally path-connected and semi-locally 1-connected;
(iii) X is an agreement space (see Definition (2.6.2));
(iv) $\pi_i(X) \otimes \mathbb{Q} = 0$, for all $i >$ some $n_1 \in \mathbb{Z}$.

Note that a paracompact space, X, is an agreement space if, in singular cohomology,
$$\varinjlim_{U(x)} H^*(U(x); \mathbb{Q}) \cong H^*(\{x\}; \mathbb{Q}),$$
for all $x \in X$, where $U(x)$ runs over the open neighbourhoods of x ([Bredon, 1967a]). In particular, any space, which is homotopy equivalent to a paracompact locally contractible space, is an agreement space.

Using Alexander-Spanier theory to define $n(X)$, $d(X)$ and $\chi''\pi(X)$, Theorem (4.3.12) is valid for any connected space, X, satisfying the following conditions:
(α) X belongs to case (A), (B) or (C) of Definition (3.2.4);
(β) $n(X) < \infty$;
(γ) $\overline{\Pi}^i_\psi(X) = 0$, for all $i >$ some n_1.

In view of Remark (4.3.7)(4), therefore, Theorem (4.3.12) is a generalization of Theorem (4.3.6).

The proof of Theorem (4.3.12) in the Alexander-Spanier theory does not follow exactly as in [Halperin, preprint], since Halperin uses the Grivel-Halperin-Thomas Theorem for $T \to ET \times X \to X_T$, and Theorem (2.6.4) does not apply directly to this fibration. It is not hard to show otherwise, however, that $\chi''\pi(X_T) = r + \chi''\pi(X)$, if $T = T^r$ is acting almost-freely on X.

(2) Let X be a path-connected space satisfying (i), (ii), (iii) and (iv) of (1), above; e.g., X a connected CW-complex. Suppose that $\pi_1(X)$ is finitely generated, and that X is nilpotent (see Definition (2.3.2)). Then, in [Halperin, preprint], it is shown that $d(X) = d_{\text{top}}(X)$, and $\chi''\pi(X) = \chi\pi_{\text{top}}(X)$. Indeed, with the notation of Definitions (4.3.9), $n(X) = d(X) + n(\tilde{X})$. Furthermore, $d(X)$ is equal to the dimension of the Lie algebra of $\pi_1(X) \otimes \mathbb{Q}$, the Mal'tsev completion of $\pi_1(X)$.

(4.3.14) Example ([Halperin, preprint])

Let
$$X = (S^1 \vee S^1) \times (\mathbb{R}P^2 \vee \mathbb{R}P^2) \times S^3 \times S^3.$$
Since the universal covering space of $\mathbb{R}P^2 \vee \mathbb{R}P^2$ is an infinite wedge of 2-spheres, $\chi\pi_{\text{top}}(X)$ and $\chi'\pi(X)$ are not defined. The minimal model of X is the same as that of $(S^1 \vee S^1) \times S^3 \times S^3$, since $\mathbb{R}P^2$ is $\mathbb{Q}$-acyclic. Hence $\mathcal{M}(X) = \mathcal{M}(S^1 \vee S^1) \otimes \mathcal{M}(S^3) \otimes \mathcal{M}(S^3)$. So $\dim_{\mathbb{Q}} \Pi^1_\psi(X) = \infty$, and $\chi\pi(X)$ is not defined. But
$$d(X) = d(S^1 \vee S^1) = n(S^1 \vee S^1) = 1.$$
Hence $\chi''\pi(X) = -3$. It follows that $\text{rank}_0(X) \leq 3$.

An examination of the spectral sequence of $X \to X_T \to BT$, however, quickly shows that $\text{rank}_0(X) = 2$.

(4.3.15) Remark

Let G be a compact connected Lie group acting almost-freely on a space X, which satisfies the usual conditions. By Theorem (3.9.3)(1), $H^*_G(X;\mathbb{Q}) \cong H^*_T(X;\mathbb{Q})^W$, where T is a maximal torus of G and W is the Weyl group. So, if $\dim_{\mathbb{Q}} H^*(X;\mathbb{Q}) < \infty$, then $\dim_{\mathbb{Q}} H^*_G(X;\mathbb{Q}) < \infty$, by Proposition (4.1.7)(2).

By using minimal models or the spectral sequence for
$$X \to X_G \to BG,$$
one can get very useful information concerning which compact connected Lie groups can act almost-freely on X. See [Allday, Halperin, 1978] and [Friedlander, Halperin, 1979] for some examples.

(4.3.16) Example

Let $f: S^3 \times S^3 \to S^6$ be a map of degree one. Let τ be the tangent bundle of S^6, and let X be the total space of the unit sphere bundle associated with $f^*(\tau)$. So we have a bundle $S^5 \to X \to S^3 \times S^3$. The minimal model $\mathcal{M}(X) = \Lambda(u,v,w)$, where $\deg(u) = \deg(v) = 3$ and $\deg(w) = 5$, with differential, d, given by $d(u) = d(v) = 0$, and

$d(w) = 2uv$. (Of course, the coefficient 2 is rationally irrelevant, but it is in keeping with the Euler class.) Theorem (4.3.6) gives $\mathrm{rank}_0(X) \leq 3$. But an examination of $R \otimes \mathcal{M}(X)$ shows quickly that $\mathrm{rank}_0(X) \leq 1$. Hence $\mathrm{rank}_0(X) = 1$.

Now consider the Lie algebra $\mathcal{L}_*(X)$. From Definition (2.3.10) and Theorem (2.3.11), we find that $\mathcal{L}_*(X)$ has a $\mathbb{Q}$-vector space basis $\{\alpha, \beta, \gamma\}$, where $\alpha, \beta \in \mathcal{L}_2(X)$, $\gamma \in \mathcal{L}_4(X)$, and the Lie bracket is given by $[\alpha, \beta] = \gamma$. Thus the centre of $\mathcal{L}_*(X)$, i.e. $\{\xi \in \mathcal{L}_*(X); [\xi, \eta] = 0$ for all $\eta \in \mathcal{L}_*(X)\}$, has dimension one, being spanned by γ. Theorem (4.3.18) gives a useful relation between $\mathrm{rank}_0(X)$ and the dimension of the centre of $\mathcal{L}_*(X)$ for simply-connected spaces X, having $\mathcal{L}_i(X) = 0$, for all odd i.

(4.3.17) Definitions
(1) Let $\mathcal{L}$ be a non-negatively graded Lie algebra. The centre of $\mathcal{L}$, denoted by $Z\mathcal{L}$, is defined by
$$Z\mathcal{L} = \{\xi \in \mathcal{L}; [\xi, \eta] = 0, \text{ for all } \eta \in \mathcal{L}\}.$$
(2) If $\lambda, \nabla : \lambda \to \lambda \otimes \lambda$, is a non-negatively graded co-Lie algebra over a field k, then there is a unique minimal k-vector subspace, $F(\lambda)$, such that $\nabla(u) \in F(\lambda) \otimes F(\lambda) \subseteq \lambda \otimes \lambda$, for all $u \in \lambda$. The cocentre of λ, denoted $\zeta(\lambda)$, is defined to be the factor space $\lambda / F(\lambda)$.

If λ is a non-negatively graded co-Lie algebra of finite type over a field k, and if $\mathcal{L} = \lambda^*$ is the graded dual Lie algebra, then it is easy to see that $Z(\mathcal{L}) = \mathrm{ann}(F(\lambda)) := \{\alpha \in \mathcal{L}; \alpha(u) = 0, \text{ for all } u \in F(\lambda)\}$. Hence $Z(\mathcal{L}) \cong \zeta(\lambda)^*$, the graded k-vector space dual of $\zeta(\lambda)$.

(4.3.18) Theorem
Suppose that X is a simply-connected finite-dimensional CW-complex of finite $\mathbb{Q}$-type; and suppose that $\pi_{2j}(X) \otimes \mathbb{Q} = 0$, for all $j \geq 1$. Then $\mathrm{rank}_0(X) \leq \dim_\mathbb{Q} Z\mathcal{L}_*(X)$.

More generally suppose that X is a connected space satisfying conditions (A), (B) or (C) of Definition (3.2.4); and suppose that
$$\overline{\Pi}_\psi^{2j}(X) = 0,$$
for all $j \geq 1$.
Then $\mathrm{rank}_0(X) \leq \dim_\mathbb{Q} \zeta(\overline{\mathcal{L}}^*(X))$.

Proof.
Let $T = T^r$ act almost-freely on X. By the Grivel-Halperin-Thomas theorem (or Corollary (2.6.6)), we have a model $R \otimes \mathcal{M}(X)$ for X_T with differential d, say. By assumption, $\mathcal{M}(X) = \Lambda(y_\beta; \beta \in \mathcal{B})$ is an exterior

algebra on a well-ordered set of elements, y_β, of odd degree. For each $\beta \in \mathcal{B}$, let $d(y_\beta) = f_\beta + u_\beta$, where $f_\beta \in R$ and u_β has no R-component. Let $J \triangleleft R$ be the ideal generated by $\{f_\beta; \beta \in \mathcal{B}\}$.

Since R is Noetherian, there exist $\beta_1, \ldots, \beta_s \in \mathcal{B}$ such that the

$$f_i := f_{\beta_i}, \quad 1 \leq i \leq s,$$

have the following properties:

(i) $\deg(f_1) \leq \ldots \leq \deg(f_s)$;
(ii) $f_1 \neq 0$, and $f_{i+1} \notin (f_1, \ldots, f_i)$, for $1 \leq i \leq s-1$;
(iii) $J = (f_1, \ldots, f_s)$.

Let $\mathcal{B}' = \mathcal{B} - \{\beta_1, \ldots, \beta_s\}$. So, for each $\beta \in \mathcal{B}'$, there exist $g_{\beta 1}, \ldots, g_{\beta s} \in R$, such that $f_\beta = \sum_{i=1}^{s} g_{\beta i} f_i$. Now let $z_{\beta_i} = y_{\beta_i}$ for $1 \leq i \leq s$; and, for each $\beta \in \mathcal{B}'$, let $z_\beta = y_\beta - \sum_{i=1}^{s} g_{\beta i} y_{\beta_i}$. Then $\mathcal{M}(X)$ is generated by $\{z_\beta; \beta \in \mathcal{B}\}$ in $R \otimes \mathcal{M}(X)$; and $d(z_\beta)$ has no R-component, for any $\beta \in \mathcal{B}'$.

Let $d_2(z_\beta)$ denote the component of $d(z_\beta)$ in $R \otimes \Lambda^2(z_\beta; \beta \in \mathcal{B})$, and set

$$d_2(z_\beta) = \sum_{i<j} r_{\beta ij} z_{\beta_i} z_{\beta_j} + \sum_{i,\gamma} r_{\beta i\gamma} z_{\beta_i} z_\gamma + \sum_{\gamma < \delta} r_{\beta\gamma\delta} z_\gamma z_\delta,$$

where $\gamma, \delta \in \mathcal{B}'$, and $r_{\beta ij}, r_{\beta i\gamma}, r_{\beta\gamma\delta} \in R$. Let $r^0_{\beta ij}, r^0_{\beta i\gamma} \in \mathbb{Q}$, denote the constant terms of $r_{\beta ij}, r_{\beta i\gamma}$, respectively. Since $d^2(z_\beta) = 0$, it follows from properties (i) and (ii), above, that $r^0_{\beta ij} = 0$ and $r^0_{\beta i\gamma} = 0$, for all β, γ, i and j.

Hence, with the notation of Definition (4.3.17)(2), $F(\mathcal{L}^*(X))$ is contained in the subspace of $\mathcal{L}^*(X)$ spanned by the elements corresponding to the z_β for $\beta \in \mathcal{B}'$. (See Section 2.3). It follows that

$$\dim_\mathbb{Q} \zeta(\mathcal{L}^*(X)) \geq s.$$

On the other hand, $\operatorname{ht}(J) \leq s$, by the Krull Height Theorem; and, clearly, $\ker(p^*) \subseteq J \subseteq \sqrt{(\ker(p)^*)}$, where $p : X_T \to BT$ is the bundle map. Since $T = T^r$ is acting almost-freely, $\operatorname{ht}(J) = r$, by Corollary (3.1.11). □

(4.3.19) Remark

Recall that the Lyusternik-Schnirel'man category of a space X, $\operatorname{cat}(X)$, is the least positive integer m such that X can be covered by $m+1$ open sets, each of which is contractible in X (or $\operatorname{cat}(X) = \infty$ if no such integer exists). For a path-connected space X there is also a notion of rational category, $\operatorname{cat}_0(X)$; and $\operatorname{cat}_0(X) \leq \operatorname{cat}(X)$: see [Félix, Halperin, 1982b]. If X is simply-connected, then $\operatorname{cat}_0(X) = \operatorname{cat}(X_0)$, where X_0 is the localization of X at (0) as in Discussion (2.3.3). Now, as pointed out in [Rachid, 1990], if one combines Theorem (4.3.18) with one of the principal results of [Félix et al, 1988], then one gets the following.

Suppose that X is a simply-connected finite-dimensional CW-complex of finite $\mathbb{Q}$-type; and suppose that $\pi_{2j}(X) \otimes \mathbb{Q} = 0$, for all $j \geq 1$. Then $\text{rank}_0(X) \leq cat_0(X)$; cf. Remark (4.4.2)(5).

We conclude this section with a useful result on realizing free torus actions. Recall the definition of a KS-extension in Section 2.2. We have the following proposition from [Halperin, 1985].

(4.3.20) Proposition
Let Y be a simply-connected topological space with $\dim_{\mathbb{Q}} H^*(Y;\mathbb{Q}) < \infty$. Let $R = \mathbb{Q}[t_1,\ldots,t_r]$ where $\deg(t_i) = 2$ for $1 \leq i \leq r$; and view R as a minimal CGDA with trivial boundary. Suppose that there exists a KS-extension
$$R \to R \otimes \mathcal{M}(Y) \to \mathcal{M}(Y)$$
with $\dim_{\mathbb{Q}} H(R \otimes \mathcal{M}(Y)) < \infty$. Then, where $G = T^r$, there is a simply-connected finite free G-CW-complex X such that X has the same rational homotopy type as Y (i.e. $\mathcal{M}(X) \cong \mathcal{M}(Y)$).

Proof.
$R \otimes \mathcal{M}(Y)$ is a 1-connected minimal CGDA with finite-dimensional cohomology. Hence, e.g., by [Tanré, 1983], Conséquence 2, p.86, there is a simply-connected finite simplicial complex Z such that
$$\mathcal{M}(Z) \cong R \otimes \mathcal{M}(Y).$$
The cocycles $t_i \otimes 1 \in R \otimes \mathcal{M}(Y) \cong \mathcal{M}(Z)$ define classes

$$\begin{array}{ccc} X & \longrightarrow & EG \\ \downarrow & & \downarrow \\ Z & \xrightarrow{\phi} & BG \end{array}$$

Diagram 4.1

$\tau_i \in H^2(Z;\mathbb{Q})$; and, hence, multiplying by integers n_i if necessary they define classes $n_i \tau_i \in H^2(Z;\mathbb{Z})$. These classes determine a map $\phi : Z \to K(\mathbb{Z}^r, 2) = BG$; ϕ is unique up to homotopy; and, without loss of generality, we may take ϕ to be simplicial. Now define X to be the pull-back of $EG \to BG$ by ϕ. So there exists the pull-back diagram shown in Diagram 4.1. Since ϕ is simplicial, it is not hard to see that X is a simply-connected finite G-CW-complex, which clearly, is free.

Finally, the fibration $G \to X \to Z$ and the Grivel-Halperin-Thomas Theorem show that $\mathcal{M}(X) \cong \mathcal{M}(Y)$. □

(4.3.21) Example
Let Y be a connected topological space with $\mathcal{M}(Y)$ an exterior algebra $\Lambda(y_i; i \in \mathcal{I})$, where the generators y_i all have odd degrees greater than 1. Suppose that $\dim_\mathbb{Q} H^*(Y;\mathbb{Q}) < \infty$. Let V be the $\mathbb{Q}$-vector space spanned by $\{y_i; i \subset \mathcal{I}\}$; and suppose that the differential δ on $\mathcal{M}(Y)$ has the property that $\delta(V) \subseteq \Lambda(W)$ where W is a subspace of V with $\dim_\mathbb{Q}(V/W) = r$. Then, by Proposition (4.3.20), there is a simply-connected finite free T^r-CW-complex X such that
$$\mathcal{M}(X) \cong \mathcal{M}(Y).$$
To see this we must put an appropriate differential d on $R \otimes \mathcal{M}(Y)$ such that $\dim_\mathbb{Q} H(R \otimes \mathcal{M}(Y)) < \infty$. Choose a splitting $V = W \oplus U$, and let $\{u_1, \ldots, u_r\}$ be a basis for U. For any $w \in W$, let $d(w) = \delta(w)$; and, for $1 \leq i \leq r$, let $d(u_i) = \delta(u_i) + t_i^{m_i}$, where $2m_i = \deg(u_i) + 1$. Since $\delta(u_i)$ is a nilpotent cycle in $R \otimes \mathcal{M}(Y)$, it follows that the radical of
$$\ker[R \to H(R \otimes \mathcal{M}(Y))]$$
is $(t_1, \ldots, t_r)$. From the Serre-type spectral sequence of $R \otimes \mathcal{M}(Y)$, $(E_2 = R \otimes H^*(Y;\mathbb{Q})$, converging to $H^*(R \otimes \mathcal{M}(Y)))$, it follows that $\dim_\mathbb{Q} H(R \otimes \mathcal{M}(Y)) < \infty$.

Now suppose further that Y, and hence X, is π-formal (coformal): i.e. $\delta(V) \subseteq \Lambda^2(W)$. Then we may take $W = F(\mathcal{L}^*(Y))$, using here the notation of Definition (4.3.17)(2). Then, if $\dim_\mathbb{Q}(V/W) = r < \infty$,
$$\dim_\mathbb{Q} \zeta(\mathcal{L}^*(Y)) = \dim_\mathbb{Q} Z\mathcal{L}_*(Y) = r.$$
By Theorem (4.3.18) then, $\mathrm{rank}_0(X) = r$.

4.4 More results on almost-free torus actions

This section is a continuation of Section 4.3; and the same basic assumptions and conventions apply.

An interesting conjecture in the study of almost-free torus actions is the following, which is generally attributed to Halperin. (The analogous conjecture for p-torus actions is stated in [Carlsson, 1987a].)

(4.4.1) Conjecture
If $\mathrm{rank}_0(X) \geq r$, then $\dim_\mathbb{Q} H^*(X;\mathbb{Q}) \geq 2^r$.

4.4. More results on almost-free torus actions

Conjecture (4.4.1) is related to a conjecture in local algebra due to Horrocks: see [Evans, Griffith, 1985], p. 60 and Exercise 14 on p. 63.

The following remarks give some cases in which the conjecture is true together with two examples which show that a tempting stronger conjecture is false.

(4.4.2) Remarks

(1) Let G be a compact connected Lie group, and let $H \subseteq G$ be a closed connected subgroup. Let $r = \text{rank}(G) - \text{rank}(H)$; so $\text{rank}_0(G/H) = r$, by Corollary (4.3.8). From the Serre spectral sequence of $H \to G \to G/H$,

$$\dim_{\mathbb{Q}} H^*(G; \mathbb{Q}) = 2^{\text{rank}(G)} \leq \dim_{\mathbb{Q}} H^*(H; \mathbb{Q}) \dim_{\mathbb{Q}} H^*(G/H; \mathbb{Q})$$
$$= 2^{\text{rank}(H)} \dim_{\mathbb{Q}} H^*(G/H; \mathbb{Q}).$$

Thus $\dim_{\mathbb{Q}} H^*(G/H; \mathbb{Q}) \geq 2^r$.

(2) X is said to be a rational cohomology Kähler space (CKS) if
  (i) there exists $\omega \in H^2(X; \mathbb{Q})$ such that $\omega^n \neq 0$ in $H^{2n}(X; \mathbb{Q}) \cong \mathbb{Q}$;
  (ii) $H^j(X; \mathbb{Q}) = 0$ for all $j > 2n$; and
  (iii) the cup-product with $\omega^j : H^{n-j}(X; \mathbb{Q}) \to H^{n+j}(X; \mathbb{Q})$ is an isomorphism for all j, $0 \leq j \leq n$.

In [Allday, Puppe, 1986a] it is shown easily that if X is a CKS, then $\text{rank}_0(X) \leq$ the maximal number of algebraically independent elements in $H^1(X; \mathbb{Q})$. So, if $\text{rank}_0(X) = r$, then $H^*(X; \mathbb{Q})$ contains an exterior algebra on r generators of degree 1. Hence $\dim_{\mathbb{Q}} H^*(X; \mathbb{Q}) \geq 2^r$. (See [Allday, Puppe, 1986a] for more on this example.)

(3) Another class of spaces for which Conjecture (4.4.1) is valid is described in [Allday, Puppe, 1986a], Section 3. This class is defined in terms of certain conditions on the rational homotopy Lie algebra and the minimal model; cf. Theorem (4.3.18).

(4) Theorem (4.4.3) shows that Conjecture (4.4.1) is valid whenever $r \leq 3$.

(5) In view of (2) above and other considerations, it is tempting to conjecture that if $\text{rank}_0(X) \geq r$, then $H^*(X; \mathbb{Q})$ must contain an exterior algebra on r generators of odd degree: but this is false as the following examples show.

Let $Y = S^3 \times S^3$ with T^2 acting freely by the standard diagonal action. Choose $x_0 \in S^3$ and let $f : T^2 \times S^3 \to Y$ be given by

$$f(z_1, z_2, x) = (z_1 x, z_2 x_0)$$

for $x \in S^3$ and $(z_1, z_2) \in T^2 = S^1 \times S^1$. Let $X = Y \cup_f (T^2 \times D^4)$. Then X is a free T^2-CW-complex; but, as is seen easily from the Mayer-Vietoris sequence for the attaching, $H^*(X; \mathbb{Q})$ does not contain an exterior algebra on two generators of odd degree. On the other hand

$\dim_{\mathbb{Q}} H^*(X; \mathbb{Q}) = 6$. (More generally let $Y = (S^{2m-1})^r$ and let $X = Y \cup_f (T^r \times D^{2m})$, where $f(z_1, \ldots, z_r, x) = (z_1 x, z_2 x_0, \ldots, z_r x_0)$.)

For the second example let $G = T^r$, and let S be the 2-skeleton of BG. So S is a wedge (pointed sum) of r copies of a sphere S^2. Let $j : S \to BG$ be the inclusion, and let X be the total space of the pullback of the universal principle G-bundle $G \to EG \to BG$ along j. So there is a principal G-bundle $G \to X \to S$. In particular X is a free G-space, and X is a 2-connected finite CW-complex. Furthermore, as is easily seen from the Serre spectral sequence of $G \to X \to S$, all cup-products of elements of positive degree in the cohomology of X (with any coefficients) are zero.

(It is interesting to compare the latter example with Remark (4.3.19), for here, by a result of [Félix, Halperin, 1982b], one has that $cat_0(X) = 1$. See also [Halperin, 1985], Section 1, for an example of a closed simply-connected manifold M with $\text{rank}_0(M) > cat_0(M)$.)

The following theorem establishes Conjecture (4.4.1) for $r \leq 3$.

(4.4.3) Theorem

Suppose that $\text{rank}_0(X) \geq r$. *Then*

(1) $\dim_{\mathbb{Q}} H^*(X; \mathbb{Q}) \geq 2r$; *and*
(2) $\dim_{\mathbb{Q}} H^*(X; \mathbb{Q}) \geq 2(r+1)$, *if* $r \geq 3$.

Proof.

(1) Recall the differential graded R-module $\mu_G^*(X; \mathbb{Q}) = R \tilde{\otimes} H^*(X; \mathbb{Q})$ of Definition (3.5.6)(1), where $G = T^r$ is acting almost-freely on X. We may assume that $r > 0$, and hence $\chi(X) = 0$. Let $\{y_1, \ldots, y_m\}$ and $\{x_1, \ldots, x_m\}$ be bases for $H^{\text{od}}(X; \mathbb{Q})$ and $H^{\text{ev}}(X; \mathbb{Q})$, respectively, where $\deg(y_1) \leq \ldots \leq \deg(y_m)$ and $\deg(x_1) < \deg(x_2) \leq \ldots \leq \deg(x_m)$, and $x_1 = 1 \in H^0(X; \mathbb{Q})$. (Clearly, without loss of generality, we may assume that X is connected.) In $\mu_G^*(X; \mathbb{Q})$ let

$$dy_i = \sum_{j=1}^m p_{ij} x_j, \text{ for } 1 \leq i \leq m, \text{ and } p_{ij} \in R;$$

and

$$dx_i = \sum_{j=1}^m q_{ij} y_j, \text{ for } 1 \leq i \leq m, \text{ and } q_{ij} \in R.$$

Then, with $p : X_G \to BG$ as usual, it follows that

$$\ker p^* \subseteq (p_{11}, \ldots, p_{m1}).$$

4.4. More results on almost-free torus actions

If $m < r$, then there exists a non-zero $\alpha = (\alpha_1, \ldots, \alpha_r) \in \bar{\mathbb{Q}}^r$ such that $p_{i1}(\alpha) = 0$ for $1 \leq i \leq m$. Then, with the notation of Definitions (3.5.6)(2),
$$S(t)^{-1} H(\mu^*_{G,\tilde{\alpha}}(X; \mathbb{Q})) \neq 0.$$
So, by Lemma (3.5.7), $X^{K(\alpha)} \neq \emptyset$. This contradicts the assumption that G is acting almost-freely. Hence $m \geq r$.

Before proving (2) of the theorem we shall give an algebraic lemma. Recall from Appendix A.5 that, if A is a commutative ring, then a sequence of elements $a_1, \ldots, a_n \in A$ is said to be a regular sequence if

(i) $a_1 \neq 0$, and $A \neq (a_1, \ldots, a_n)$, the ideal generated by $a_1, \ldots, a_n$; and

(ii) for $1 \leq j \leq n$, the image of a_j in A/J_j is not a zero-divisor, where J_j is ideal $(a_1, \ldots, a_{j-1})$ and $J_1 = (0)$.

In general the regularity of a sequence depends upon its order.

(4.4.4) Lemma

Let k be any field and let $R = k[t_1, \ldots, t_r]$ be a polynomial ring in r generators. Suppose that $f_1, \ldots, f_m \in R$ is a regular sequence of homogeneous polynomials of positive degree. Then:

(1) *Any permutation of $f_1, \ldots, f_m$ is again a regular sequence.*

(2) *Suppose that $g_1, \ldots, g_m \in R$ are homogeneous, and that*
$$f_1 g_1 + \ldots + f_m g_m = 0.$$
Then, as ideals, $(g_1, \ldots, g_m) \subset (f_1, \ldots, f_m)$. In particular, if $g_1, \ldots, g_m$ is also regular, then $(g_1, \ldots, g_m) = (f_1, \ldots, f_m)$.

(3) *Suppose that $m = r$, and set $f_i = \sum_{j=1}^{r} f_{ij} t_j$, for $1 \leq i \leq r$. So $F = (f_{ij})$ is a $r \times r$ matrix of homogeneous polynomials in R. Then $\det(F) \notin (f_1, \ldots, f_r)$.*

We also have:

(4) *Let $h_1, \ldots, h_m \in R$ be homogeneous polynomials of positive degree, and let $J = (h_1, \ldots, h_m) \triangleleft R$. Suppose that $\mathrm{ht}(J) = m$. Then $h_1, \ldots, h_m$ is a regular sequence in R; (cf. [Baum, 1968], Proposition 3.9.)*

Proof.

For (1) see Theorem (A.5.4); and for (4) see Proposition (A.6.9).

For (2), we have $f_j g_j \in (f_1, \ldots, f_{j-1}, f_{j+1}, \ldots, f_m)$. By (1),
$$f_1, \ldots, f_{j-1}, f_{j+1}, \ldots, f_m, f_j$$
is a regular sequence. Hence
$$g_j \in (f_1, \ldots, f_{j-1}, f_{j+1}, \ldots, f_m).$$

For (3), let $J = (f_1, \ldots, f_r)$, and consider the two Koszul complexes $K_1 = R \otimes \Lambda(s_1, \ldots, s_r)$ with boundary given by $\partial(s_i) = f_i$ for $1 \leq i \leq r$, and $K_2 = R \otimes \Lambda(\sigma_1, \ldots, \sigma_r)$ with boundary given by $\partial(\sigma_i) = t_i$ for $1 \leq i \leq r$; K_1 is a free resolution of R/J, and K_2 is a free resolution of k. (See Theorem (A.5.5)(1).) The matrix F defines a multiplicative map $\phi_F : K_1 \to K_2$ covering the quotient map $R/J \to k$ and given by $\phi_F(s_i) = \sum_{j=1}^r f_{ij}\sigma_j$ for $1 \leq i \leq r$.

ϕ_F induces a homomorphism

$$\phi_F^* : \mathrm{Ext}_R^r(k, R) \to \mathrm{Ext}_R^r(R/J, R).$$

$$\mathrm{Ext}_R^r(k, R) \cong k \quad \text{and} \quad \mathrm{Ext}_R^r(R/J, R) \cong R/J;$$

and $\phi_F^*(1) = \det(F) + J$. Thus it is enough to show that $\phi_F^* \neq 0$. Let $\mathfrak{m} = (t_1, \ldots, t_r)$. From the long Ext sequence coming from the short exact sequence

$$0 \to \mathfrak{m}/J \to R/J \to k \to 0$$

it is enough to show that $\mathrm{Ext}_R^{r-1}(\mathfrak{m}/J, R) = 0$.

If $\mathfrak{p} \in \mathrm{Spec}(R)$ and $\mathfrak{p} \notin V(J) = \{\mathfrak{p} \in \mathrm{Spec}(R) : \mathfrak{p} \supseteq J\}$, then $(\mathfrak{m}/J)_\mathfrak{p} = 0$. Thus $\mathrm{Supp}(\mathfrak{m}/J) \subseteq V(J)$. Hence $\mathrm{Ext}_R^{r-1}(\mathfrak{m}/J, R) = 0$, by [Matsumura, 1986], Theorem 16.6. □

Proof of Theorem (4.4.3)(2).

We shall use the same notation as used in the proof of (1). We shall suppose that $m = r$, and, deduce that $r \leq 2$.

$\ker(p^*) \subset (p_{11}, \ldots, p_{r1})$; and, since T^r is acting almost-freely on X, $\sqrt{(\ker(p^*))} = (t_1, \ldots, t_r) = \mathfrak{m}$, the maximal homogeneous ideal in R, by Corollary (3.1.11). So $\sqrt{(p_{11}, \ldots, p_{r1})} = \mathfrak{m}$; and hence $p_{11}, \ldots, p_{r1}$ is a regular sequence by Lemma (4.4.4)(4).

For $1 \leq i \leq r$, put $p_{i1} = \sum_{j=1}^r r_{ij}t_j$. Let $P = (r_{ij})$. By Lemma (4.4.4)(3), $\det(P) \notin (p_{11}, \ldots, p_{r1})$. Hence $\det(P) \notin \ker(p^*)$.

Let $n(X) = \max\{j : H^j(X; \mathbb{Q}) \neq 0\} = \max\{\deg(y_r), \deg(x_r)\}$. Then $\deg(\det(P)) \leq n(X) - r$. (See Example (4.3.5)(2): from the Leray-Serre spectral sequence for $G \to X \times EG \to X_G$, $n(X) = n(X_G) + r$.) But

$$\deg(r_{ij}) = \deg(p_{i1}) - 2 = \deg(y_i) - 1.$$

Hence $\deg(\det(P)) = \sum_{i=1}^r \deg(y_i) - r$. Thus $\sum_{i=1}^r \deg(y_i) \leq n(X)$; and, hence, if $r > 1$, x_r has maximal degree in $H^*(X; \mathbb{Q})$.

Now $dx_r = \sum_{j=1}^r q_{rj}y_j$. If $q_{r1}, \ldots, q_{rr}$ is not regular, then, by Lemma (4.4.4)(4), $\mathrm{ht}(q_{r1}, \ldots, q_{rr}) < r$; and so there exists a non-zero

$$\alpha = (\alpha_1, \ldots, \alpha_r) \in \overline{\mathbb{Q}}^r,$$

such that $q_{rj}(\alpha) = 0$, for $1 \leq j \leq r$. It follows that x_r represents a non-zero class in $S(t)^{-1}H(\mu^*_{G,\tilde{\alpha}}(X;\mathbb{Q}))$. So $q_{r1}, \ldots, q_{rr}$ is regular.

4.4. More results on almost-free torus actions

Since $d(dx_r) = 0$, $q_{r1}p_{11} + \ldots + q_{rr}p_{r1} = 0$. Hence
$$(q_{r1}, \ldots, q_{rr}) = (p_{11}, \ldots, p_{r1}),$$
by Lemma (4.4.4)(2). For $1 \leq i \leq r$, put
$$q_{ri} = \sum_{j=1}^{r} s_{ij}t_j;$$
and let $C = (s_{ij})$. As above $\det(C) \notin \ker p^*$; and so
$$\deg(\det(C)) \leq n(X) - r = \deg(x_r) - r.$$
But
$$\deg(s_{ij}) = \deg(q_{ri}) - 2 = \deg(x_r) - \deg(y_i) - 1.$$
So
$$\deg(\det(C)) = r\deg(x_r) - \sum_{i=1}^{r}\deg(y_i) - r.$$
Hence
$$(r-1)\deg(x_r) \leq \sum_{i=1}^{r}\deg(y_i) \leq \deg(x_r).$$
Thus $r \leq 2$. $\square$

The second theorem of this section concerns the number of distinct non-zero cohomology groups of X in relation to $\operatorname{rank}_0(X)$.

(4.4.5) Theorem
If each connected component of X is open in X, and if $\operatorname{rank}_0(X) \geq r$, then $H^j(X;\mathbb{Q}) \neq 0$ for at least $r+1$ different values of j.

Proof.
The restriction homomorphism $H^*(X;\mathbb{Q}) \to H^*(c;\mathbb{Q})$ is surjective for any component c of X, since c is open in X. Thus, without loss of generality, we may assume that X is connected. Also if
$$n(X) := \sup\{i : H^i(X;\mathbb{Q}) \neq 0\}$$
is infinite, then there is nothing to prove. So we shall assume that $n(X) < \infty$. Let $G = T^r$ act almost-freely on X. Then $n(X_G) = n(X) - r$, as in the proof of Theorem (4.4.3)(2). The proof is now analogous to the proof of Theorem (1.4.14). By Definition (3.5.6)(1) we have a minimal Hirsch-Brown model $\mu_G^*(X;\mathbb{Q}) = R\tilde{\otimes} H^*(X;\mathbb{Q})$ with differential δ, say. One obtains a filtration of $\mu_G^*(X;\mathbb{Q})$ as follows.

Let $\mathcal{F}_0 H^*(X;\mathbb{Q}) = \{x \in H^*(X;\mathbb{Q}); \delta(1 \otimes x) = 0\}$. Inductively for $q > 0$, let $\mathcal{F}_q H^*(X;\mathbb{Q}) = \{x \in H^*(X;\mathbb{Q}); \delta(1 \otimes x) \in \mathcal{F}_{q-1}\mu_G^*(X;\mathbb{Q})\}$;

and for all $q \geq 0$ let $\mathcal{F}_q \mu_G^*(X; \mathbb{Q}) = R \otimes \mathcal{F}_q H^*(X; \mathbb{Q})$. Since $n(X_G) < \infty$ by above, it follows from Proposition (1.4.16) that

$$\mathcal{F}_{r-1} H^*(X; \mathbb{Q}) \neq H^*(X; \mathbb{Q}).$$

Now suppose that $H^j(X; \mathbb{Q}) \neq 0$ only for $j = j_0, \ldots, j_m$, where $0 = j_0 < \ldots < j_m$. By induction it is immediate that

$$H^{j_i}(X; \mathbb{Q}) \subseteq \mathcal{F}_i H^*(X; \mathbb{Q})$$

for $0 \leq i \leq m$. Hence $m > r - 1$. $\square$

(4.4.6) Remarks
(1) The proof of Corollary (1.4.21) can be imitated in the present context to give an alternative proof of Theorem (4.4.3)(1).
 (2) $\operatorname{rank}_0(T^r) = r$, and $H^j(T^r; \mathbb{Q}) \neq 0$ for exactly $r+1$ different values of j.

We shall conclude this section by pointing out the analogues of Theorems (4.4.3)(1) and (4.4.5) for free p-torus actions. The usual convention concerning cohomology theories applies. For more details and further results see Section 1.4, Theorem (4.6.42) and Discussion (4.6.43).

(4.4.7) Theorem
Suppose that $G = (\mathbb{Z}_p)^r$ and that X is a free G-space satisfying condition (LT) of the introduction to Section 3.10. Then:
(1) *if $p = 2$, $\dim_{\mathbb{F}_2} H^*(X; \mathbb{F}_2) \geq 2r$; and*
(2) *if each connected component of X is open in X, and G is acting trivially on $H^*(X; \mathbb{F}_p)$, then $H^j(X; \mathbb{F}_p) \neq 0$ for at least $r + 1$ different values of j.*

Proof.
The results follow at once from Corollary (1.4.21) and Theorem (1.4.14) provided $n(X_G) := \sup\{i; H_G^i(X; \mathbb{F}_p) \neq 0\}$ is finite. In case (B),

$$n(X_G) = n(X/G) \leq \operatorname{cd}_{\mathbb{F}_p}(X/G) \leq \operatorname{cd}_{\mathbb{F}_p}(X) < \infty.$$

In case (C) there is nothing to prove unless $n(X) < \infty$: and in that case $n(X_G) = n(X/G) \leq n(X)$ by Proposition (3.10.9) and Lemma (3.10.16)(2). $\square$

4.5 The method of Browder and Gottlieb

In this section we shall describe a method for analysing elementary abelian p-group actions due to Browder ([Browder, 1983] and [Browder, 1988]) and Gottlieb ([Gottlieb, 1986]). p may be any prime number

4.5. The method of Browder and Gottlieb

(i.e. we shall work in cases (2) and (odd p)); and we shall use the convention regarding cohomology theories which is stated in the introduction to Section 3.10.

To save writing we make the following definition.

(4.5.1) Definition
A topological pair (X, A) will be called a $(n, \mathbb{Z})$-pair if $H^j(X, A; \mathbb{Z}) = 0$ for all $j > n$, and $H^n(X, A; \mathbb{Z}) \cong \mathbb{Z}$. If $A = \emptyset$, we shall say that X is a $(n, \mathbb{Z})$-space.

We can now summarize Corollaries (3.10.21) and (3.10.23) as follows.

(4.5.2) Proposition
Let $G = (\mathbb{Z}_p)^r$ and let (X, A) be a pair of G-spaces satisfying condition (LT) of the introduction to Section 3.10. Suppose that (X, A) is a $(n, \mathbb{Z})$-pair; and that G acts trivially on $H^n(X, A; \mathbb{Z})$ if $p = 2$. Then:

(1) $(X/G, A/G)$ is a $(n, \mathbb{Z})$-pair; and
(2) if G acts freely on $X - A$, if $\pi : (X, A) \to (X/G, A/G)$ is the orbit map, and if $u \in H^n(X, A; \mathbb{Z})$ and $v \in H^n(X/G, A/G; \mathbb{Z})$ are generators, then $\pi^*(v) = \pm p^r u$.

(4.5.3) Definition
Let $G = (\mathbb{Z}_p)^r$ and let (X, A) be a $(n, \mathbb{Z})$-pair of G-spaces. The G-exponent of (X, A), denoted $e(X, A; G)$, is defined to be the order of $\mathrm{coker}[i^* : H^n_G(X, A; \mathbb{Z}) \to H^n(X, A; \mathbb{Z})]$. If $u \in H^n(X, A; \mathbb{Z})$ is a generator, then $e(X, A; G)$ is the least positive integer ν such that $\nu u \in \mathrm{im}(i^*)$. Here i is the inclusion of the fibre in the bundle

$$(X, A) \xrightarrow{i} (X_G, A_G) \xrightarrow{p} BG.$$

We shall see below that $i^* \neq 0$, and, hence, $e(X, A; G)$ is always defined. If $A = \emptyset$, we shall denote the G-exponent by $e(X; G)$. (In [Gottlieb, 1986], $e(X, A; G)$ is called the fibre number of the bundle p.)

(4.5.4) Lemma
Let $G = (\mathbb{Z}_p)^r$ and let (X, A) be a $(n, \mathbb{Z})$-pair of G-spaces with X paracompact and A closed in X. If $p = 2$ suppose that G acts trivially on $H^n(X, A; \mathbb{Z})$. Then:

(1) $e(X, A; G) | p^r$; and
(2) if G acts trivially on $H^*(X, A; \mathbb{Z})$, then $e(X, A; G) | p^c$, where c is the number of integers j such that $j < n$ and $H^j(X, A; \mathbb{Z}) \neq 0$.

Proof.

(1) Consider Diagram 4.2

$$\begin{array}{ccc} (X,A) & \xrightarrow{\text{id}} & (X,A) \\ \downarrow{i_0} & & \downarrow{i} \\ EG \times (X,A) & \xrightarrow{j} & (X_G, A_G) \end{array}$$

Diagram 4.2

where i_0 is the inclusion $x \mapsto (z_0, x)$ for a base point $z_0 \in EG$, and j is the orbit map. Since EG is contractible, i_0^* is an isomorphism. So, given a generator $u \in H^n(X, A; \mathbb{Z})$, there exists $y \in H^n(EG \times X, EG \times A; \mathbb{Z})$ such that $i_0^*(y) = u$. Let μ be the transfer of Theorem (3.10.15) corresponding to j. (Note that $EG \times X$ is paracompact: see Section 3.2.) Then

$$i^*\mu(y) = i_0^* j^* \mu(y) = i_0^* \left(\sum_{g \in G} g^*(y) \right) = i_0^*(p^r y) = p^r u.$$

(2) If G acts trivially on $H^*(X, A; \mathbb{Z})$, then, in the Leray-Serre spectral sequence of $(X_G, A_G) \to BG$ with coefficients in $\mathbb{Z}$,

$$pE_2^{a,b} = pH^a(BG; H^b(X, A; \mathbb{Z})) = 0$$

wherever $a > 0$. It follows that $e(X, A; G) | p^c$ by considering the differentials $d_m : E_m^{0,n} \to E_m^{m, n-m+1}$. □

(4.5.5) Proposition

Let $G = (\mathbb{Z}_p)^r$, and let (X, A) be a $(n, \mathbb{Z})$-pair of G-spaces which satisfies condition (LT). If $p = 2$, suppose that G acts trivially on $H^n(X, A; \mathbb{Z})$. If G acts freely on $X - A$, then $e(X, A; G) = p^r$.

Proof.

Let $u \in H^n(X, A; \mathbb{Z})$ be a generator. By Proposition (4.5.2)(2), there is a generator $v \in H^n(X/G, A/G; \mathbb{Z})$ such that $\pi^*(v) = p^r u$. Now $\pi = \psi i$, where $\psi : (X_G, A_G) \to (X/G, A/G)$ is the standard map: and ψ^* is an isomorphism by Proposition (3.10.9). □

(4.5.6) Corollary

Under the conditions of Proposition (4.5.5), if G acts trivially on $H^*(X, A; \mathbb{Z})$, then:

(1) $H^j(X, A; \mathbb{Z}) \neq 0$ for at least $r+1$ different values of j; (cf. Theorem (1.4.14); see also Corollary (4.6.37) and Discussion (4.6.43));
and

4.5. The method of Browder and Gottlieb

(2) there are exactly r different non-zero differentials
$$d_m : E_m^{0,n} \to E_m^{m,n-m+1}$$
in the Leray-Serre spectral sequence of $(X_G, A_G) \to BG$ with integer coefficients.

Proof.
This is immediate from Proposition (4.5.5) and from Lemma (4.5.4)(2) and its proof. $\square$

(4.5.7) Corollary
If X satisfies condition (LT), if
$$H^*(X; \mathbb{Z}) \cong H^*((S^n)^m; \mathbb{Z})$$
for $n, m > 0$, and if $G = (\mathbb{Z}_p)^r$ acts freely on X and trivially on $H^*(X; \mathbb{Z})$, then $r \leq m$.
(See Discussion (4.6.43).)

We come now to a theorem of [Browder, 1988], which can be considered as a partial generalization of Proposition (3.10.9). Note that, by means of the mapping cylinder, $\psi : X_G \to X/G$ can be viewed as an inclusion up to homotopy: hence one may consider the relative cohomology $H^*(X/G, X_G; \mathbb{Z})$.

(4.5.8) Theorem
Let $G = (\mathbb{Z}_p)^r$ and let X be any paracompact G-space. Let ν be the number of distinct orders of non-trivial isotropy subgroups: i.e. ν is the order of $\{|G_x|; x \in X \text{ and } G_x \neq \{1\}\}$. Then $p^\nu u = 0$ for any $u \in H^*(X/G, X_G; \mathbb{Z})$.

Before proving the theorem we have the following lemma which eliminates the trivial isotropy subgroups (i.e. the free part of the action).

(4.5.9) Lemma
Let $G = (\mathbb{Z}_p)^r$ and let X be any paracompact G-space. Then
$$H^*(X/G, X_G; \mathbb{Z}) \cong H^*(X_1/G, (X_1)_G; \mathbb{Z})$$
where $X_1 = \{x \in X; G_x \neq \{1\}\}$, the singular set.

Proof.
We shall use integer coefficients throughout and we shall omit them from the notation. X_1 is closed, since $X - X_1$, is open by Proposition (4.1.11). Let W be a closed invariant neighbourhood of X_1 in X, let V

be the complement of the interior of W, and let $U = V \cap W$. Then there is a Mayer-Vietoris sequence
$$\ldots \to H^j(X/G, X_G) \to H^j(V/G, V_G) \oplus H^j(W/G, W_G)$$
$$\to H^j(U/G, U_G) \to \ldots$$
But $V_G \simeq V/G$ and $U_G \simeq U/G$ by Proposition (3.10.9). Hence
$$H^*(X/G, X_G) \cong H^*(W/G, W_G).$$
So the result follows by tautness. $\square$

Proof of Theorem (4.5.8).

By Lemma (4.5.9) we may assume that $X = X_1$, and we shall proceed by induction on ν.

If $\nu = 1$, let $K_1, \ldots, K_n$ be the distinct isotropy subgroups. For $1 \leq i \leq n$, let $X(i) = \{x \in X; G_x = K_i\}$. By Proposition (4.1.11), each $X(i)$ is open, and hence also closed. So
$$H^*(X/G, X_G) \cong \bigoplus_{i=1}^{n} H^*(X(i)/G, X(i)_G).$$
(Again we shall use only integer coefficients, and we shall omit them from the notation.) Thus, without loss of generality, $n = 1$, and $K = K_1$ is the only isotropy subgroup. Now
$$X_G \approx BK \times X_{G/K} \simeq BK \times X/G/K \approx BK \times X/G.$$
Hence, by the Künneth theorem,
$$H^j(X_G) \cong \bigoplus_{i=0}^{j} H^{j-i}(BK) \otimes H^i(X/G) \oplus \bigoplus_{i=0}^{j+1} H^{j+1-i}(BK) * H^i(X/G)$$
$$\cong H^j(X/G) \oplus A^j,$$
where $pA^j = 0$.

By the naturality of the Künneth exact sequence,
$$H^{j+1}(X/G, X_G) \cong A^j.$$
So $pH^*(X/G, X_G) = 0$.

Now suppose that $\nu > 1$. Let $n_1, \ldots, n_\nu$, with $n_1 < \ldots < n_\nu$, be the distinct orders of the isotropy subgroups. Let $Y = \{x \in X; |G_x| > n_1\}$. Again $X - Y$ is open by Proposition (4.1.11); and so Y is closed. Let $f^* : H^*(X/G, X_G) \to H^*(Y/G, Y_G)$ be the restriction; and let
$$u \in H^j(X/G, X_G).$$
By the induction hypothesis, $p^{\nu-1} f^*(u) = 0$.

By tautness, $H^*(Y/G, Y_G) \cong \lim_{\overrightarrow{U}} H^*(U/G, U_G)$, where U ranges over the closed invariant neighbourhoods of Y in X. So there is a closed

4.5. The method of Browder and Gottlieb

invariant neightbourhood W of Y in X such that $p^{\nu-1} f_W^*(u) = 0$ where $f_W^* : H^*(X/G, X_G) \to H^*(W/G, W_G)$ is the restriction.

Let V be the complement of the interior of W, and let $U = V \cap W$. From the Mayer-Vietoris sequence

$$\ldots \xrightarrow{\delta} H^j(X/G, X_G) \xrightarrow{\alpha} H^j(V/G, V_G) \oplus H^j(W/G, W_G)$$
$$\to H^j(U/G, U_G) \to \ldots,$$

since $V \subseteq X - Y = \{x \in X; |G_x| = n_1\}$, $p^{\nu-1}\alpha(u) = 0$. So there exists $v \in H^{j-1}(U/G, U_G)$ such that $p^{\nu-1}u = \delta(v)$. Since $U \subseteq X - Y$, $pv = 0$; and so $p^\nu u = 0$. □

(4.5.10) Corollary
Let $G = (\mathbb{Z}_p)^r$ and let X be any paracompact G-space. Then

$$p^s H^*(X/G, X_G; \mathbb{Z}) = 0,$$

where $p^s = \max\{|G_x|; x \in X\}$.

Recall from Definition (4.1.5) that if $\Phi : G \times X \to X$ is an action of $G = (\mathbb{Z}_p)^r$ on X, then $\text{rank}(\Phi) = \min\{\text{rank}_p(G/G_x); x \in X\}$. So, if $\text{rank}(\Phi) = \rho$, then p^ρ is the order of the smallest orbit. By means of a series of lemmata we shall prove the following theorem of [Browder, 1988] and [Gottlieb, 1986]. Afterwards, briefly, we shall discuss various generalizations.

(4.5.11) Theorem
Let M be a connected closed orientable topological manifold. Let Φ . $G \times M \to M$ be an action of $G = (\mathbb{Z}_p)^r$ on M, and suppose that the action is orientation preserving if $p = 2$. Then $e(M; G)$ is equal to the order of the smallest orbit: i.e. $e(M; G) = p^\rho$, where $\rho = \text{rank}(\Phi)$.

From now until the theorem has been proved we shall assume that M is a connected closed orientable topological n-manifold and that $\Phi : G \times M \to M$ is an action of $G = (\mathbb{Z}_p)^r$ on M, which is orientation preserving if $p = 2$. Let $[M] \in H_n(M; \mathbb{Z})$ denote a top homology class, and let $\langle M \rangle \in H^n(M; \mathbb{Z})$ be the corresponding top cohomology class. M/G is a $(n, \mathbb{Z})$-space by Proposition (4.5.2)(1); and we shall let $\langle M/G \rangle \in H^n(M/G; \mathbb{Z})$ be a generator chosen so that $\pi^*\langle M/G \rangle$ is a positive multiple of $\langle M \rangle$, where $\pi : M \to M/G$ is the orbit map.

(4.5.12) Lemma
If Φ is effective, then $\pi^*\langle M/G \rangle = p^r \langle M \rangle$.

Proof.

If ϕ is effective, then $\text{cd}_{\mathbb{Z}} M_1 \leq n-2$, where $M_1 = \{x \in M; G_x \neq \{1\}\}$, the singular set. If Φ is smooth or locally smooth ([Bredon, 1972]), then this is clear by considering linear slices. More generally it follows from [Borel et al., 1960], Chapter V, 2.6. The lemma now follows by applying Proposition (4.5.2)(2) with $X = M$ and $A = M_1$. □

(4.5.13) Corollary
If Φ is effective and $\text{rank}(\Phi) = \rho$, then $e(M; G) \geq p^\rho$.

Proof.

Consider Diagram 4.3 (with integer coefficients).

$$H^n(M/G) \xrightarrow{\psi^*} H^n_G(M) \xrightarrow{\delta} H^{n+1}(M/G, M_G)$$
$$\searrow{\pi^*} \quad \downarrow{i^*}$$
$$H^n(M)$$

Diagram 4.3

Let $e(M; G) = e$ and choose $y \in H^n_G(M)$ so that $i^*(y) = e\langle M\rangle$. By Theorem (4.5.8), $p^{r-\rho}\delta(y) = 0$; and so there is an integer a such that $p^{r-\rho}y = \psi^*(a\langle M/G\rangle)$. Hence

$$p^{r-\rho}e\langle M\rangle = \pi^*(a\langle M/G\rangle) = ap^r\langle M\rangle,$$

by Lemma (4.5.12). $e = ap^\rho$, therefore. □

(4.5.14) Lemma
If $\text{rank}(\Phi) = \rho$, then $e(M; G) \leq p^\rho$.

Proof.

Let N be an integer very much larger than n, and let E^N_G be a compact N-dimensional manifold approximating EG. Let $M^N_G = (E^N_G \times M)/G$; and for any subgroup $K \subseteq G$ let $B^N_K = E^N_G/K$. We may take E^N_G to be closed and orientable, so that M^N_G and B^N_K are also closed and orientable. As above $[M^N_G]$, $[B^N_K]$, $\langle M^N_G\rangle$ and $\langle B^N_K\rangle$ will denote the top homology and cohomology classes. Let $p_N : M^N_G \to B^N_G$ be the bundle map and let $\phi_K : B^N_K \to B^N_G$ be the standard map which is the orbit map for the free G/K-action on B^N_K. In particular, $\phi^*_K\langle B^N_G\rangle = |G/K|\langle B^N_K\rangle$ by Proposition (4.5.2)(2).

Now let $K = G_x$ for some $x \in M$. Then inclusion of the orbit, $G(x) \to M$, induces a map

$$\sigma_K : B^N_K \approx G(x)^N_G \to M^N_G;$$

4.5. The method of Browder and Gottlieb

and $p_N \sigma_K = \phi_K$. By Poincaré duality there is a uniquely determined class $\theta_K \in H^n(M_G^N; \mathbb{Z})$ such that
$$[M_G^N] \cap \theta_K = \sigma_{K*}[B_K^N].$$

Now
$$\sigma_K^* p_N^* \langle B_G^N \rangle = |G/K|\langle B_K^N \rangle,$$
by above; and so
$$[M_G^N] \cap \theta_K p_N^* \langle B_G^N \rangle = \sigma_{K*}[B_K^N] \cap p_N^* \langle B_G^N \rangle$$
$$= \sigma_{K*}([B_K^N] \cap \sigma_K^* p_N^* \langle B_G^N \rangle) = |G/K|.$$

In the Leray-Serre spectral sequence for p_N with integer coefficients, $E_2^{*,n} \cong H^*(B_G^N; \mathbb{Z}) \otimes H^n(M; \mathbb{Z})$, and $E_\infty^{*,n}$ is a $H^*(B_G^N; \mathbb{Z})$-submodule of $E_2^{*,n}$. Let $i : M \to M_G^N$ be the inclusion of a fibre, and let $i^*(\theta_K) = a\langle M \rangle$.

Let $\pi_n : H^*(M_G^N; \mathbb{Z}) \to E_\infty^{*,n}$ be the projection associated with the increasing filtration on E_∞. Then
$$\pi_n(\theta_K) = 1 \otimes a\langle M \rangle.$$

Hence
$$p_N^* \langle B_G^N \rangle \pi_n(\theta_K) = \pi_n(p_N^* \langle B_G^N \rangle \theta_K) = \langle B_G^N \rangle \otimes a\langle M \rangle$$
in $E_\infty^{N,n} = E_2^{N,n}$. So $p_N^* \langle B_G^N \rangle \theta_K = a \langle M_G^N \rangle$.

From above, then $a = \pm |G/K|$. Hence $e(M;G) \leq |G/K|$. $\square$

Proof of Theorem (4.5.11).

By Corollary (4.5.13) and Lemma (4.5.14), the result holds if Φ is effective. If Φ is not effective, let $\mathbb{Z}_p$ act on S^2 by rotations, and let G act on $W = (S^2)^r$ diagonally. Let $\Psi : G \times M \times W \to M \times W$ be the

$$\begin{array}{ccccc} M & \xleftarrow{\pi_1} & M \times W & \xrightarrow{\pi_2} & W \\ \downarrow{i} & & \downarrow{j} & & \downarrow{i_W} \\ M_G & \xleftarrow{\overline{\pi}_1} & (M \times W)_G & \xrightarrow{\overline{\pi}_2} & W_G \end{array}$$

Diagram 4.4

diagonal action. Since $W^G \neq \emptyset$, $\text{rank}(\Psi) = \text{rank}(\Phi) = \rho$, say; and Ψ is effective, since G acts effectively on W. So $e(M \times W; G) = p^\rho$.

Now consider Diagram 4.4.

There exist $y \in H_G^n(M; \mathbb{Z})$ and $z \in H_G^{2r}(W; \mathbb{Z})$ such that $i^*(y) = e\langle M \rangle$, where $e = e(M;G)$, and $i_W^*(z) = \langle W \rangle$, since $e(W;G) = 1$, by Lemma (4.5.14), because $W^G \neq \emptyset$. Then
$$j^*(\overline{\pi}_1^*(y)\overline{\pi}_2^*(z)) = e\pi_1^*\langle M \rangle \pi_2^* \langle W \rangle = e\langle M \times W \rangle.$$

Thus $e \geq e(M \times W; G) = p^\rho$: and the theorem follows since effectiveness was not assumed in Lemma (4.5.14). $\square$

We shall now indicate briefly how to get a relative version of Theorem (4.5.8), which will then be applied to get Theorem (4.5.11) for compact orientable manifolds with boundary.

(4.5.15) Theorem
Let $G = (\mathbb{Z}_p)^r$, let X be any paracompact G-space and let $A \subset X$ be any closed invariant subspace. Let ν be the number of distinct orders of non trivial isotropy subgroups in $X - A$: i.e. ν is the order of $\{|G_x|; x \in X - A$ and $G_x \neq \{1\}\}$. Then $p^\nu u = 0$, for any $u \in H^*(X/G, X_G \cup A/G; \mathbb{Z})$.

(If $\psi : X_G \to X/G$ is the standard map and if $M(\psi)$ is the mapping cylinder of ψ, then $M(\psi|A_G) \subseteq M(\psi)$, where $\psi|A_G : A_G \to A/G$; and the pair $(X/G, X_G \cup A/G)$ is to interpreted as $(M(\psi), X_G \cup M(\psi|A_G))$.)

Proof.
First, $H^*(X/G, X_G \cup A/G; \mathbb{Z}) \cong H^*(X_1/G, (X_1)_G \cup A_1/G; \mathbb{Z})$, where $A_1 = X_1 \cap A$. This follows directly from Lemma (4.5.9) by considering the homomorphism from the long exact sequence of the triple $(X/G, X_G \cup A/G, X_G)$ to the long exact sequence of the triple
$$(X_1/G, (X_1)_G \cup A_1/G, (X_1)_G)$$
together with the excision isomorphism
$$H^*(X_G \cup A/G, X_G; \mathbb{Z}) \xrightarrow{\cong} H^*(A/G, A_G; \mathbb{Z}).$$
Secondly the argument for the case where $\nu = 1$ goes through as before. One has that
$$H^*(X/G, X_G \cup A/G; \mathbb{Z}) \cong \bigoplus_{i=1}^n H^*(Y(i)/G, Y(i)_G \cup A/G; \mathbb{Z})$$
where $Y(i) = X(i) \cup A$, and $X(i) = \{x \in X; G_x = K_i\}$, as before. Clearly $Y(i)$ is closed by Proposition (4.1.11). If K is the only isotropy subgroup on $X - A$, then $H^*(X_G, A_G; \mathbb{Z}) \cong H^*(BK \times (X/G, A/G); \mathbb{Z})$; and the rest of the argument for the case where $\nu = 1$ proceeds by means of the Künneth theorem as before.

For the induction argument when $\nu > 1$, we let $n_1 < \ldots < n_\nu$ be the distinct orders of isotropy subgroups in $X - A$, and let
$$Y = \{x \in X; |G_x| > n_1\}$$
as before. Again Y is closed by Proposition (4.1.11). By considering the restriction
$$H^*(X/G, X_G \cup A/G; \mathbb{Z}) \to H^*(Y/G, Y_G \cup (Y \cap A)/G; \mathbb{Z})$$
we can argue as before provided that we can show that
$$H^*(Y/G, Y_G \cup (Y \cap A)/G; \mathbb{Z}) \cong \varinjlim_U H^*(U/G, U_G \cup (U \cap A)/G; \mathbb{Z}).$$

4.5. The method of Browder and Gottlieb

where U ranges over the closed invariant neighbourhoods of Y. By looking at the long exact sequence for $(U/G, U_G \cup (U \cap A)/G, U_G)$, the latter follows provided that we can show the following: if N is any closed neighbourhood of $(Y \cap A)/G$ in A/G, then there exists a closed neighbourhood W of Y/G in X/G such that $W \cap A/G \subseteq N$. Well, given such a neighbourhood N, its interior in A/G has the form $U \cap A/G$ where U is open in X/G. Since X/G is paracompact, it is regular, and hence there is an open neighbourhood V of Y/G in X/G such that $V \cap A/G = U \cap A/G$. Since X/G is paracompact, it is normal, and hence there is a closed neighbourhood W of Y/G in X/G such that $W \subseteq V$. Hence $W \cap A/G \subseteq N$. □

(4.5.16) Theorem
Let M be a connected compact orientable topological manifold with boundary ∂M. Let $\Phi : G \times M \to M$ be an action of $G = (\mathbb{Z}_p)^r$ on M, and suppose that the action is orientation preserving if $p = 2$. Then $e(M, \partial M; G)$ is equal to the order of the smallest orbit: i.e. $e(M, \partial M; G) = p^\rho$, where $\rho = \mathrm{rank}(\Phi)$.

Proof.
The proof is essentially the same. If G is effective on M, then, with the notation of [Bredon, 1967a], p.74,
$$\dim_{c,\mathbb{Z}}(M - \partial M)_1 = \dim_{\mathbb{Z}}(M - \partial M)_1 \leq n - 2.$$
Hence $H^j(M_1, (\partial M)_1) = 0$ for all $j \geq n - 1$; and so
$$H^j(M_1/G, (\partial M)_1/G) = 0$$
for all $j \geq n - 1$, by Lemma (3.10.20). Let
$$\pi : (M, \partial M) \to (M/G, \partial M/G)$$
and
$$\overline{\pi} : (M, \partial M \cup M_1) \to (M/G, (\partial M \cup M_1)/G)$$
be the orbit maps. From the long exact sequence of $(M, \partial M \cup M_1, \partial M)$ and the excision
$$(\partial M \cup M_1, \partial M) \sim (M_1, (\partial M)_1),$$
it follows that π^* and $\overline{\pi}^*$ are isomorphic maps in degree n. Hence
$$\pi^* \langle M/G, \partial M/G \rangle = p^r \langle M, \partial M \rangle$$
by Proposition (4.5.2). It now follows easily from Theorem (4.5.15) that $e(M, \partial M; G)$ is no less than the order of the smallest orbit in $M - \partial M$.

From essentially the same duality argument as before, one gets that $e(M, \partial M; G)$ is at most the order of the smallest orbit in M.

Again the argument with $W = (S^2)^r$ removes the need to assume that Φ is effective. One concludes that $e(M, \partial M; G)$ is equal to the order of the smallest orbit in M, which is equal to the order of the smallest orbit in $M - \partial M$. (Note that the last equality, which follows from the inequalities above, is clear for smooth actions; but it is not so clear for arbitrary continuous actions.) □

Using cohomology with compact supports there is a second generalization of Theorem (4.5.8), namely the following.

(4.5.17) Theorem
Let $G = (\mathbb{Z}_p)^r$, let X be any locally compact paracompact G-space, and let A by any closed invariant subspace of X. Let ν be the number of distinct orders of non-trivial isotropy subgroups in $X - A$. Then, for any $j \geq 0$, for any N very much larger than j, and for any $u \in H_c^j(X/G, X_G^N \cup A/G; \mathbb{Z})$, $p^\nu u = 0$.

We shall leave the details of the proof (i.e. adapting the proof of Theorem (4.5.15) to work with compact supports) as an exercise: and the same goes for the following consequence. Before stating the next theorem, however, we note that if G is acting on a non-compact connected orientable topological n-manifold without boundary, then

$$i^* : H_c^n(M_G^N; \mathbb{Z}) \to H_c^n(M; \mathbb{Z})$$

is independent of N, for all N very much larger than n. This follows from comparing the Leray spectral sequences of $M_G^N \to B_G^N$ with compact supports for different values of N. Thus we may define $e_c(M; G)$ to be the order of $\mathrm{coker}[i^* : H_c^n(M_G^N; \mathbb{Z}) \to H_c^n(M; \mathbb{Z})]$, for any N very much larger than n.

(4.5.18) Theorem
Let M be a connected orientable paracompact topological manifold without boundary. Let $\Phi : G \times M \to M$ be an action of $G = (\mathbb{Z}_p)^r$ on M, and suppose that Φ is orientation preserving if $p = 2$. Then $e_c(M; G)$ is equal to the order of the smallest orbit: i.e. $e_c(M; G) = p^\rho$, where $\rho = \mathrm{rank}(\Phi)$.

(4.5.19) Remarks
(1) The case of non-compact manifolds with boundary is essentially contained in Theorem (4.5.18), since

$$H_c^*(M, \partial M; \mathbb{Z}) \cong H_c^*(M - \partial M; \mathbb{Z}).$$

(Note that $(\partial M)_G^N \approx \partial(M_G^N)$. Also one finds that the size of the smallest

4.5. The method of Browder and Gottlieb

orbit in M is the same as that in $M - \partial M$, as in the proof of Theorem (4.5.16).)

(2) All the results of this section hold if one replaces $\mathbb{Z}$ by $\mathbb{Z}_{(p)}$, the integers localized at p.

(3) Topological manifolds do not form a convenient class of spaces on which to study continuous group actions, since fixed point sets may not be topological manifolds. For example, there is an involution on a sphere the fixed point set of which is not even an absolute neighbourhood retract. Cohomology manifolds, on the other hand, form a much more convenient class of spaces with good duality properties. (See [Borel et al., 1960] and [Bredon, 1967a].) Owing to the good duality properties of cohomology manifolds, it follows at once that Theorem (4.5.11) is valid if M is any connected compact orientable cohomology manifold over $\mathbb{Z}$ (or $\mathbb{Z}_{(p)}$) without boundary, and that Theorem (4.5.18) is valid if M is any connected paracompact orientable cohomology over $\mathbb{Z}$ (or $\mathbb{Z}_{(p)}$) without boundary. (Cohomology manifolds are locally compact by definition.)

(4) The parts of Theorems (4.5.11), (4.5.16) and (4.5.18) which assert that the G-exponent is no less than the order of the smallest orbit use only the fact that the singular set has codimension at least two when the action is effective. This allows one to conclude that the G-exponent is no less than the order of the smallest orbit (without assuming effectiveness, by the arguments with $W = (S^2)^r$ in the proof of Theorem (4.5.11)) whenever M is a manifold (or cohomology manifold) except for some singular set of codimension at least two. This is useful for studying certain complex algebraic varieties. See the discussion of near-manifolds in [Browder, 1988] for details.

(5) Let $G = \mathbb{Z}_p$ act on S^4 by standard rotations fixing the north and south poles, n and s. Let X be obtained from S^4 by attaching a 1-cell with one end at n and the other at s, and extend the action by letting G act trivially on the 1-cell. By the Localization Theorem, since X^G is just the 1-cell, $e(X;G) = p$. So Theorem (4.5.11) does not hold for near-manifolds in the sense of [Browder, 1988].

(6) It might be useful in studying actions on non-orientable manifolds (or cohomlgy manifolds) to consider coefficients in the orientation sheaf. Otherwise one might study the lifted action on the orientable double covering. (See [Bredon, 1972], Chapter I, Corollary 9.4. For the case of cohomology manifolds, see [Bredon, 1960].)

(7) See [Browder, 1988] and [Gottlieb, 1986] for further results.

We shall conclude this section with some applications due to Adem and Browder ([Adem, Browder, 1988]). These applications are along the lines of Corollary (3.11.16) and Corollary (4.5.7). In Corollary (3.11.16) we assumed that $H^*(X;\mathbb{F}_p)$ and $H^*((S^n)^m;\mathbb{F}_p)$ were isomorphic as graded $\mathbb{F}_p$-algebras; and in Corollary (4.5.7) we only needed to assume that $H^*(X;\mathbb{Z})$ and $H^*((S^n)^m;\mathbb{Z})$ were isomorphic as graded abelian groups. Adem and Browder assume that $H^*(X;\mathbb{Z})$ and $H^*((S^n)^m;\mathbb{Z})$ are isomorphic as graded rings, and they get much stronger results.

For any finite group G, commutative ring k, and $k[G]$-module M, the k-module of coinvariants is defined to be $M_G := k \otimes_{k[G]} M$, where k is viewed as a trivial $k[G]$-module. Thus $M_G \cong H_0(G;M) \cong M/IM$, where $I \triangleleft k[G]$ is the augmentation ideal.

Now the first main result of [Adem, Browder, 1988] is the following.

(4.5.20) Proposition
Suppose X satisfies condition (LT), and $H^*(X;\mathbb{Z})$ and $H^*((S^n)^m;\mathbb{Z})$ are isomorphic as graded rings, where n, $m > 0$. Let $G = (\mathbb{Z}_p)^r$, where p is any odd prime number, and let $H = \{g \in G;\, g$ acts trivially on $H^*(X;\mathbb{F}_p)\}$. Let L be an isotropy subgroup of maximal p-rank for the action of H on X. Suppose either:

(1) *G is acting freely on X (and so, in particular, $L = \{1\}$); or*
(2) *X is a compact orientable cohomology manifold over $\mathbb{Z}$ (without boundary).*

Then
$$\mathrm{rank}_p(H) - \mathrm{rank}_p(L) \leq \dim_{\mathbb{F}_p} H^n(X;\mathbb{F}_p)_G.$$

Before starting the proof of the proposition we shall prove the following simple lemma.

(4.5.21) Lemma

(1) *Let $G = (\mathbb{Z}_p)^r$, where p is an odd prime, and let M be a $\mathbb{Z}[G]$-module. Suppose that M is a finitely generated free abelian group and that $M \otimes_{\mathbb{Z}} \mathbb{F}_p$ is a trivial $\mathbb{F}_p[G]$-module. Then M is a trivial $\mathbb{Z}[G]$-module.*

(2) *Let M be a finitely generated free abelian group and let V be a $\mathbb{F}_p$-vector-space. Let $\phi : M \to V$ be a homomorphism of abelian groups. Then there is a basis $\{u_1, \ldots, u_n\}$ for M, where $n = \mathrm{rank}(M)$, and $s \leq n$, such that $\{\phi(u_1), \ldots, \phi(u_s)\}$ is a $\mathbb{F}_p$-vector-space basis for $\mathrm{im}(\phi)$, and $\phi(u_i) = 0$ for $i > s$.*

Proof.

(1) Let $\{v_1, \ldots, v_n\}$ be a basis for M as a free abelian group. Let $g \in G$, and let $gv_i = v_i + \sum_{j=1}^{n} a_{ij} v_j$ for $1 \leq i \leq n$. Since $g \otimes 1$ is trivial

on $M \otimes_{\mathbb{Z}} \mathbb{F}_p$, $p|a_{ij}$ for all i, j. Let A be the $n \times n$ integer matrix (a_{ij}) and let I be the identity matrix. Since $g^p = 1$, $(I + A)^p = I$. Hence

$$pA = -\left[\binom{p}{2}A^2 + \ldots + A^p\right].$$

Choose a_{ij} with least p-divisibility: i.e. for some integer $m \geq 1$, $p^m|a_{ij}$, $p^{m+1} \not| a_{ij}$ and $p^m|a_{st}$ for all s, t. But p^{2m+1} divides all entries of $\binom{p}{2}A^2 + \ldots + A^p$, whereas $p^{2m+1} \not| pa_{ij}$.

(2) Let $\{v_1, \ldots, v_n\}$ be any basis for M. From $\{\phi(v_1), \ldots, \phi(v_n)\}$ choose a basis $\{\phi(v_{i_1}), \ldots, \phi(v_{i_s})\}$ for im(ϕ) as a $\mathbb{F}_p$-vector-space, where $i_1 < \ldots < i_s$. Let $N = \mathbb{Z}^s$ with basis $\{w_1, \ldots, w_s\}$, and define

$$\pi : N \to \text{im}(\phi)$$

by

$$\pi(w_j) = \phi(v_{i_j})$$

for $1 \leq j \leq s$. Define

$$\overline{\phi} : M \to N$$

by

$$\overline{\phi}(v_{i_j}) = w_j$$

for $1 \leq j \leq s$, and for $i \notin \{i_1, \ldots, i_s\}$, let $\overline{\phi}(v_i) = w'_i$, where $\pi(w'_i) = \phi(v_i)$. Then $\pi\overline{\phi} = \phi$, $\overline{\phi}$ is surjective, and $M \cong N \oplus \ker \overline{\phi}$. □

Proof of Proposition (4.5.20).

Consider the Leray-Serre spectral sequence for $X \xrightarrow{i} X_H \to BH$ with integer coefficients. By Lemma (4.5.21)(1), H acts trivially on $H^*(X; \mathbb{Z})$. Consider

$$d_{n+1} : E_{n+1}^{0,n} \cong H^n(X; \mathbb{Z}) \to E_{n+1}^{n+1,0} \cong H^{n+1}(BH; \mathbb{Z}).$$

$H^{n+1}(BH; \mathbb{Z})$ is a $\mathbb{F}_p$-vector-space; and so, by Lemma (4.5.21)(2), there is a basis $\{u_1, \ldots, u_m\}$ for $H^n(X; \mathbb{Z})$ such that $\{d_{n+1}(u_1), \ldots, d_{n+1}(u_s)\}$ is a basis for im(d_{n+1}) as a $\mathbb{F}_p$-vector-space, and $d_{n+1}(u_i) = 0$ for $i > s$. Thus $pu_1, \ldots, pu_s, u_{s+1}, \ldots, u_m$ all survive to infinity: i.e. they are all in im(i^*). Hence $p^s u_1 \ldots u_m \in \text{im}(i^*)$. So $e(X; H) \leq p^s$.

Now in case (1), by Proposition (4.5.5), rank$_p(H) \leq s$. And, in case (2), by Theorem (4.5.11), and Remark (4.5.19)(3), rank$_p(H)$ − rank$_p(L) \leq s$.

The group G acts on $X_H = (EH \times X)/H$ by $gH(z,x) = H(z, gx)$, for all $g \in G$, $z \in EH$ and $x \in X$. So G acts on the spectral sequence, trivially on the base $E_q^{*,0}$. Thus there is an epimorphism of $\mathbb{Z}[G]$-modules, $d_{n+1} : H^n(X; \mathbb{Z}) \to \text{im}(d_{n+1})$; and hence an epimorphism of $\mathbb{F}_p[G]$-modules $H^n(X; \mathbb{F}_p) \to \text{im}(d_{n+1})$. But im$(d_{n+1})$ is a trivial $\mathbb{F}_p[G]$-module; and so the latter epimorphism factors through $H^n(X;\mathbb{F}_p)_G$. Hence

$$s = \dim_{\mathbb{F}_p} \operatorname{im}(d_{n+1}) \leq \dim_{\mathbb{F}_p} H^n(X;\mathbb{F}_p)_G.$$

□

(4.5.22) Remarks
(1) If $G = (\mathbb{Z}_p)^r$ is generated by $g_1, \ldots, g_r$, if k is any field of characteristic p, and if M is a $k[G]$-module, then $M_G \cong M/\mathfrak{m}M$, where $\mathfrak{m} = (1-g_1, \ldots, 1-g_r)$, the unique prime ideal in $k[G]$. Since $\mathfrak{m}^{(p-1)r+1} = (0)$, if M is finitely generated, then $\dim_k M_G$ is the minimum possible number of elements in a generating set for M as a $k[G]$-module.

Also $\operatorname{Hom}_k(M_G, k) \cong \operatorname{Hom}_k(M; k)^G$; and so, in Proposition (4.5.20), we could replace $H^n(X;\mathbb{F}_p)_G$ by $H_n(X;\mathbb{F}_p)^G$, where homology is understood as the dual of cohomology when X is not an agreement space.
(2) See Discussion (4.6.43) for some refinements of Proposition (4.5.20), especially for ways in which condition (2) in the statement of the proposition can be relaxed.

Using some ingenious, but not very difficult, representation theory, Adem and Browder ([Adem, Browder, 1988]) carry Proposition (4.5.20) a big step further, arriving at the following result.

(4.5.23) Theorem
Suppose that X satisfies condition (LT), and $H^*(X;\mathbb{Z})$ and $H^*((S^n)^m;\mathbb{Z})$ are isomorphic as graded rings, where $m, n > 0$. Let $G = (\mathbb{Z}_p)^r$, where p is any odd prime number, and let $H = \{g \in G;\ g\ \text{acts trivially on } H^*(X;\mathbb{F}_p)\}$. Let L be an isotropy subgroup of maximal p-rank for the action of H on X. Suppose either

(1) *G is acting freely on X (and so, in particular, $L = \{1\}$); or*
(2) *X is a compact orientable cohomology manifold over $\mathbb{Z}$ (without boundary).*

Then

$$\operatorname{rank}_p(G) - \operatorname{rank}_p(L)$$
$$\leq \frac{1}{p-2} \dim_{\mathbb{F}_p} H^n(X;\mathbb{F}_p) + \frac{p-3}{p-2} \dim_{\mathbb{F}_p} H^n(X;\mathbb{F}_p)_G.$$

In particular, $\operatorname{rank}_p(G) - \operatorname{rank}_p(L) \leq \dim_{\mathbb{F}_p} H^n(X;\mathbb{F}_p) = m$. Thus, in case (1), $r \leq m$.
See [Adem, Browder, 1988] for the proof.

(4.5.24) Remarks

(1) Proposition (4.5.20) and Theorem (4.5.23) hold with $\mathbb{Z}$ replaced by $\mathbb{Z}_{(p)}$, the integers localized at p, throughout.
(2) Adem and Browder use rational representation theory. Thus they must assume that X has the same $\mathbb{Z}$- or $\mathbb{Z}_{(p)}$-cohomology as $(S^n)^m$ despite Discussion (4.6.43)(4).
(3) By Discussion (4.6.43)(3)(b), Theorem (4.5.23) holds if condition (2) is replaced by the much weaker assumption that X satisfies condition (LT). In case (B) one need only assume that $\mathrm{cd}_{\mathbb{F}_p}(X) < \infty$.
(4) When $p = 2$, under the other conditions of Theorem (4.5.23), and assuming that X is homotopy equivalent to $(S^n)^m$, Adem and Browder obtain the inequality

$$\mathrm{rank}_p(G) - \mathrm{rank}_p(L) \leq \dim_{\mathbb{F}_p} H^n(X; \mathbb{F}_p) = m$$

provided that $n \neq 1$, 3 or 7. The case when $p = 2$ and $n = 1$, 3 or 7 is left open. (See also Discussion (4.6.43)(3)(c).)
(5) See [Adem, 1987] for some further results concerning actions of $\mathbb{Z}_p$ on spaces X with $H^*(X; \mathbb{Z}) \cong H^*((S^n)^m; \mathbb{Z})$.

4.6 Equivariant Tate cohomology

In this section we shall describe Swan's construction of equivariant Tate cohomology ([Swan, 1960]), prove Swan's main theorem (Theorem (4.6.15)), and give another form of the Browder-Gottlieb exponent (Definition (4.5.3)). To begin with we shall let G be any finite group, and we shall let k be any commutative ring with identity. We shall maintain our usual convention regarding cohomology theories, as stated in the introduction to Section 3.10.

Recall that a complete resolution ([Brown, K.S., 1982], Chapter VI, Section 3) of a finite group G over k is an acyclic complex

$$\ldots \to P_n \to P_{n-1} \to \ldots \to P_1 \to P_0 \to P_{-1} \to P_{-2} \to \ldots$$
$$\to P_{-n} \to P_{-n-1} \to \ldots$$

of projective left $k[G]$-modules together with the factoring shown in Diagram 4.5,

$$\begin{array}{ccc} P_0 & \longrightarrow & P_{-1} \\ {}_\varepsilon \searrow & & \nearrow {}_\eta \\ & k & \end{array}$$

Diagram 4.5

where k is viewed as a trivial $k[G]$-module, ε is surjective and η is injective. We shall assume, as we may, that each P_i, $i \in \mathbb{Z}$, is finitely generated.

(4.6.1) Example
Let $G = (\mathbb{Z}_p)^r$, and for $i \geq 0$, let $P_i = \mathcal{E}_i(G)$ as in Recollections (3.11.1)(2). Let $\varepsilon : \mathcal{E}_0(G) \to k$ also be as in Recollections (3.11.1)(2). For $i \geq 1$, let
$$P_{-i} = P_{i-1}^* = \mathrm{Hom}_{k[G]}(\mathcal{E}_{i-1}(G), k[G]);$$
and define
$$\eta : k \to P_{-1} = \mathrm{Hom}_{k[G]}(\mathcal{E}_0(G), k[G]) = \mathrm{Hom}_{k[G]}(k[G], k[G])$$
by $\eta(\alpha) = \alpha\nu$, where $\nu : k[G] \to k[G]$ is defined by $\nu(1) = \sum_{g \in G} g$.

If M is a left $k[G]$-module, then the Tate cohomology of G with coefficients in M, $\hat{H}^*(G;M)$, is defined to be the homology of the cochain complex $\mathrm{Hom}_{k[G]}(P_*, M)$. Since we are assuming that each P_i is finitely generated, we have in degree n,
$$\mathrm{Hom}_{k[G]}(P_*, M)^n = \mathrm{Hom}_{k[G]}(P_n, M) \cong P^n \otimes_{k[G]} M,$$
where P^n is the dual $P_n^* = \mathrm{Hom}_{k[G]}(P_n, k[G])$. (See, e.g., [Brown, K.S., 1982], Chapter I, Proposition (8.3); cf. (1.2.2).) Now we wish to define the Tate cohomology of G with coefficients in a cochain complex of $k[G]$-modules. So suppose that C^* is a cochain complex of left $k[G]$-modules. One forms $\mathrm{Hom}_{k[G]}(P_*, C^*)$: but there is some ambiguity as to how one should define the cochains of degree n. It is conventional to define them as products; but it is more convenient here to use sums. We shall consider both definitions.

(4.6.2) Definition
Let G be a finite group, let k be a commutative ring with identity and let P_* be a complete resolution for G over k consisting of finitely generated projective left $k[G]$-modules. Let C^* be any cochain complex of left $k[G]$-modules. Let $A^n(C^*) = \prod_{i+j=n} \mathrm{Hom}_{k[G]}(P_i, C^j)$ and let
$$B^n(C^*) = \bigoplus_{i+j=n} \mathrm{Hom}_{k[G]}(P_i, C^j).$$
Let $A(C^*) = \bigoplus_{n=-\infty}^{\infty} A^n(C^*)$ and let $B(C^*) = \bigoplus_{n=-\infty}^{\infty} B^n(C^*)$ with the standard differentials coming from P_* and C^*. Note that
$$A^n(C^*) \cong \prod_{i+j=n} P^i \otimes_{k[G]} C^j$$

and
$$B^n(C^*) \cong \bigoplus_{i+j=n} P^i \otimes_{k[G]} C^j$$
where
$$P^i = P_i^* = \mathrm{Hom}_{k[G]}(P_i, k[G]).$$
Also let
$$A^{p,q}(C^*) = B^{p,q}(C^*) = \mathrm{Hom}_{k[G]}(P_p, C^q) \cong P^p \otimes_{k[G]} C^q.$$

Define $\hat{H}_G^*(C^*)$ to be the homology of the cochain complex $B(C^*)$: i.e. $\hat{H}_G^*(C^*) = H(B(C^*))$.

Note that there is a natural inclusion $B(C^*) \to A(C^*)$; and clearly we have the following.

(4.6.3) Lemma
The natural inclusion $B(C^*) \to A(C^*)$ is an isomorphism if C^* is bounded: i.e. if there exist integers M, N such that $C^j = 0$ for all $j < M$ and all $j > N$.

(4.6.4) Definition
We shall call the first filtration the decreasing filtration on $B(C^*)$ defined by
$$\mathcal{F}^p B^n(C^*) = \bigoplus_{i \geq p} \mathrm{Hom}_{k[G]}(P_i, C^{n-i}) \cong \bigoplus_{i \geq p} P^i \otimes_{k[G]} C^{n-i}.$$
And we shall call the second filtration the decreasing filtration on $A(C^*)$ defined by
$$\mathcal{F}^q A^n(C^*) = \prod_{j \geq q} \mathrm{Hom}_{k[G]}(P_{n-j}, C^j) \cong \prod_{j \geq q} P^{n-j} \otimes_{k[G]} C^j.$$

(4.6.5) Lemma
If C^* is bounded below (i.e. there is an integer N such that $C^j = 0$ for all $j < N$), then

(1) *the first filtration is regular in the sense of [Cartan, Eilenberg, 1956], Chapter XV, §4; and*
(2) *the second filtration is complete in the sense of [Eilenberg, Moore, 1961].*

Proof.
Clearly $B(C^*) = \bigcup_p \mathcal{F}^p B(C^*)$, and $\mathcal{F}^p B^n(C^*) = 0$ for $p > n - N$.
For the second filtration we have that $\mathcal{F}^N A(C^*) = A(C^*)$, which shows I-completeness; and P-completeness is also clear. $\square$

(4.6.6) Remark
The first filtration on $B(C^*)$ is the completion (I-completion) of the corresponding filtration on $A(C^*)$ in the sense of [Eilenberg, Moore, 1961].

(4.6.7) Corollary
If C^* is bounded below, then there are the following spectral sequences (which we call the first spectral sequence and the second spectral sequence respectively):
(1) $E_2^{p,q} \cong \hat{H}^p(G; H^q(C^*))$ *converging to* $\hat{H}_G^*(C^*)$; *and*
(2) $E_1^{p,q} \cong \hat{H}^q(G; C^p)$ *converging to* $H(A(C^*))$.

(4.6.8) Remarks
(1) The convergence of the first spectral sequence has the following sense. For any p, q and $r > q + 1 - N$, where $C^j = 0$ for all $j < N$, there are epimorphisms
$$E_r^{p,q} \to E_{r+1}^{p,q} \to \cdots$$
and
$$E_\infty^{p,q} = \varinjlim_{s \geq r} E_s^{p,q}.$$
And there is a filtration $\mathcal{F}^p \hat{H}_G^*(C^*)$ on $\hat{H}_G^*(C^*)$ such that
$$\mathcal{F}^p \hat{H}_G^{p+q}(C^*)/\mathcal{F}^{p+1} \hat{H}_G^{p+q}(C^*) \cong E_\infty^{p,q};$$
and
$$\hat{H}_G^n(C^*) \cong \varinjlim_p \mathcal{F}^p \hat{H}_G^n(C^*),$$
where the latter is the direct limit of the sequence of inclusions
$$\cdots \subseteq \mathcal{F}^p \hat{H}_G^n(C^*) \subseteq \mathcal{F}^{p-1} \hat{H}_G^n(C^*) \subseteq \cdots.$$

(2) The convergence of the second spectral sequence should be interpreted as being in the sense of [Eilenberg, Moore, 1961], Theorem 7.4. In particular, if $f : C_1^* \to C_2^*$ is a $k[G]$-cochain map of $k[G]$-cochain complexes which are bounded below, and if f induces an isomorphism on the E_r-level of the second spectral sequences for some $r \geq 1$, then f induces an isomorphism $H(A(C_1^*)) \to H(A(C_2^*))$. Thus (and this is all that we shall need), if $E_r^{*,*} = 0$ for some $r \geq 1$ in the second spectral sequence for C^* (bounded below), then $H(A(C^*)) = 0$. Actually we shall only apply the second spectral sequence when C^* is bounded (above and below), in which case the convergence is clearly much stronger.

(4.6.9) Definition
Let G be a finite group, let X be a paracompact G-space, let k be a commutative ring with identity and let Λ be a k-module. Let $\overline{C}^*(X;\Lambda) = \varinjlim_{\mathcal{U}} C^*(\overline{\mathcal{U}};\Lambda)$, where $\mathcal{U}$ ranges over the locally finite Čech-G-coverings of X (see Definitions and Comments (3.11.2)), $\overline{\mathcal{U}}$ is the Vietoris nerve of $\mathcal{U}$, and $C^*(\overline{\mathcal{U}};\Lambda)$ is the ordered cochain complex of $\overline{\mathcal{U}}$ with coefficients in Λ. Then the equivariant Tate cohomology of X with coefficients in Λ is defined to be $\hat{H}_G^*(X;\Lambda) = \hat{H}_G^*(\overline{C}^*(X;\Lambda))$ where $\overline{C}^*(X;\Lambda)$ is taken with the standard $k[G]$-module structure.

If A is a closed invariant subspace of X, then let $\overline{C}^*(X,A;\Lambda)$ be the kernel of the epimorphism $\overline{C}^*(X;\Lambda) \to \overline{C}^*(A;\Lambda)$. Now the equivariant Tate cohomology of the pair (X,A) with coefficients in Λ is defined to be $\hat{H}_G^*(X,A;\Lambda) := \hat{H}_G^*(\overline{C}^*(X,A;\Lambda))$.

(4.6.10) Proposition
Let G, X, k and Λ be as in Definition (4.6.9). Then:

(1) for any closed invariant subspace $A \subseteq X$, there is a natural long exact sequence
$$\ldots \to \hat{H}_G^n(X,A;\Lambda) \to \hat{H}_G^n(X;\Lambda) \to \hat{H}_G^n(A;\Lambda)$$
$$\to \hat{H}_G^{n+1}(X,A;\Lambda) \to \ldots;$$

(2) for any closed invariant subspace $A \subseteq X$
$$\hat{H}_G^*(A;\Lambda) \cong \varinjlim_V \hat{H}_G^*(V;\Lambda)$$
where V ranges over the closed invariant neighbourhoods of A; and

(3) for any closed invariant subspaces $A, B \subseteq X$ with $A \cup B = X$, there is a long exact Mayer-Vietoris sequence
$$\ldots \to \hat{H}_G^n(X;\Lambda) \to \hat{H}_G^n(A;\Lambda) \oplus \hat{H}_G^n(B;\Lambda) \to \hat{H}_G^n(A \cap B;\Lambda)$$
$$\to \hat{H}_G^{n+1}(X;\Lambda) \to \ldots.$$

Proof.
(1) For fixed G, k and complete resolution P_* it is clear that the functor B of Definition (4.6.2) is exact on $k[G]$-cochain compelexes. So (1) follows from the exact sequence
$$0 \to \overline{C}^*(X,A;\Lambda) \to \overline{C}^*(X;\Lambda) \to \overline{C}^*(A;\Lambda) \to 0.$$

(2) $\varinjlim_V \overline{C}^*(V;\Lambda) \to \overline{C}^*(A;\Lambda)$ is a weak equivalence by the tautness of A. Hence $\hat{H}_G^*(A;\Lambda) \cong \hat{H}_G^*\left(\varinjlim_V \overline{C}^*(V;\Lambda)\right)$ by the first spectral sequence (Corollary (4.6.7)(1)). So the result follows since direct limits commute with tensors, sums and homology; cf. Remark (3.11.4)(3).

(3) Let V and W be closed invariant neighbourhoods of A and B respectively. Let $\mathcal{U}$ be any open covering of X which refines the covering $\{\operatorname{int} V, \operatorname{int} W\}$. Then we have an exact sequence of ordered chain complexes (with coefficients in k):

$$0 \to C_*(\overline{\mathcal{U}}_{V\cap W}) \to C_*(\overline{\mathcal{U}}_V) \oplus C_*(\overline{\mathcal{U}}_W) \to C_*(\overline{\mathcal{U}}) \to 0,$$

where, for example, $\mathcal{U}_V = \{U \cap V; U \in \mathcal{U}\}$. Taking cochains and direct limits over $\mathcal{U}$, there is an exact sequence:

$$0 \to \overline{C}^*(X; \Lambda) \to \overline{C}^*(V; \Lambda) \oplus \overline{C}^*(W; \Lambda) \to \overline{C}^*(V \cap W; \Lambda) \to 0,$$

as can also be seen directly.

By the exactness of the functor B, as in (1), we then have the long exact sequence

$$\ldots \to \hat{H}^n_G(X; \Lambda) \to \hat{H}^n_G(V; \Lambda) \oplus \hat{H}^n_G(W; \Lambda) \to \hat{H}^n_G(V \cap W; \Lambda) \to \ldots$$

The result now follows by taking direct limits over pairs of closed invariant neighbourhoods directed by $(V', W') \geq (V, W)$ if $V' \subseteq V$ and $W' \subseteq W$. □

It is convenient to have some other ways of computing $\hat{H}^*_G(X; \Lambda)$.

(4.6.11) Lemma
Let G, X and Λ be as in Definition (4.6.9). Then there is an isomorphism $\hat{H}^*_G(X; \Lambda) \cong \varinjlim_{\mathcal{U}} \hat{H}^*_G(S^*(|\check{\mathcal{U}}|; \Lambda))$, where S^* indicates singular cochains and $\mathcal{U}$ ranges over the locally finite Čech-G-coverings of X. Furthermore this isomorphism is natural with respect to equivariant maps of G-spaces for fixed G.

Proof.
Remark (3.11.4)(1) shows that $\varinjlim_{\mathcal{U}} \hat{H}^*_G(S^*(|\check{\mathcal{U}}|; \Lambda))$ is well defined, and that there is an isomorphism $\hat{H}^*_G(C^*(\overline{\mathcal{U}}; \Lambda)) \cong \hat{H}^*_G(S^*(|\check{\mathcal{U}}|; \Lambda))$ which is natural with respect to refinements. The result follows since

$$\hat{H}^*_G(X; \Lambda) \cong \varinjlim_{\mathcal{U}} \hat{H}^*_G(C^*(\overline{\mathcal{U}}; \Lambda)).$$

□

(4.6.12) Lemma
Let G and Λ be as in Definition (4.6.9). Let X be a G-CW-complex. Then there are isomorphisms

$$\hat{H}^*_G(X; \Lambda) \cong \hat{H}^*_G(S^*(X; \Lambda)) \cong \hat{H}^*_G(W^*(X; \Lambda)),$$

where $W^*(X; \Lambda)$ is the cellular cochain complex of X with coefficients in

4.6. Equivariant Tate cohomology

Λ: i.e. $W^n(X;\Lambda) = H^n(X^n, X^{n-1};\Lambda)$, where X^n is the $(G-)n$-skeleton of X.

Proof.
Let $\mathcal{U}$ be an open covering of X. Then there are chain maps
$$S_*(X) \xleftarrow{\alpha(\mathcal{U})} S_*(\mathcal{U}) \longrightarrow C_*(\overline{\mathcal{U}}),$$
where $S_*(X)$ is the singular chain complex of X with coefficients in k, $S_*(\mathcal{U})$ is the subcomplex of small chains, i.e. $S_n(\mathcal{U})$ is the free k-module generated by singular simplexes $\sigma : \Delta^n \to X$ such that $\sigma(\Delta^n) \subseteq U$ for some $U \in \mathcal{U}$, $\alpha(\mathcal{U})$ is the inclusion, and $C_*(\overline{\mathcal{U}})$ is the ordered chain complex of $\overline{\mathcal{U}}$; $\alpha(\mathcal{U})$ is a chain homotopy equivalence by [Spanier, 1966], Chapter 4, Theorem 4.14. Thus we have cochain maps
$$S^*(X;\Lambda) \xrightarrow{\alpha} \varinjlim_{\mathcal{U}} \text{Hom}_k(S_*(\mathcal{U}), \Lambda) \xleftarrow{\beta} \bar{C}^*(X;\Lambda),$$
and α is a weak equivalence. The natural transformation,
$$\bar{H}^*)X;\Lambda) \to H^*(X;\Lambda),$$
from Alexander-Spanier theory to singular theory is $H(\alpha)^{-1}H(\beta)$: and it is an isomorphism if X is a CW-complex. Now α and β induce maps of the first spectral sequences: when $\mathcal{U}$ ranges over the cofinal system of Čech-G-coverings and if X is a G-CW-complex one gets isomorphisms on the E_2-level. Thus $\hat{H}^*_G(X;\Lambda) \cong \hat{H}^*_G(S^*(X;\Lambda))$, if X is a G-CW-complex.

To show the second isomorphism, again thanks to the first spectral sequence, all we need is a natural weak-equivalence $W^*(X;\Lambda) \to S^*(X;\Lambda)$. This exists as follows.

By [Dold, 1980], Chapter V, §1, there is a chain homotopy equivalence $\theta : W_*(X) \to S_*(X)$ which is natural up to chain homotopy. By acyclic models ([Dold, 1980], Chapter VI, Proposition 11.7) there is a natural transformation $\psi : S_*(X) \to W_*(X)$ which induces $H(\theta)^{-1}$ in degree 0. Thus $\theta\psi$ induces a natural transformation of singular cohomology theory with coefficients in Λ to itself. The induced map in degree 0 is the identity; and hence it is an isomorphism on a point. By the Comparison Theorem of cohomology theories ([Dyer, 1969], Chapter I, Section B3), $\theta\psi$ induces an isomorphism of cohomology theories (for CW-complexes). Thus ψ induces a natural weak equivalence $W^*(X;\Lambda) \to S^*(X;\Lambda)$.

(Alternatively see Remark (1.2.7)(2) and its proof.) $\square$

It is now an easy matter to deduce a relative version of Lemma (4.6.12). (See, e.g., [Dold, 1980], Chapter V, Proposition 4.7.)

(4.6.13) Corollary

Let G and Λ be as in Definition (4.6.9). Let X be a G-CW-complex, and let $A \subseteq X$ be a G-CW-subcomplex. Then there are isomorphisms
$$\hat{H}_G^*(X, A; \Lambda) \cong \hat{H}_G^*(S^*(X, A; \Lambda)) \cong \hat{H}_G^*(W^*(X, A; \Lambda)),$$
where $W^n(X, A; \Lambda) = H^n(X^n \cup A, X^{n-1} \cup A; \Lambda)$.

(4.6.14) Corollary

If X is a finite-dimensional G-CW-complex, and if $A \subseteq X$ is a G-CW-subcomplex such that G acts freely on $X - A$, then $\hat{H}_G^*(X, A; \Lambda) = 0$.

Proof.

By Lemma (4.6.3) and Corollary (4.6.13) the second spectral sequence associated with $A(W^*(X, A; \Lambda))$ converges to $\hat{H}_G^*(X, A; \Lambda)$. The E_1-term of this spectral sequence is given by $E_1^{p,*} \cong \hat{H}^*(G; W^p(X, A; \Lambda))$. But $W^p(X, A; \Lambda) = H^p(X^p \cup A, X^{p-1} \cup A; \Lambda)$, which is a product of copies of the $k[G]$-module $\Lambda[G] = \Lambda \otimes_k k[G]$, one copy for each G-p-cell in $X - A$. Thus $E_1^{p,*}$ is a product of copies of $\hat{H}^*(G; \Lambda[G])$. The latter group is zero by Shapiro's Lemma ([Brown, K.S., 1982], Chapter VI, (5.3)). □

We are now in a position to prove Swan's main theorem ([Swan, 1960]). We shall give the proof for finite-dimensional G-CW-complexes and, more generally, for paracompact finitistic spaces (case (C)) from section 3.2). Indeed the notion of a finitistic space seems to have been invented by Swan as a general kind of space for which the theorem holds. Note, however, that the theorem also holds for pairs (X, A), where $A \subseteq X$ is a closed invariant subspace, X is paracompact and $\mathrm{cd}_k(X) < \infty$, where k is as in the statement of the theorem below (case (B) of Section 3.2): see Exercise (4.10). Recall that for a G-space X the singular set is defined to be $X_1 := \{x \in X; |G_x| > 1\}$.

(4.6.15) Theorem

Let G be any finite group, k any commutative ring with identity and Λ any k-module. Let (X, A) be a pair of G-spaces, and suppose that either:

(1) X is a finite-dimensional G-CW-complex and A is a G-CW-subcomplex; or

(2) X is paracompact and finitistic and A is a closed invariant subspace.

Then restriction induces an isomorphism
$$\hat{H}_G^*(X, A; \Lambda) \xrightarrow{\cong} \hat{H}_G^*(X_1, A_1; \Lambda).$$

4.6. Equivariant Tate cohomology

Proof.
By the long exact sequence of Proposition (4.6.10)(1) it is enough to prove the result when $A = \emptyset$.

If X is a finite-dimensional G-CW-complex, then $\hat{H}^*_G(X, X_1; \Lambda) = 0$ by Corollary (4.6.14). So the long exact sequence completes the proof in this case.

Now suppose that X is paracompact and finitistic. Thanks to Proposition (4.6.10)(2) and (3), by the now familiar argument (see the proofs of Lemma (3.2.3), Theorem (3.11.13) and Lemma (4.5.9)), it is enough to show that $\hat{H}^*_G(X; \Lambda) = 0$ if G acts freely on X.

Suppose that G acts freely on X. Recall the notion of a Vietoris-G-covering from Definitions and Comments (3.11.2). These exist if X is free by considering slices or the covering map $X \to X/G$. Clearly Vietoris-G-coverings are cofinal if they exist. So finite-dimensional Vietoris-G-coverings are cofinal here. Let $\mathcal{U}$ be a finite-dimensional Vietoris-G-covering. Then $|\check{\mathcal{U}}|$ is a finite-dimensional free G-CW-complex. Thus $\hat{H}^*_G(|\check{\mathcal{U}}|; \Lambda) = 0$ by Corollary (4.6.14). The result now follows from Lemmas (4.6.11) and (4.6.12). $\square$

As a simple application of Swan's Theorem we give some inequalities from [Swan, 1960], which are a variant of those given in Theorems (3.10.6) and (3.10.18) and Remark (3.10.19)(1).

(4.6.16) Corollary
Let $G = \mathbb{Z}_p$ and let (X, A) be any pair of G-spaces satisfying the conditions of Theorem (4.6.15), Then for any integers m and n,

$$\sum_{i=0}^{\infty} \dim_{\mathbb{F}_p} H^{n+i}(X^G, A^G; \mathbb{F}_p) \leq \sum_{i=0}^{\infty} \dim_{\mathbb{F}_p} \hat{H}^{m-n-i}(G; H^{n+i}(X, A; \mathbb{F}_p))$$

$$\leq \sum_{i=0}^{\infty} \dim_{\mathbb{F}_p} H^{n+i}(X, A; \mathbb{F}_p)^G.$$

Proof.
Since $G = \mathbb{Z}_p$, $(X_1, A_1) = (X^G, A^G)$; and so the inclusion

$$\phi : (X^G, A^G) \to (X, A)$$

induces an isomorphism

$$\phi^* : \quad \hat{H}^*_G(X, A; \mathbb{F}_p) \to \hat{H}^*_G(X^G, A^G; \mathbb{F}_p).$$

Thus, using the first filtration as in Remarks (4.6.8)(1), there is an epimorphism

$$\hat{H}^m_G(X, A; \mathbb{F}_p)/\mathcal{F}^{m-n+1}\hat{H}^m_G(X, A; \mathbb{F}_p) \to \hat{H}^m_G(X^G, A^G; \mathbb{F}_p)/$$
$$\mathcal{F}^{m-n+1}\hat{H}^m_G(X^G, A^G; \mathbb{F}_p),$$

for any integers m and n.

Let $E^{i,j}_r(X, A)$ denote the first spectral sequence for (X, A). Since G acts trivially on (X^G, A^G),

$$E^{i,j}_\infty(X^G, A^G) = E^{i,j}_2(X^G, A^G) \cong \hat{H}^i(G; \mathbb{F}_p) \otimes H^j(X^G, A^G; \mathbb{F}_p).$$

Since $\hat{H}^i(G; \mathbb{F}_p) = \mathbb{F}_p$ for all i,

$$\sum_{i=0}^\infty \dim_{\mathbb{F}_p} H^{n+i}(X^G, A^G; \mathbb{F}_p)$$

$$= \sum_{i=0}^\infty \dim_{\mathbb{F}_p} E^{m-n-i, n+i}_\infty(X^G, A^G)$$
$$= \dim_{\mathbb{F}_p} \hat{H}^m_G(X^G, A^G; \mathbb{F}_p)/\mathcal{F}^{m-n+1}\hat{H}^m_G(X^G, A^G; \mathbb{F}_p)$$
$$\leq \dim_{\mathbb{F}_p} \hat{H}^m_G(X, A; \mathbb{F}_p)/\mathcal{F}^{m-n+1}\hat{H}^m_G(X, A; \mathbb{F}_p)$$
$$= \sum_{i=0}^\infty \dim_{\mathbb{F}_p} E^{m-n-i, n+i}_\infty(X, A)$$
$$\leq \sum_{i=0}^\infty \dim_{\mathbb{F}_p} E^{m-n-i, n+i}_2(X, A)$$
$$= \sum_{i=0}^\infty \dim_{\mathbb{F}_p} \hat{H}^{m-n-i}(G; H^{n+i}(X, A; \mathbb{F}_p)).$$

The second inequality in the statement of the theorem follows since, for any G-module M, $\hat{H}^i(G; M) \cong M^G/\nu M$ if i is even, and isomorphic to $\ker \nu / \tau M$ if i is odd, where $\tau = 1 - g$, $\nu = 1 + g + \ldots + g^{p-1}$ and g is a generator of G: thus if M is a $\mathbb{F}_p[G]$-module which is finite-dimensional over $\mathbb{F}_p$, then $\dim_{\mathbb{F}_p} \hat{H}^i(G; M) = \dim_{\mathbb{F}_p}(M^G/\nu M)$ for all i. (Note too that if M is a $\mathbb{F}_p[G]$-module such that $\dim_{\mathbb{F}_p} M^G < \infty$, then $\dim_{\mathbb{F}_p} M < \infty$.) □

Next we give some simple general inequalities due basically to Heller ([Heller, 1959], but see also [Adem, 1988]). Recall that if G is a finite-group and if k is a field of characteristic 0 or p where $p \nmid |G|$, then $\hat{H}^*(G; M) = 0$ for any $k[G]$-module M ([Brown, K.S., 1982], Chapter VI, (5.6)).

(4.6.17) Proposition

Let G be any finite group and let k be a field of characteristic p where $p \mid |G|$, Let C^* be a cochain complex of $k[G]$-modules such that $C^i = 0$ for all $i < 0$ and $H^j(C^*) = 0$ for all $j > N$, where N is some positive integer.

Then, for any integer m,

$$\dim_k \hat{H}^{m+1}(G; H^0(C^*)) \leq \dim_k \hat{H}_G^{m+1}(C^*) + \sum_{j=1}^{N} \dim_k \hat{H}^{n-j}(G; H^j(C^*)).$$

Proof.
Let $E_r^{i,j}$ denote the first spectral sequence for C^*. Let $e_r^{i,j} = \dim_k E_r^{i,j}$. Then the inequality to be proven is

$$e_2^{m+1,0} \leq \sum_{i=-m-1}^{\infty} e_\infty^{-i,m+1+i} + \sum_{j=1}^{\infty} e_2^{m-j,j}.$$

Hence it is enough to show that

$$e_2^{m+1,0} \leq e_\infty^{m+1,0} + \sum_{j=1}^{N} e_2^{m-j,j} = e_{N+2}^{m+1,0} + \sum_{j=1}^{N} e_2^{m-j,j}.$$

Now there is an exact sequence

$$E_r^{m+1-r,r-1} \xrightarrow{d_r} E_r^{m+1,0} \to E_{r+1}^{m+1,0} \to 0$$

for all $r \geq 2$. Hence $e_r^{m+1,0} - e_{r+1}^{m+1,0} \leq e_r^{m+1-r,r-1}$.
Adding these inequalities for $2 \leq r \leq N+1$, yields

$$e_2^{m+1,0} - e_{N+2}^{m+1,0} \leq \sum_{r=2}^{N+1} e_r^{m+1-r,r-1} = \sum_{j=1}^{N} e_{j+1}^{m-j,j}.$$

and clearly $e_{j+1}^{m-j,j} \leq e_2^{m-j,j}$ for all $j \geq 1$. □

As a simple application of Proposition (4.6.17) we give Heller's original proof of his result previously proven in Example (3.10.17).

(4.6.18) Example

Let $G = (\mathbb{Z}_p)^3$ and suppose that X is either a finite-dimensional G-CW-complex or that X is paracompact and finitistic. Suppose that $H^*(X; \mathbb{F}_p) \cong H^*(S^a \times S^b; \mathbb{F}_p)$ where a and b are integers with $a > b > 0$. Then G cannot act freely on X.

To see this one simply takes $m = a+b$ in the inequality of Proposition (4.6.17), noting that $\hat{H}_G^*(X; \mathbb{F}_p) = 0$ if X is a free G-space, by Theorem (4.6.15).

Now recall from Definition (4.1.5) or the discussion following Corollary (4.5.10) that if $\Phi : G \times X \to X$ is an action of $G = (\mathbb{Z}_p)^r$ or a space X, then rank(Φ) is defined to be ρ where p^ρ is the minimum of the orders of the orbits of Φ. By analogy with Definition (4.3.1) one can make the following definition.

(4.6.19) Definition
$\text{rank}_p(X) = \sup \text{rank}(\Phi)$, where Φ ranges over all continuous actions of p-tori (elementary abelian p-groups) on X.

A major problem in studying $\text{rank}_p(X)$ is that the proof of Lemma (4.3.3) is not valid in this context since $\mathbb{F}_p$ is a finite field. For example, in Example (4.6.18) above, $(\mathbb{Z}_p)^3$ cannot act freely on X: but this does not immediately imply that $\text{rank}_p(X) \leq 2$. One way around this difficulty is to consider shifted subgroups: when $G = (\mathbb{Z}_p)^r$ a shifted subgroup is a special kind of subgroup of the group ring $E[G]$, where E is an extension field of $\mathbb{F}_p$, typically the algebraic closure $\bar{\mathbb{F}}_p$. We shall give a precise definition of a shifted subgroup shortly.

Let $G = (\mathbb{Z}_p)^r$ and let $g_1, \ldots, g_r$ be elements which generate G. If E is any field of characteristic p, then there is an isomorphism

$$E[x_1, \ldots, x_r]/(x_1^p, \ldots, x_r^p) \to E[G]$$

sending x_i to $1 - g_i$ for $1 \leq i \leq r$. Identifying the two rings by means of the isomorphism it is clear that the ideal $(x_1, \ldots, x_r)$ is the one and only prime ideal in $E[G]$. We shall let $\mathfrak{m}(G; E)$ or just $\mathfrak{m}$ denote this unique prime ideal in $E[G]$.

(4.6.20) Definition
Using the notation above we shall say that an element $u \in E[G]$ is a non-trivial unit if there exist $\alpha_1, \ldots, \alpha_r \in E$, not all zero, such that $u = 1 - \sum_{i=1}^{r} \alpha_i(1 - g_i)$, modulo $\mathfrak{m}^2$. Clearly $u^p = 1$.

Now let $u_1, \ldots, u_s$ be non-trivial units in $E[G]$. Let $\Gamma(u_1, \ldots, u_s)$ denote the elementary abelian p-subgroup of the group of units of $E[G]$ generated by $u_1, \ldots, u_s$. The inclusion $\Gamma = \Gamma(u_1, \ldots, u_s) \to E[G]$ induces a homomorphism $i_\Gamma : E[\Gamma] \to E[G]$.

If $\alpha = (\alpha_1, \ldots, \alpha_r) \in E^r$ is non-zero, let $\Gamma(\alpha) = \Gamma(u)$ where $u = 1 - \sum_{i=1}^{r} \alpha_i(1 - g_i)$.

Suppose that $u_1, \ldots, u_s$ are non-trivial units in $E[G]$, and suppose that $u_j = 1 - \sum_{i=1}^{r} \alpha_{ji}(1 - g_i)$, modulo $\mathfrak{m}^2$, for $1 \leq j \leq s$. Now $\Gamma(u_1, \ldots, u_s)$ is said to be a shifted subgroup of rank s if the vectors $\alpha_j = (\alpha_{j1}, \ldots, \alpha_{jr}) \in E^r$, for $1 \leq j \leq s$, are linearly independent over E. It is easy to see that in this case $\Gamma = \Gamma(u_1, \ldots, u_s)$ is an elementary

4.6. Equivariant Tate cohomology

abelian p-group of rank s, i.e. $\Gamma \cong (\mathbb{Z}_p)^s$; and it follows from Theorem (4.6.21)(1) that i_Γ is injective.

Furthermore we shall write $u \sim v$ for $u, v \in E[G]$ if $u - v \in \mathfrak{m}^2$.

We shall now list some of the fundamental properties of shifted subgroups, the proofs of which may be found in [Carlson, 1983]. (We shall include below a proof of the first part of (1), which is easy. And it is this part which is most used subsequently.)

(4.6.21) Theorem

Let $G = (\mathbb{Z}_p)^r$ and let E be a field of characteristic p. Then:

(1) If $\Gamma \subseteq E[G]$ is a shifted subgroup of rank s, then $E[G]$ is a free $E[\Gamma]$-module (via $i_\Gamma : E[\Gamma] \to E[G]$). If $s = r$, then the i_Γ is an isomorphism. ([Carlson, 1983], Lemma 2.9.)

If E is algebraically closed, and if $u_1, \ldots, u_s$ are non-trivial units in $E[G]$ such that $\Gamma = \Gamma(u_1, \ldots, u_s)$ has order p^s, then $E[G]$ is a free $E[\Gamma]$-module if and only if $1 - u_1, \ldots, 1 - u_s$ are linearly independent modulo $\mathfrak{m}^2$ (i.e. if and only if $u_1, \ldots, u_s$ generate Γ as a shifted subgroup). ([Carlson, 1983], Theorem 6.2.)

(2) Suppose that E is algebraically closed and that $u, v \in E[G]$ are non-trivial units. Let M be a finitely generated $E[G]$-module; and suppose that $u \sim v$. Then M is a free $E[\Gamma(u)]$-module if and only if M is a free $E[\Gamma(v)]$-module. In particular, if

$$u = 1 - \sum_{i=1}^{r} \alpha_i(1 - g_i),$$

modulo $\mathfrak{m}^2$, where $\alpha = (\alpha_1, \ldots, \alpha_r) \neq 0$, then M is a free $E[\Gamma(u)]$-module if and only if M is a free $E[\Gamma(\alpha)]$-module. ([Carlson, 1983], Lemma 6.4.)

(3) Dade's Lemma. If E is algebraically closed, and if M is a finitely generated $E[G]$-module, then M is a free $E[G]$-module if and only if M is a free $E[\Gamma(\alpha)]$-module for all non-zero $\alpha \in E^r$. ([Carlson, 1983], Theorem 4.4, or [Dade, 1978].)

Proof of the first part of (1).

Let $G = \langle g_1, \ldots, g_r \rangle$ and let $\tau_i = 1 - g_i \in E[G]$, $1 \leq i \leq r$. Also let $\Gamma = \Gamma(u_1, \ldots, u_s)$, where $u_j = 1 - \sum_{i=1}^{r} \alpha_{ji}\tau_i$ modulo $\mathfrak{m}^2$. Let

$$\sigma_j = 1 - u_j \in E[\Gamma], \text{ and let } \sigma'_j = \sum_{i=1}^{r} \alpha_{ji}\tau_i, \quad 1 \leq j \leq s.$$

So

$$\sigma'_j - \sigma_j \in \mathfrak{m}^2.$$

Since Γ is a shifted subgroup of rank s, $\sigma'_1, \ldots, \sigma'_s$ are linearly independent. Let $\{\sigma'_1, \ldots, \sigma'_r\}$ be a basis for the E-vector subspace of $E[G]$ spanned by $\tau_1, \ldots, \tau_r$. Let $u_i = 1 - \sigma'_i$ for $s+1 \leq i \leq r$, and let $\Gamma' = \Gamma(u_{s+1}, \ldots, u_r)$. Then

$$i_\Gamma : E[\Gamma] \to E[G] \text{ and } i_{\Gamma'} : E[\Gamma'] \to E[G]$$

induce (via the direct sum of E-algebras) a homomorphism

$$\phi : E[\Gamma] \otimes E[\Gamma'] \to E[G].$$

Since $\sigma'_j - \sigma_j \in \mathfrak{m}^2$ for $1 \leq j \leq s$, and since $\mathfrak{m}^{(p-1)r+1} = 0$, it is clear that ϕ is surjective. Hence ϕ is an isomorphism since its domain and codomain are finite-dimensional E-vector-spaces. $\square$

(4.6.22) Definition
Let $G = (\mathbb{Z}_p)^r$ and let $\alpha = (\alpha_1, \ldots, \alpha_r) \in \overline{\mathbb{F}}_p^r$. Recall from Recollection (3.11.7)(1) that α defines a homomorphism, also denoted by α, from $R = \mathbb{F}_p[t_1, \ldots, t_r]$, the polynomial part of $H^*(BG; \mathbb{F}_p)$, to $\overline{\mathbb{F}}_p$, by $\alpha(t_i) = \alpha_i$ for $1 \leq i \leq r$. With the notation of Definition (3.1.4), there is a subgroup $K(\alpha) \subseteq G$, defined by $PK(\alpha) = \sigma \ker(\alpha)$. Now let X be a G-space. We define the G-variety of X to be

$$V_G(X) := \{\alpha \in \overline{\mathbb{F}}_p^r; X^{K(\alpha)} \neq \emptyset\}.$$

If M is a $\overline{\mathbb{F}}_p[G]$-module, then one defines the G-variety or rank variety of M to be

$$V_G(M) := \{0\} \cup \{\alpha \in \overline{\mathbb{F}}_p^r; M \text{ is not a free } \overline{\mathbb{F}}_p[\Gamma(\alpha)]\text{-module}\}$$

(See [Carlson, 1983]).

(4.6.23) Lemma
(1) *If $G = (\mathbb{Z}_p)^r$, if X is a G-space, and if*

$$K_1, \ldots, K_n$$

are the isotropy subgroups of G on X, then

$$V_G(X) = \bigcup_{i=1}^n V(PK_i),$$

where, for $1 \leq i \leq n$, $V(PK_i)$ is the variety in $\overline{\mathbb{F}}_p^r$ of the homogeneous linear prime ideal $PK_i \subseteq R$.
(2) *If $G = (\mathbb{Z}_p)^r$, if X is a G-CW-complex, and if $W_*(X; \overline{\mathbb{F}}_p)$ is the cellular chain complex of X with coefficients in $\overline{\mathbb{F}}_p$, so that*

$$W_n(X; \overline{\mathbb{F}}_p) = H_n(X^n, X^{n-1}; \overline{\mathbb{F}}_p),$$

where X^n is the n-skeleton of X, then

$$V_G(X) = V_G(W_*(X; \overline{\mathbb{F}}_p)).$$

4.6. Equivariant Tate cohomology

Proof.
(1)
$$X^{K(\alpha)} \neq \emptyset \Leftrightarrow K(\alpha) \subseteq K_i \text{ for some } i, 1 \leq i \leq n$$
$$\Leftrightarrow \ker(\alpha) \supseteq PK_i \text{ for some } i, \text{ since } PK(\alpha) = \sigma\ker(\alpha)$$
$$\Leftrightarrow \alpha \in V(PK_i) \text{ for some } i.$$

(2) Let $\mathcal{E}_n$ be the set of G-n-cells of X. A G-n-cell σ has the form $G/G_\sigma \times D^n$, where G_σ is the isotropy group of σ. So
$$W_n(X; \overline{\mathbb{F}}_p) \cong \bigoplus_{\sigma \in \mathcal{E}_n} \overline{\mathbb{F}}_p[G/G_\sigma].$$

As $\overline{\mathbb{F}}_p[\Gamma(\alpha)]$ is a local ring, any finitely generated projective $\overline{\mathbb{F}}_p[\Gamma(\alpha)]$-module is free (see, e.g., [Kunz, 1985], p. 112). So $W_*(X; \overline{\mathbb{F}}_p)$ is a free $\overline{\mathbb{F}}_p[\Gamma(\alpha)]$-module if and only if each $\overline{\mathbb{F}}_p[G/G_\sigma]$ is a free $\overline{\mathbb{F}}_p[\Gamma(\alpha)]$-module. If $X^{K(\alpha)} = \emptyset$, then it follows from the proofs of Lemma (3.11.11) and Proposition (3.11.24) that each $\overline{\mathbb{F}}_p[G/G_\sigma]$ is free. On the other hand, if $X^{K(\alpha)} \neq \emptyset$, then $K(\alpha) \subseteq G_\sigma$ for some σ. By Recollections (3.11.7)(3), there is a non-trivial unit $u \in \overline{\mathbb{F}}_p[G]$ such that $u = 1 - \sum_{i=1}^r \alpha_i(1 - g_i)$ modulo $\mathfrak{m}^2$, and $u \in \overline{\mathbb{F}}_p[K(\alpha)]$. Clearly, then, $\overline{\mathbb{F}}_p[G/G_\sigma]$ is not a free $\overline{\mathbb{F}}_p[\Gamma(u)]$-module. So by Theorem (4.6.21)(2), $\overline{\mathbb{F}}_p[G/G_\sigma]$ is not a free $\overline{\mathbb{F}}_p[\Gamma(\alpha)]$-module. □

Now let $G = (\mathbb{Z}_p)^r$, let E be a field of characteristic p, and let $\Gamma \subseteq E[G]$ be a shifted subgroup. By Example (4.6.1), there is a complete resolution, P_*, of G over E consisting of finitely generated free $E[G]$-modules. Hence, by Theorem (4.6.21)(1), P_* is a complete resolution of Γ over E consisting of finitely generated free $E[\Gamma]$-modules. For a cochain complex of $E[G]$-modules, C^*, we can form $A(C^*)$ and $B(C^*)$ as in Definition (4.6.2) but using $\operatorname{Hom}_{E[\Gamma]}(P_i, C^j)$ instead of $\operatorname{Hom}_{E[G]}(P_i, C^j)$. In particular, for a G-space X, we can define $\hat{H}^*_\Gamma(X; E)$; and we have an obvious map $\hat{H}^*_G(X; E) \to \hat{H}^*_\Gamma(X; E)$. It is easy to see that the results (4.6.3)-(4.6.13) remain valid using the induced $E[\Gamma]$ structures instead of the standard $E[G]$ structures. The same goes for the inequalities of Proposition (4.6.17). Now the main result on shifted subgroups in this context is the following (see [Adem, 1988]).

(4.6.24) Proposition
Let $G = (\mathbb{Z}_p)^r$, and let $\Phi : G \times X \to X$ be an action of G on X. Let $\operatorname{rank}(\Phi) = \rho$. Suppose that X is paracompact and finitistic: e.g., X could be a finite-dimensional G-CW-complex. Then there is a finite-dimensional extension field E over $\mathbb{F}_p$, and a shifted subgroup $\Gamma \subseteq E[G]$ such that Γ has rank ρ and $\hat{H}^*_\Gamma(X; E) = 0$.

Before proving the proposition we note that, thanks to the shifted version of Proposition (4.6.17), there is the following immediate corollary.

(4.6.25) Corollary
Let X be as in Example (4.6.18). Then $\mathrm{rank}_p(X) \leq 2$.

We shall prove Proposition (4.6.24) first when X is a finite-dimensional G-CW-complex.

(4.6.26) Lemma
Proposition (4.6.24) holds when X is a finite-dimensional G-CW-complex.

Proof.
Let G be generated by $g_1, \ldots, g_r$, and let $\tau_i = 1 - g_i$ for $1 \leq i \leq r$. Let F denote the algebraic closure $\overline{\mathbb{F}}_p$, and let $\mathfrak{m} = (\tau_1, \ldots, \tau_r)$ be the augmentation ideal of $F[G]$. Let $K_1, \ldots, K_n$ be the maximal isotropy subgroups. Let K_i be generated by $h_{i1}, \ldots, h_{ir_i} \in G$ where $r_i = \mathrm{rank}(K_i)$. Let $\tau_{ij} = 1 - h_{ij}$, $1 \leq i \leq n$, $1 \leq j \leq r_i$. Then, as in Recollection (3.11.7)(3), for each i and j, one can write $\tau_{ij} = \sum_{s=1}^{r} \mu_{ijs} \tau_s$ modulo $\mathfrak{m}^2$, where $\mu_{ijs} \in \mathbb{F}_p$; and setting $\tau'_{ij} = \sum_{s=1}^{r} \mu_{ijs} \tau_s$, for each i, $\tau'_{i1}, \ldots, \tau'_{ir_i}$ are linearly independent over F, since the $r_i \times r$ matrix $\mu_i = (\mu_{ijs})$ has a right inverse.

Now let $V \subseteq F[G]$ be the F-vector-space spanned by $\tau_1, \ldots, \tau_r$; and, for $1 \leq i \leq n$, let $V_i \subseteq V$ be the subspace spanned by $\tau'_{i1}, \ldots, \tau'_{ir_i}$. In particular, $\dim_F V = r$, $\dim_F V_i = r_i$; and

$$\max\{r_i; 1 \leq i \leq n\} = r - \rho.$$

Since F is an infinite field, there is a subspace $W \subseteq V$ such that $\dim_F W = \rho$ and $W \cap V_i = 0$ for $1 \leq i \leq n$. Let $\{\sigma_1, \ldots, \sigma_\rho\}$ be a basis for W. So $\sigma_1, \ldots, \sigma_\rho$ define a shifted subgroup $\Gamma \subseteq F[G]$ of rank ρ.

Consider the homomorphisms $\pi_i : F[G] \to F[G/K_i]$, induced by the quotient maps $G \to G/K_i$ for $1 \leq i \leq \rho$. Since $\tau_{ij} = \tau'_{ij}$ modulo $\mathfrak{m}^2$, it follows that $\pi_i(\sigma_1), \ldots, \pi_i(\sigma_\rho)$ are linearly independent over F; and each $1 - \pi_i(\sigma_j)$ is a non-trivial unit in $F[G/K_i]$. Thus $\pi_i | F[\Gamma]$ is a monomorphism; and, by Theorem (4.6.21)(1), $F[G/K_i]$ is a free $F[\Gamma]$-module.

Let $H \subseteq G$ be any isotropy subgroup. So $H \subseteq K_i$ for some i. Let

$$\pi : F[G] \to F[G/H]$$

be induced by the quotient map $G \to G/H$. So $\pi_i = q\pi$ where q is induced by $G/H \to G/K_i$. Hence $\pi | F[\Gamma]$ is a monomorphism; and so $F[G/H]$ is a free $F[\Gamma]$-module as before.

4.6. Equivariant Tate cohomology

Let $\sigma_i = \sum_{j=1}^{r} \alpha_{ij}\tau_j$ for $1 \leq i \leq \rho$, where each $\alpha_{ij} \in F$. Let E be the finite extension field of $\mathbb{F}_p$ generated by $\{\alpha_{ij}; 1 \leq i \leq \rho, 1 \leq j \leq r\}$. So we may view Γ as a shifted subgroup of $E[G]$.

Let M be any $E[G]$-module, and suppose that $F \otimes_E M$ is a free $F[\Gamma] = F \otimes_E E[\Gamma]$-module: i.e. M is a $E[\Gamma]$-module via the inclusion $E[\Gamma] \to E[G]$, and $F \otimes_E M$ is a free $F[\Gamma]$-module via extension of the coefficients. Then $\hat{H}^*(\Gamma; F \otimes_E M) = 0$ by [Brown, K.S., 1982], Chapter VI, Theorem (8.5). But since there are compete resolutions for $E[\Gamma]$ of finite type, it is clear that $\hat{H}^*(\Gamma; F \otimes_E M) \cong F \otimes_E \hat{H}^*(\Gamma; M)$. So $\hat{H}(\Gamma; M) = 0$; and by the same theorem in [Brown, K.S., 1982], M is a free $E[\Gamma]$-module.

Thus we have shown that $E[G/H]$ is a free $E[\Gamma]$-module for every isotropy subgroup $H \subseteq G$. So each $W^j(X; E)$ is a product of free $E[\Gamma]$-modules. Hence $\hat{H}^*_\Gamma(X; E) = 0$ by the second spectral sequence. (This also shows that each $W^j(X; E)$ is a free $E[\Gamma]$-module.) □

In order to extend Proposition (4.6.24) and other results to paracompact finitistic spaces the following notion is useful.

(4.6.27) Definition
Let G be a finite group and let X be a G-space. Let $\mathcal{U}$ be an invariant open covering of X. For each $U \in \mathcal{U}$, let
$$G_U = \{g \in G; gU = U\}.$$
We shall say that $\mathcal{U}$ is faithful if for any $U \in \mathcal{U}$ there is a $x \in X$ such that $G_U \subseteq G_x$.

If $\mathcal{U}$ is a faithful Čech-G-covering, then clearly the maximal elements of $\{G_U; U \in \mathcal{U}\}$ are the same as the maximal elements of $\{G_x; x \in X\}$.

(4.6.28) Lemma
Let G be a finite group and let X be a G-space. If X is paracompact (resp., paracompact and finitistic), then faithful Čech-G-coverings (resp., faithful finite-dimensional Čech-G-coverings) are cofinal.

Proof.

Clearly any invariant covering which refines a faithful Čech-G-covering is faithful. So, by [Bredon, 1972], Chapter III, Theorem 6.1, it is enough to show that there exists a faithful Čech-G-covering.

Since X is paracompact, slices exist; and, since G is finite, slices are open. By choosing one slice per orbit and translating one has an invariant covering by slices. Clearly this is a faithful Čech-G-covering by, e.g., [Bredon, 1972], Chapter II, Theorem 4.4. □

Proof of Proposition (4.6.24).

Let X be a paracompact finitistic G-space. Choose Γ just as in the proof of Lemma (4.6.26). Let $\mathcal{U}$ be a faithful finite-dimensional Čech-G-covering of X. Then $\hat{H}^*(|\mathcal{U}|;E) = 0$ by Lemma (4.6.26). Hence, $\hat{H}^*_\Gamma(X;E) = 0$ by Lemmas (4.6.11), (4.6.12) and (4.6.28). □

Now we turn to the equivariant Tate cohomology version of the G-exponent of Browder and Gottlieb that was defined in Definition (4.5.3). If A is an abelian group, then let $\exp(A)$ be the least positive integer n such that $na = 0$ for all $a \in A$.

(4.6.29) Definition
Let G be a finite group and let X be a G-space. Let P be a one-point space with trivial G-action. Then the map $X \to P$ induces a map
$$\varepsilon(X) : \hat{H}^*(G;\mathbb{Z}) \cong \hat{H}^*_G(P;\mathbb{Z}) \to \hat{H}^*_G(X;\mathbb{Z}).$$
Let
$$\mathrm{im}(\varepsilon(X)) = \mathrm{im}[\varepsilon(X) : \hat{H}^0(G;\mathbb{Z}) \to \hat{H}^0_G(X;\mathbb{Z})].$$
Then the Tate G-exponent of X, $\hat{e}(X;G)$, is defined by
$$\hat{e}(X;G) := |G|/\exp(\mathrm{im}(\varepsilon(X))).$$
Note that $\hat{H}^0(G;\mathbb{Z}) \cong \mathbb{Z}/(|G|)$; and so $\exp(\mathrm{im}(\varepsilon(X))) \mid |G|$.

The main theorem concerning $\hat{e}(X;G)$ is the following which is due to Adem ([Adem, 1989]).

(4.6.30) Theorem
Let $G = (\mathbb{Z}_p)^r$ and let $\Phi : G \times X \to X$ be an action of G on X. Let $\mathrm{rank}(\Phi) = \rho$. Suppose either that X is a finite-dimensional G-CW-complex or that X is paracompact and finitistic. Then $\hat{e}(X;G) = p^\rho$.

Note that no Poincaré duality assumptions are made on X - cf. Theorem (4.5.11).

We shall not give the whole proof of Theorem (4.6.30): instead we shall state a general result from [Adem,1989] and show how to deduce Theorem (4.6.30) and other results from it. The general result is as follows: the proof, which depends heavily on results and methods from [Carlson, 1989], [Benson, Carlson, 1987] and [Carlson, 1984], can be found in [Adem, 1989]. For a finite group G let $\mathrm{rank}_p(G)$ be the maximum of all $\mathrm{rank}_p(T)$ where T ranges over all elementary abelian p-subgroups of G.

(4.6.31) Theorem
Let G be a finite group and let $\Phi : G \times X \to X$ be an action of G on a space X. Suppose either that X is a finite-dimensional G-CW-complex

4.6. Equivariant Tate cohomology

or that X is paracompact and finitistic. Let κ be the maximum of all $\operatorname{rank}_p(G_x)$ where x ranges over all points in X and p ranges over all prime numbers. Then there are homogeneous elements of positive degree in $H^*(G;\mathbb{Z})$, $\xi_1,\ldots,\xi_\kappa$, such that

$$\exp(\hat{H}^*_G(X;\mathbb{Z})) \mid \prod_{i=1}^{\kappa}\exp(\xi_i),$$

where, for $1 \leq i \leq \kappa$, $\exp(\xi_i)$ is the least positive integer n_i such that $n_i \xi_i = 0$.

(4.6.32) Remark
Adem proves the result only when X is a finite-dimensional G-CW-complex: but the paracompact finitistic case follows easily. In Adem's proof one can see that $\xi_1,\ldots,\xi_\kappa$ depend only on the maximal isotropy subgroups. Thus, if X is paracompact and finitistic, then the same $\xi_1,\ldots,\xi_\kappa$ will work for all $|\check{\mathcal{U}}|$, where $\mathcal{U}$ ranges over the faithful finite-dimensional Čech-G-coverings. Let $N = \prod_{i=1}^{\kappa}\exp(\xi_i)$. Since

$$\hat{H}^*_G(X;\mathbb{Z}) \cong \varinjlim_{\mathcal{U}} \hat{H}^*_G(|\check{\mathcal{U}}|;\mathbb{Z})$$

$Na = 0$ for all $a \in \hat{H}^*_G(X;\mathbb{Z})$. So $\exp(\hat{H}^*_G(X;\mathbb{Z}))|N$.

If $G = (\mathbb{Z}_p)^r$, then $p\xi = 0$ for any ξ of positive degree in $H^*(G;\mathbb{Z})$. So the next corollary follows easily.

(4.6.33) Corollary
Let $G = (\mathbb{Z}_p)^r$ and let $\Phi : G \times X \to X$ be an action of G on a space X. Suppose either that X is a finite-dimensional G-CW-complex or that X is paracompact and finitistic. Let $\kappa = r - \operatorname{rank}(\Phi)$: i.e. p^κ is the order of a largest isotropy subgroup. Then $p^\kappa \hat{H}^*_G(X;\mathbb{Z}) = 0$.

To prove Theorem (4.6.30) we want the following lemma.

(4.6.34) Lemma
Let G be a finite group, and let X be a paracompact G-space. Let K be any isotropy subgroup. Then $\hat{e}(X;G) \mid |G/K|$.

Proof.

The inclusion of the orbit G/K into X induces a homomorphism $\psi^*_K : \hat{H}^*_G(X;\mathbb{Z}) \to \hat{H}^*_G(G/K;\mathbb{Z})$.

Now let P_* be a complete resolution of G over $\mathbb{Z}$; and, for each i, $-\infty < i < \infty$, consider the Diagram 4.6,

$$\text{Hom}_{\mathbf{Z}[G]}(P_i, \text{Hom}_{\mathbf{Z}}(\mathbf{Z}[G/K], \mathbf{Z})) \longrightarrow \text{Hom}_{\mathbf{Z}[G]}(P_i, \text{Hom}_{\mathbf{Z}[K]}(\mathbf{Z}[G], \mathbf{Z}))$$
$$\uparrow \qquad\qquad\qquad\qquad\qquad\qquad \downarrow$$
$$\text{Hom}_{\mathbf{Z}[G]}(P_i, \mathbf{Z}) \longrightarrow \text{Hom}_{\mathbf{Z}[K]}(P_i, \mathbf{Z})$$

Diagram 4.6

where the left-hand vertical map is induced by the map $\mathbf{Z} \to \text{Hom}_{\mathbf{Z}}(\mathbf{Z}[G/K], \mathbf{Z})$ given by $n \mapsto f_n$, where $f_n(gK) = n$, for all $g \in G$; the upper horizontal map is induced by the map $\text{Hom}_{\mathbf{Z}}(\mathbf{Z}[G/K], \mathbf{Z}) \to \text{Hom}_{\mathbf{Z}[K]}(\mathbf{Z}[G], \mathbf{Z})$ given by $f \mapsto f'$, where $f'(g) = f(g^{-1}K)$; the right hand vertical map is induced by the map $\text{Hom}_{\mathbf{Z}[K]}(\mathbf{Z}[G], \mathbf{Z}) \to \mathbf{Z}$ given by $f \mapsto f(1)$; and the lower horizontal map is the standard inclusion. Clearly the diagram commutes.

The upper horizontal map is an isomorphism with inverse induced by the map $\text{Hom}_{\mathbf{Z}[K]}(\mathbf{Z}[G], \mathbf{Z}) \to \text{Hom}_{\mathbf{Z}}(\mathbf{Z}[G/K], \mathbf{Z})$ given by $\phi \mapsto \overline{\phi}$, where $\overline{\phi}(gK) = \phi(g^{-1})$. The right hand vertical map is an isomorphism by, e.g., [Brown, K.S., 1982], Chapter III, (3.5) and (3.6). Finally P_* is a complete resolution of K over $\mathbf{Z}$, since $\mathbf{Z}[G]$ is a free $\mathbf{Z}[K]$-module by, e.g., [Brown, K.S., 1982], Chapter I, Section 3.

So $\psi_K^* \varepsilon(X)$ is isomorphic to the restriction $\hat{H}^*(G; \mathbf{Z}) \to \hat{H}^*(K; \mathbf{Z})$ and for $1 \in \hat{H}^0(G; \mathbf{Z})$, $\psi_K^* \varepsilon(X)(1) = 1 \in \hat{H}^0(K; \mathbf{Z})$. It follows that $|K| \mid \exp(\text{im}(\varepsilon(X)))$, since $\hat{H}^0(K; \mathbf{Z}) \cong \mathbf{Z}/(|K|)$. Hence $\hat{e}(X; G) \mid |G/K|$. $\square$

Proof of Theorem (4.6.30).

$\hat{e}(X; G) \mid p^\rho$ by Lemma (4.6.34), and $p^\rho \mid \hat{e}(X; G)$ by Corollary (4.6.33). $\square$

We shall now give some immediate corollaries.

(4.6.35) Corollary
Let G, X, Φ and ρ be as in Theorem (4.6.30). Let $\kappa = r - \rho$, so that $p^\kappa = \max\{|G_x|; x \in X\}$. Then
$$\exp(\text{im}(\varepsilon(X))) = \exp(\hat{H}_G^0(X; \mathbf{Z})) = \exp(\hat{H}_G^*(X; \mathbf{Z})) = p^\kappa.$$

The following lemma combined with Theorem (4.6.30) is interesting.

(4.6.36) Lemma
Let $G = (\mathbb{Z}_p)^r$ and let X be a paracompact G-space. Suppose that X is connected. Then

$$\hat{e}(X;G) \mid \prod_{j=1}^{\infty} \exp(\hat{H}^{-j-1}(G; H^j(X;\mathbb{Z}))).$$

Proof.
It is clear (from the first spectral sequence, for example) that $\varepsilon(X)$ can be factored as

$$\hat{H}^0(G;\mathbb{Z}) \to \hat{H}^0(G; H^0(X;\mathbb{Z})) \to \hat{H}^0_G(X;\mathbb{Z}).$$

Letting $E_n^{*,*}$ denote the first spectral sequence, the image of the second map is $E_\infty^{0,0}$. So, if X is connected, $\text{im}(\varepsilon(X)) = E_\infty^{0,0}$. Now the differentials which map into $E_n^{0,0}$ are $d_n : E_n^{-n,n-1} \to E_n^{0,0}$. The result follows since

$$E_2^{-n,n-1} \cong \hat{H}^{-n}(G; H^{n-1}(X;\mathbb{Z})).$$

□

(4.6.37) Corollary
Let $G = (\mathbb{Z}_p)^r$, and let Φ be an action of G on a connected space X. Suppose either that X is a finite-dimensional G-CW-complex, or that X is paracompact and finitistic. Suppose that G acts trivially on $H^*(X;\mathbb{Z})$, and let $\text{rank}_p(\Phi) = \rho$. Then $H^j(X;\mathbb{Z}) \neq 0$ for at least $\rho+1$ different values of j.

Proof.
Since G acts trivially on $H^*(X;\mathbb{Z})$, $\exp(\hat{H}^{-j-1}(G; H^j(X;\mathbb{Z}))) = p$ for $j \geq 1$. So the corollary follows from Lemma (4.6.36) and Theorem (4.6.30).
□

To obtain the next corollary we need a simple lemma and a major theorem from [Quillen, 1971a]. The lemma is the following.

(4.6.38) Lemma
Let G be a finite group, let X be any paracompact G-space, let $A \subseteq X$ be a closed invariant subspace, and let k be any commutative ring with identity. Then there is a natural homomorphism $H^j_G(X, A; k) \to \hat{H}^j_G(X, A; k)$ which is an isomorphism for all $j > \max\{i; H^i(X, A; k) \neq 0\}$.

Proof.
Let P_* be a complete resolution for G over k, and let $C^* = \overline{C}^*(X, A; k)$. Let

$$\check{B}^n(C^*) = \bigoplus_{i=0}^{n} \text{Hom}_{k[G]}(P_i, C^{n-i}).$$

Then clearly there is an inclusion

$$\check{B}(C^*) \to B(C^*);$$

and this induces the desired homomorphism by Definition (4.6.9) and Lemma (3.11.5), which holds clearly for any finite group G.

If $\max\{i; H^i(X, A; k) \neq 0\} = N$, say, is finite, then define C_t^* by $C_t^j = C^j$ for $j < N$, $C_t^N = Z^N$, the cocycles of degree N, and $C_t^j = 0$ for $j > N$. Clearly the inclusion of C_t^* into C^* is a weak-equivalence; and hence, by the first spectral sequence and its analogue for equivariant cohomology (the Leray-Serre spectral sequence), $\hat{H}_G^*(C_t^*) \cong \hat{H}_G^*(C^*)$ and $H(\check{B}(C_t^*)) \cong H(\check{B}(C^*))$. On the other hand, the inclusion

$$\check{B}(C_t^*) \to B(C_t^*)$$

is the identity in degrees greater than or equal to N. Hence

$$H_G^j(X, A; k) \to \hat{H}_G^j(X, A; k)$$

is an isomorphism for $j > N$. $\square$

Now let G be a finite group and let X be a paracompact G-space. Let $H_G(X; \mathbb{F}_p) = \bigoplus_{i=0}^{\infty} H_G^{2i}(X; \mathbb{F}_p)$ if p is odd, and let $H_G(X; \mathbb{F}_p) = H_G^*(X; \mathbb{F}_p)$ if $p = 2$. If $\dim_{\mathbb{F}_p} H^*(X; \mathbb{F}_p) < \infty$, then $H_G(X; \mathbb{F}_p)$ is a commutative Noetherian ring ([Quillen, 1971a], Corollary 2.2; cf. Proposition (3.10.1).) In the latter situation the Krull dimension of $H_G(X; \mathbb{F}_p)$, $\dim H_G(X; \mathbb{F}_p)$, is the same as the order of pole of the Poincaré series of $H_G^*(X; \mathbb{F}_p)$, $P(H_G^*(X; \mathbb{F}_p), z)$, at $z = 1$; cf. Lemma (4.1.3). The following is Theorem 7.7 of [Quillen, 1971a]. Note that case (C) can be included thanks to the general form of the Localization Theorem (Theorem (3.2.6)).

(4.6.39) Theorem
Let G be a finite group and let X be a G-space satisfying condition (LT) of the introduction to Section 3.10. Suppose further that $\dim_{\mathbb{F}_p} H^*(X; \mathbb{F}_p) < \infty$. Let $\kappa_p(X; G)$ be the largest integer r such that there is an elementary abelian p-subgroup $H \subseteq G$ having rank r (i.e. $H \cong (\mathbb{Z}_p)^r$) such that $X^H \neq \emptyset$. (I.e. $\kappa_p(X; G) = \max\{\text{rank}_p(G_x); x \in X\}$.) Then $\dim H_G(X; \mathbb{F}_p) = \kappa_p(X; G)$.

Now the following comes immediately from Corollary (4.6.35).

(4.6.40) Corollary
Let G be a finite group. Let X be a G space which is either a finite-dimensional G-CW-complex or paracompact and finitistic. Suppose that $\dim_{\mathbb{F}_p} H^*(X;\mathbb{F}_p) < \infty$. If $p^\lambda = \max\{\exp(\hat{H}_K^0(X;\mathbb{Z}))\}$, where K ranges over all elementary abelian p-subgroups of G, then $\dim H_G(X;\mathbb{F}_p) = \lambda$.

(4.6.41) Remark
Corollary (4.6.40) shows the curious fact that the order of pole of the Poincaré series of $H_G^*(X;\mathbb{F}_p)$ is determined by the exponents of the groups $\hat{H}_K^0(X;\mathbb{Z})$.

We shall give next a generalization of Theorem (1.4.14) which is closely related to Corollary (4.6.37). The proof of the theorem below does not use exponents or, in an essential way, equivariant Tate cohomology: it does, however, use shifted subgroups together with the filtration method of Theorem (1.4.14). We shall then finish the section with a summary of some of the results on filtrations and the problem of p-tori acting on products of spheres of the same dimension.

(4.6.42) Theorem
Let $G = (\mathbb{Z}_p)^r$, where p is any prime number; and let $\Phi : G \times X \to X$ be an action of G on a space X. Suppose that X is paracompact and finitistic and that every connected component of X is open in X: e.g., X could be a finite-dimensional G-CW-complex. Then

$$\sum_{j=0}^{\infty} \ell(H^j(X;\mathbb{F}_p)) \geq \operatorname{rank}(\Phi) + 1,$$

where ℓ indicates the length of the G-filtration as defined immediately above Theorem (1.4.14). (See Definition (4.1.5) for the definition of $\operatorname{rank}(\Phi)$.)

In particular, if G acts trivially on $H^*(X;\mathbb{F}_p)$, then $H^j(X;\mathbb{F}_p)$ must be non-zero for at least $\operatorname{rank}(\Phi) + 1$ different values of j.

If $\operatorname{rank}(\Phi) = r$, i.e. G is acting freely on X, then the result holds also if X is paracompact with $\operatorname{cd}_{\mathbb{F}_p}(X) < \infty$ (and if every connected component of X is open in X).

Proof.
Since every connected component of X is open it is clear that it is enough to prove the result when X is connected. So we shall assume that X is connected. Also there is nothing to prove if

$$n(X) := \sup\{j; H^j(X;\mathbb{F}_p) \neq 0\}$$

is infinite. So we shall assume that $n(X) < \infty$.

Let rank(Φ) = ρ. Then, by Proposition (4.6.24), there is a finite extension field E of $\mathbb{F}_p$ and a shifted subgroup $\Gamma \subseteq E[G]$ such that Γ has rank ρ and $\hat{H}_\Gamma^*(X;E) = 0$.

Now we can define $\overline{\beta}_\Gamma^*(X;E)$ and $H_\Gamma^*(X;E)$ just as in Definitions (3.11.8). (We are writing H instead of $\overline{H}$ for Alexander-Spanier cohomology here.) And clearly we can construct a minimal Hirsch-Brown model for $\overline{\beta}_\Gamma^*(X;E)$ exactly as was done in Section 1.4. So thanks to Proposition (1.4.16) and the rest of the proof of Theorem (1.4.14) it remains only to show that $H_\Gamma^i(X;E) = 0$, for all sufficiently large i. But, by the obvious analogue of Lemma (4.6.38) for shifted subgroups, $H_\Gamma^i(X;E) \cong \hat{H}_\Gamma^i(X;E) = 0$ for all $i > n(X)$. Note that because the augmentation ideal of $E[\Gamma]$ is contained in the augmentation ideal of $E[G]$ it follows that, for all j, the length of the Γ-filtration on $H^j(X;E)$ is at most the length of the G-filtration. Furthermore, it suffices to consider $\ell(H^j(X;\mathbb{F}_p))$ instead of $\ell(H^j(X;E))$ since $H^*(X;E) \cong H^*(X;\mathbb{F}_p) \otimes_{\mathbb{F}_p} E$. (See Remarks (3.11.14)(2).)

If rank(Φ) = r, then the result in case (B) (i.e. X paracompact with $\mathrm{cd}_{\mathbb{F}_p}(X) < \infty$) follows directly from Theorem (1.4.14) if every connected component of X is open in X. □

(4.6.43) Discussion

Let $G = (\mathbb{Z}_p)^r$ where p is a prime number, and let $\Phi: G \times X \to X$ be an action of G on a paracompact space X. To simplify things a little we shall assume that X is connected.

(1) Consider the following two statements.

(F1) $\sum_{i=0}^{\infty} \ell(H^i(X;k)) \geq \mathrm{rank}(\Phi) + 1$, where the coefficients k will be specified below, and ℓ denotes the length of the G-filtration.

(F2) If G acts trivially on $H^*(X;k)$, then $H^j(X;k) \neq 0$ for at least rank(Φ) + 1 different values of j.

The statements (F1) and (F2) have been established under the following conditions.

(a) (F1) and hence (F2) hold if $k = \mathbb{F}_p$ and if X is finitistic (e.g., a finite-dimensional G-CW-complex) by Theorem (4.6.42).

(b) (F2) holds if $k = \mathbb{Z}$ and if X is finitistic by Corollary (4.6.37).

(c) If rank(Φ) = r, i.e. G is acting freely on X, then (F1) and hence (F2) hold if $k = \mathbb{F}_p$ and if $\mathrm{cd}_{\mathbb{F}_p}(X) < \infty$ by Theorem (1.4.14).

(d) If rank(Φ) = r, then (F2) holds if $k = \mathbb{Z}$ and if $\mathrm{cd}_{\mathbb{Z}}(X) < \infty$ by Theorem (4.6.15) and Exercise (4.10).

(2) Note that an analogue of (F2) in case (0) appears as Theorem (4.4.5).

(3) Consider the following statement.

(S) Suppose that there is an isomorphism
$$\phi : H^*(X;k) \to H^*((S^n)^m;k).$$
Then rank(Φ) $\leq m$.

Statement (S) has been established under the following conditions.

(a) G is acting trivially on $H^*(X;k)$, ϕ is an isomorphism of graded k-modules, $k = \mathbb{Z}$ or $\mathbb{F}_p$, and X is finitistic (or $\operatorname{cd}_k(X) < \infty$ if rank(Φ) $= r$). This follows from statement (F2) above.

(b) G is not necessarily acting trivially on $H^*(X;k)$, p is odd, $k = \mathbb{Z}$, ϕ is an isomorphism of graded rings, and X is finitistic or $\operatorname{cd}_{\mathbb{F}_p}(X) < \infty$. In fact, under these conditions, $\operatorname{rank}_p(G) - \operatorname{rank}_p(L) \leq m$, where L is an isotropy subgroup of maximal rank for the action of
$$H := \{g \in G; g \text{ acts trivially on } H^*(X;\mathbb{F}_p)\}$$
on X.

This rather more general version of the Adem-Browder theorem (Theorem (4.5.23)) follows by an obvious alteration of the argument given in the proof of Proposition (4.5.2) using the result of Carlsson given in (4) below.

(c) The same conditions as in (b) above, but with $p = 2$, and with the additional conditions that X be homotopy equivalent to $(S^n)^m$ and that $n \neq 1$, 3 or 7. Again one has the refinement that $\operatorname{rank}_p(G) - \operatorname{rank}_p(L) \leq m$, where L is defined as before. And again the technical improvement to somewhat more general spaces than those considered in Section 4.5 is owing to Carlsson: see (5) below.

(4) Suppose that p is odd and that, in statement (S), $k = \mathbb{F}_p$ and ϕ is an isomorphism of graded $\mathbb{F}_p$-algebras. Suppose too that G is acting trivially on $H^*(X;k)$. Let $\pi : X_G \to BG$ be the bundle map. Let $\{u_1, \ldots, u_m\}$ be a basis for $H^n(X;k)$; and in the Leray-Serre spectral sequence of π with coefficients in k let $v_1, \ldots, v_m \in H^{n+1}(BG;k)$ be the transgressions of $u_1, \ldots, u_m$, respectively. Let $f_1, \ldots, f_m \in R$ be the polynomial parts of $v_1, \ldots, v_m$, respectively. If one follows the method of [Carlsson, 1982] using Alexander-Spanier cochains instead of cellular chains, then it follows that the two ideals in R, $(f_1, \ldots, f_m)$ and $\ker(\pi^*|R)$, have the same radical.

So if X is finitistic or if $\operatorname{cd}_k(X) < \infty$, then rank($\Phi$) $\leq m$ by the Localization Theorem. (Under condition (LT), $\ker(\pi^*|R)$ is a Hsiang-Serre ideal. Hence under the conditions here $(f_1, \ldots, f_m)$ is a Hsiang-Serre ideal; and by the Krull Height Theorem, there must exist an isotropy subgroup of corank at most m.)

(5) In [Carlsson, 1980] the result of (4) when $p = 2$ was proved differently using the theorem of Serre given in Remark (3.7.19). Since $p = 2$, of course, $f_i = v_i$ for $1 \leq i \leq m$. Let $J = (f_1, \ldots, f_m)$. One begins by showing that J is stable under the Steenrod algebra. By the Cartan formula it is enough to show that $\mathrm{Sq}^i(f_j) \in J$ for $1 \leq i \leq n+1$ and $1 \leq j \leq m$. Clearly $\mathrm{Sq}^{n+1}(f_j) = f_j^2 \in J$. So suppose that $i < n+1$. Since transgression commutes with the Steenrod operations,

$$\mathrm{Sq}^i(f_j) = \mathrm{Sq}^i(d_{n+1}(u_j)) = d_{n+1+i}(\mathrm{Sq}^i(u_j)) = 0$$

in $E_{n+1+i}^{n+1+i,0}$, since $\mathrm{Sq}^i(u_j) = 0$ by degree considerations or the fact that $u_j^2 = 0$. But after d_{n+1} the next possible non-zero differential is d_{2n+1}. Hence $\mathrm{Sq}^i(f_j) \in J$ as desired.

So, by Serre's theorem, J is a Hsiang-Serre ideal. Let PK be a minimal prime of J. So $\mathrm{cork}(K) = \mathrm{ht}(PK) \leq m$ by the Krull Height Theorem. Looking at the spectral sequence for $X_K \to BK$ and the map of spectral sequences induced by $X_K \to X_G$, it follows that $u_1, \ldots, u_m$ all transgress to zero in the spectral sequence of $X_K \to BK$. Hence the latter spectral sequence collapses; and it follows from the Localization Theorem that $X^K \neq \emptyset$ if X satisfies condition (LT). (Note that, just as in (4), it follows that $\sqrt{J} = \sqrt{(\ker(\pi^*))}$ whether X satisfies condition (LT) or not.)

(6) One might ask if in (4) above the transgressions $v_1, \ldots, v_m$ necessarily lie in R. This is not so. An example due to Browder ([Browder, 1989]) shows that $v_1, \ldots, v_m$ can have non-zero exterior parts.

(7) See Corollaries (4.5.6) and (4.5.7) for another version of statement (F2) and proof of a case of statement (S).

(8) The main references for the results above are [Adem, 1988], [Adem, Browder, 1988], [Baumgartner, 1990], [Benson, Carlson, 1987], [Browder, 1983, 1988], [Carlsson, 1980, 1982, 1983], and [Gottlieb, 1986].

(9) See also [Davis, Milgram, preprint] and [Pardon, 1980]. In the latter reference, for example, it is shown that given any finite group G and any $n \equiv 3 \pmod 4$, $n \geq 7$, there is a smooth closed simply-connected manifold M such that M has the same rational homotopy type as S^n and such that G can act freely and smoothly on M. In particular, taking $G = (\mathbb{Z}_p)^r$ with $r > n$, G must act non-trivially on $H^*(M; \mathbb{F}_p)$ by Theorem (4.6.42).

4.7 Two theorems on topological symmetry

In this section we shall give a brief account of two of the main results of [Hauschild, 1986]. We begin by recalling the following definitions from [Hsiang, 1975].

(4.7.1) Definitions
Let X be a topological space.
  (1) The topological degree of symmetry of X is defined to be $N_t(X) :=$ $\max\{\dim G; G$ can act effectively on $X\}$, where G ranges over all compact Lie groups.
  (2) The topological toric degree of symmetry of X is defined to be $T_t(X) := \max\{\dim T; T$ can act effectively on $X\}$, where T ranges over all tori. (The subscript t indicates that any topological, i.e. continuous, action is allowed. Clearly one could consider only smooth or locally smooth actions instead, but that would make no difference in Theorem (4.7.2) below.)
  The first main result of this section is the following.

(4.7.2) Theorem
Let G be a compact connected semisimple Lie group, and let T be a maximal torus of G. Then
$$T_t(G/T) = \operatorname{rank}(G) := \dim T.$$
In order to prove Theorem (4.7.2) it is convenient to study torus actions which are cohomologically effective: this notion is defined as follows.

(4.7.3) Definition
Let X be a topological space, and let G be a compact Lie group acting on X. The action is said to be cohomologically effective if the restriction homomorphism $H^*(X; \mathbb{Q}) \to H^*(X^K; \mathbb{Q})$ is not an isomorphism for any non-trivial subtorus $K \subseteq G$.

(4.7.4) Remark
An action $\Phi : G \times X \to X$ is said to be almost-effective if the ineffective kernel, $\ker(\Phi) := \{g \in G : gx = x$, for all $x \in X\}$, is finite. By dividing by $\ker(\Phi)$, an almost-effective action gives rise to an effective action by a group of the same dimension. If G is a compact Lie group and X is a compact orientable manifold, then, clearly, any smooth or locally smooth action of G on X is cohomologically effective if and only if it is almost-effective: indeed this is still true, though less clearly, for continuous actions on compact orientable manifolds. More generally, if X is a compact connected orientable cohomology manifold over $\mathbb{Q}$, then any continuous action of a compact Lie group on X is cohomologically effective if and only if it is almost-effective. (See [Borel et al, 1960], Chapter I, Corollary 4.6, and Chapter V, Theorem 3.2.)
  The following is a technical definition in order to save writing.

(4.7.5) Definition
A topological space X is said to be of q-type if
$$H^*(X;\mathbb{Q}) \cong \mathbb{Q}[x_1,\ldots x_n]/I,$$
where $\deg(x_i) = 2$ for $1 \leq i \leq n$, if the ideal of relations, I, is generated by elements of degree at least four, and if there exists a homogeneous element of degree four, $q = q(x_1,\ldots,x_n)$, in I, such that for any rational numbers $\alpha_1,\ldots,\alpha_n$, if $q(\alpha_1,\ldots,\alpha_n) = 0$, then $\alpha_1 = \ldots = \alpha_n = 0$.

Note that q is a quadratic form in $x_1,\ldots,x_n$; and, hence, by the diagonalizability of quadratic forms (see, e.g., [Jacobson, 1974], Theorem 6.5), without loss of generality one can choose the generators $x_1,\ldots,x_n$ so that $q(x_1,\ldots,x_n) = c_1 x_1^2 + \ldots + c_n x_n^2$, for some $c_1,\ldots,c_n \in \mathbb{Q}$.

(4.7.6) Example
Let G be a compact connected semisimple Lie group and let T be a maximal torus of G. Then G/T is a space of q-type. For, by [Borel, 1967], Theorem 20.3, one has that
$$H^*(G/T;\mathbb{Q}) \cong H^*(BT;\mathbb{Q})/I,$$
where I is the ideal generated by elements of positive degree which are invariant under the action of the Weyl group of G, $W = W(G) := NT/T$. To see that I contains an appropriate quadratic form one can either use the negative definitive Killing form (as in [Hauschild, 1986]), or, using less Lie theory, one can embed G into some unitary group $U(n)$, extend T to a maximal torus $\overline{T} \subseteq U(n)$, and observe that $H^*(B\overline{T};\mathbb{Q})$ contains a positive definitive quadratic form invariant under $W(U(n))$ - see, e.g., [Hsiang, 1975], Chapter III, §1(B), Example 1 - which then maps to a positive definitive form in $H^*(BT;\mathbb{Q})$, which is invariant under $W(G)$.

(4.7.7) Proposition
Let G be a torus of rank $r \geq 1$, let X be a G-space satisfying condition (LT) of the introduction to Section 3.10, suppose that X is a space of q-type with $H^*(X;\mathbb{Q}) \cong \mathbb{Q}[x_1,\ldots,x_n]/I$, as in Definition (4.7.5), and suppose that the action of G on X is cohomologically effective. Then

(1) $r \leq n$; and

(2) X^G is not connected; and each component of X^G is a space of q-type whose rational cohomology can be generated by fewer than n elements of degree 2.

4.7. Two theorems on topological symmetry

Proof.
Clearly X is TNHZ in $X_G \to BG$, and, letting $H^*(BG;\mathbb{Q}) = R$, there is a presentation of $H_G^*(X;\mathbb{Q})$ as a R-algebra of the form

$$0 \to J \to R[X_1, \ldots, X_n] \xrightarrow{\pi} H_G^*(X;\mathbb{Q}) \to 0$$

where $i^*\pi(X_j) = x_j$ for $1 \leq j \leq n$, where $i: X \to X_G$ is the inclusion of the fibre. Let F be a component of X^G: then, using the notation of Section 3.8, we may assume without loss of generality, that $X_i \in P(G,F)$ for $1 \leq i \leq n$; and hence J contains an element of the form

$$q(X_1, \ldots, X_n) + p_1 X_1 + \ldots + p_n X_n,$$

where $p_i \in H^2(BG;\mathbb{Q})$ for $1 \leq i \leq n$.

If $n < r$, then there is a non-trivial subtorus $K \subsetneq G$ such that

$$PK = (p_1, \ldots, p_n).$$

Let c be a component of X^K. By Lemma (3.8.4)(1), there exist $b_1, \ldots, b_n \in R$ such that

$$q(X_1, \ldots, X_n) \in PK + (X_1 - b_1, \ldots, X_n - b_n) = P(K,c).$$

So $q(b_1, \ldots, b_n) \in PK$. Let $\bar{b}_i$ be the image of b_i in $R_K := H^*(BK;\mathbb{Q}) \cong R/PK$. Then

$$q(\bar{b}_1, \ldots, \bar{b}_n) = 0$$

in R_K. By the condition imposed on q in Definition (4.7.5), it follows that

$$\bar{b}_1 = \ldots = \bar{b}_n = 0.$$

Hence $P(K,c) = PK + (X_1, \ldots, X_n)$. Since distinct components of X^K give rise to distinct ideals $P(K,c)$, by Proposition (3.6.16), it follows that X^K is connected. It now follows readily from the Localization Theorem that $H^*(X;\mathbb{Q}) \to H^*(X^K;\mathbb{Q})$ is surjective since $H^*(X;\mathbb{Q})$ is generated by elements of degree two. Since X is TNHZ in $X_K \to BK$, and since $\dim_{\mathbb{Q}} H^*(X;\mathbb{Q}) < \infty$ (see (3.8.18)), the restriction

$$H^*(X;\mathbb{Q}) \to H^*(X^K;\mathbb{Q})$$

is an isomorphism. Thus $r \leq n$; and also X^G is not connected.

Now let F be a component of X^G. By Theorem (3.8.12) $H^*(F;\mathbb{Q})$ can be generated by n or fewer elements, which, by the Localization Theorem, must have degree two. Suppose that n generators, $y_1, \ldots, y_n$, say, are needed. Without loss of generality, under the restriction

$$\phi^* : H^*(X;\mathbb{Q}) \to H^*(F;\mathbb{Q})$$

we may assume that $\phi^*(x_i) = y_i$, for $1 \leq i \leq n$. Hence $q(y_1, \ldots, y_n) + p_1 y_1 + \ldots p_n y_n = 0$. Since this equation is in $H_G^*(F;\mathbb{Q}) \cong R \otimes H^*(F;\mathbb{Q})$, $p_1 = \ldots = p_n = 0$. As before this would imply that X^G is connected.

Finally let $H^*(F; \mathbf{Q})$ be generated by $y_1, \ldots, y_m$, with $m < n$. Let $\phi^*(x_i) = h_i(y_1, \ldots, y_m)$, for $1 \leq i \leq n$. Then $q(h_1, \ldots, h_n) = 0$. Writing $q(h_1, \ldots, h_n) = \tilde{q}(y_1, \ldots, y_m)$, it is clear that $\tilde{q}$ satisfies the desired condition. □

Proof of Theorem (4.7.2).
By Remark (4.7.4), Example (4.7.6) and Proposition (4.7.7),

$$T_t(G/T) \leq \dim T.$$

On the other hand G acts on G/T by left translation: i.e. for $g, h \in G$, $(g, hT) \mapsto ghT$. $g \in G$ acts trivially if and only if $g \in \bigcap_{h \in G} hTh^{-1}$. But the latter group, being the intersection of all the maximal tori of G, is the centre of G, $Z(G)$. Since G is semisimple, $Z(G)$ is finite. □

The second theorem from [Hauschild, 1986] which we shall give is the following.

(4.7.8) Theorem
Let G be a compact connected semisimple Lie group acting cohomologically effectively on a space X of q-type satisfying condition (LT) of the introduction to Section 3.10. Then the Euler characteristic of X, $\chi(X)$, is divisible by the order of the Weyl group of G, $|W(G)|$. (Note that $\chi(X) = \dim_{\mathbf{Q}} H^*(X; \mathbf{Q})$ here.)

Proof.
As before let $H^*(X; \mathbf{Q}) = \mathbf{Q}[x_1, \ldots, x_n]/I$, and let

$$0 \to J_G \to R_G[X_1, \ldots, X_n] \to H_G^*(X; \mathbf{Q}) \to 0$$

be a presentation, where $R_G = H^*(BG; \mathbf{Q})$. For certain $p_0, p_1, \ldots, p_n \in R_G$,

$$q(X_1, \ldots, X_n) + p_1 X_1 + \ldots + p_n X_n + p_0 \in J_G.$$

As remarked in Definition (4.7.5), we may assume that

$$q(X_1, \ldots, X_n) = c_1 X_1^2 + \ldots + c_n X_n^2.$$

By definition of q-type, each $c_i \neq 0$; and so, by completing the square and putting

$$Y_i = X_i + p_i/2c_i,$$

$q(Y_1, \ldots, Y_n) - \beta \in J_G$, where

$$\beta = p_1^2/4c_1 + \ldots p_n^2/4c_n - p_0 \in R_G.$$

4.7. Two theorems on topological symmetry 331

Let T be a maximal torus of G and let $R = H^*(BT; \mathbf{Q})$. R is a free R_G-module (see, e.g., Lemma (3.9.1)(1)); and so, by Theorem (3.9.3)(2), there is a presentation

$$0 \to J \to R[Y_1, \ldots, Y_n] \to H_T^*(X; \mathbf{Q}) \to 0,$$

where $J = J_G \otimes_{R_G} R$. Furthermore, the action of the Weyl group $W := W(G)$ on $H_T^*(X; \mathbf{Q})$ can via its action on R be lifted to one of W on

$$R[Y_1, \ldots Y_n].$$

Identify β with its image under the inclusion $R_G \to R$.

Now let F be a component of X^T, and $P(T, F) = (Y_1 - a_1, \ldots, Y_n - a_n)$. Then $\beta = q(a_1, \ldots, a_n)$. Let $PK = (a_1, \ldots, a_n)$, and let c be a component of X^K. Let $P(K, c) = PK + (Y_1 - b_1, \ldots, Y_n - b_n)$. Now $\beta \in PK$; and so $q(b_1, \ldots, b_n) \in PK$. As in the proof of Proposition (4.7.7) this implies that X^K is connected; and this contradicts the assumption that G is acting cohomologically effectively unless K is trivial. Thus $\{a_1, \ldots, a_n\}$ spans the $\mathbf{Q}$-vector space $H^2(BT; \mathbf{Q})$.

W acts on the set of components of X^T; and we shall show that this action is free. Suppose that $g \in W$ fixes the component F. Then $g^*P(T, F) = P(T, F)$. So, for each i, $1 \leq i \leq n$,

$$g^*(Y_i - a_i) = Y_i - g^*(a_i) = \sum_{j=1}^n \lambda_{ij}(Y_j - a_j),$$

for some $\lambda_{ij} \in \mathbf{Q}$. Since this is an equation in $R[Y_1, \ldots, Y_n]$,

$$Y_i = \sum_{j=1}^n \lambda_{ij} Y_j,$$

for $1 \leq i \leq n$. Thus $\lambda_{ij} = \delta_{ij}$, the Kronecker δ; and so $g^*(a_i) = a_i$, for $1 \leq i \leq n$. Since $\{a_1, \ldots, a_n\}$ spans $H^2(BT; \mathbf{Q})$, it follows that g^* is the identity on $H^*(BT; \mathbf{Q})$. From the fibration $G/T \to BT \to BG$, it follows that g acts trivially on $H^*(G/T; \mathbf{Q})$; and so g is the identity element of W. (See, e.g., [Borel, 1967], Proposition 20.2 and its proof.)

Finally let $F_1, \ldots, F_m$ be a complete set of representatives of the orbits of W on the set of components of X^T. Then

$$\dim_{\mathbf{Q}} H^*(X; \mathbf{Q}) = \dim_{\mathbf{Q}} H^*(X^T; \mathbf{Q}) = \sum_{i=1}^m |W| \dim_{\mathbf{Q}} H^*(F_i; \mathbf{Q})$$

$$= |W| \sum_{i=1}^m \dim_{\mathbf{Q}} H^*(F_i; \mathbf{Q}) \qquad \square$$

(4.7.9) Example

The orders of the Weyl groups of the compact connected simple Lie groups of types $A_n (n \geq 1)$, $B_n (n \geq 2)$, $C_n (n \geq 3)$, $D_n (n \geq 4)$, G_2, F_4,

E_6, E_7 and E_8 are $(n+1)!$, $n!2^n$, $n!2^n$, $n!2^{n-1}$, 12, $2^7 \times 3^2$, $2^7 \times 3^4 \times 5$, $2^{10} \times 3^4 \times 5 \times 7$ and $2^{14} \times 3^5 \times 5^2 \times 7$, respectively. (See, e.g., [Samuelson, 1969], pp. 79-86.) Thus, for example, if E_8 can act cohomologically effectively on a space X of q-type satisying condition (LT), then $\chi(X) = \dim_{\mathbb{Q}} H^*(X;\mathbb{Q})$ must be divisible by $2^{14} \times 3^5 \times 5^2 \times 7 = 696\,729\,600$.

See [Hauschild, 1986] for more results like those of this section. See also [Hauschild, 1985].

4.8 Localization and the Steenrod algebra

Throughout this section G will be a p-torus, i.e. $G = (\mathbb{Z}_p)^r$ for some $r \geq 1$, and X will be a G-space satisying condition (LT) of the introduction to Section 3.10. Recall from Examples (3.1.1)(2) and (3) that $R = H^*(BG;\mathbb{F}_2)$ if $p = 2$, and R is the polynomial part of $H^*(BG;\mathbb{F}_p)$ for odd p. Recall also from Example (3.1.5) that if $K \subseteq G$ is a subgroup then $S(K)$ denotes the multiplicative subset of R generated by the homogeneous linear polynomials which are not in PK. The purpose of this section is to state and prove the main theorem of [Dwyer, Wilkerson, 1988], which, by means of the Steenrod algebra, identifies precisely those elements of $S(K)^{-1}H_G^*(X^K;\mathbb{F}_p)$ which belong to $H_G^*(X^K;\mathbb{F}_p)$. Of course $S(K)^{-1}H_G^*(X^K;\mathbb{F}_p) \cong S(K)^{-1}H_G^*(X;\mathbb{F}_p)$ by the Localization Theorem. Indeed, the inclusion $H_G^*(X^K;\mathbb{F}_p) \to S(K)^{-1}H_G^*(X^K;\mathbb{F}_p)$, followed by the inverse of the localized restriction homomorphism, gives a natural homomorphism $i_K : H_G^*(X^K;\mathbb{F}_p) \to S(K)^{-1}H_G^*(X;\mathbb{F}_p)$. By Lemma (3.7.5), i_K is a monomorphism. The theorem of this section uses the Steenrod algebra to identify the image of i_K.

Let $\mathcal{A}_p$ denote the modulo p Steenrod algebra. For odd p the Steenrod operations will be denoted $\mathcal{P}^i$, $i \geq 0$, and the Bockstein operation will be denoted β.

(4.8.1) Definition
Let M be a $\mathbb{Z}$-graded module over $\mathcal{A}_p$, which is not assumed to satisfy the dimension condition. Then $Un(M)$ denotes the $\mathbb{Z}$-graded $\mathbb{F}_p$-vector subspace of (weakly) unstable elements defined as follows for each $n \in \mathbb{Z}$.

(1) If $p = 2$, then $Un(M_n) = \{x \in M_n : \mathrm{Sq}^i(x) = 0, \text{ if } i > n\}$.
(2) If p is odd, then $Un(M_n) = \{x \in M_n : \mathcal{P}^i(x) = 0 \text{ if } 2i > n\}$.
(In [Dwyer, Wilkerson, 1988] the unstable elements are defined by
$$Un_0(M_n) = \{x \in M_n; \mathrm{Sq}^i\,\mathrm{Sq}^I(x) = 0 \text{ for all multi-indices}$$
$$I \text{ such that } i > n + \deg(\mathrm{Sq}^I)\},$$

4.8. Localization and the Steenrod algebra

when $p = 2$, where for a multi-index $I = \{i_1, \ldots, i_m\}$, $\text{Sq}^I = \text{Sq}^{i_1} \ldots \text{Sq}^{i_m}$; and

$$Un_0(M_n) = \{x \in M_n; \beta^j \mathcal{P}^i \mathcal{P}^I(x) = 0 \text{ for all multi-indices}$$
$$I \text{ such that } 2i + j > n + \deg(\mathcal{P}^I)\},$$

when p is odd, where for a multi-index $I = \{\varepsilon_0, i_1, \varepsilon_1, \ldots, i_m, \varepsilon_m\}$, $\varepsilon_i = 0$ or 1 for $0 \leq i \leq m$, $\mathcal{P}^I = \beta^{\varepsilon_0} \mathcal{P}^{i_1} \beta^{\varepsilon_1} \ldots \mathcal{P}^{i_m} \beta^{\varepsilon_m}$.

Thanks to the Adem relations, $Un_0(M) = Un(M)$ when $p = 2$; and $Un_0(M) = Un(M)$ when p is odd, if the Bockstein is omitted from the definition of Un_0. It is an immediate consequence of the proof below, however, that, in the notation of Theorem (4.8.4),

$$Un_0(S(K)^{-1} H_G^*(X; \mathbb{F}_p)) = Un(S(K)^{-1} H_G^*(X, \mathbb{F}_p)).)$$

(4.8.2) Definition

For any topological space Y, let $H^*(Y; \mathbb{F}_p)[[\xi]]$ be the power series ring in ξ with coefficients in $H^*(Y; \mathbb{F}_p)$, where ξ is assumed to commute with every element of $H^*(Y; \mathbb{F}_p)$. Define

$$\mathcal{P}_\xi : H^*(Y; \mathbb{F}_p) \to H^*(Y; \mathbb{F}_p)[[\xi]]$$

by the formula

$$\mathcal{P}_\xi(x) = \sum_{i=0}^\infty \mathcal{P}^i(x) \xi^i$$

if p is odd, and similarly if $p = 2$. By the Cartan formula $\mathcal{P}_\xi$ is a ring homomorphism.

(4.8.3) Lemma

The action of $\mathcal{A}_p$ on $H_G^*(X; \mathbb{F}_p)$ can be extended to $S(K)^{-1} H_G^*(X; \mathbb{F}_p)$ by the formulas

$$\mathcal{P}_\xi(x/f) = \mathcal{P}_\xi(x)/\mathcal{P}_\xi(f),$$

for any $x \in H_G^*(X; \mathbb{F}_p)$ and $f \in S(K)$, and, in case (odd p), $\beta(x/f) = \beta(x)/f$.

Proof.

If $x/f = y/g$ for $x, y \in H_G^*(X; \mathbb{F}_p)$ and $f, g \in S(K)$, then there is $h \in S(K)$ such that $hgx = hfy$. So

$$\mathcal{P}_\xi(h) \mathcal{P}_\xi(g) \mathcal{P}_\xi(x) = \mathcal{P}_\xi(h) \mathcal{P}_\xi(f) \mathcal{P}_\xi(y).$$

But, for any $a \in S(K)$,

$$\mathcal{P}_\xi(a) = a + \mathcal{P}^1(a)\xi + \ldots$$

is clearly a unit in $S(K)^{-1}H_G^*(X;\mathbb{F}_p)[[\xi]]$. Hence
$$\mathcal{P}_\xi(x)/\mathcal{P}_\xi(f) = \mathcal{P}_\xi(y)/\mathcal{P}_\xi(g).$$
For odd p, $\beta(a) = 0$, for any $a \in S(K)$. So $hg\beta(x) = hf\beta(y)$. Hence $\beta(x)/f = \beta(y)/g$. □

It now makes sense to state the Dwyer-Wilkerson theorem.

(4.8.4) Theorem
$\operatorname{im}[i_K : H_G^*(X^K;\mathbb{F}_p) \to S(K)^{-1}H_G^*(X;\mathbb{F}_p)] = Un(S(K)^{-1}H_G^*(X;\mathbb{F}_p))$.
I.e. i_K gives an isomorphism
$$H_G^*(X^K;\mathbb{F}_p) \overset{\cong}{\to} Un(S(K)^{-1}H_G^*(X;\mathbb{F}_p)).$$
Before proving the theorem we need the following lemma.

(4.8.5) Lemma
$Un(S(K)^{-1}H_G^*(X;\mathbb{F}_p))$ is a R-submodule of $S(K)^{-1}H_G^*(X;\mathbb{F}_p)$.

Proof.
Clearly
$$\mathcal{P}_\xi : S(K)^{-1}H_G^*(X;\mathbb{F}_p) \to S(K)^{-1}H_G^*(X;\mathbb{F}_p)[[\xi]]$$
is, by definition, a ring homomorphism, and so the Cartan formula holds in $S(K)^{-1}H_G^*(X;\mathbb{F}_p)$.

Now let $z \in S(K)^{-1}H_G^*(X;\mathbb{F}_p)$ and let $a \in R$. Then $\mathcal{P}^j(az) = \sum_{i=0}^{j}\mathcal{P}^i(a)\mathcal{P}^{j-i}(z)$. Suppose that $2j > \deg(az)$. If $2i \leq \deg(a)$, then $2(j-i) > \deg(z)$. Thus $\mathcal{P}^j(az) = 0$ if $z \in Un(S(K)^{-1}H_G^*(X;\mathbb{F}_p))$. □

Proof of Theorem (4.8.4).
Clearly the image of i_K is contained in $Un(S(K)^{-1}H_G^*(X;\mathbb{F}_p))$. Let
$$x/f \in Un(S(K)^{-1}H_G^*(X;\mathbb{F}_p))$$
where $f \in S(K)$ is linear. As before we shall consider only case (odd p): case (2) is exactly similar.

Let $\nu = [\frac{1}{2}\deg(x)]$, the greatest integer $\leq \frac{1}{2}\deg(x)$. Then
$$\mathcal{P}_\xi(x) = \sum_{i=0}^{\nu} x_i\xi^i,$$
where $x_0 = x$ and $x_i = \mathcal{P}^i(x)$, and
$$\mathcal{P}_\xi(f) = f + f^p\xi = f(1 + f^{p-1}\xi).$$
Also
$$\mathcal{P}_\xi(f)\mathcal{P}_\xi(x/f) = \mathcal{P}_\xi(x);$$

4.8. Localization and the Steenrod algebra

and so

$$fP_\xi(x/f) = (x_0 + x_1\xi + \ldots + x_\nu\xi^\nu)(1 - f^{p-1}\xi + f^{2(p-1)}\xi^2 + \ldots$$
$$+ (-1)^j f^{j(p-1)}\xi^j + \ldots)$$
$$= \sum_{m=0}^{\infty} \sum_{i+j=m} (-1)^j f^{j(p-1)} x_i \xi^m.$$

Set $\Phi(x) = \sum_{i+j=\nu} (-1)^j f^{j(p-1)} x_i$. Since

$$x/f \in Un(S(K)^{-1} H_G^*(X; \mathbb{F}_p))$$

and $\deg(x/f) = \deg(x) - 2$, $\Phi(x) = 0$.

Let $\{f_1, \ldots, f_h, \ldots, f_r = f\}$ be a basis for the $\mathbb{F}_p$-vector-space of linear elements of R such that $PK = (f_1, \ldots, f_h)$. Let $(f_1, \ldots, f_{r-1}) = PL$. Hence $L \subseteq K$, and R_L, the polynomial part of $H^*(BL; \mathbb{F}_p)$, is isomorphic to $\mathbb{F}_p[f]$. Also $H_G^*(X^K; \mathbb{F}_p) \cong R_L \otimes \wedge(s) \otimes H_{G/L}^*(X^K; \mathbb{F}_p)$, where s generates $H^1(BL; \mathbb{F}_p)$. Identifying $S(K)^{-1} H_G^*(X; \mathbb{F}_p)$ with $S(K)^{-1} H_G^*(X^K; \mathbb{F}_p)$ we may suppose that x is an element of

$$H_G^*(X^K; \mathbb{F}_p),$$

and hence put

$$x = 1 \otimes y_0 + f \otimes y_1 + \ldots + f^n \otimes y_n,$$

where each $y_i \in \wedge(s) \otimes H_{G/L}^*(X^K; \mathbb{F}_p)$, $0 \leq i \leq n$.

Now, in general, if $y \in Un\, H_G^*(X^K; \mathbb{F}_p)$ and $\deg(fy) = \deg(x)$, then

$$\Phi(fy) := \sum_{i+j=\nu} (-1)^j f^{j(p-1)} \mathcal{P}^i(fy)$$
$$= f\mathcal{P}^\nu(y) + f^p \mathcal{P}^{\nu-1}(y) - f^{p-1}\{f\mathcal{P}^{\nu-1}(y) + f^p \mathcal{P}^{\nu-2}(y)\}$$
$$+ f^{2(p-1)}\{f\mathcal{P}^{\nu-2}(y) + f^p \mathcal{P}^{\nu-3}(y)\}$$
$$- \ldots + (-1)^\nu f^{\nu(p-1)} fy$$
$$= f\mathcal{P}^\nu(y) = 0,$$

since $2\nu > \deg(y)$.

So $\Phi(1 \otimes y_0) = 0$, since $\Phi(x) = 0$: i.e.

$$1 \otimes \mathcal{P}^\nu(y_0) - f^{p-1} \otimes \mathcal{P}^{\nu-1}(y_0) + \ldots (-1)^\nu f^{\nu(p-1)} \otimes y_0 = 0.$$

Hence $y_0 = 0$; and so x/f is in the image of i_K.

Finally suppose that

$$x/f \in Un(S(K)^{-1} H_G^*(X; \mathbb{F}_p))$$

where $f = f_1 \ldots f_n$ is a product of linear elements $f_i \in S(K)$. By Lemma (4.8.5),

$$x/f_n \in Un(S(K)^{-1} H_G^*(X; \mathbb{F}_p)).$$

By the above, x/f_n is in the image of i_K. So, for some $y_1 \in H_G^*(X^K; \mathbb{F}_p))$,

336   4. General results on torus and p-torus actions

$x/f = y_1/f_1 \ldots f_{n-1}$. We can now proceed inductively to deduce that x/f is in the image of i_K. □

(4.8.6) Corollary
There is a natural isomorphism
$$H^*(X^G; \mathbb{F}_p) \xrightarrow{\cong} \mathbb{F}_p \otimes_{H_G^*} Un(S^{-1}H_G^*(X; \mathbb{F}_p)),$$
where $H_G^* = H^*(BG; \mathbb{F}_p)$, and $S = S(G)$ is the multiplicative set generated by all non-zero homogeneous linear polynomials in R.

We have already seen an example when the Steenrod algebra was used to gain additional information about the fixed point set of a 2-torus action, namely Example (3.8.11). A much more complicated example is the main theorem of [Back, 1977]. Here now is a very simple example.

(4.8.7) Example
Suppose that X satisfies condition (LT), let $G = \mathbb{Z}_2$, $k = \mathbb{F}_2$, suppose that $H^*(X;k) \cong H^*(S^n;k)$, and let G act on X in such a way that X^G is non-empty and connected. So $H^*(X^G;k) \cong H^*(S^m;k)$ for some m, $1 \leq m \leq n$. From Hsiang's Fundamental Fixed Point Theorem, Theorem (3.8.7), $H_G^*(X;k) \cong R[x]/(x^2)$, where x has degree n and maps to the generator of $H^n(X;k)$ via the inclusion $i: X \to X_G$, and, as usual, $R = H^*(BG;k) = k[t]$ and $\deg(t) = 1$.

Let $\phi: X^G \to X$ be the inclusion, and let $y \in H^m(X^G;k)$ be the generator. Consider $\phi^*: H_G^*(X;k) \to H_G^*(X^G,k) = R \otimes H^*(X^G,k)$. Clearly, by the Localization Theorem, $\phi^*(x) = t^{n-m} \otimes y$. Since ϕ^* is injective here, and since $\mathrm{Sq}^i y = 0$ for all $i > 0$, it follows that $n - m$ is the largest integer j such that $\mathrm{Sq}^j x \neq 0$.

(4.8.8) Remark
Somewhat related to the results of this section is the work of Lannes. See, for example, [Lannes, 1987]; and for a direct connection between Theorem (4.8.4) and Lannes' functor T, see [Dwyer, Wilkerson, 1991]. See also some of the work on the Segal and Sullivan conjectures: for example, see the survey article by Carlsson on the Segal conjecture ([Carlsson, 1987b]), the work of Miller on the Sullivan conjecture ([Miller, 1984]), and the many related papers cited in the works listed above.

4.9 The rational homotopy Lie algebra of a fixed point set

In this short section we shall state without proof (but cf. Remark (4.9.3)(7)) the main results of [Allday, Puppe, 1986b]. The first theorem is concerned with a torus T acting on a space X with the fixed point set $X^T \neq \emptyset$. Under certain conditions, the theorem compares the number of generators and relations in a minimal presentation of $\mathcal{L}_*(F)$, the rational homotopy Lie algebra (see Section 2.3) of a component F of X^T, with the corresponding numbers for $\mathcal{L}_*(X)$. This theorem is an analogue of Theorem (3.8.12). The conditions needed are as follows.

(4.9.1) Conditions
Let T be a torus, and let X be a T-space with $X^T \neq \emptyset$. Let F be a component of X^T. Suppose that

(1) $\dim_{\mathbb{Q}} H^*(X; \mathbb{Q}) < \infty$;
(2) X satisfies conditions (A), (B) or (C) of Definition (3.2.4);
(3) (i) X and F are 1-connected,
 (ii) X and F are agreement spaces (see Definition (2.6.2)), and
 (iii) F is full (see Definition (3.3.13), Proposition (3.3.15) and the concluding remarks of Section 3.4).

(4.9.2) Theorem
Let T be a torus, and let X be a T-space with $X^T \neq \emptyset$. Let F be a component of X^T. Suppose that X and F satisfy Conditions $(4.9.1)(1)$, (2) and (3). Suppose further that $\mathcal{L}_*(X)$ is finitely presented as a graded Lie algebra, and let $L(W_X) \to L(V_X) \to \mathcal{L}_*(X) \to 0$ be a minimal presentation, where L is the free graded Lie algebra functor.

Then $\mathcal{L}_*(F)$ is finitely presented; and if

$$L(W_F) \to L(V_F) \to \mathcal{L}_*(F) \to 0$$

is a minimal presentation, then

(1) $\dim_{\mathbb{Q}}(V_F) \leq \dim_{\mathbb{Q}}(V_X)$,
(2) $\dim_{\mathbb{Q}}(W_F) \leq \dim_{\mathbb{Q}}(W_X)$, and, indeed,
(3) $\mathrm{cid}(\mathcal{L}_*(F)) \leq \mathrm{cid}(\mathcal{L}_*(X))$,
where $\mathrm{cid}(\mathcal{L}_*(X)) = \dim_{\mathbb{Q}}(W_X) - \dim_{\mathbb{Q}}(V_X)$.

(4.9.3) Remarks
(1) As in Theorem (3.8.12) parts (1), (2) and (3) of Theorem (4.9.2) can be refined into the corresponding inequalities for generators, relations and complete intersection defects of even degrees and odd degrees separately.

(2) If Conditions (4.9.1)(3)(i) and (ii) are dropped, then some results can be obtained using the Alexander-Spanier theory of Section 2.6 and co-Lie algebra structures. This is clear from the proof in [Allday, Puppe, 1986b].

(3) Condition (4.9.1)(3)(iii) cannot be dropped: see [Allday, Puppe, 1986b], Examples 4.4 and 4.5.
(4) If X is a finite-dimensional T-CW-complex with FMCOT, then Conditions (4.9.1)(2) and (3)(ii) are satisfied automatically.
(5) Under Conditions (4.9.1)(2)(B) or (C), Condition (4.9.1)(1) could be replaced by the conditions that both $H^*(X;\mathbb{Q})$ and $H^*(F;\mathbb{Q})$ have finite type. This is not so clear in case (A) unless $T = S^1$; see Lemma (4.2.1) and the first paragraph of Section 4 of [Allday, Puppe, 1986b], where an easily corrected (because of Lemma (4.2.1)) mistake is made concerning the reduction of the proof to the case where $T = S^1$.
(6) In Section 4 of [Allday, Puppe 1986b] the quantities $\dim_\mathbb{Q}(V_X) - \dim_\mathbb{Q}(V_F), \dim_\mathbb{Q}(W_X) - \dim_\mathbb{Q}(W_F)$ and $\text{cid}(\mathcal{L}_*(X)) - \text{cid}(\mathcal{L}_*(F))$ are computed from the E_2-term of a certain spectral sequence.
(7) Theorem (4.9.2) can be proved along the lines of the argument used to obtain Theorem (1.3.17) at least if $\dim_\mathbb{Q} \mathcal{L}_*(X) < \infty$. See Exercise (4.12) and also [Allday, Puppe, 1984].
(8) Under Conditions (4.9.1) if $\dim_\mathbb{Q} \mathcal{L}_*(X) < \infty$, it follows also that the degree of nilpotency of $\mathcal{L}_*(F)$ is greater than or equal to the degree of nilpotency of $\mathcal{L}_*(X)$: see [Allday, Puppe, 1986b], Theorem 4.6.

The theorem cited in Remarks (4.9.3)(8) has a cohomology analogue which is as follows. (See [Allday, Puppe, 1986b], Remarks 4.7(ii) and cf. (1.3.16) for case (2). The proof of (1.3.16) can be imitated here.)

(4.9.4) Theorem
Let X and F satisfy Conditions (4.9.1)(1) and (2). Suppose that X is connected. Suppose also that X is TNHZ in $X_T \to BT$ (with respect to rational cohomology). Then, for any integer $q \geq 0$,
$$\dim_\mathbb{Q} \overline{H^*(F,\mathbb{Q})}/\overline{H^*(F,\mathbb{Q})}^q \leq \dim_\mathbb{Q} \overline{H^*(X,\mathbb{Q})}/\overline{H^*(X,\mathbb{Q})}^q,$$
where $\overline{H^*(-;\mathbb{Q})}$ denotes reduced (singular or Alexander-Spanier, as the case may be) cohomology.

In particular, if X^T is connected, then $n_0(X^T) \geq n_0(X)$, where n_0 is the rational cup length - i.e. $n_0(X) = \max\{q; \overline{H^*(X,\mathbb{Q})}^q \neq 0\}$.

In [Tomter, 1974] there is an example of the circle group acting on $X = S^4 \times S^{2n}$ with fixed point set $X^T = \mathbb{C}P^3$. So $n_0(X^T) > n_0(X)$ in this example.

Exercises

In the following exercises, unless stated otherwise, all G-spaces should be assumed to satisfy condition (LT) of the introduction to Section 3.10.

Exercises for Chapter 4

(4.1) Let G be a torus (case (0)), and let X be a G-space with
$$\dim_{\mathbb{Q}} H^*(X;\mathbb{Q}) < \infty.$$
Let $M = H_G^*(X;\mathbb{Q})$, and let $P(M,z)/P(R,z) = f(z) \in \mathbb{Z}[z]$ as in Lemma (4.1.1). Show that $f(-1) = \chi(X)$.

(4.2) (Skjelbred) With the conditions and notation of Exercise (4.1) and Theorem (4.1.16), let $P(M,z) = \sum_{m=-\infty}^{\infty} c_m (1+z)^m$ be the Laurent expansion about $z = -1$. Show that

(i) $c_m = 0$ for $m < -s$; and
(ii) $c_{-s} = 2^{-s} \chi(X_s/G) = 2^{-s} \sum_{j=1}^{\sigma} \chi(X^{K_j}/G)$.

(4.3) Let $f : S^3 \times S^3 \to S^6$ be a map of degree one. Let θ be the pull-back of the tangent bundle of S^6 along f, and let X be the total space of the unit sphere bundle associated with θ. Show that the only non-trivial compact connected Lie group which can act almost-freely on X is S^1.

(4.4) Let X be the total space of the principal S^1-bundle over $T^2 = S^1 \times S^1$ which is classified by a top class (generator) in
$$H^2(T^2;\mathbb{Z}) = [T^2, BS^1].$$
Let $A = S^1 \times S^2$ with S^1 acting trivially on the first factor and in the standard way on the second factor. Let Y be the connected sum $(A - \overset{\circ}{D}) \cup_{\partial D} (A - \overset{\circ}{D})$ where $D \subseteq A$ is a closed invariant 3-disc centred on a fixed point (e.g., a closed slice). Show that
(1) $H^*(X;\mathbb{Q}) \cong H^*(Y;\mathbb{Q})$;
(2) Y inherits an action of $G = S^1$ such that Y^G consists of three disjoint circles; and
(3) the only possible non-trivial effective compact connected Lie group actions on X are almost-free S^1-actions.

(4.5) (Skjelbred) Let G be a torus (case (0)) and let X be a path-connected G-space. Let $H_G(X) = H_G^{ev}(X;\mathbb{Q})$ as in Section 3.6. Let $I \triangleleft H_G(X)$ be the ideal generated by all nilpotent and all R-torsional elements. Suppose that $\pi_{2m}(X) \otimes \mathbb{Q} = 0$ for $1 \leq m \leq N$, where $N \geq 1$ is some integer; and suppose that $H_G(X)/I$ is generated as a R-algebra by some elements of degree $\leq 2N$ and a finite number, α, say, of elements of degree $> 2N$. If X is not compact and not a G-CW-complex and if $\dim_{\mathbb{Q}} H^*(X;\mathbb{Q})$ is not finite, then assume that the natural map $H_G^*(X^L;\mathbb{Q}) \to \prod_c H_G^*(c;\mathbb{Q})$, where c ranges over all the components of X^L, is surjective for all subtori $L \subseteq G$. Suppose that $X^G \neq \emptyset$, and let F_1 and F_2 be two components of X^G. Let K be a maximal connecting subtorus of F_1 and F_2. (See Remark (3.8.6).)

Show that $\text{cork}(K) \leq \alpha$.

(4.6) suppose that $G = T^r$ acts almost-freely on X. Let
$$n(X) = \sup\{j : H^j(X; \mathbb{Q}) \neq 0\},$$
and suppose that $n(X) < \infty$. Show that $n(X_G) < \infty$, and hence that $n(X_G) = n(X) - r$. And so, in particular, $r \leq n(X)$. (Cf. Example (4.3.5)(2).)

(4.7) Verify that $\text{rank}_0((S^1 \vee S^1) \times S^3 \times S^3) = 2$. (See Example (4.3.14).)

(4.8) Let $G = T^r$, and let $R = H^*(BG; \mathbb{Q}) = \mathbb{Q}[t_1, \ldots, t_r]$. In projective geometry a homogeneous ideal $I \triangleleft R$ is said to be irrelevant if $\sqrt{I} = (t_1, \ldots, t_r)$. We shall say that an irrelevant ideal $I \triangleleft R$ is completely irrelevant if given any homogeneous $f_1, \ldots, f_r \in I$ and setting $f_i = \sum_{j=1}^{r} f_{ij} t_j$ for $1 \leq i \leq r$, then $\det(f_{ij}) \in I$.

Now suppose that G is acting almost-freely on a connected space X, and that $\ker(p^*)$ is not completely irrelevant, where $p^* : R \to H_G^*(X; \mathbb{Q})$ is induced by the bundle map $p : X_G \to BG$. Show that $H^*(X; \mathbb{Q})$ contains an exterior algebra on r homogeneous elements of odd degree. In particular, $\dim_{\mathbb{Q}} H^*(X; \mathbb{Q}) \geq 2^r$.

(4.9) Let $G = (\mathbb{Z}_p)^r$ and let k be a field of characteristic p. Let X be any paracompact free G-space, and let $\overline{C}^j(X; k)$, $j \geq 0$, denote the Alexander-Spanier cochains of degree j with coefficients in k. Show that (with the standard module structure) $\overline{C}^j(X; k)$ is a free $k[G]$-module for all $j \geq 0$.

Similarly suppose that X is any paracompact G-space and let E, α, $K(\alpha)$, γ and Γ be as in Recollections (3.11.7) chosen so that $X^{K(\alpha)} = \emptyset$. Show that $\overline{C}^j(X; E)$ is a free $E[\Gamma]$-module for all $j \geq 0$.

Also if X is a paracompact finitistic G-space and if E and Γ are chosen as in Proposition (4.6.24) and its proof, then show that $\overline{C}^j(X; E)$ is a free $E[\Gamma]$-module for all $j \geq 0$.

(4.10) The purpose of this exercise is to give a proof of Swan's theorem (Theorem (4.6.15)) in case (B).

Let G be a finite group, k a commutative ring and Λ a k-module. Let X be a free G-space, and suppose that X is paracompact with $\text{cd}_k(X) < \infty$. Let $\text{cd}_k(X) = n$. Also let $C^j = \overline{C}^j(X; \Lambda)$, the Alexander-Spanier cochains of X of degree j with coefficients in Λ, for $j \geq 0$. So $C^j \cong \varinjlim_{\mathcal{U}} \text{Hom}_k(C_j(\overline{\mathcal{U}}; k), \Lambda)$, where $\mathcal{U}$ ranges over the faithful Čech-G-coverings (i.e. Vietoris-G-coverings) of X, and $C_j(\overline{\mathcal{U}}; k)$ denotes the ordered j-chains of $\overline{\mathcal{U}}$ with coefficients in k.

For any $k[G]$-module M we shall say that M is acyclic, resp. Tate acyclic, if $H^i(G; M) = 0$ for all $i > 0$, resp. $\hat{H}^i(G; M) = 0$ for all i.
(1) Show that C^j is Tate acyclic for all $j \geq 0$.
(2) Define the cochain complex C_t^* as follows. $C_t^j = C^j$ for $j < n$, $C_t^n = Z^n$, the module of cycles in C^n, and $C_t^j = 0$ for $j > n$. Let the differential on C_t^* be the restriction of the differential on C^*. Show that the inclusion $i : C_t^* \to C^*$ is a weak-equivalence. Deduce that $H_G^*(C_t^*) \cong H_G^*(C^*) \cong H_G^*(X; \Lambda)$ and that
$$\hat{H}_G^*(C_t^*) \cong \hat{H}_G^*(C^*) = \hat{H}_G^*(X; \Lambda).$$
(3) Show that Z^n is acyclic in either of the following two ways.
(a) Look at the second spectral sequence for $H_G^*(C_t^*)$ and use the fact that $\mathrm{cd}_k(X/G) \leq n$.
(b) Let $Q_i = C^{n+i}$ for $i \geq 0$. Show that
$$0 \to Z^n \to Q_0 \to Q_1 \to Q_1 \to \cdots$$
is a coresolution of Z^n by acyclic $k[G]$-modules. Deduce that $H^*(G; Z^n)$ is the homology of the cochain complex
$$0 \to Q_0^G \to Q_1^G \to \cdots$$
(Just as in sheaf theory, injective coresolutions may be replaced by acyclic coresolutions. See [Bredon, 1967a], Chapter II, Section 4, pp. 33-4.)
Now use the second spectral sequence for $H_G^*(C^*) \cong H_G^*(X; \Lambda)$ and the fact that $\mathrm{cd}_k(X/G) \leq n$.
(4) Deduce that Z^n is Tate acyclic. (See [Brown, K.S., 1982], Chapter VI, Theorem (8.7) and Proposition (8.8).)
(5) Conclude that $\hat{H}_G^*(X; \Lambda) = 0$.
(6) Show that Theorem (4.6.15) is valid as stated if (X, A) is any pair of G-spaces such that X is paracompact, $\mathrm{cd}_k(X) < \infty$ and A is a closed invariant subspace.

(4.11) Let $G = (\mathbb{Z}_p)^r$, where p is any prime number. Let X be a finite G-CW-complex; and suppose that the rank of the action is ρ. Show that $p^\rho \mid \chi(X)$.

(4.12) Suppose that, in addition to Conditions (4.9.1), X has FDRH (i.e $\dim_\mathbb{Q} \Pi_\psi^*(X) < \infty$). Give a proof of Theorem (4.9.2) by the method of the proof of Theorem (1.3.17). (With $R = \mathbb{Q}[t] = H^*(BT; \mathbb{Q})$ as usual, where $T = S^1$, show that the R-co-Lie algebra structure on $Q_{(T,F)}(X)$ can be dualized to give a graded R-Lie algebra, $R \tilde{\otimes} \mathcal{L}_*(X) = L$, say, where $t \in R$ has degree -2, $L_0 \cong \mathcal{L}_*(X)$ and $L_1 \cong \mathcal{L}_*(F)$. Here L_0, resp. L_1, indicates L evaluated at 0, resp. 1; and the isomorphism $L_1 \cong \mathcal{L}_*(F)$ is $\mathbb{Z}_2$-graded. See Section A.7.)

5

Actions on Poincaré duality spaces

If M is a compact, oriented n-dimensional manifold then the cohomology $H^*(M;k)$ with coefficients in a field k is a Poincaré algebra, i.e. the orientation of M (resp. the corresponding fundamental class $[M] \in H_n(M;k)$) gives a k-linear map $\mathcal{O}_M : H^*(M;k) \to k$ defined on $H^n(M;k)$ as the evaluation on the fundamental class of M and to be zero on $H^i(M;k)$ for $i \neq n$, such that the bilinear form obtained by composing the cup-product on $H^*(M;k)$ with $\mathcal{O}_M$

$$H^*(M;k) \times H^*(M;k) \overset{\cup}{\to} H^*(M;k) \overset{\mathcal{O}_M}{\to} k$$

is non-singular. This means that the induced map

$$H^*(M;k) \overset{\mathcal{D}}{\to} \mathrm{Hom}_k(H^*(M;k), k)$$

sending $a \in H^*(M;k)$ to $\mathcal{D}_a \in \mathrm{Hom}_k(H^*(M;k), k)$, where

$$\mathcal{D}_a : H^*(M;k) \to k$$

is given by $\mathcal{D}_a(x) := \mathcal{O}_M(a \cup x)$, is an isomorphism.

In the situation above the duality actually holds already over $\mathbb{Z}$ (as coefficients), but in view of the material we are going to describe we restrict ourselves to field coefficients here. For a detailed exposition of duality on manifolds see, e.g., [Dold, 1980].

We will study certain implications for the cohomology of fixed point sets of group actions on spaces X for which the above duality holds. Of the many different aspects of group actions on manifolds (see, e.g., [Schultz, 1985] and the references given there, and the references in the references...) we discuss only a small section for which the purely algebraic duality properties of the cohomology $H^*(X;k)$ of X mostly suffice and no geometric reason behind it (i.e. X being a manifold) is needed.

5.1 Algebraic preliminaries

Since the passage from the cohomology of a G-space to that of its fixed point set has little control on the grading, cf. (1.3.5), (1.4.5), (3.5.7), we define a notion of Poincaré duality which is not restricted to graded algebras.

Let A be an algebra over a field k.

(5.1.1) Definition
(1) An orientation of A is a non-trivial (and hence surjective) k-linear map $\mathcal{O}_A : A \to k$.
(2) $(A, \mathcal{O}_A)$ is called a Poincaré algebra if the k-bilinear form

$$A \times A \xrightarrow{\mu_A} A \xrightarrow{\mathcal{O}_A} k$$

given by the composition of the multiplication $\mu_A : A \times A \to A$ and the orientation $\mathcal{O}_A$ is non-singular, i.e. induces a k-isomorphism

$$\mathcal{D} : A \to \text{Hom}_k(A, A) \to \text{Hom}_k(A, k),$$

where $\mathcal{D}_a := \mathcal{D}(a)$ is defined by $\mathcal{D}_a(x) := \mathcal{O}_A(ax)$ (using the notation $ax := \mu_A(a,x)$), between A and its k-dual $\text{Hom}_k(A,k)$. (Note that A necessarily has to be of finite dimension as a k-vector-space.)
(3) If $\bigoplus_{i=0}^n A^i$ is a graded algebra and fulfils (2) with respect to a graded orientation $\mathcal{O}_A : A \to k$, i.e. $\mathcal{O}_A$ is trivial on all A^i with $0 \leq i < n$, we call $(A, \mathcal{O}_A)$ a graded Poincaré algebra of formal dimension n ($\text{fd}(A) = n$). If one defines a grading on $\text{Hom}_k(A,k)$ by $\text{Hom}_k(A,k)^i := \text{Hom}_k(A^{n-i}, k)$, the isomorphism $\mathcal{D} : A \to \text{Hom}_k(A,k)$ becomes an isomorphism of graded k-vector-spaces.
(4) Let A be a filtered algebra with filtration

$$0 = \mathcal{F}_{-1}A \subset \mathcal{F}_0 A \subset \ldots \subset \mathcal{F}_n A = A$$

of length $n+1$ such that $\mu_A(\mathcal{F}_i A \times \mathcal{F}_j A) \subset \mathcal{F}_{i+j} A$ (we always assume this compatibility of the filtration with the product structure for a filtered algebra) and $\mathcal{O}_A$ an orientation on A which is trivial on $\mathcal{F}_{n-1}(A)$. We call $(A, \mathcal{O}_A)$ a filtered Poincaré algebra of length $n+1$, if

$$\mathcal{D} : A \to \text{Hom}_k(A,k)$$

is an isomorphism of filtered k-vector-spaces, where the filtration on

$\operatorname{Hom}_k(A,k)$ is defined by
$$\mathcal{F}_q \operatorname{Hom}_k(A,k) := \{\phi \in \operatorname{Hom}_k(A,k); \phi|\mathcal{F}_{n-q-1}(A) = 0\}$$
$$\cong \operatorname{Hom}_k(A/\mathcal{F}_{n-q-1}A, k).$$
(We consider the filtration as part of the structure even if it might not be explicitly indicated in the notation.)

(5.1.2) Remark
(1) The most important examples for graded Poincaré algebras of formal dimension n are the cohomology algebras of n-dimensional, compact, oriented manifolds. In fact, Poincaré's results on the cohomology of such manifolds motivated the defiintion of a Poincaré algebra. But, in particular, if one restricts to specific coefficients k the cohomology algebra $H^*(X;k)$ might be a Poincaré algebra for a space X which need not be a manifold. We will call such a space X a Poincaré duality space with respect to the coefficients k ('k-PD-space' for short) and by the formal dimension, $\operatorname{fd}(X)$, we mean the formal dimension $\operatorname{fd}(H^*(X;k))$ of $H^*(X;k)$.

(2) Clearly a graded Poincaré algebra $(A, \mathcal{O}_A)$ of formal dimension n becomes a filtered Poincaré algebra if one defines the filtration of A by $\mathcal{F}_q A = \bigoplus_{i=0}^{q} A^i$.

If A is a filtered algebra with filtration
$$\mathcal{F}_{-1} A = 0 \subseteq \mathcal{F}_0 A \subset \ldots \subset \mathcal{F}_n A = A$$
of length $n+1$ and $\mathcal{O}_A : A \to k$ an orientation of A which vanishes on $\mathcal{F}_{n-1}$, then clearly $\mathcal{O}_A$ induces an orientation $\mathcal{O}_{gr A}$ of the associated graded algebra $gr A := \bigoplus_{i=0}^{n} \mathcal{F}_i A / \mathcal{F}_{i-1} A$.

(5.1.3) Proposition
Under the above assumption $(A, \mathcal{O}_A)$ is a filtered Poincaré algebra if and only if $(gr A, \mathcal{O}_{gr A})$ is a graded Poincaré algebra.

Proof.
The map $\mathcal{D} : A \to \operatorname{Hom}_k(A,k)$ defined by $\mathcal{D}_a(x) := \mathcal{O}_A(ax)$ for $a \in A$ respects the filtrations, where
$$\mathcal{F}_q \operatorname{Hom}_k(A,k) := \operatorname{Hom}_k(A/\mathcal{F}_{n-q-1}A, k)$$
(see (5.1.1)(4)), since
$$\mathcal{O}_A(\mu_A(\mathcal{F}_i A \times \mathcal{F}_j A)) \subset \mathcal{O}_A(\mathcal{F}_{i+j}A) = 0$$
for $i + j < n$. The associated graded vector-space $gr \operatorname{Hom}_k(A,k)$ is canonically isomorphic (including the grading) to $\operatorname{Hom}(gr A, k)$ equipped

5.1. Algebraic preliminaries

with the grading
$$\operatorname{Hom}_k(grA, k)^i := \operatorname{Hom}_k((grA)^{n-i}, k)$$
(see (5.1.1)(3)). Hence the map
$$gr\mathcal{D} : grA \to gr\operatorname{Hom}_k(A, k) \cong \operatorname{Hom}_k(grA, k)$$
is an isomorphism of graded vector-spaces if and only if
$$\mathcal{D} : A \to \operatorname{Hom}_k(A, k)$$
is an isomorphism of filtered vector-spaces. □

If $A = \bigoplus_{i=0}^{n} A^i$ is a connected (i.e. $A^0 = k$), graded k-algebra and $\mathcal{O}_A : A \to k$ is a graded orientation such that $(A, \mathcal{O}_A)$ is a graded Poincaré algebra of formal dimension n, then A^0 is isomorphic to
$$\operatorname{Hom}_k(A, k)^0 = \operatorname{Hom}_k(A^n, k)$$
and the unit in $k = A^0$ corresponds to $\mathcal{O}_A | A^n$ under this isomorphism.

In particular $\mathcal{O}_A | A^n : A^n \to k$ is an isomorphism ($\mathcal{O}_A(A^i) = 0$ for $0 \le i < n$, since $\mathcal{O}_A$ is assumed to be graded) and it is easy to check that any isomorphism between A^n and k could serve as an orientation. In other words if there is at all a way to choose an orientation $\mathcal{O}_A$ on A such that $(A, \mathcal{O}_A)$ is a graded Poincaré algebra, then any isomorphism between A^n and k, extended trivially to A^i, $0 \le i < n$, will do.

The following characterizes those connected graded k-algebras
$$A = \bigoplus_{i=0}^{n} A^i,$$
which by some choice of an orientation become graded Poincaré algebras, purely in terms of the multiplicative structure of A, i.e. without referring to a particular orientation.

(5.1.4) Proposition
Let $A = \bigoplus_{i=0}^{n} A^i$, $n > 0$, $A^n \ne 0$, be a connected, graded k-algebra which has finite dimension as a k-vector-space.

(1) If $\mathcal{O}_A : A \to k$ is an orientation (not necessarily fulfilling $\mathcal{O}(A^i) = 0$ for $0 \le i < n$) such that $(A, \mathcal{O}_A)$ is a Poincaré algebra, then $\dim_k \ker(\overline{A} \xrightarrow{\hat{\mu}} \operatorname{Hom}_k(\overline{A}, \overline{A})) = 1$, where $\overline{A} := \bigoplus_{i=1}^{n} A^i$ and the map $A \xrightarrow{\hat{\mu}} \operatorname{Hom}_k(\overline{A}, \overline{A})$ is given by $a \mapsto \hat{\mu}_a$ with $\hat{\mu}_a(\overline{x}) = a\overline{x}$ for $a \in A$.

(2) If $\dim \ker \hat{\mu} = 1$, then $(A, \mathcal{O}_A)$ is a graded Poincaré algebra of formal dimension n for any graded orientation
$$\mathcal{O}_A : A \to k \quad (\mathcal{O}_A(A^i) = 0 \text{ for } 0 \le i < n).$$

Proof.
(1) By assumption $\mathcal{D} : A \to \mathrm{Hom}_k(A, A) \to \mathrm{Hom}_k(A, k)$ is an isomorphism. Restricting to $\overline{A}$ one gets the commutative diagram shown in Diagram 5.1.

$$\begin{array}{ccc} A & \xrightarrow{\mathcal{D}} & \mathrm{Hom}_k(A, k) \\ \uparrow & & \uparrow \\ \Lambda & \xrightarrow{\mathcal{D}|\overline{A}} & \mathrm{Hom}_k(A, k) \ni \phi \\ \downarrow & & \downarrow \quad\quad\quad \downarrow \\ \overline{A} & \xrightarrow{\overline{\mathcal{D}}} & \mathrm{Hom}_k(\overline{A}, k) \ni \phi|\overline{A} \end{array}$$

Diagram 5.1

Clearly $\dim_k \ker(\mathrm{Hom}_k(A, k) \to \mathrm{Hom}_k(\overline{A}, k)) = 1$ and hence

$$\dim_k \ker \hat{\mu} \leq 1,$$

since $\mathcal{D}$ is injective.

On the other hand $A^n \subset \ker \hat{\mu}$, since $A^n \overline{A} = 0$ for degree reasons, and therefore part (1) follows.

(2) Since $0 \neq A^n \subset \ker \hat{\mu}$ and $\dim_k \ker \hat{\mu} = 1$, one gets $A^n = \ker \hat{\mu} \cong k$. Choose any orientation $\mathcal{O}_A : A \to k$ with $\mathcal{O}_A(A^i) = 0$ for $0 \leq i < n$. It suffices to show that $\mathcal{D} : A \to \mathrm{Hom}_k(A, A) \to \mathrm{Hom}_k(A, k)$ is injective (since $\dim_k A < \infty$). One only needs to check injectivity on the homogeneous parts A^i of A, for $\mathcal{D}$ preserves degree. If $i = 0, n$ this clear since $A^0 A^n = A^n A^0 = A^n$ and $\mathcal{O}_A | A^n$ is an isomorphism. Because $A^n = \ker \hat{\mu}$ one has that $\hat{\mu}|A^i$ is injective for $0 < i < n$. This means that for any non-zero element $a \in A^i$, $0 < i < n$, there exists an $x \in A^j$, $1 \leq j$, such that $ax \neq 0$. Since $ax \in A^{i+j}$, $i < i + j \leq n$, iterating the argument (if necessary) gives an element $\overline{a} \in A^{n-i}$ such that $a\overline{a} \neq 0$ in A^n. Hence $\mathcal{O}_A(a\overline{a}) \neq 0$ and $\mathcal{D}_a \in \mathrm{Hom}_k(A, k)$ is non-trivial, i.e. the map $\mathcal{D}$ is injective. □

As long as the specific choice of an orientation does not matter we refer to A as a graded Poincaré algebra, if $(A, \mathcal{O}_A)$ is a graded Poincaré algebra for some graded orientation $\mathcal{O}_A$.

(5.1.5) Remark
In commutative algebra there is the well-known notion of a Gorenstein ring (algebra) (e.g., [Kunz, 1985]). For a connected, graded, strictly commutative k-algebra $A = \bigoplus_{i=0}^{n} A^i$, with $\dim_k A < \infty$, this amounts precisely to the condition $\dim_k \ker \hat{\mu} = 1$ in (5.1.4). Hence part of (5.1.4)

can be restated as: a connected, graded, strictly commutative algebra $A = \bigoplus_{i=0}^{n} A^i$ with $\dim_k A < \infty$ is a Gorenstein ring (algebra) if and only if it is a Poincaré algebra.

Gorenstein rings (algebras) are usually considered in the strict commutative setting but it is simple to extend at least some basic definitions and properties to the graded commutative case.

5.2 Poincaré duality for the fixed point set

If G is a compact Lie group acting differentiably and in an orientation preserving way on a compact, orientable, differentiable manifold M, then the fixed point set M^G is again a compact, orientable, differentiable manifold, which may consist of different components having not necessarily the same dimension (see e.g., [Bredon, 1972] or [Jänich, 1968] for basic material on differentiable G-manifolds). In particular the cohomology $H^*(F_i; k)$ of each component $F_i \subset M^G$ with coefficients in a field k forms a graded Poincaré algebra.

The following result (see [Chang, Skjelbred, 1972], and [Bredon, 1973]) shows that one can deduce Poincaré duality for the fixed point set of a torus or p-torus action on a space X under weaker assumptions on X.

(5.2.1) Theorem
Let G be a torus or p-torus and $k = \mathbb{Q}$ or $\mathbb{F}_p$ respectively. If X is a finite-dimensional G-CW-complex which is also a k-PD-space (i.e. $H^*(X; k)$ is a Poincaré algebra) then each component F_i of X^G is a k-PD-space. The formal dimension $\text{fd}_k(F_i)$ of F_i is smaller or equal to the formal dimension of the component of X which contains F_i, and the formal dimensions coincide modulo 2 in the cases (0) (i.e. $(G, k) = ((S^1)^r, \mathbb{Q})$) and (odd p).

Proof.
It suffices to prove the theorem for groups of rank equal to 1. The general case then follows by induction. We also can assume that X is connected and $X^G \neq \emptyset$. To simplify the notation we suppress the coefficients, i.e. 'everything' (cohomology, cochain complexes, etc.) is understood to be taken with coefficients in k without this being mentioned explicitly.

We first consider the case $p = 2$. The cup-product on $H^*(X)$ is induced by a chain map $\mu_X : S^*(X) \otimes S^*(X) \to S^*(X)$, where $S^*(X)$ denotes the singular cochain complex of X with coefficients in k, which by naturality can be chosen to be compatible with the G-action, i.e. μ_X is a morphism of cochain complexes over the group ring $k[G]$, where $S^*(X) \otimes S^*(X)$ carries the diagonal action and the usual tensor product boundary. If $\mathrm{fd}(X) = n$ we define a $k[G]$-subcomplex $\overline{S}^*(X)$ of $S^*(X)$ by

$$\overline{S}^i(X) = \begin{cases} S^i(X) & \text{for } i < n \\ Z^n(X) & \text{for } i = n \\ 0 & \text{for } i > n \end{cases}$$

where $Z^n(X)$ denotes the n-cycles of $S^*(X)$. Since $H^i(X) = 0$ for $i > n$ the inclusion $\overline{S}^*(X) \subset S^*(X)$ induces an isomorphism in homology; in fact, it is a homotopy equivalence of complexes over k. Let

$$\mathcal{O}_{H^*(X)} : H^*(X) \to k$$

be a graded orientation for the Poincaré algebra $H^*(X)$, i.e.

$$\mathcal{O}_{H^*(X)} : H^n(X) \to k$$

is an isomorphism and $\mathcal{O}_{H^*(X)}$ is trivial on $H^i(X)$, $i \neq n$. Then $\mathcal{O}_{H^*(X)}$ can be lifted to a morphism

$$\mathcal{O}_{\overline{S}^*(X)} : \overline{S}^*(X) \to k$$

of cochain complexes over $k[G]$ by putting $\mathcal{O}_{\overline{S}^*(X)}$ to be zero on $\overline{S}^i(X)$ for $i \neq n$ and to be the composition of the projection $Z^n(X) \to H^n(X)$ with $\mathcal{O}_{H^*(X)}$ on $\overline{S}^n(X)$ (the $k[G]$-structure and the coboundary on k is defined to be trivial). Using the product and the orientation on the chain level we would like to define a morphism of $k[G]$-complexes between $S^*(X)$ and $\mathrm{Hom}(S^*(X), k)$ (or $\overline{S}^*(X)$ and $\mathrm{Hom}(\overline{S}^*(X); k)$) which induces the duality isomorphism in homology.

There is the following difficulty: on one hand it may not be possible to extend $\mathcal{O}_{\overline{S}^*(X)} : \overline{S}^*(X) \to k$ to $S^*(X)$ as a morphism of $k[G]$-complexes, and on the other hand μ_X does not restrict to an (internal) multiplication on $\overline{S}^*(X)$. (Of course, since $\overline{S}^*(X)$ is homotopy equivalent to $S^*(X)$ one can obtain a multiplication on $\overline{S}^*(X)$, which corresponds to μ_X up to homotopy. But this multiplication may not be compatible with the G-action.) We consider the following sequence of morphisms of $k[G]$-complexes

$$(*) \quad \overline{S}^*(X) \xrightarrow{\phi} \mathrm{Hom}(\overline{S}^*(X), S^*(X)) \xleftarrow{\psi} \mathrm{Hom}(\overline{S}^*(X), \overline{S}^*(X)) \xrightarrow{\eta} \mathrm{Hom}(\overline{S}^*(X), k).$$

The map ϕ is adjoint to μ_X (restricted to $\overline{S}^*(X) \otimes \overline{S}^*(X)$), the map ψ is induced by the inclusion $\overline{S}^*(X) \to S^*(X)$ and the map η by the

5.2. Poincaré duality for the fixed point set

orientation $\mathcal{O}_{\overline{S}^*(X)} : \overline{S}^*(X) \to k$. (For two right $k[G]$-modules A, B the right $k[G]$-module structure on $\mathrm{Hom}(A, B)$ is defined by

$$(\alpha g)(a) := (\alpha(ag^{-1}))g,$$

$\alpha \in \mathrm{Hom}(A, B)$.) Passing to homology the sequence $(*)$ gives the duality isomorphism

$$H^*(X) \xrightarrow{\mathcal{D}} \mathrm{Hom}(H^*(X), k),$$

since $H(\overline{S}^*(X))$ can be identified with $H(S^*(X)) = H^*(X)$, and the map ψ in the above sequence, which is a homotopy equivalence of complexes over k, induces the 'identity' in homology.

We now apply the functor $\beta_{G,1}$ (Section 1.3) to the above sequence $(*)$ and get a sequence of morphisms of differential k-vector-spaces

$$\beta_{G,1}(*) \ \beta_{G,1}(\overline{S}^*(X)) \to \beta_{G,1} \mathrm{Hom}(\overline{S}^*(X), S^*(X))$$
$$\leftarrow \beta_{G,1} \mathrm{Hom}(\overline{S}^*(X), \overline{S}^*(X)) \to \beta_{G,1} \mathrm{Hom}(\overline{S}^*(X), k).$$

For a $k[G]$-cochain complex C^* (bounded below) the differential k-vector-space $\beta_{G,1}(C^*)$ is nothing but the direct sum $\bigoplus_q C^q$ equipped with the differential $\tilde{\delta}_1$ defined by $\tilde{\delta}_1(c) = \delta c + c(1+g)$, where δ is the coboundary of C^* and g is the generator of $G = \mathbb{Z}_2$ (since here k is a field of characteristic 2, we need not worry about signs). It is easy to check (see Lemma (5.2.2)) that $\beta_{G,1} \mathrm{Hom}(\overline{S}^*(X)), k)$ is isomorphic to $\mathrm{Hom}(\beta_{G,1}(\overline{S}^*(X)), k)$.

We claim that - taking homology - $\beta_{G,1}(*)$ induces an isomorphism $H^{(*)}(X^G) \to \mathrm{Hom}(H^{(*)}(X^G), k)$. The sequence $\beta_{G,1}(*)$ is obtained from $(*)$ by first applying the functor $\beta_G^*(-)$ and then evaluating at $t = 1$. The inclusion $\overline{S}^*(X) \to S^*(X)$ gives a homotopy equivalence

$$\beta_G^*(\overline{S}^*(X)) \to \beta_G^*(S^*(X))$$

(see (1.2.3)(3), (1.2.4)(3)). The induced map in homology is an isomorphism of $k[t]$-modules. Hence the evaluation at $t = 1$ (which commutes with taking homology of objects in $\delta gk[t]$-Mod) gives an isomorphism

$$H(\beta_{G,1}(\overline{S}^*(X))) = H_{G,1}(\overline{S}^*(X)) \cong H_{G,1}(S^*(X)) = H(\beta_{G,1} S^*(X)).$$

Using (1.3.5) we therefore get $H(\beta_{G,1}(\overline{S}^*(X))) \cong H^{(*)}(X^G)$ and by the Universal Coefficients Theorem furthermore

$$H(\beta_{G,1} \mathrm{Hom}(\overline{S}^*(X), k)) \cong H(\mathrm{Hom}(\beta_{G,1}(\overline{S}^*(X)), k))$$
$$\cong \mathrm{Hom}(H(\beta_{G,1}(\overline{S}^*(X))), k)$$
$$\cong \mathrm{Hom}(H^{(*)}(X^G), k).$$

It follows from Corollary (1.2.5) that $\beta_G^*(\psi)$ and $\beta_G^*(\eta)\beta_G^*(\psi)^{-1}\beta_G^*(\phi)$ induce isomorphisms in homology (where $\beta_G^*(\psi)^{-1}$ is a homotopy inverse of $\beta_G^*(\psi)$ in δgk-Mod).

Using again the fact that the evaluation at $t = 1$ commutes with taking homology (of objects in $\delta gk[t]$-Mod) (see (A.7.2)) one obtains the desired isomorphism

$$H(\beta_{G,1}(\eta))H(\beta_{G,1}(\psi))^{-1}H(\beta_{G,1}(\phi)) :$$
$$H(\beta_{G,1}(S^*(X))) \cong H^{(*)}(X^G) \to H(\beta_{G,1}\operatorname{Hom}(\overline{S}^*(X), k))$$
$$\cong \operatorname{Hom}(H^{(*)}(X^G), k).$$

It remains to show that this isomorphism is indeed induced by the cup-product on $H^{(*)}(X^G)$ and an appropriate orientation of $H^{(*)}(X^G)$ (this follows - not quite obviously - by naturality of the construction), and that it implies Poincaré duality for the cohomology of each component F_i of X^G.

We consider the commutative diagram (Diagram 5.2) in $\delta gk[G]$-Mod. We use the following abbreviations: S^* for $S^*(X)$, S_G^* for $S_G^*(X^G)$, $\overline{S}^*$ for $\overline{S}^*(X)$, $\overline{S}_G^*$ for $\overline{S}^*(X^G)$ and $H_{S^*}^{\overline{S}^*}$ for $\operatorname{Hom}(\overline{S}^*, S^*)$, and so on.

$$\begin{array}{ccccccc}
\overline{S} & \to & H_{S^*}^{\overline{S}^*} & \leftarrow & H_{\overline{S}^*}^{\overline{S}^*} & \to & H_k^{\overline{S}^*} \\
& \searrow & & \swarrow {\scriptstyle (T_1)} & & & \\
\downarrow & & H_{S_G^*}^{\overline{S}^*} & & & & \uparrow \\
& \nearrow & & \nwarrow {\scriptstyle (T_2)} & & & \\
\overline{S}_G^* & \to & H_{S_G^*}^{\overline{S}_G^*} & \leftarrow & H_{\overline{S}_G^*}^{\overline{S}_G^*} & & H_k^{\overline{S}_G^*}
\end{array}$$

Diagram 5.2

The top line is the sequence (*) above, and the bottom line is an analogous sequence for X^G instead of X, e.g.,

$$\overline{S}^i(X^G) = \begin{cases} S^i(X^G) & \text{for } i < n \\ Z^n(X^G) & \text{for } i = n \\ 0 & \text{for } i > n. \end{cases}$$

Corollary (1.3.8) implies that the inclusion $\overline{S}^*(X^G) \to S^*(X^G)$ is a homotopy equivalence in δgk-Mod.

The other maps in the diagram are given by the canonical morphisms $S^*(X) \to S^*(X^G)$, $\overline{S}^*(X) \to \overline{S}^*(X^G)$, $\overline{S}^*(X^G) \to S^*(X^G)$. Note that the bottom line is 'incomplete', i.e. there is no obvious orientation for $\overline{S}^*(X^G)$ that is compatible with the orientation for $\overline{S}^*(X)$.

We now apply the functor $H_{G,1}(-) = H(\beta_{G,1}(-))$ to Diagram 5.2 and by Lemma (5.2.2) and Theorem (1.3.5) we get that all maps in the triangles (T_1), (T_2) and also the vertical maps in Diagram 5.2 become isomorphisms. In particular the composition (extracted from the diagram $H_{G,1}$ Diagram 5.2)

$$H_{G,1}(\overline{S}^*(X^G)) \to H_{G,1}(\text{Hom}(\overline{S}^*(X^G), S^*(X^G)))$$
$$\xrightarrow{\cong} H_{G,1}(\text{Hom}(\overline{S}^*(X), S^*(X^G)))$$
$$\xleftarrow{\cong} H_{G,1}(\text{Hom}(\overline{S}^*(X), \overline{S}^*(X)))$$
$$\to H_{G,1}(\text{Hom}(\overline{S}^*(X), k))$$
$$\xleftarrow{\cong} H_{G,1}(\text{Hom}(\overline{S}^*(X^G), k))$$

is an isomorphism. By (5.2.2) and (1.3.5) this composition is equivalent to

$$H^{(*)}(X^G) \to \text{Hom}_k(H^{(*)}(X^G), H^{(*)}(X^G))$$
$$= \text{Hom}_k(H^{(*)}(X^G), H^{(*)}(X^G))$$
$$= \text{Hom}_k(H^{(*)}(X^G), H^{(*)}(X^G))$$
$$\to \text{Hom}(H^{(*)}(X^G), k)$$
$$= \text{Hom}(H^{(*)}(X^G), k),$$

where the first map in the composition is adjoint to the cup-product of $H^{(*)}(X^G)$ and the last map is induced by the map

$$H_{G,1}(\mathcal{O}_{\overline{S}^*(X)}) : H_{G,1}(\overline{S}^*(X)) = H^{(*)}(X^G) \to k.$$

(We here identify $H_{G,1}(\overline{S}^*(X))$, $H_{G,1}(\overline{S}^*(X^G))$ and $H^{(*)}(X^G)$ according to Theorem (1.3.5).)

Hence $H_{G,1}(\mathcal{O}_{\overline{S}^*(X)}) : H^{(*)}(X^G) \to k$ is the desired orientation of $H^{(*)}(X^G)$ which - together with the cup-product on $H^{(*)}(X^G)$ - gives Poincaré duality for $H^{(*)}(X^G)$.

Let $F_1, \ldots, F_\nu$ be the components of X^G; hence

$$H^{(*)}(X^G) \cong \prod_{i=1}^{\nu} H^{(*)}(F_i).$$

Since, for $i \neq j$, $H^{(*)}(F_i)$ and $H^{(*)}(F_j)$ are orthogonal with respect to the non-singular bilinear form on $H^{(*)}(X^G)$ given by the cup-product and the orientation, the restriction of this bilinear form to any $H^{(*)}(F_i)$ is also non-singular. This means that for any $i = 1, \ldots, \nu$, $H^{(*)}(F_i)$ together with restriction of the orientation of $H^{(*)}(X^G)$ to $H^{(*)}(F_i)$ fulfils Poincaré duality.

By Proposition (5.1.4) we then have that $H^*(F_i)$ is a graded Poincaré algebra for $i = 1, \ldots, \nu$.

Clearly the formal dimension of $H^*(F_i)$ cannot exceed the formal dimension of $H^*(X)$ (see (1.3.8)).

This finishes the proof of (5.2.1) for the case $p = 2$ (except for Lemma (5.2.2)).

The proof in case (odd p) is similar and we only point out the necessary alterations. The more significant differences between the two cases are in the proof of Lemma (5.2.2). We recall that for odd p the functor $H_{G,1}(-)$ is defined as

$$H_{G,1}(C^*) := (H(\beta_G^*(C^*) \otimes_{k[t]} k_1)) \otimes_{\Lambda(s)} k_0,$$

(see Section 1.4), and for a finite-dimensional G-CW-complex X one has natural isomorphisms

$$H(\beta_G^*(S^*(X)) \otimes_{k[t]} k_1) \cong H^{(*)}(X^G) \otimes \Lambda(s)$$

and $H_{G,1}(S^*(X)) \cong H^{(*)}(X^G)$ (see Section 1.4).

One now again applies the functor $H_{G,1}(-)$ to the sequence $(*)$ and Diagram 5.2. Using (1.2.5), (1.4.5) (instead of (1.3.5)), and (5.2.2) one obtains again that $H_{G,1}(-)$ applied to the sequence $(*)$ above gives an isomorphism $H^{(*)}(X^G) \to \text{Hom}(H^{(*)}(X^G), k)$ and $H_{G,1}(-)$ applied to Diagram 5.2 shows that this isomorphism is indeed induced by the cup-product on $H^{(*)}(X^G)$ and the orientation

$$H_{G,1}(\mathcal{O}_{\overline{S}^*(X)}) : H^{(*)}(X^G) \to k.$$

It remains to show that the formal dimension $\text{fd}_k(F_i)$ of a component F_i of X^G coincides modulo 2 with $\text{fd}_k(X)$ in case (odd p). But this is an immediate consequence of the fact that $H_{G,1}(\mathcal{O}_{\overline{S}^*(X)})$ is $\mathbb{Z}_2$-graded and changes degree by n mod 2, i.e. $H_{G,1}(\mathcal{O}_{\overline{S}^*(X)})$ restricted to $H^j(X^G)$ must vanish for $j \neq n$ mod 2. This finishes the argument in case (odd p).

For the case (0) we make use of the minimal Hirsch-Brown model $H^*(X)\tilde{\otimes}R$, $R = k[t]$, $(H^*(X)\tilde{\otimes}R \cong \mu_G^*(X;\mathbb{Q})$ in the notation of Section 3.5) which inherits a R-bilinar product from the multiplication of the Sullivan minimal model $\mathcal{M}^*(X) \otimes R$. The evaluation of the product at $t = 0$ gives the cup-product of $H^*(X)$. We consider the following sequence of in δgR-Mod:

$(*_0)$
$$H^*(X)\tilde{\otimes}R \xrightarrow{\phi} \text{Hom}_R(H^*(X)\tilde{\otimes}R, H^*(X)\tilde{\otimes}R)$$
$$\xrightarrow{\eta} \text{Hom}_R(H^*(X)\tilde{\otimes}R, R),$$

where the map ϕ is adjoint to the multiplication

$$\tilde{\mu}_X : (H^*(X)\tilde{\otimes}R) \otimes (H^*(X)\tilde{\otimes}R) \to H^*(X)\tilde{\otimes}R,$$

and η is induced by the (graded) orientation

$$\mathcal{O}_{H^*(X)} : H^*(X) \to k$$

$(\mathcal{O}_{H^*(X)} : H^n(X) \xrightarrow{\cong} k$, $\mathcal{O}_{H^*(X)}|H^j(X) = 0$ for $i \neq n$) extended to a R-linear map $\mathcal{O}_{H^*(X)\tilde{\otimes}R} : H^*(X)\tilde{\otimes}R \to R$ in the canonical way. (Note

5.2. Poincaré duality for the fixed point set

that although the differential on $H^*(X)\tilde\otimes R$ is twisted, in general, this map is a morphism in δgR-Mod.)

One has canonical isomorphisms of R-modules

$$\mathrm{Hom}_R(H^*(X)\otimes R, R) \cong \mathrm{Hom}_k(H^*(X), R) \cong \mathrm{Hom}_k(H^*(X), k) \otimes R,$$

and the twisted differential on $H^*(X)\tilde\otimes R$ gives a corresponding twisted differential on $\mathrm{Hom}_k(H^*(X), k) \otimes R$, which vanishes if one evaluates at $t = 0$. Using the above isomorphism the composition $\eta\phi$ gives a morphism in δgR-Mod

$$H^*(X)\tilde\otimes R \xrightarrow{\tilde{\mathcal{D}}} \mathrm{Hom}_k(H^*(X), k)\tilde\otimes R,$$

which evaluated at $t = 0$ is just the duality isomorphism

$$H^*(X) \xrightarrow{\mathcal{D}} \mathrm{Hom}_k(H^*(X), k)$$

of the Poincaré algebra $(H^*(X), \mathcal{O}_{H^*(X)})$. Hence, see (A.7.3), $\tilde{\mathcal{D}}$ is an isomorphism and applying $H(-)_1$, i.e. evaluating at $t = 1$ and taking homology (or the other way around which gives the same), gives an isomorphism

$$H^{(*)}(X^G) \cong H(H^*(X)\tilde\otimes R)_1 \to H(\mathrm{Hom}_k(H^*(X), k)\tilde\otimes R)_1$$
$$\cong \mathrm{Hom}(H^{(*)}(X^G), k)$$

(see Lemma (5.2.2)).

Again it remains to show that this isomorphism is induced by the cup-product on $H^{(*)}(X^G)$ and an appropriate orientation of $H^{(*)}(X^G)$. (The assertion for the components F_i of X^G, i.e. $H^*(F_i)$ being a Poincaré algebra of formal dimension $\mathrm{fd}_k(F_i) \leq \mathrm{fd}_k(X)$ with $\mathrm{fd}_k(F_i) \equiv \mathrm{fd}_k(X)$ mod 2, follows as in the case (odd p).).

We consider Diagram 5.3, a homotopy commutative diagram in δgR-Mod, in which we abbreviate $H^*(X)\tilde\otimes R$ by $\tilde H_R$, and $H^*(X^G)\otimes R$ (resp. $H^*(X^G)\tilde\otimes R$) by H_R^G (resp. $\tilde H_R^G$).

$$
\begin{array}{ccccc}
\tilde H_R & \longrightarrow & \mathrm{Hom}_R(\tilde H_R, \tilde H_R) & \longrightarrow & \mathrm{Hom}_R(\tilde H_R) \\
& & \downarrow & & \\
\downarrow & & \mathrm{Hom}_R(\tilde H_R, \tilde H_R^G) & & \uparrow \\
& & \uparrow & & \\
H_R^G & \longrightarrow & \mathrm{Hom}_R(H_R^G, H_R^G) & & \mathrm{Hom}_R(H_R^G, R)
\end{array}
$$

Diagram 5.3

Diagram 5.3 corresponds to the diagram obtained from Diagram 5.2 by applying the minimal Hirsch-Brown model in the cases (odd p) and (2), which can be considered as a functor up to homotopy from $\delta gk[G]$-Mod to δgR-Mod. Note, though, that even in case $p = 2$ the multiplication induced on $H^*(X)\tilde{\otimes}R$ need not be $k[t]$-bilinear, in general (cf. Example (3.11.18)).

Applying the functor $H(-)_1$ to Diagram 5.3 gives a commutative diagram with all vertical maps being isomorphisms (see (3.5.7) and (5.2.2)). In particular the composition (extracted from the diagram $H(\text{Diagram } 5.3)_1$)

$$H(H^*(X^G) \otimes R)_1 \to H(\text{Hom}_R(H^*(X^G) \otimes R, H^*(X^G) \otimes R))_1$$
$$\xrightarrow{\cong} H(\text{Hom}_R(H^*(X)\tilde{\otimes}R, H^*(X^G) \otimes R))_1$$
$$\xleftarrow{\cong} H(\text{Hom}_R(H^*(X)\tilde{\otimes}R, H^*(X)\tilde{\otimes}R))_1$$
$$\to H(\text{Hom}_R(H^*(X)\tilde{\otimes}R, R))_1$$
$$\xleftarrow{\cong} H(\text{Hom}_R(H^*(X^G) \otimes R, R))_1$$

is an isomorphism. By (5.2.2) and (3.5.7) this composition is equivalent to

$$H^{(*)}(X^G) \to \text{Hom}_k(H^{(*)}(X^G), H^{(*)}(X^G)) \to \text{Hom}_k(H^{(*)}(X^G), k),$$

where the first map is adjoint to the cup-product in $H^{(*)}(X^G)$ and the last map is induced by the map

$$H(\mathcal{O}_{H^*(X)\tilde{\otimes}R})_1 : H(H^*(X)\tilde{\otimes}R)_1 \cong H^{(*)}(X^G) \to k.$$

The rest of the proof is analogous to the other cases. □

(5.2.2) Lemma

Let X, Y, Z be finite-dimensional G-CW-complexes, $G = \mathbb{Z}_p$, resp. S^1, $k = \mathbb{F}_p$ resp. $\mathbb{Q}$. Then - in cases (2) and (odd p) one has the following natural (in X, Y, Z) isomorphisms, which essentially are induced by the inclusions of the respective fixed point sets:

(a) $H_{G,1}(S^*(X) \otimes S^*(Y)) \cong H^{(*)}(X^G) \otimes H^{(*)}(Y^G)$
(b) $H_{G,1}(\text{Hom}(\overline{S}^*(X), S^*(Y))) \cong \text{Hom}(H^{(*)}(X^G), H^*(Y^G))$, where

$$\overline{S}^i(X) := \begin{cases} S^i(X) & \text{for } i < n \\ Z^n(X) & \text{for } i = n \\ 0 & \text{for } i > n \end{cases}$$

with $n = \max\{q; H^q(X; k) \neq 0\}$.

The $k[G]$-complexes $S^*(X)$, $S^*(Y)$ in part (a), and the $k[G]$-complex $S^*(Y)$ in part (b) can be replaced by the $k[G]$-complexes $\overline{S}^*(X)$, $\overline{S}^*(Y)$,

5.2. Poincaré duality for the fixed point set

and $\overline{S}^*(Y)$ respectively. In particular one gets for $Y = \mathrm{pt}$:
$$H_{G,1}(\mathrm{Hom}(\overline{S}^*(X), k)) \cong \mathrm{Hom}(H^{(*)}(X^G), k).$$

Using the isomorphisms from part (a) and (b) one gets, abbreviating, in Diagrams 5.4–5.7 as follows: S_X for $S^*(X)$, $\overline{S}_X$ for $\overline{S}^*(X)$, and similarly with X replaced by Y, Z, H_{X^G} for $H^{(*)}(X^G)$ etc., and $\mathrm{Hom}(-,-)$ denoted by Hom^-_-:

(c)

$$\begin{array}{ccc}
H_{G,1}(\mathrm{Hom}^{\overline{S}_X \otimes \overline{S}_Y}_{S_Z}) & \cong & \mathrm{Hom}^{H_{X^G} \otimes H_{Y^G}}_{H_{Z^G}} \\
\downarrow \cong & & \downarrow \cong \\
H_{G,1}(\mathrm{Hom}(\overline{S}_X, \mathrm{Hom}^{\overline{S}_Y}_{S_Z})) & \cong & \mathrm{Hom}(H_{X^G}, \mathrm{Hom}^{H_{Y^G}}_{H_{Z^G}}),
\end{array}$$

Diagram 5.4

where the horizontal maps are analogous to those in (a) and (b), and the vertical maps are given by the adjointness of the functors $- \otimes -$ and $\mathrm{Hom}(-,-)$.

(d)

$$\begin{array}{ccc}
H_{G,1}(\mathrm{Hom}^{\overline{S}_X}_{\overline{S}_Y} \otimes \mathrm{Hom}^{\overline{S}_Y}_{S_Z}) & \cong & \mathrm{Hom}^{H_{X^G}}_{H_{Y^G}} \otimes \mathrm{Hom}^{H_{Y^G}}_{H_{Z^G}} \\
\downarrow & & \downarrow \\
H_{G,1}(\mathrm{Hom}^{\overline{S}_X}_{S_Z}) & \cong & \mathrm{Hom}^{H_{X^G}}_{H_{Z^G}},
\end{array}$$

Diagram 5.5

where the vertical maps are given by composition of morphisms and the horizontal isomorphisms are again as in (a) and (b).

In case (0) - denoting the minimal Hirsch-Brown models of X, Y, Z by $H^*(X) \tilde{\otimes} R$, $H^*(Y) \tilde{\otimes} R$, $H^*(Z) \tilde{\otimes} R$, resp., $R = k[t]$, - one has natural isomorphisms:

(a_0) $H((H^*(X) \tilde{\otimes} R) \otimes_R (H^*(Y) \tilde{\otimes} R))_1 \cong H^{(*)}(X^G) \otimes H^{(*)}(Y^G)$;

(b_0) $H(\mathrm{Hom}_R(H^*(X) \tilde{\otimes} R, H^*(Y) \tilde{\otimes} R))_1 \cong \mathrm{Hom}(H^{(*)}(X^G), H^{(*)}(Y^G))$,
(in particular, $H(\mathrm{Hom}_R(H^*(X) \tilde{\otimes} R, R)) \cong \mathrm{Hom}(H^{(*)}(X^G), k)$);

and the commutative diagrams, where H^*_X etc denotes $H^*(X)$:

(c_0)

$$\begin{array}{ccc}
H(\mathrm{Hom}_R(H^*_X \tilde{\otimes} R) \otimes_R (H^*_Y \tilde{\otimes} R), H^*_Z \tilde{\otimes} R))_1 & \cong & \mathrm{Hom}^{H_{X^G} \otimes H_{Y^G}}_{H_{Z^G}} \\
\downarrow \cong & & \downarrow \cong \\
H(\mathrm{Hom}_R(H^*_X \tilde{\otimes} R, \mathrm{Hom}_R(H^*_Y \tilde{\otimes} R, H^*_Z \tilde{\otimes} R)))_1 & \cong & \mathrm{Hom}(H_{X^G}, \mathrm{Hom}^{H_{X^G}}_{H_{Z^G}})
\end{array}$$

Diagram 5.6

(d₀)

$$H(\operatorname{Hom}_R(H_X\tilde\otimes R, H_Y^*\tilde\otimes R))\otimes_R \operatorname{Hom}_{H_Z^*\tilde\otimes R}^{H_Y^*\tilde\otimes R})_1 \cong \operatorname{Hom}_{H_{YG}}^{H_{XG}} \otimes \operatorname{Hom}_{H_{ZG}}^{H_{YG}}$$

$$\downarrow \qquad\qquad\qquad\qquad \downarrow$$

$$H(\operatorname{Hom}_R(H_X^*\tilde\otimes R, H_Z^*\tilde\otimes R))_1 \qquad \cong \qquad \operatorname{Hom}_{H_{ZG}}^{H_{XG}}$$

Diagram 5.7

Proof.

Part (a) follows from (1.2.4)(3), and (1.3.5) in case (2) resp. (1.4.5) in case (odd p), since $S^*(X)\otimes S^*(Y)$ is homotopy equivalent to $S^*(X\times Y)$ in δgk-Mod and the homotopy equivalence can be chosen to be a morphism in $\delta gk[G]$-Mod, where $S^*(X)\otimes S^*(Y)$ and $S^*(X\times Y)$ are equipped with the diagonal G-action. The fixed point set of $X\times Y$ (with respect to the diagonal action) is $(X\times Y)^G = X^G \times Y^G$ and hence

$$H_{G,1}(S^*(X)\otimes S^*(Y)) \cong H_{G,1}(S^*(X\times Y)) \cong H^{(*)}(X^G\times Y^G)$$
$$\cong H^{(*)}(X^G)\otimes H^{(*)}(Y^G).$$

To prove part (b) we consider the following morphism in $\delta gk[G]$-Mod, induced by $\overline{S}^*(X)\to \overline{S}^*(X^G)$ and $S^*(Y)\to S^*(Y^G)$ respectively,

$$\operatorname{Hom}(\overline{S}^*(X),S^*(Y))\xrightarrow{\Phi}\operatorname{Hom}(\overline{S}^*(X),S^*(Y^G))\xleftarrow{\Psi}\operatorname{Hom}(\overline{S}^*(X^G),S^*(Y^G)),$$

and show that $H_{G,1}(\Phi)$ and $H_{G,1}(\Psi)$ are isomorphisms. We want to apply the Comparison Theorem (1.1.3) (see also (1.1.4) and (1.1.5)). The functor $H_{G,1}(\operatorname{Hom}(\overline{S}^*(X),S^*(Y)))$ can be viewed as a G-cohomology theory in the variable Y (for X fixed). Hence in order to show that $H_{G,1}(\Phi)$ is an isomorphism it suffices to calculate that

$$H_{G,1}(\operatorname{Hom}(\overline{S}^*(X),S^*(Y))) = 0$$

if $Y = \coprod G$. As in the proof of (1.3.5) and (1.4.5) one can reduce this to the case $Y = G$

$$\operatorname{Hom}\left(\overline{S}^*(X),S^*\left(\coprod G\right)\right) \cong \operatorname{Hom}\left(\overline{S}^*(X),\prod S^*(G)\right)$$
$$\cong \prod \operatorname{Hom}(\overline{S}^*(X),k[G]),$$

identifying $k[G]$ with its k-dual, using the canonical k-basis given by the elements of G). The complex $\operatorname{Hom}(\overline{S}^*(X),k[G])$ is bounded and free over $k[G]$ (even if the action of G on X is not free), and hence by induction (using the 5-lemma) it is enough to show that $H_{G,1}(M) = 0$ where M is a free $k[G]$-module (with trivial boundary). This is a simple calculation (see the proof of (1.3.5) and (1.4.5)). The argument for $H_{G,1}(\Psi)$ being an isomorphism is analogous. On the other hand, since $\operatorname{Hom}(\overline{S}^*(X^G),S^*(Y^G))$ carries the trivial G-action, one has by the

Universal Coefficients Theorem
$$H_{G,1}(\operatorname{Hom}(\overline{S}^*(X^G), S^*(Y^G))) \cong \operatorname{Hom}(H^{(*)}(X^G), H^{(*)}(Y^G)),$$
which finishes the proof of part (b).

The proof of (c) and (d) is analogous. One first shows that one can replace X, Y, Z by their respective fixed point sets X^G, Y^G, Z^G and then - these having trivial G-action - the result follows from the standard universal coefficients business.

The proofs in case (0) are in a sense similar (applying (3.5.7) instead of (1.3.5) and (1.4.5)), but simpler. One uses the fact that evaluation at $t = 1$ commutes with taking homology plus some standard isomorphisms in homological algebra.

For part (a_0) one only needs to observe that
$$((H^*(X)\tilde{\otimes} R) \otimes_R (H^*(Y)\tilde{\otimes} R)) \otimes_R k_1$$
$$\cong ((H^*(X)\tilde{\otimes} R) \otimes_R k_1) \otimes ((H^*(Y)\tilde{\otimes} R) \otimes_R k_1),$$
and to apply (3.5.7) and the Künneth formula.

For part (b_0) one uses the isomorphism
$$(\operatorname{Hom}_R(H^*(X)\tilde{\otimes} R, H^*(Y)\tilde{\otimes} R)) \otimes_R k_1$$
$$\cong \operatorname{Hom}((H^*(X)\tilde{\otimes} R) \otimes_R k_1, (H^*(Y)\tilde{\otimes} R) \otimes_R k_1)$$
and the Universal Coefficients Theorem.

The proof of (c_0) and (d_0) is again analogous. □

(5.2.3) Remark
Since finite p-groups are solvable (cf. Exercise (1.5) (b)), Theorem (5.2.1) also holds for any finite p-group G (and $k = \mathbb{F}_p$) by induction on the order of G.

(5.2.4) Remark
The hypothesis in (5.2.1) that X is a finite-dimensional G-CW-complex can be weakened. If one uses Alexander-Spanier cohomology then it suffices to assume instead that X is a G-space satisfying the condition (LT) (see Section 3.10). The proof given above can be imitated in this more general case, replacing the singular cochain complex by the Alexander-Spanier cochain complex. We sketch another proof of (5.2.1), which relies on the Localization Theorem (3.2.6) rather than the Evaluation Theorems (1.3.5), (1.4.5) and (3.5.7), and which also applies to the more general case.

The cup-product of the H_G^*-algebra $H_G^*(X)$ ($H_G^* = H^*(BG)$) together with the orientation $H(\beta_G^*(\mathcal{O}_{\overline{S}^*(X)})) : H_G^*(X) \to H_G^*$ gives an $H^*(BG)$-linear map
$$H_G^*(X) \to \operatorname{Hom}_{H_G^*}(H_G^*(X), H_G^*).$$

Localizing this map with respect to $S = \{1, t, t^2, \ldots\} \subset H_G^*$ gives a morphism
$$S^{-1}H_G^*(X) \to \mathrm{Hom}_{S^{-1}H_G^*}(S^{-1}H_G^*(X), S^{-1}H_G^*),$$
which by the Localization Theorem (3.2.6) is equivalent to the corresponding morphism for X^G instead of X
$$S^{-1}H_G^*(X^G) \to \mathrm{Hom}_{S^{-1}H_G^*}(S^{-1}H_G^*(X^G), S^{-1}H_G^*)$$
$$\cong S^{-1}H_G^* \otimes \mathrm{Hom}(H^*(X^G), k)$$
(using $S^{-1}H_G^*(X^G) = S^{-1}H_G^* \otimes H^*(X^G)$). In order to show that the fixed point set satifies Poincaré duality it suffices to prove that the above map is an isomorphism (which can be interpreted as Poincaré duality for the $S^{-1}H_G^*$-algebra $S^{-1}H_G^*(X)$). This assertion follows from the Leray-Serre spectral sequence of $X \to X_G \to BG$, localized with respect to $S \subset H_G^*$. The multiplicative properties of the spectral sequence give morphisms of $\delta g S^{-1}H_G^*$-modules
$$S^{-1}E_r^{*,*} \to \mathrm{Hom}_{S^{-1}H_G^*}(S^{-1}E_r^{*,*}, S^{-1}H_G^*).$$

The Poincaré duality for $H^*(X)$ implies that for $r = 2$ one has an isomorphism of $S^{-1}H_G^*$-complexes which, in particular, induces an isomorphism in homology. To perform an induction (on r) one has to check that $H(\mathrm{Hom}_{S^{-1}H_G^*}(S^{-1}E_r^{*,*}, S^{-1}H_G^*))$ is isomorphic to
$$\mathrm{Hom}_{S^{-1}H_G^*}(S^{-1}H(E_r^{*,*}), S^{-1}H_G^*)$$
(where homology is taken with respect to the boundaries given by $d_r : E_r \to E_r$). In cases (0) and (2) one has that $S^{-1}H_G^*$ is a 'graded field', i.e. any homogeneous element different from zero has a multiplicative inverse, and hence the desired isomorphism follows easily. The case (odd p) is a little more delicate. One has $S^{-1}H_G^* \cong \Lambda(s) \otimes k[t, t^{-1}]$ for $G = \mathbb{Z}_p$, and not every (finitely generated) $S^{-1}H_G^*$-module needs to be free. But in addition to the $S^{-1}H_G^*$-module structure on $S^{-1}H_G^*(X)$, $S^{-1}E_r^{*,*}$, etc., one has the action of the Bockstein operator β, which on $S^{-1}H_G^*$ is given by $\beta(s) = t$, $\beta(t) = 0$, $\beta(t^{-1}) = 0$ (extended to $S^{-1}H_G^*$ as a derivation). Now it is easy to see that if M^* is a H_G^*-module with a (compatible) Bockstein operator then $S^{-1}M^*$ is free as a $S^{-1}H_G^*$-module. (For any element $m \in M^*$ with $sm = 0$ one has:
$$0 = \beta(sm) = \beta(s)m - s\beta(m) = tm - s\beta(m).$$
Therefore $m = s\beta(m)/t$ in $S^{-1}M^*$, and this implies that the kernel of the multiplication by s, $\mu_s : S^{-1}M^* \to S^{-1}M^*$, coincides with the image $\mu_s(S^{-1}M^*)$. One gets a $S^{-1}H_G^*$-basis of $S^{-1}M^*$ by choosing a set of elements in $S^{-1}M^*$ which is mapped to a $k[t, t^{-1}]$-basis of $\mu_s(S^{-1}M^*)$ under μ_s.) Hence we get the above desired isomorphism also in case (odd p). We therefore get in all three cases that

$S^{-1}E_r^{*,*} \to \operatorname{Hom}_{S^{-1}H_G^*}(S^{-1}E_r^{*,*}, S^{-1}H_G^*)$ is an isomorphism and this implies that $S^{-1}H^*(X_G) \to \operatorname{Hom}_{S^{-1}H_G^*}(S^{-1}H^*(X_G), S^{-1}H_G^*)$, is an isomorphism, too, finishing the argument.

Instead of using the spectral sequence argument sketched in Remark (5.2.4) above one could also imitate - using the Localization Theorem rather than evaluation - the argument given for (5.2.1) and (5.2.2) to obtain the following result:

(5.2.5) Theorem
Let G be a torus or p-torus and $k = \mathbb{Q}$ or $\mathbb{F}_p$ respectively. If X is a G-space which fulfils (LT) and which also is a k-PD-space, then $S^{-1}H_G^*(X)$ fulfils Poincaré duality considered as an $S^{-1}H_G^*$-algebra, where S is the multiplicative set of homogeneous non-zero polynomials in $H_G^* = H^*(BG)$.

More precisely: if $\operatorname{fd}_k(X) = n$ and $\mathcal{O}_{\overline{S}^*(X)} : \overline{S}^*(X) \to k$ is the chosen orientation for $\overline{S}^*(X)$ (see above), then the cup-product on $S^{-1}H_G^*(X)$ and the orientation

$$S^{-1}H_G(\mathcal{O}_{\overline{S}^*(X)}) : S^{-1}H_G^*(X) \to S^{-1}H_G^{*-n}$$

induce an isomorphism of $S^{-1}H_G^*$-modules

$$S^{-1}H_G^*(X) \to \operatorname{Hom}_{S^{-1}H_G^*}(S^{-1}H_G^*(X), S^{-1}H_G^*(X))$$
$$\to \operatorname{Hom}_{S^{-1}H_G^*}(S^{-1}H_G^*(X), S^{-1}H_G^*).$$

Under the isomorphism given by the Localization Theorem this isomorphism corresponds to

$$H^*(X^G) \otimes S^{-1}H_G^*$$
$$\to \operatorname{Hom}_{S^{-1}H_G^*}(H^*(X^G) \otimes S^{-1}H_G^*, H^*(X^G) \otimes S^{-1}H_G^*)$$
$$\to \operatorname{Hom}_{S^{-1}H_G^*}(H_G^*(X^G) \otimes S^{-1}H_G^*, S^{-1}H_G^*)$$
$$\cong \operatorname{Hom}(H_G^*(X^G), k) \otimes S^{-1}H_G^*.$$

It is, of course, easy to deduce Theorem (5.2.1) from Theorem (5.2.5), e.g., by evaluating the last line in (5.2.5) at $t = 1$ ($s = 0$) in case of a group of rank 1, and then applying induction on the rank to get the general case. But for the proof of (5.2.5) one does not need to go through an induction on the rank, in fact, the proof along the line of the argument given for (5.2.1) and (5.2.2) in case of $\operatorname{rank}(G) = 1$ using the functor $H_{G,1}(-)$, works well for G a torus or p-torus of any rank using the functor $S^{-1}H_G(-)$ (instead of $H_{G,1}(-)$).

The statement of Theorem (5.2.5) can be interpreted as Poincaré duality for the localized cohomology of the total space X_G of the Borel fibration $X \to X_G \to BG$ 'over the base BG' (assuming that X is a

PD-space over k). The original proofs of Chang, Skjelbred [1972] and Bredon [1973] use instead Poincaré duality (over k) of an approximation of X_G (replacing BG by an appropriate manifold) in order to obtain Poincaré duality for the fixed point set.

5.3 Equivariant Gysin homomorphism, Euler classes and a formula of A. Borel

In the case of differentiable manifolds one defines the Gysin homomorphism (Euler class, etc.) of an embedding (and more generally for any map between closed, oriented differentiable manifolds) using the corresponding normal bundle (see, e.g., [Milnor, Stasheff, 1974]), and this can be carried over to an equivariant setting (see, e.g., [Kawakubo, 1980], where in a rather general frame this machinery is used to discuss relations between global and local invariants of G-manifolds with respect to some G-cohomology theory: 'global' refers to the G-manifold itself, and 'local' to its fixed point set).

The Euler class, for example, gives some information about the neighbourhood of the embedded manifold in the ambient manifold. For singular cohomology the Gysin homomorphism can be defined purely algebraically as the Poincaré dual of the induced map in cohomology (if one knows that the spaces involved fulfil Poincaré duality). We will use a similar definition and the results of the preceding section to get equivariant Gysin homomorphisms, Euler classes, etc. in the case of G-spaces, fulfilling the hypotheses (LT) of the Localization Theorem (3.2.6), which are k-PD-spaces, $G =$ torus or p-torus, and $k = \mathbb{Q}$ or $\mathbb{F}_p$ respectively. If not indicated otherwise we will always assume in this section that the spaces considered fulfil these assumptions.

(5.3.1) Definition

Let $f : X \to Y$ be a G-map between G-spaces that fulfil the above assumptions, $\mathrm{fd}(X) = m$, $\mathrm{fd}(Y) = n$ (actually for this definition the hypothesis (LT) is not necessary). We fix graded orientations $\mathcal{O}_{H^*(X)}$: $H^*(X) \to k$ and $\mathcal{O}_{H^*(Y)} : H^*(Y) \to k$ and define the Gysin homomorphism

$$f_! : H_G^{*+m}(X) \to H_G^{*+n}(Y)$$

induced by $f : X \to Y$ by the following procedure. In case (2) and (odd p) we consider the sequence

$$S^*(X) \leftarrow \overline{S}^*(X) \to \mathrm{Hom}(\overline{S}^*(X), S^*(X)) \leftarrow \mathrm{Hom}(\overline{S}^*(X), \overline{S}^*(X))$$
$$\to \mathrm{Hom}(\overline{S}^*(X), k) \leftarrow \mathrm{Hom}(S^*(X), k)$$

5.3. Equivariant Gysin homomorphism

and the corresponding sequence for Y, which both induce isomorphisms if the functor

$$H_G^*(-) = H(\beta_G^*(-))$$

is applied (see (1.2.5)). Of course, $f : X \to Y$ induces a morphism $f^* : S^*(Y) \to S^*(X)$ of $k[G]$-complexes and hence a map

$$f^* : H_G^*(Y) \to H_G^*(X).$$

(We use the same symbol f^* for different maps induced by f if it is clear from the context, which map we actually mean.) On the other hand $f^* : S^*(Y) \to S^*(X)$ also induces a morphism of $k[G]$-complexes $f_* : \mathrm{Hom}(S^*(X), k) \to \mathrm{Hom}(S^*(Y), k)$. Applying $H(\beta_G^*(-))$ and using the above isomorphisms, which come from the Poincaré duality of $H^*(X)$ and $H^*(Y)$, respectively, (see Section 5.2), one obtains the Gysin homomorphism

$$f_! : H_G^{*+m}(X) \to H_G^{*+n}(Y).$$

Note that $f_!$ increases degree by $n - m = \mathrm{fd}(Y) - \mathrm{fd}(X)$. This comes from the fact that the orientations $\mathcal{O}_{\overline{S}^*(X)} : \overline{S}^*(X) \to k$ and

$$\mathcal{O}_{\overline{S}^*(Y)} : \overline{S}^*(Y) \to k$$

decrease degree by m and n, respectively.

In case (0) we use the minimal Hirsch-Brown models and the sequence

$$H^*(X) \tilde{\otimes} R \to \mathrm{Hom}_R(H^*(X) \tilde{\otimes} R, H^*(X) \tilde{\otimes} R) \to \mathrm{Hom}_R(H^*(X) \tilde{\otimes} R, R),$$

where $R = H^*(BG) = k[t_1, \ldots, t_r]$ for $G = T^r = S^1 \times \ldots \times S^1$ a torus of rank r, and proceed in an analogous fashion.

It should be noted that the Gysin homomorphism depends on the chosen orientations.

(5.3.2) Remark
An alternative definition of the Gysin homomorphism under the above assumptions can be given using the Leray-Serre spectral sequence of the Borel fibration $X \to X_G \to BG$ (cf. also (5.3.19)).

The following properties of the Gysin homomorphism are analogous to those in the differentiable case.

(5.3.3) Proposition
Let $(X, \mathcal{O}_{H^*(X)})$, $(Y, \mathcal{O}_{H^*(Y)})$ and $(Z, \mathcal{O}_{H^*(Z)})$ be oriented k-PD-spaces of formal dimension l, m and n respectively, on which the group G acts. Let $f : X \to Y$ and $g : Y \to Z$ be G-maps. Then

(1) $f_! : H_G^*(X) \to H_G^{*+(m-l)}(Y)$ is a morphism of H_G^*-modules, which depends only on the G-homotopy class of f.

(2) (a) $(\mathrm{id}_X)_! = \mathrm{id}_{H_G^*(X)}$
 (b) $(g \circ f)_! = g_! \circ f_!$, i.e. the Gysin homomorphism is natural in X.
 (c) For $pr: X \to pt$ one has: $pr_! = H_G^*(\mathcal{O}_{S^*(X)}) : H_G^*(X) \to H_G^{*-l}$.
(3) If $\gamma : K \to G$ is a group morphism, then Diagram 5.8 commutes.

$$\begin{array}{ccc} H_G^*(X) & \xrightarrow{f_!^G} & H_G^*(Y) \\ \downarrow \gamma^* & & \downarrow \gamma^* \\ H_K^*(X) & \xrightarrow{f_!^K} & H_K^*(Y), \end{array}$$

Diagram 5.8

In Diagram 5.8 $f_!^G$ and $f_!^K$ denote the equivariant Gysin homomorphisms with respect to the groups G and K (K acts on X and Y via γ and the action of G), and γ^* is the natural transformation between the equivariant cohomology theories $H_G^*(-)$ and $H_K^*(-)$ induced by $\gamma : K \to G$. (This may be expressed by saying that the Gysin homomorphism is natural in G.)

(4) $f_!(x \cup f^*(y)) = f_!(x) \cup y$ for $x \in H_G^*(X)$ and $y \in H_G^*(Y)$, i.e. $f_! : H_G^*(X) \to H_G^*(Y)$ is a morphism of $H_G^*(Y)$-modules if one defines the $H_G^*(Y)$-module structure of $H_G^*(Y)$ via the cup-product, and of $H_G^*(X)$ by first applying $f^* : H_G^*(Y) \to H_G^*(X)$ and then taking the cup-product.

Proof.
Using the fact that $H(\beta_G^*(-)) : \delta gk[G]\text{-Mod} \to gH_G^*\text{-Mod}$ is a G-homotopy invariant functor and $\gamma^* : H(\beta_G^*(-)) \to H(\beta_K^*(-))$ a natural transformation parts (1)-(3) are immediate consequences of the definition of the Gysin homomorphism in case G is a finite group.

In case (0) one uses the fact that the minimal Hirsch-Brown model $H^*(X) \tilde{\otimes} H^*(BG)$ is functorial up to homotopy in X and in G. Part (4) follows by a calculation on the chain level. We handle the case of a finite group G and leave the proof for case (0), which - using the minimal Hirsch-Brown models - is similar, but simpler, to the reader.

Under duality the cup-product $S^*(X) \otimes S^*(X) \xrightarrow{\cup} S^*(X)$ corresponds to the map $\mathrm{Hom}(S^*(X), k) \otimes S^*(X) \xrightarrow{\cap} \mathrm{Hom}(S^*(X), k)$ given by

$$\phi \otimes x_1 \mapsto \phi \cap x_1,$$

with $(\phi \cap x_1)(x_2) = \phi(x_1 \cup x_2)$ for $\phi \in \mathrm{Hom}(S^*(X), k); x_1, x_2 \in S^*(X)$. More precisely, one has a commutative diagram, Diagram 5.9, in $\delta gk[G]$-Mod

5.3. Equivariant Gysin homomorphism

$$
\begin{array}{ccc}
S^*(X) \otimes S^*(X) & \xrightarrow{\cup} & S^*(X) \\
\downarrow & & \downarrow \\
\mathrm{Hom}(S^*(X), S^*(X)) \otimes S^*(X) & \longrightarrow & \mathrm{Hom}(S^*(X), S^*(X)) \\
\uparrow & & \uparrow \\
\mathrm{Hom}(S^*(X), \overline{S}^*(X)) \otimes S^*(X) & \longrightarrow & \mathrm{Hom}(S^*(X), \overline{S}^*(X)) \\
\downarrow & & \downarrow \\
\mathrm{Hom}(S^*(X), k) \otimes S^*(X) & \xrightarrow{\cap} & \mathrm{Hom}(S^*(X), k),
\end{array}
$$

Diagram 5.9

where the maps are the 'canonical' ones, i.e. the vertical maps to the right give the duality isomorphism in cohomology, the vertical maps to the left are obtained by taking the identity $\mathrm{id}_{S^*(X)}$ on the second factor in the tensor product, and the horizontal maps are given by the cup-product (see above for the map $\cap$). A similar diagram holds of course for Y instead of X, and a straightforward calculation shows that Diagram 5.10, where S_X, resp. S_Y denotes $S^*(X)$ resp. $S^*(Y)$,

$$
\begin{array}{ccccc}
\mathrm{Hom}(S_X, k) \otimes S_Y & \xrightarrow{\mathrm{id} \otimes f^*} & \mathrm{Hom}(S_X, k) \otimes S_X & \xrightarrow{\cap} & \mathrm{Hom}(S_X, k) \\
\downarrow{\mathrm{id}} & & & & \downarrow{f_*} \\
\mathrm{Hom}(S_X, k) \otimes S_Y & \xrightarrow{f_* \otimes \mathrm{id}} & \mathrm{Hom}(S_Y, k) \otimes S_Y & \xrightarrow{\cap} & \mathrm{Hom}(S_Y, k)
\end{array}
$$

Diagram 5.10

is commutative for $\phi \in \mathrm{Hom}(S^*(X), k), y_1, y_2 \in S^*(Y)$ one has:

$$
\begin{aligned}
(f_*(\phi \cap f^*(y_1)))(y_2) &= (\phi \cap f^*(y_1)) f^*(y_2) \\
&= \phi(f^*(y_1) \cup f^*(y_2)) \\
&= \phi(f^*(y_1 \cup y_2)),
\end{aligned}
$$

and

$$
((f_*\phi) \cap y_1)(y_2) = (f_*\phi)(y_1 \cup y_2) = \phi(f^*(y_1 \cup y_2))).
$$

5. Actions on Poincaré duality spaces

Applying the functor $H_G^*(-) = H(\beta_G^*(-))$ to the Diagrams 5.9 and 5.10 gives, of course, again commutative diagrams. In particular one gets that Diagram 5.11, abbreviating $S^*(X)$, $S^*(Y)$ by S_X, S_Y respectively,

$$\begin{array}{ccccc} H_G^*(S_X \otimes S_Y) & \xrightarrow{H_G(\mathrm{id} \otimes F^*)} & H_G^*(S_X \otimes S_X) & \xrightarrow{H_G^*(\cup)} & H_G^*(S_X) \\ \downarrow {\scriptstyle \mathrm{id}} & & & & \downarrow {\scriptstyle f_!} \\ H_G^*(S_X \otimes S_Y) & \xrightarrow{H_G^*(f_! \otimes \mathrm{Id})} & H_G^*(S_Y \otimes S_Y) & \xrightarrow{H_G^*(\cup)} & H_G^*(S_Y) \end{array}$$

Diagram 5.11

is commutative, where - by a slight abuse of notation - $H_G^*(f_! \otimes \mathrm{id})$ denotes the map obtained by tensoring with $\mathrm{id}_{S^*(Y)}$ the sequence of chain maps that give $f_!$ and then applying $H_G^*(-)$.

The cup-product on $H_G^*(X)$ is given by the composition

$$H_G^*(X) \otimes H_G^*(X) \cong H(\beta_G^*(S^*(X)) \otimes \beta_G^*(S^*(X)))$$
$$\longrightarrow H(\beta_G^*(S^*(X) \otimes S^*(X)))$$
$$\xrightarrow{H_G^*(\cup)} H(\beta_G^*(S^*(X))) = H_G^*(X)$$

where the first arrow is induced by the diagonal

$$\triangle : \mathcal{E}_*(G) \to \mathcal{E}_*(G) \otimes \mathcal{E}_*(G)$$

of $\mathcal{E}_*(G)$, i.e. for any two $k[G]$-complexes C_1^*, C_2^* one gets a morphism

$$\beta_G^*(C_1^*) \otimes \beta_G^*(C_2^*) = \mathrm{Hom}_{k[G]}(\mathcal{E}_*(G), C_1^*) \otimes \mathrm{Hom}_{k[G]}(\mathcal{E}_*(G), C_2^*)$$
$$\to \mathrm{Hom}_{k[G]}(\mathcal{E}_*(G), C_1^* \otimes C_2^*)$$
$$= \beta_G^*(C_1^* \otimes C_2^*)$$

by defining $\phi_1 \otimes \phi_2 \mapsto \phi$, where $\phi(a) := (\phi_1 \otimes \phi_2)(\triangle a)$, $a \in \mathcal{E}_*(G)$ ($C_1^* \otimes C_2^*$ is equipped with the diagonal action); and for $C_1^* = C_2^* = S^*(X)$ one obtains the desired map (see Section 1.2). Since this construction is natural in C_1^* and C_2^* one gets the commutative Diagrams 5.12 and 5.13

$$\begin{array}{ccc} H_G^*(S^*(X) \otimes S^*(Y)) & \xrightarrow{H_G^*(\mathrm{id} \otimes f^*)} & H_G^*(S^*(X) \otimes S^*(X)) \\ \uparrow & & \uparrow \\ H_G^*(X) \otimes H_G^*(Y) & \xrightarrow{\mathrm{id} \otimes f^*} & H_G^*(X) \otimes H_G^*(X) \end{array}$$

Diagram 5.12

5.3. Equivariant Gysin homomorphism

$$\begin{array}{ccc}
H_G^*(X) \otimes H_G^*(Y) & \xrightarrow{f_! \otimes \mathrm{id}} & H_G^*(Y) \otimes H_G^*(Y) \\
\downarrow & & \downarrow \\
H_G(S^*(X) \otimes S^*(Y)) & \xrightarrow{H_G^*(f_! \otimes \mathrm{id})} & H_G^*(S^*(Y) \otimes S^*(Y)),
\end{array}$$

Diagram 5.13

where the vertical maps are induced by the diagonal of $\mathcal{E}_*(G)$.

Combining these two diagrams with Diagram 5.11 one gets (5.3.3)(4). □

(5.3.4) Corollary
Let X be a k-PD-space of formal dimension $\mathrm{fd}_k(X) = n$ on which a group G acts. If $X^G \neq \emptyset$, then for any generator $\langle X \rangle \in H^n(X)$ there is a class $u \in H_G^n(X)$ which maps to $\langle X \rangle$ under the map induced by the inclusion $X \to X_G$ of the fibre into the total space of the Borel fibration $X \to X_G \to BG$.

Proof.
We choose the orientation $\mathcal{O}_{H^*(X)} : H^*(X) \to k$ in such a way that $\mathcal{O}_{H^*(X)}\langle X \rangle = 1$, where $\langle X \rangle \in H^*(X)$ is a generator.

Define $u := f_!^G(1)$, where $f : pt \to X$ is the inclusion of a fixed point, $f_!^G : H_G^*(pt) = H^*(BG) \to H_G^{*+n}(X)$ is the corresponding Gysin homomorhism, and $1 \in H^*(BG)$ is the unit.

The restriction to the fibre, i.e. the map $H_G^*(X) = H^*(X_G) \to H^*(X)$ induced by the inclusion $X \to X_G$, is nothing but the map obtained by restricting the group action to the unit element e in G, i.e. the map

$$\gamma^* : H_G^*(X) \to H_{\{e\}}^*(X) = H^*(X)$$

for $\gamma : \{e\} \to G$. By (5.3.3)(3) one has: $\gamma^*(f_!^G(1)) = f_!^{\{e\}}(\gamma^*(1))$, which is the element in $H^*(X)$ that is dual to $1 \in H^*(X)$ under Poincaré duality in $H^*(X)$. Hence $\gamma^*(f_!^G(1)) = \langle X \rangle$. □

(5.3.5) Remark
For the above corollary one does not need the assumption (LT) for X nor that G is a p-torus or torus and $k = \mathbb{F}_p$ or $\mathbb{Q}$ respectively. One just needs the existence of an equivariant Gysin homomorphism which fulfils (5.3.3)(3). Our approach gives this for G any finite group (and G a torus, $k = \mathbb{Q}$). The construction using the Leray-Serre spectral sequence (see Remark (5.3.2)) would even work for any compact Lie group. In the next section we give a converse of (5.3.4) (see (5.4.1), (5.4.2)), i.e.

we deduce the existence of fixed points from the existence of a class $u \in H_G^n(X)$ which restricted to the fibre in the Borel fibration gives a generator in the top degree of $H^*(X)$. For that we have to assume that X fulfils (LT), G a p-torus or torus, and $k = \mathbb{F}_p$ or $\mathbb{Q}$ respectively; but we do not need the assumption that X is a k-PD-space.

(5.3.6) Definition
(1) Let $f : X \to Y$ be as above, then
$$e_G(f) := f^* f_!(1) \in H_G^{n-m}(X),$$
$m := \text{fd}(X)$, $n := \text{fd}(Y)$, is called the equivariant Euler class of $f : X \to Y$. (In case of an inclusion $X \hookrightarrow Y$ we use the notation $e_G(X, Y)$ for the equivariant Euler class.)
(2) If X is as above then the Gysin homomorphism
$$pr_! : H_G^*(X) \to H_G^*$$
induced by the projection $pr : X \to pt$ is also called the index of X and denoted by $\text{ind}_G(X)$.

(Note that $\text{ind}_G(X) = H_G^*(\mathcal{O}_{\overline{S}^*(X)})$ (see (5.3.3)(2)(c)).

Clearly the above definitions and general properties carry over in a natural way to the localized equivariant cohomology. We will discuss the behaviour of the Gysin homomorphism and the Euler class under localization and apply the results to relate global information (on X) to local information (on X^G).

(5.3.7) Proposition
Let X be a G-space as above, G a torus or p-torus and $k = \mathbb{Q}$ or $\mathbb{F}_p$ respectively. Let $\phi_i : F_i \to X$ denote the inclusion of the component F_i of X^G into X.

(1) The Gysin homomorphism $(\phi_i)_! : H_G^{*+m_i}(F_i) \to H_G^{*+m}(X)$, $m_i = \text{fd}(F_i)$, $m = \text{fd}(X)$, becomes an inclusion of a direct summand (over $S^{-1} H_G^*$) after localization with respect to the multiplicative set S of non-zero, homogeneous polynomials in H_G^*. In fact, for the localized equivariant cohomology one has an isomorphism
$$\phi_i^*(\phi_i)_! : S^{-1} H_G^{*+m_i}(F_i) \to S^{-1} H_G^{*+m}(F_i)$$
of $S^{-1} H_G^*$-modules.
(2) The composition $\phi_i^*(\phi_i)_! : H_G^{*+m_i}(F_i) \to H_G^{*+m}(F_i)$ is given by the multiplication with the equivariant Euler class
$$e_G(F_i, X) \in H_G^{m-m_i}(F_i),$$
and $e_G(F_i, X)$ is a unit in $S^{-1} H_G^*(F_i)$ (i.e. invertible with respect to the cup-product).

Proof.
(1) The localized Gysin homomorphism
$$(\phi_i)_! : S^{-1}H_G^{*+m_i}(F_i) \to S^{-1}H_G^{*+m}(X)$$
is the dual (adjoint) of
$$\phi_i^* : S^{-1}H_G^*(X) \to S^{-1}H_G^*(F_i),$$
using Poincaré duality of $S^{-1}H_G^*(X)$ and $S^{-1}H_G^*(F_i)$ over $S^{-1}H_G^*$ (see (5.2.5)). Since - by the Localization Theorem - the inclusions $\phi_i : F_i \to X$ of the components $F_1, \ldots, F_\nu$ of X^G induce an isomorphism
$$S^{-1}H_G^*(X) \cong \prod_{i=1}^{\nu} S^{-1}H_G^*(F_i),$$
and Poincaré duality of $S^{-1}H_G^*(X)$ respects this decomposition (being induced by the cup-product), the assertion follows.

(2) Since $H_G^*(F_i)$ injects into $S^{-1}H_G^*(F_i)$ (and $e_G(F_i, X)$ is already defined in $H_G^*(F_i)$) it suffices to prove part (2) after localization. By (5.3.3)(4) one has
$$\phi_i^*(\phi_i)_!(1 \cup \phi_i^*(x)) = \phi_i^*(\phi_i)_!(1) \cup \phi_i^*(x)$$
for $x \in S^{-1}H_G^*(X)$. Since $\phi_i^* : S^{-1}H_G^*(X) \to S^{-1}H_G^*(F_i)$ is surjective one obtains $\phi_i^*(\phi_i)_!(y_i) = e_G(F_i, X) \cup y_i$ for $y_i \in S^{-1}H_G^*(F_i)$. By part (1)
$$\phi_i^*(\phi_i)_! : S^{-1}H_G^{*+m_i}(F_i) \to S^{-1}H_G^{*+m}(F_i)$$
is an isomorphism. Hence $e_G(F_i, X)$ must be invertible in $S^{-1}H_G^*(F_i)$. □

We define $e_G(X^G, X) := \sum_{i=1}^{\nu} e_G(F_i, X)$ to be the equivariant Euler class of the inclusion $\phi : X^G \to X$, and the corresponding Gysin homomorphism $\phi_! : H_G^*(X^G) \to H_G^*(X)$ by $\phi_!|H_G^*(F_i) = (\phi_i)_!$ using the decomposition $H_G^*(X^G) = \prod_{i=1}^{\nu} H_G^*(F_i)$. (Note that $e_G(X^G, X)$ need not be homogeneous and $\phi_!$ may not have a fixed degree, but - in the cases (odd p) and (0) the Euler class $e_G(X^G, X)$ consists of terms of even degree, see (5.2.1).) The Euler class $e_G(X^G, X^K)$ for $K < G$ is defined analogously.

(5.3.8) Corollary
Under the hypothesis of (5.3.7) one has:

(1) *The fixed point set X^G of X is non-empty if and only if $\text{ind}_G(X)$ is non-zero.*

(2) *If $X^G \neq \emptyset$ and F_j is a component of X^G, then $\text{fd}(F_j) = \text{fd}(X)$ if and only if the restriction homomorphism $H^*(X) \to H^*(F_j)$ is an isomorphism. (Note that $\text{fd}(F_j) \leq \text{fd}(X)$ by Theorem (5.2.1).)*

Proof.

(1) Suppose $X^G = \emptyset$. Then, by the Localization Theorem,
$$p^* : R \to H_G^*(X)$$
is not injective, where $p : X_G \to BG$ is the bundle projection. Choose a non-zero $a \in R$ such that $p^*(a) = 0$. Now by Proposition (5.3.3)(4), for any $v \in H_G^*(X)$, $0 = \text{ind}_G(X)(p^*(a)v) = a\,\text{ind}_G(X)(v)$. Hence $\text{ind}_G(X)(v) = 0$.

Now suppose that $X^G \neq \emptyset$. Let $x \in X^G$, and let $j : \{x\} \to X$ be the inclusion. By Proposition (5.3.3)(2), $\text{ind}_G(X)j_!$ is the identity on H_G^*. Hence $\text{ind}_G(X)$ is surjective.

(2) Let $\overline{\phi}_j^* : H^*(X) \to H^*(F_j)$ be the restriction homomorphism in non-equivariant cohomology. Clearly $\text{fd}(F_j) = \text{fd}(X)$ if $\overline{\phi}_j^*$ is an isomorphism. So suppose that $\text{fd}(F_j) = \text{fd}(X)$. By Proposition (5.3.7)(2), $e_G(F_j, X) \in H_G^0(F_j)$ is non-zero. Let $e_G(F_j, X) = \alpha \in k$. Let $f_j \in H^n(F_j)$, $n = \text{fd}(X) = \text{fd}(F_j)$, be a generator.

Let $u \in H^*(F_j)$ and let $w = \alpha^{-1}(\phi_j)_!(u)$. By Proposition (5.3.7)(2) $\phi_j^*(w) = u$, where ϕ_j^* is the restriction homomorphism in equivariant cohomology (as usual). Thus $\overline{\phi}_j^*(i^*(w)) = u$, where $i : X \to X_G$ is the inclusion. Hence $\overline{\phi}_j^*$ is surjective. And, in particular, there exists $x \in H^n(X)$ such that $\overline{\phi}_j^*(x) = f_j$. So $\overline{\phi}_j^*$ is injective by Poincaré duality. $\square$

Corollary (5.3.8) can be used to detect fixed points from global information on X. We will discuss results in this direction in the following sections.

The equivariant Euler class has the following multiplicative property.

(5.3.9) Proposition
Let X be as above and let K be a subtorus of G. If $\psi : X^G \to X^K$ denotes the inclusion of the fixed point set X^G of the G-action on X into the fixed point set X^K of the action restricted to $K \subset G$, then
$$e_G(X^G, X) = e_G(X^G, X^K) \cup \psi^*(e_G(X^K, X)).$$

Proof.
By definition (and (5.3.3)(2))
$$e_G(X^G, X) = \psi^* \phi^* \phi_! \psi_!(1),$$
where $\psi : X^G \to X^K$, $\phi : X^K \to X$ are the inclusions.

It suffices to prove the above equation between the Euler classes in the localized theory since $H_G^*(X^G) \subset S^{-1}H_G^*(X^G)$. After localization the maps ψ^*, ϕ^*, $\phi_!$, $\psi_!$ become isomorphisms. Therefore there exists a $x \in S^{-1}H_G^*(X)$ such that $\phi^*(x) = \psi_!(1)$. Hence by (5.3.3)(4) one has:

$$\begin{aligned}
e_G(X^G, X) &= \psi^*\phi^*\phi_!\psi_!(1) = \psi^*\phi^*\phi_!\phi^*(x) = \psi^*\phi^*\phi_!(1 \cup \phi^*(x)) \\
&= \psi^*(\phi^*(\phi_!(1) \cup x)) = \psi^*(\phi^*\phi_!(1) \cup \phi^*(x)) \\
&= \psi^*\phi^*\phi_!(1) \cup \psi^*\psi_!(1) \\
&= \psi^*(e_G(X^K, X)) \cup e_G(X^G, X^K) \\
&= e_G(X^G, X^K) \cup \psi^*(e_G(X^K, X))
\end{aligned}$$

(since - in the cases (odd p) and (0) - $e_G(X^G, X^K)$ and $e_G(X^K, X)$ consist of terms of even degree (cf. (5.2.1)). $\square$

The preceding proposition and the following remark will be used to prove the important Borel formula which gives a relation between the dimensions of a G-space X, its fixed point set X^G and the fixed point sets X^K of subtori $K \subset G$ of corank one (see [Borel et al, 1960]).

(5.3.10) Remark
Under the hypothesis of (5.3.9) let

$$\rho_K : G \to G/K =: L$$

denote the canonical projection and $\rho_K^* : H_L^*(Y) \to H_G^*(Y)$ the map induced in equivariant cohomology for any L-space Y. As in Chapter 3 we denote by $S(K) \subset H_G^*$ the multiplicative subset generated by the linear polynomials in H_G^*, which are not in the kernel of $H_G^* = H^*(BG) \to H^*(BK) = H_K^*$. Then the Euler class $e_G(X^G, X^K)$ is contained in

$$\rho_G^*(H_L^*(X^G)) \cong H^*(X^G) \otimes \rho_G^* H_L^* \subset H^*(X^G) \otimes H_G^* = H_G^*(X^G),$$

and $\psi^* e_G(X^K, X)$ is invertible in

$$S(K)^{-1}H_G^*(X^G) \cong H^*(X^G) \otimes S(K)^{-1}H_G^*.$$

Proof.
Since K acts trivially on X^G and X^K these spaces can be viewed as L-spaces (the original G-action is recovered via ρ). The naturality (with respect to G) of the Gysin homomorphism and the Euler class (see (5.3.3)) gives:

$$\rho_K^* e_L(X^G, X^K) = e_G(X^G, X^K).$$

By the Localization Theorem (see (3.2.6)) one has an isomorphism $S(K)^{-1}H_G^*(X) \cong S(K)^{-1}H_G^*(X^K)$. Therefore - by the same argument as in the proof of (5.3.7) - the Euler class $e_G(X^K, X)$ is invertible in $S(K)^{-1}H_G^*(X^K)$. Applying $\psi^* : S(K)^{-1}H_G^*(X^K) \to S(K)^{-1}H_G^*(X^G)$ gives that $\psi^* e_G(X^K, X)$ is invertible in

$$S(K)^{-1}H_G^*(X^G) \cong H^*(X^G) \otimes S(K)^{-1}H_G^*.$$

□

Let X be a G-space fulfilling the usual assumptions for this section and let θ_1 be the set of subtori of G of corank one (i.e. $\operatorname{rank}(G/K) = 1$ for $K \in \theta_1$). Furthermore F denotes a component of X^G and $F(K)$ the component of X^K which contains F ($K \in \theta_1$).

(5.3.11) Theorem (The Borel formula)
$$\operatorname{fd}(X) - \operatorname{fd}(F) = \sum_{K \in \theta_1} (\operatorname{fd}(F(K)) - \operatorname{fd}(F)),$$
where $\operatorname{fd}(Y)$ denotes the formal dimension of the k-PD-space Y.

Proof.
For a given G-space X as above we may replace the set θ_1 by the subset $LW(F) := \{K \in \theta_1; \operatorname{fd}(F(K)) > \operatorname{fd}(F)\}$ without changing the right side of the above formula. As the arguments below will show, the set $LW(F)$ is finite even in case (0). (This is completely obvious in the other cases, since there θ_1 is already finite). Note that

$$\operatorname{fd}(X) - \operatorname{fd}(F) \; (\text{resp. } \operatorname{fd}(F(K)) - \operatorname{fd}(F))$$

is just the degree of the Euler class $e_G(F, X)$ (resp. $e_G(F, F(K))$). We are going to use the product formula for the Euler classes (see (5.3.9)) to relate these degrees. Rather than performing the necessary calculations in $H_G^*(X^G)$ (resp. $S^{-1}H_G^*(X^G)$) we restrict to a point in X^G, i.e. map to H_G^* (resp. $S^{-1}H_G^*$), and - in case (odd p) - we project further to $k[t_1, \ldots, t_r]$ (resp. $S^{-1}k[t_1, \ldots, t_r]$), $r := \operatorname{rank}(G)$, by dividing by the radical $\sqrt{(0)} \subset H_G^*$ (cf. Section 3.6).

We denote by $\bar{e}_G(F, X)$ ($\bar{e}_G(F, F(K)), \bar{e}(F(K), X), \ldots$) the image of $e_G(F, X)$ ($e_G(F, F(K)), \psi^* e_G(F(K), X), \ldots$) under the above restriction.

The product formula (5.3.9) then implies:

$$\bar{e}_G(F, X) = \bar{e}_G(F, F(K))\bar{e}_G(F(K), X)$$

in $R := k[t_1, \ldots, t_r]$. We claim that

$$\bar{e}_G(F, X) = c \prod_{K \in LW(F)} \bar{e}_G(F, F(K))$$

5.3. Equivariant Gysin homomorphism

with $c \neq 0$, $c \in k$. We prove this claim in three steps:

(i) $\bar{e}_G(F, F(K))$ divides $\bar{e}_G(F, X)$ in R for all $K \in \theta_1$.

This is clear from the above product formula.

(ii) $\prod_{K \in LW(F)} \bar{e}_G(F, F(K))$ divides $\bar{e}_G(F, X)$ in R.

For this it suffices to show that $\bar{e}_G(F, F(K))$ and $\bar{e}_G(F, F(H))$ are relatively prime in R if $K, H \in LW(F)$ and $K \neq H$. It follows from (5.3.10) that $\bar{e}_G(F, F(K)) \in \rho_K^* R_{G/K}$ and $\bar{e}_G(F, F(H)) \in \rho_H^* R_{G/H}$ (where $R_L := H^*(BL)/\sqrt{(0)}$). Since $\operatorname{rank}(G/K) = 1$ the ring $R_{G/K}$ is a polynomial ring in one generator and its image $\rho_K^* R_{G/K}$ in R is generated by a linear polynomial ℓ_K. In particular, $\bar{e}_G(F, F(K)) = c_K \ell_K^{m_K}$, where $c_K \in k$, $c_K \neq 0$ and $m_K = \operatorname{fd}(F(K)) - \operatorname{fd}(F)$ in case (2), $m_K = \frac{1}{2}(\operatorname{fd}(F(K)) - \operatorname{fd}(F))$ in the cases (odd p) and (0). The same holds for H instead of K, but since $K \neq H$ the linear polynomials ℓ_K and ℓ_H are linearly independent over k. Hence ℓ_K and ℓ_H, and therefore also $\bar{e}_G(F, F(K))$ and $\bar{e}_G(F, F(H))$, are relatively prime. Since $m_K > 0$ for $K \in LW(F)$, the set $LW(F)$ must be finite. Finally:

(iii) $\bar{e}_G(F, X) = c \prod_{K \in LW(F)} \bar{e}_G(F, F(K))$, where $c \in k$, $c \neq 0$.

It only remains to show that the factor c is indeed a non-zero element in k (i.e. of degree zero). Clearly c is non-zero since $\bar{e}_G(F, X)$ is invertible in $S^{-1}R$. But $\bar{e}_G(F, X)$ is invertible already in $S(G)^{-1}R$, where $S(G)$ is the multiplicative set generated by $\{\ell_K; K \in \theta_1\}$, which we may consider as a subset of H_G^*. This follows as in (5.3.7) using the Localization Theorem in the form $S(G)^{-1} H_G^*(X) \cong S(G)^{-1} H_G^*(X^G)$ (see (3.2.6) and (3.1.5)). (In the notation of Chapter 3 one has to show that $X^{S(G)} := \{x \in X; j_x^* p^*(f) \neq 0$ for all $f \in S(G)\}$ is equal to X^G, where $j_x^* p^* : H_G^* \to H_{G_x}^*$ is induced by the inclusion $G_x \hookrightarrow G$. Clearly $X^G \subset X^{S(G)}$. If $x \in X$ is not a fixed point then there exists a $K \in \theta_1$ such that $G_x < K$ in case G is a p-torus, and $G_x^0 < K$ in the case (0). Since, by definition, ℓ_K is an element of positive degree in the image of the map $H_{G/K}^* \to H_G^*$ it is in the kernel of $H_G^* \to H_K^*$ and hence of $H_G^* \to H_K^* \to H_{G_x}^*$. (Note that in case (0) one has: $H_{G_x}^* \cong H_{G_x^0}^*$ since k has characteristic 0.) Therefore $x \notin X^{S(G)}$, and we get $X^G = X^{S(G)}$.)

Since the units (i.e. invertible elements) in $S(G)^{-1}R$ are the elements in $S(G)$ up to a non-zero factor in k one has $\bar{e}_G(F, X) \in k S(G)$. So c must be an element in $k S(G)$ also.

We now fix a $K \in LW(F)$ and localize with respect to the subset S_K of $S(G)$ which is generated by $\{\ell_H; H \in \theta_1, H \neq K\}$. The same argument as above shows that $\bar{e}_G(F(K), X)$ becomes a unit already in

$S_K^{-1}R$ and hence is contained in kS_K. The above product formula
$$\bar{e}_G(F, F(K))\bar{e}_G(F(K), X) = \bar{e}_G(F, X)$$
$$= \bar{e}_G(F, F(K))c \prod_{H \in LW(F)-\{K\}} \bar{e}_G(F, F(H))$$
then implies that c is not divisible by ℓ_K, and since this holds for any $K \in LW(F)$, on the one hand, and c is an element in $kS(G)$, on the other hand, c must have degree zero. Since $c \neq 0$ it is a unit in k. □

(5.3.12) Definition
In the notation of the proof of Theorem (5.3.11),
$$LW(F) := \{K \in \theta_1 | \operatorname{fd}(F(K)) > \operatorname{fd}(F)\}.$$
The elements of $LW(F)$ are called the local weights (of the given G-action) at F. Note that with the notation of the proof of Theorem (5.3.11), $PK = (\ell_K)$ for each $K \in LW(F)$ (see Definition (3.1.4) for PK); ℓ_K is a non-zero homogeneous linear polynomial in R, uniquely determined by K up to non-zero scalar multiples in k. In some works (e.g., [Hsiang, 1975]) the set of local weights is defined to be $\{\ell_K | K \in LW(F)\}$.

We shall give some applications of local weights and the Borel formula (Theorem (5.3.11)) below. First, however, we shall give a few more properties of the Gysin homomorphism and the index.

(5.3.13) Definition
Let X, G, $F_i \subseteq X^G$, $\phi : X^G \to X$ and $\phi_i : F_i \to X$ be as above. Let $\operatorname{fd}(X) = n$, $\operatorname{fd}(F_i) = n_i$, and let $f_i \in H^{n_i}(F_i)$ be a generator. Let $\theta_i := (\phi_i)_!(1)$, where $1 \in H_G^0(F_i)$ is the identity, and let $U_i := (\phi_i)_!(f_i)$. So $\theta_i \in H_G^{n-n_i}(X)$ and $U_i \in H_G^n(X)$.

Also abbreviate $e_G(F_i, X) = e_i$, and let $\Omega_i = \psi^*(e_i)$, where
$$\psi^* : H_G^*(F_i) \to H_G^*$$
is induced by the inclusion of a point into F_i. Note that, by definition, $e_i = \phi_i^*(\theta_i)$.

(5.3.14) Lemma
With the above notation
(1) $(\phi_i)_! \phi_i^*(v) = \theta_i v$ *for any* $v \in H_G^*(X)$; *and*
(2) *for any* $u \in H_G^*(F_i)$, $\phi^*(\phi_i)_!(u) = \phi_i^*(\phi_i)_!(u) = e_i u$ *in* $H_G^*(X^G)$. *In particular,* $\phi_j^*(\phi_i)_!(u) = 0$, *for any component* F_j *of* X^G *distinct from* F_i.

5.3. Equivariant Gysin homomorphism

Proof.
(1) This follows immediately from Proposition (5.3.3)(4).
(2) Given $u \in H_G^*(F_i) \subseteq H_G^*(X)$, by the Localization Theorem, there exists $v \in H_G^*(X)$ and non-zero $a \in R$ such that $\phi^*(v) = au$. So
$$a\phi^*(\phi_i)_!(u) = \phi^*(\phi_i)_!\phi_i^*(v) = \phi^*(\theta_i v),$$
by (1), $= a\phi^*(\theta_i)u$. Since $H_G^*(X^G)$ is a free R-module,
$$\phi^*(\phi_i)_!(u) = \phi^*(\theta_i)u \in H_G^*(F_i),$$
cf. Proposition (5.3.7). □

(5.3.15) Lemma
With the notation above (see Definitions (5.3.12) and (5.3.13)), $\Omega_i = c_i \prod_{K \in LW(F_i)} \ell_K^{m_K}$, where $c_i \in k$, $c_i \neq 0$, and (see the proof of Theorem (5.3.11)) $m_K = \mathrm{fd}(F_i(K)) - \mathrm{fd}(F_i)$ in case (2), and $m_K = \frac{1}{2}[\mathrm{fd}(F_i(K)) - \mathrm{fd}(F_i)]$ in cases (0) and (odd p). ($F_i(K)$ is the component of X^K which contains F_i, and F_i is any component of X^G.)
In particular $\Omega_i \in R$ even in case (odd p).

Proof.
Looking at steps (i) and (ii) of the proof of Theorem (5.3.11), but projecting only onto H_G^*, and not onto $R = H_G^*/\sqrt{(0)}$ in case (odd p), it follows as before that $\prod_{K \in LW(F_i)} \ell_K^{m_K}$ divides Ω_i. (Note that ℓ_K comes from the generator of $H_{G/K}^2$ in case (odd p); and the latter is purely polynomial since G/K has rank one.) So the result follows since the degrees agree by Theorem (5.3.11). □

Recall from Example (3.7.16)(1) that for any $u \in H_G^*(X^G)$,
$$I_u := \{a \in R \mid au \in \mathrm{im}(\phi^*)\}.$$
The following, combined with Lemma (5.3.15), is referred to as the topological splitting principle in [Allday, Skjelbred, 1974] and [Chang, Skjelbred, 1974]. (See [Allday, Skjelbred, 1974] for a somewhat different approach.)

(5.3.16) Proposition
With the notation above (see Definition (5.3.13)), for any component F_j of X^G
(1) $i^*(U_j)$ generates $H^n(X)$, where $i : X \to X_G$ is the inclusion;
(2) $\phi^*(U_j) = \Omega_j f_j$; and
(3) $I_{f_j} = (\Omega_j)$.

Proof.

(1) $\operatorname{ind}_G(X)(U_j) = \operatorname{ind}_G(X)(\phi_j)_!(f_j)$
$= \operatorname{ind}_G(F_j)(f_j)$ by Proposition (5.3.3)(2)
$\neq 0$, clearly.

Now the result follows from Proposition (5.3.3)(3) with Y a one-point space and $K = \{1\}$.

(2) From Lemma (5.3.14)(2), $\phi^*(U_j) = \phi^*(\phi_j)_!(f_j) = e_j f_j$. But $e_j f_j = \Omega_j f_j$, since $f_j u = 0$ for any u of positive degree in $H^*(F_j)$.

(3) $\Omega_j \in I_{f_j}$ by (2). Now suppose $a \in I_{f_j}$, and choose $v \in H^*_G(X)$ such that $\phi^*(v) = a f_j$. Then $\phi^*(aU_j - \Omega_j v) = 0$; and so, by the Localization Theorem, there is a non-zero $b \in R$ such that $abU_j = \Omega_j bv$. Hence $ab\operatorname{ind}_G(X)(U_j) = \Omega_j b\operatorname{ind}_G(X)(v)$. By the proof of (1), $\operatorname{ind}_G(X)(U_j)$ is non-zero in H^0_G; and b can be cancelled in H^*_G. So Ω_j divides a. □

(5.3.17) Remarks

(1) Lemma (5.3.15) and Proposition (5.3.16)(3) show that $LW(F_j)$ is determined by the ideal I_{f_j} in R.

(2) Example (3.7.18)(1) gives another way of characterizing $LW(F_j)$. Namely $K \in LW(F_j)$ if and only if $K \in \theta_1$ and $(K, F_j(K))$ belongs to $H^*_G(X, F_j) \otimes_{H_G} H^*_G(F_j)$. (See Definition (3.7.11) for the meanings of 'belongs'.)

The second part of the next proposition is called the integration formula in [Atiyah, Bott, 1984]. (See (3.8) there.)

(5.3.18) Proposition

With the above notation and assumptions let $F_1, \ldots, F_\nu$ be the components of X^G. Let S be the multiplicative subset of R generated by the non-zero homogeneous linear polynomials. Then

(1) *for any $u \in S^{-1} H^*_G(X)$,*

$$u = \sum_{i=1}^{\nu} S^{-1}(\phi_i)_!(S^{-1}\phi_i^*(u)/e_i); \text{ and}$$

(2) *Diagram 5.14 commutes:*

$$\begin{array}{ccc}
S^{-1} H^*_G(X) & \xrightarrow{\sum_{i=1}^{\nu} S^{-1}\phi_i^*/e_i} & S^{-1} H^*_G(X^G) \\
{\scriptstyle S^{-1}\operatorname{ind}_G(X)} \searrow & & \swarrow {\scriptstyle \sum_{i=1}^{\nu} S^{-1}\operatorname{ind}_G(F_i)} \\
& S^{-1} H^*_G &
\end{array}$$

Diagram 5.14

5.3. Equivariant Gysin homomorphism

Proof.

(2) This follows from (1) since $\text{ind}_G(F_i) = \text{ind}_G(X)(\phi_i)_!$, by Proposition (5.3.3)(2).

(1) This follows since

$$S^{-1}\phi^*(\sum_{i=1}^{\nu} S^{-1}(\phi_i^*)_!(S^{-1}\phi_i^*(u)/e_i)) = \sum_{i=1}^{\nu} S^{-1}\phi_i^*(u),$$

by Lemma (5.3.14)(2), $= S^{-1}\phi^*(u)$; and $S^{-1}\phi^*$ is an isomorphism by the Localization Theorem. □

When $X^G \neq \emptyset$ there is another way of defining the index. Consider the Leray-Serre spectral sequence of $X_G \to BG$ with coefficients in k, E_r, say, and let $\mathcal{F}_q H_G^*(X)$ be the associated increasing (row-wise) filtration. Let $n = \text{fd}(X)$ and let π be the projection onto $E_\infty^{*,n}$ in the following short exact sequence:

$$0 \to \mathcal{F}_{n-1} H_G^*(X) \to H_G^*(X) \xrightarrow{\pi} E_\infty^{*,n} \to 0.$$

By Proposition (5.3.16)(1), $E_\infty^{*,n} = E_2^{*,n}$ if $X^G \neq \emptyset$; and $E_2^{*,n}$ is a free H_G^*-module generated by any generator $x \in H^n(X)$. Now define

$$i_G(X) : H_G^*(X) \to H_G^*$$

by the formula $\pi(v) = i_G(X)(v)x$ for any $v \in H_G^*(X)$. (We have not included the dependence of $i_G(X)$ on the choice of x in the notation: $i_G(X)$ is uniquely determined up to a non-zero scalar multiple in k.) The next lemma shows that $i_G(X)$ and $\text{ind}_G(X)$ are essentially the same.

(5.3.19) Lemma

With the notation above the two homomorphisms $\text{ind}_G(X)$ and $i_G(X) : H_G^*(X) \to H_G^*$, both of degree $-n$, are the same except perhaps for a non-zero scalar multiple in k depending on the choices of orientation.

Proof.

Choose $x \in H^n(X)$ and the orientation so that $\text{ind}_{\{1\}}(X)(x) = 1$. Choose $f_i \in H^{n_i}(F_i)$ so that $i^*(U_i) = x$ for $1 \leq i \leq \nu$, where $F_1, \ldots, F_\nu$ are the components of X^G, $n_i = \text{fd}(F_i)$, and $U_i = (\phi_i)_!(f_i)$ as in Definition (5.3.13); see Proposition (5.3.16) (1). By Proposition (5.3.3)(3) with Y a one-point space and $K = \{1\}$, it follows that $\text{ind}_G(X)(U_i) = 1$ for $1 \leq i \leq \nu$. Hence also $\text{ind}_G(F_i)(f_i) = 1$ for $1 \leq i \leq \nu$. Also using x to define $i_G(X)$ and f_i to define $i_G(F_i)$ we have that $i_G(X)(U_i) = 1$ and $i_G(F_i)(f_i) = 1$ for $1 \leq i \leq \nu$.

Let $u \in H_G^*(F_i)$. Set $u = af_i + \sum_j b_j y_j$ where $a, b_j \in H_G^*$ and each $y_j \in H^*(F_i)$ has degree less than n_i. So $(\phi_i)_!(y_j)$ has degree less than n;

and hence $i_G(X)(\phi_i)_!(y_j) = 0$. Thus $i_G(X)(\phi_i)_!(u) = i_G(X)(aU_i) = a$. So

$$i_G(X)(\phi_i)_! = i_G(F_i) = \mathrm{ind}_G(F_i) = \mathrm{ind}_G(X)(\phi_i)_!$$

for $1 \leq i \leq \nu$.

Hence by Proposition (5.3.18)(1), $S^{-1}i_G(X) = S^{-1}\mathrm{ind}_G(X)$. It follows that $i_G(X) = \mathrm{ind}_G(X)$ since $H_G^* \to S^{-1}H_G^*$ is injective. □

As a simple application of the Borel formula we shall give some upper bounds for the ranks of torus or p-torus groups which can act effectively on Poincaré duality spaces. These upper bounds follow from little more than the existence of local weights, and yet in many cases they are surprisingly strong. Recall from Definition (4.1.5) the definition of rank(Φ) where $\Phi : G \times X \to X$ is an action of a torus or p-torus. Recall also the definition of $\mathrm{rank}_0(X)$ from Definition (4.3.1): $\mathrm{rank}_0(X) = \sup \mathrm{rank}(\Phi)$ where Φ ranges over all torus actions on X. Similarly, as in Definition (4.6.19), we set $\mathrm{rank}_p(X) := \sup \mathrm{rank}(\Phi)$ where Φ ranges over all p-torus actions on X. Also recall from Definitions (4.7.1)(2) that the topological toric degree of symmetry of a space X is $T_t(X) := \sup \mathrm{rank}(T)$ where T is a torus which can act effectively on X. Similarly we define the topological p-toric degree of symmetry of X to be $T_{t,p}(X) := \sup \mathrm{rank}(G)$ where G is a p-torus which can act effectively on X. Note that in the definitions of $\mathrm{rank}_0(X)$ and $T_t(X)$ we shall require that the torus actions have FMCOT if X is not compact.

Applying the Borel formula to actions on PD-spaces will, in general, yield information only about cohomologically effective actions. The notion of a cohomologically effective action in case (0) was defined in Definition (4.7.3). If in Definition (4.7.3) one considers p-tori instead of tori and one uses coefficients in $\mathbb{F}_p$ instead of $\mathbb{Q}$, then one gets the analogous definition of cohomologically effective actions in cases (2) and (odd p).

It is clear that a cohomologically effective action is always almost-effective, resp. effective, in case (0), resp. cases (2) and (odd p). On the other hand, as pointed out in Remark (4.7.4), any almost-effective torus action on a compact orientable topological manifold, M, is cohomologically effective: and this is true more generally if M is a compact orientable $\mathbb{Q}$-cohomology manifold. Similarly any effective p-torus action on a compact orientable topological manifold, M, is cohomologically effective: and this is true more generally if M is a compact orientable $\mathbb{F}_p$-cohomology manifold. The key reference for these facts is [Borel et al, 1960], Chapter I, Corollary 4.6, together with Chapter V, Theorem 3.2, resp. 2.2, in the same work, in case (0), resp. cases (2) and (odd p).

5.3. Equivariant Gysin homomorphism

Note that T^n and $(\mathbb{Z}_p)^n$ can act effectively on $\mathbb{R}^{2n}$; but no such action can be cohomologically effective.

For convenience we shall make the following definitions.

(5.3.20) Definitions
(1) Let $T^c_t(X)$, resp. $T^c_{t,p}(X)$, be sup rank(G) where G is a torus, respectively p-torus, which can act cohomologically effectively on X. In case (0), as usual, we shall suppose that the torus actions have FMCOT if X is not compact.

Note that by the discussion above one always has $T^c_t(X) \leq T_t(X)$ and $T^c_{t,p}(X) \leq T_{t,p}(X)$: but $T^c_t(X) = T_t(X)$ and $T^c_{t,p}(X) = T_{t,p}(X)$ if X is a compact orientable topological manifold, or, more generally, a compact orientable k-cohomology manifold, where $k = \mathbb{Q}$ or $\mathbb{F}_p$ as the case may be.

(2) Let X be a k-PD-space, where, as usual $k = \mathbb{Q}$, resp. $\mathbb{F}_p$, when considering torus, resp. p-torus, actions. We shall let $\mu d_0(X)$, respectively $\mu d_p(X)$, denote the minimal possible formal dimension (over k) of a component of the fixed point set of a torus, resp. p-torus, action on X with non-empty fixed point set. For example $\mu d_0(S^{2n+1}) = 1$, and $\mu d_p(S^{2n+1}) = 1$ for odd p; $\mu d_0(S^{2n}) = 0$, $\mu d_p(S^{2n}) = 0$, $\mu d_2(S^n) = 0$, $\mu d_0(\mathbb{C}P^n) = 0$, $\mu d_p(\mathbb{C}P^n) = 0$ and $\mu d_0((S^{2n+1})^m) = m$.

Now a simple application of the Borel Formula is the following.

(5.3.21) Proposition
Let X be a k-PD-space satisfying condition (LT), where $k = \mathbb{Q}$, resp. $\mathbb{F}_p$, in case (0), resp. case (2) or (odd p). Then:
(1) $T^c_t(X) \leq \text{rank}_0(X) + \frac{1}{2}[\text{fd}(X) - \mu d_0(X)]$;
(2) $T^c_{t,2}(X) \leq \text{rank}_2(X) + \text{fd}(X) - \mu d_2(X)$; and
(3) $T^c_{t,p}(X) \leq \text{rank}_p(X) + \frac{1}{2}[\text{fd}(X) - \mu d_p(X)]$ for odd p.

Proof.

We shall give the proof in case (odd p), the other cases being similar. Let G be a p-torus acting cohomologically effectively on X, and let K be an isotropy group of maximal rank. (In case (0) K would be the identity component of an isotropy group of maximal rank.) If $\Phi : G \times X \to X$ denotes the action, then, by definition, rank$(\Phi) = \text{rank}(G/K)$. So

$$\text{rank}(G) = \text{rank}(\Phi) + \text{rank}(K) \leq \text{rank}_p(X) + \text{rank}(K).$$

$X^K \neq \emptyset$. Let F be a component of X^K. Since K is acting cohomologically effectively, fd$(F) < $ fd(X), unless $K = \{1\}$, by Corollary (5.3.8)(2). And if fd$(F) < $ fd(X), then, by the Borel Formula, there is a subgroup

$K_1 \subseteq K$ of corank 1 such that $\mathrm{fd}(F) < \mathrm{fd}(F(K_1))$. If $K_1 \neq \{1\}$, then, by considering the Borel formula for the action of K_1 on X, there is a subgroup $K_2 \subseteq K_1$ of corank one such that $\mathrm{fd}(F(K_1)) < \mathrm{fd}(F(K_2))$. Thus, continuing in this way, if $\mathrm{rank}(K) = r$, we get a sequence of subgroups $\{1\} = K_r \subset \ldots \subset K_2 \subset K_1 \subset K$, with each corank one in the next, such that

$$\mathrm{fd}(F) < \mathrm{fd}(F(K_1)) < \mathrm{fd}(F(K_2)) < \ldots < \mathrm{fd}(F(K_r)) = \mathrm{fd}(X).$$

Since each difference in the latter sequence is at least two, it follows that $\mathrm{fd}(X) - \mathrm{fd}(F) \geq 2r$. So

$$\mathrm{rank}(K) \leq \frac{1}{2}[\mathrm{fd}(X) - \mathrm{fd}(F)] \leq \frac{1}{2}[\mathrm{fd}(X) - \mu d_p(X)].$$

□

(5.3.22) Examples
(1) The proposition gives the estimate $T_t(\mathbb{C}P^n) \leq n$. But the standard action of $SU(n+1)$ on $\mathbb{C}P^n$ has ineffective kernel $\mathbb{Z}_{n+1}$: hence $T_t(\mathbb{C}P^n) = n$.
(2) The proposition gives

$$T_t((S^{2n+1})^m) \leq m + \frac{1}{2}[(2n+1)m - m] = (n+1)m.$$

But the standard action of $SO(2n+2)$ on S^{2n+1} is effective. Hence $T_t((S^{2n+1})^m) = (n+1)m$.
(3) The proposition gives $T_{t,2}(S^n) \leq n+1$, since $\mathrm{rank}_2(S^n) = 1$. The standard action of $O(n+1)$ on S^n is effective. Hence $T_{t,2}(S^n) = n+1$.

We shall now give some further results about local weights, concluding the section with a non-trivial application (Theorem (5.3.27)). We begin by recalling the conditions under which we are working and adding a little more notation.

(5.3.23) Conditions
Let G be a torus, respectively p-torus, let $k = \mathbb{Q}$, resp. $\mathbb{F}_p$, and let X be a Poincaré duality space over k on which G acts. Suppose that condition (LT) is satisfied. Suppose too that $X^G \neq \emptyset$, and let F be a component of X^G. For any subtorus $L \subseteq G$, as above, let $F(L)$ be the component of X^L which contains F. Also let $\theta_1(L)$ denote the set of corank one subtori which contain L: i.e.

$$\theta_1(L) = \{K \subseteq T : \mathrm{cork}(K) = 1 \text{ and } K \supseteq L\}.$$

(5.3.24) Lemma
Under Conditions (5.3.23) let $L \subseteq G$ be a subtorus. Then the set of local weights at F for the G-action on $F(L)$ is $\theta_1(L) \cap LW(F)$.

5.3. Equivariant Gysin homomorphism

In particular, by the Borel formula,

$$\mathrm{fd}(F(L)) - \mathrm{fd}(F) = \sum_{H \in \theta_1(L) \cap LW(F)} \{\mathrm{fd}(F(H)) - \mathrm{fd}(F)\}.$$

Proof.
If $H \in \theta_1(L) \cap LW(F)$, then clearly H is a local weight at F for the G-action on $F(L)$.

If H has a corank one and H does not contain L, then $G = HL$. Hence the component of $F(L)^H$ which contains F is F: and so H is not a local weight. □

(5.3.25) Definitions

Under Conditions (5.3.23):

(1) let $\mathcal{I}(F)$ denote the set of all intersections of members of $LW(F)$ in cases (2) and (odd p), and let $\mathcal{I}(F)$ denote the set of all identity components of intersections of members of $LW(F)$ in case (0);

(2) say that a proper subtorus $L \subset G$ is essential at F if for all subtori K which properly contain L, $\mathrm{fd}(F(K)) < \mathrm{fd}(F(L))$.

Note that a typical element of $\mathcal{I}(F)$ has the form $H_1 \cap \ldots \cap H_n$ in cases (2) and (odd p), resp. $(H_1 \cap \ldots \cap H_n)^0$ in case (0), where $H_i \in LW(F)$ for $1 \leq i \leq n$.

Note also that $L \subset G$ is essential at F if and only if the restriction homomorphism $H^*(F(L); k) \to H^*(F(K); k)$ is not an isomorphism for any $K \supset L$. (See Corollary (5.3.8)(2).)

Now the following theorem is a very useful consequence of the Borel formula. It is a version of [Hsiang, 1975], Theorem (V.I').

(5.3.26) Theorem

Under Conditions (5.3.23), a subtorus $L \subset G$ is essential at F if and only if $L \in \mathcal{I}(F)$.

Proof.
Suppose that L is essential at F. Let $\theta_1(L) \cap LW(F) = \{H_1, \ldots, H_n\}$; and let $K = H_1 \cap \ldots \cap H_n$ in cases (2) and (odd p), resp. $(H_1 \cap \ldots \cap H_n)^0$ in case (0). Then $K \supseteq L$; and $\theta_1(K) \cap LW(F) = \theta_1(L) \cap LW(F)$. By Lemma (5.3.24), $\mathrm{fd}(F(K)) = \mathrm{fd}(F(L))$. Since L is essential at F, $K = L$; and so $L \in \mathcal{I}(F)$.

Now let $L \in \mathcal{I}(F)$. Let $\theta_1(L) \cap LW(F) = \{H_1, \ldots, H_n\}$. So

$$L = H_1 \cap \ldots \cap H_n$$

in cases (2) and (odd p), respectively $(H_1 \cap \ldots \cap H_n)^0$ in case (0). Suppose $K \supset L$. Then
$$\theta_1(K) \cap LW(F) \subset \theta_1(L) \cap LW(F).$$
Hence, by Lemma (5.3.24), $\mathrm{fd}(F(K)) < \mathrm{fd}(F(L))$. □

As a further application of local weights and the Borel formula we now give the following theorem of [Skjelbred, 1978b].

(5.3.27) Theorem
Let $G = T^r$ be a torus of rank r. Let X be a rational Poincaré duality space, and suppose that G is acting on X so that condition (LT) is satisfied. Suppose that X is TNHZ in $X_G \to BG$ with respect to rational cohomology, suppose that X^G has exactly two components, F_1 and F_2, say, and suppose that the restriction homomorphism $H^*(X;\mathbb{Q}) \to H^*(F_1;\mathbb{Q})$ is surjective. Suppose further that the action of G is cohomologically effective. Then:
1. *if $\mathrm{fd}(F_1) > \mathrm{fd}(F_2)$, then $r = 1$; and*
2. *if $\mathrm{fd}(F_1) < \mathrm{fd}(F_2)$, then $r \leq 3$.*

(5.3.28) Remarks
(1) There is no conclusion if $\mathrm{fd}(F_1) = \mathrm{fd}(F_2)$. For example T^n can act effectively on S^{2n} with two isolated fixed points.
(2) In [Skjelbred, 1978b] there is given an example where $G = (\mathbb{Z}_2)^4$ is acting effectively on the Cayley projective plane, X with $X^G = S^1 +$ an isolated point. Thus the theorem does not hold for 2-tori and coefficients in $\mathbb{F}_2$.

The proof will result from the following three lemmata. For a component F of X^G we shall view the set of local weights at F, $LW(F)$, both as a set of corank one subtori or as a set of non-zero elements of $H^2(BG;\mathbb{Q})$: i.e. we view $LW(F)$ as either the set of all subtori K of corank one such that $\mathrm{fd}(F(K)) > \mathrm{fd}(F)$ or as the set of generators $\omega \in H^2(BG;\mathbb{Q})$ for the ideals PK where K has corank one and $\mathrm{fd}(F(K)) > \mathrm{fd}(F)$. Of course, in the latter view where $PK = (\omega)$, ω is defined only up to non-zero rational multiples. (See Definition (5.3.12).)

(5.3.29) Lemma
Under the conditions of Theorem (5.3.27),
$$LW(F_1) \subseteq LW(F_2),$$
and X^K is connected for all $K \in LW(F_1)$. (The cohomological effectiveness of the action is not needed here.)

5.3. Equivariant Gysin homomorphism

Proof.
Let $K \in LW(F_1)$. Then the restriction homomorphism
$$H^*(F_1(K); \mathbb{Q}) \to H^*(F_1; \mathbb{Q})$$
is not an isomorphism. It is, however, surjective because of the composition
$$H^*(X; \mathbb{Q}) \to H^*(F_1(K); \mathbb{Q}) \to H^*(F_1; \mathbb{Q}).$$
So $\dim_{\mathbb{Q}} H^*(F_1(K); \mathbb{Q}) > \dim_{\mathbb{Q}} H^*(F_1; \mathbb{Q})$. But X is TNHZ, and so
$$\dim_{\mathbb{Q}} H^*(X^K; \mathbb{Q}) = \dim_{\mathbb{Q}} H^*(F_1; \mathbb{Q}) + \dim_{\mathbb{Q}} H^*(F_2; \mathbb{Q}).$$
It follows that X^K is connected and $K \in LW(F_2)$. $\square$

The next lemma is due to T.Chang.

(5.3.30) Lemma
Under the conditions of Theorem (5.3.27), suppose that $r \geq 2$. Then $\mathrm{fd}(F_1) \leq \mathrm{fd}(F_2)$. Furthermore, if $\mathrm{fd}(F_1) < \mathrm{fd}(F_2)$ then:

(1) *for any two distinct H_1 and H_2 in $LW(F_1)$, there is a*
$$K \in LW(F_2) - LW(F_1)$$
such that $K \supseteq (H_1 \cap H_2)^0$; and
(2) *for any $H \in LW(F_1)$ and $K \in LW(F_2) - LW(F_1)$, there is a $H' \in LW(F_1)$ such that $H' \neq H$ and $H' \supseteq (H \cap K)^0$.*

Proof.
Let $\Omega = LW(F_1)$ and let $N = LW(F_2) - LW(F_1)$. Since $r \geq 2$ and the action is cohomologically effective, by considering the Borel formula
$$\mathrm{fd}(X) - \mathrm{fd}(F_1) = \sum_{H \in \Omega} \{\mathrm{fd}(F(H)) - \mathrm{fd}(F_1)\},$$
where
$$F(H) = F_1(H) = F_2(H) = X^H,$$
it follows that Ω contains at least two elements. Let $H_1, H_2 \in \Omega$ be distinct, and let $L = (H_1 \cap H_2)^0$. By Lemma (5.3.24) one has
$$\mathrm{fd}(F(L)) - \mathrm{fd}(F_1) = \sum_{H \in \theta_1(L) \cap \Omega} \{\mathrm{fd}(F(H)) - \mathrm{fd}(F_1)\},$$
and
$$\mathrm{fd}(F(L)) - \mathrm{fd}(F_2) = \sum_{H \in \theta_1(L) \cap \Omega} \{\mathrm{fd}(F(H)) - \mathrm{fd}(F_2)\}$$
$$+ \sum_{K \in \theta_1(L) \cap N} \{\mathrm{fd}(F_2(K)) - \mathrm{fd}(F_2)\}.$$

Subtracting, one gets

$$\text{fd}(F_2) - \text{fd}(F_1) = \sum_{H \in \theta_1(L) \cap \Omega} \{\text{fd}(F_2) - \text{fd}(F_1)\}$$
$$- \sum_{K \in \theta_1(L) \cap N} \{\text{fd}(F_2(K)) - \text{fd}(F_2)\}.$$

Since $\theta_1(L) \cap \Omega$ contains H_1 and H_2, and since $\text{fd}(F_2(K)) - \text{fd}(F_2) \geq 0$, it follows that $\text{fd}(F_2) \geq \text{fd}(F_1)$; and $\theta_1(L) \cap N \neq \emptyset$ if $\text{fd}(F_2) > \text{fd}(F_1)$.

This proves all but (2). To see (2) let $H \in \Omega$ and $K \in N$, and let $L = (H \cap K)^0$. By Lemma (5.3.24) applied at F_2, $\text{fd}(F_2(L)) > \text{fd}(F(H))$. But $F_2(L) = X^L = F_1(L)$, since $L \subseteq H$ and X^H is connected. And so, by Lemma (5.3.24) applied at F_1, H is not the only member of $\theta_1(L) \cap \Omega$. □

Theorem (5.3.27)(1) is already contained in Lemma (5.3.30): and so, to complete the proof of the theorem, it is enough to show that the conclusions (1) and (2) of Lemma (5.3.30) cannot be satisfied if $r \geq 4$. Without loss of generality we may assume that $r = 4$. View $\Omega = LW(F_1)$ and $N = LW(F_2) - LW(F_1)$ as sets of vectors in the rational vector space $H^2(BG; \mathbb{Q}) \cong \mathbb{Q}^4$: indeed, more correctly, Ω and N should then be viewed as sets of lines through the origin in $\mathbb{Q}^4$. Let V be a three-dimensional subspace of $\mathbb{Q}^4$ which does not contain any of the lines in Ω or N. Let W be a three-dimensional affine subspace of $\mathbb{Q}^4$ parallel to V and not containing the origin. Let A be the set of points in which the lines in Ω meet W, and let B be the set of points in which the lines in N meet W.

For any two subtori $L_1, L_2 \subseteq G$, it is easy to see that

$$P(L_1 \cap L_2)^0 = PL_1 + PL_2.$$

Thus, e.g., conclusion (1) of Lemma (5.3.30) becomes, in the context above, the following: for any two points $a_1, a_2 \in A$, there is a point $b \in B$ such that b lies on the line joining a_1 and a_2. Also letting $\Omega = \{H_1, \ldots, H_n\}$, since G is acting cohomologically effectively, $(H_1 \cap \ldots \cap H_n)^0$ must be the trivial subtorus, by Theorem (5.3.26). Thus, letting $PH_i = (\omega_i)$, for $1 \leq i \leq n$, $\{\omega_1, \ldots, \omega_m\}$ must span $H^2(BG; \mathbb{Q})$, i.e. A must contain a set of points in general position. Thus we are reduced to the following purely combinatorial lemma.

(5.3.31) Lemma
In $\mathbb{R}^3$ (or $\mathbb{Q}^3$) there cannot exist two non-empty finite disjoint sets of points, A and B, such that

(1) A contains a subset of four points in general position;
(2) for any two points a_1, $a_2 \in A$, there is a point $b \in B$ which lies on the line joining a_1 and a_2; and
(3) for any $a \in A$ and $b \in B$, there is a point $a' \in A$ such that $a' \neq a$, and a' lies on the line joining a and b.

Proof.
Let $a_1, a_2, a_3, a_4 \in A$ be in general position. Let P' be a plane through a_1 which does not contain any of the lines $a_1 a$ for $a \in A$, $a \neq a_1$. Let P be a plane not through a_1 and parallel to P'. Let $\pi : A - \{a_1\} \to P$ be stereographic projection from a_1.

Now the theorem of Sylvester and Grünwald says that if C is any finite set of non-collinear points in $\mathbb{R}^n$, then there is a line in $\mathbb{R}^n$ which contains exactly two points of C. (See, e.g., [Hadwiger, Debrunner, Klee, 1964].) Since a_1, a_2, a_3, a_4 are not coplanar, $\pi(a_2)$, $\pi(a_3)$ and $\pi(a_4)$ are not collinear. So there is a line L in P containing exactly two points in the image of π, $\pi(x)$ and $\pi(y)$, say.

Let $b \in B$ be on the line $a_1 x$; and let $a \in A$ be on the line yb with $a \neq y$. Then $\pi(a)$ is on L, but $\pi(a) \neq \pi(x)$ or $\pi(y)$. $\square$

5.4 Torus actions and Pontryagin classes

In this section we shall give some simple but very useful results concerning the existence of fixed points of torus actions on manifolds. Some of the results, especially Proposition (5.4.1), apply more generally; and the main result on Pontryagin classes for torus actions (Theorem (5.4.5)) also has an analogue concerning Stiefel-Whitney classes for 2-torus actions (Theorem (5.4.9)).

First there is the following simple fact.

(5.4.1) Proposition
Let G be a torus (resp., p-torus) and let $k = \mathbb{Q}$ (resp., $\mathbb{F}_p$). Let X be a G-space satisfying condition (LT) of the introduction to Section 3.10. Suppose that X is a Poincaré duality space over k of formal dimension n; and let $\langle X \rangle \in H^n(X;k)$ be a generator. Let $X \xrightarrow{i} X_G \xrightarrow{p} BG$ be the fibre bundle of the Borel construction.

Then $X^G \neq \emptyset$ if and only if there is a class $u \in H_G^n(X;k)$ such that $i^*(u) = \langle X \rangle$.

Proof.

If $X^G \neq \emptyset$, then the existence of u is shown in Corollary (5.3.4) and again in Proposition (5.3.16)(1).

If u exists, then $E_\infty^{*,n}$ is a free $H^*(BG;k)$-module in the Leray-Serre spectral sequence of $X_G \to BG$. Using the increasing filtration (as in the proof of Corollary (3.1.14)), there is an exact sequence

$$0 \to \mathcal{F}_{n-1} H_G^*(X;k) \to H_G^*(X;k) \to E_\infty^{*,n} \to 0.$$

Thus $S^{-1} H_G^*(X;k) \neq 0$ for any multiplicative set $S \subseteq R$, the polynomial part of $H^*(BG;k)$. So $X^G \neq \emptyset$ by the Localization Theorem. □

(5.4.2) Remarks
(1) If X is a compact manifold which is orientable over k, and if $X^G \neq \emptyset$, then u can be seen to exist at once by considering the equivariant inclusion $X \to (X, X - \{x_0\})$, where $x_0 \in X^G$.

(2) The part of the proposition which says that $X^G \neq \emptyset$ if $\langle X \rangle$ survives to ∞ in the spectral sequence does not require Poincaré duality. Indeed $X^G \neq \emptyset$ if any class of top degree survives to ∞.

Some immediate corollaries are as follows.

(5.4.3) Corollary
Let G and X be as in Proposition (5.4.1). If $X^G \neq \emptyset$, then X^G contains at least two points. (Cf. [Bredon, 1972], Chapter IV, Corollary 2.3.)

Proof.

Both $E_\infty^{*,n}$ and $E_\infty^{*,0}$ are free $H^*(BG;k)$-modules. □

(5.4.4) Corollary
If $H^j(X;\mathbf{Q}) = 0$ for $j > 2n$, if $H^1(X;\mathbf{Q}) = 0$, and if there exists $w \in H^2(X;\mathbf{Q})$ such that $w^n \neq 0$, then any torus action on X satisfying condition (LT) has a fixed point.

The next result, especially when combined with Proposition (5.4.1), is useful for showing that certain torus actions must have fixed points. If M is a topological manifold, then let, as usual, $p(M) \in H^*(M;\mathbf{Q})$ denote the total rational Pontryagin class of M. So $p(M) = \sum_{i=0}^{\infty} p_i(M)$ where $p_i(M) \in H^{4i}(M;\mathbf{Q})$ is the ith Pontryagin class of M. (For the existence of topological rational Pontryagin classes see [Milnor, Stasheff, 1974], Epilogue, and [Kirby, Siebenmann, 1977], Essay IV, §8.) Note that we always assume that a topological manifold is paracompact.

(5.4.5) Theorem
Let G be a torus acting on a topological manifold M. Let
$$M\xrightarrow{i}M_G\xrightarrow{\pi}BG$$
be the fibre bundle of the Borel construction. Then there is a class $u \in H_G^*(M;\mathbb{Q})$ such that $i^*(u) = p(M)$.

Thus all the rational Pontryagin classes of M survive to ∞ in the Leray-Serre spectral sequence of π in rational cohomology.

Proof.
Choose an integer N very much larger than $\dim M$. Let E_G^N be a N-connected compact approximation to EG so that $B_G^N = E_G^N/G$ is a compact manifold, and let $M_G^N = (E_G^N \times M)/G$. Consider the fibre bundle $M\xrightarrow{i_N}M_G^N\xrightarrow{\pi_N}B_G^N$. Let $T(M)$, resp. $T(M_G^N)$, denote the tangent microbundle of M, resp. M_G^N. Since π_N is a fibre bundle, M has a trivial normal microbundle in M_G^N. And so, as can be seen directly or by [Milnor, 1964], Theorem (5.9), $i_N^*T(M_G^N) \cong T(M) \oplus \nu$, where ν is a trivial microbundle.

Hence - see, e.g. [Milnor, Stasheff, 1974], Epilogue - $i_N^*p(M_G^N) = p(M)$: and so $p(M) \in \operatorname{im} i_N^*$. Since N can be arbitrarily large, by comparing the Leray-Serre spectral sequence of π_N with that of π, it follows that $p(M) \in \operatorname{im} i^*$. $\square$

(5.4.6) Remark
In ascribing Pontryagin classes to microbundles we implicitly use Kister's theorem ([Kister, 1964]) for which the manifold is assumed to be separable. But any component of a topological manifold is separable (see, e.g., [Munkres, 1975], §6-5, Exercises 2 and 3), and any component is invariant under a torus action.

(5.4.7) Corollary ([Bott, 1967])
Let M be a compact orientable topological manifold which has a non-zero Pontryagin number. Then any torus group action on M has a non-empty fixed point set.

Proof.
The result is immedaite from Theorem (5.4.5) and Proposition (5.4.1).
$\square$

(5.4.8) Example
Let M and N be two compact connected oriented smooth manifolds of the same dimension. Then the connected sum $M \sharp N$ is oriented cobor-

dant to the disjoint union $M + N$. Hence, where σ denotes signature, $\sigma(M \sharp N) = \sigma(M) + \sigma(N)$, by Thom's theorem. (See, e.g., [Milnor, Stasheff, 1974], Lemma 19.3.) Let
$$X = (S^1 \times S^3) \sharp (S^1 \times S^3) \sharp \mathbb{C}P^2 \sharp \mathbb{C}P^2.$$
So $\sigma(X) \neq 0$; and so X has a non-zero Pontryagin number by the Signature Theorem ([Milnor, Stasheff, 1974], Theorem 19.4). But the Euler characteristic $\chi(M \sharp N) = \chi(M) + \chi(N) - \chi(S^n)$, where $n = \dim M = \dim N$, as is easily seen by the Mayer-Vietoris sequence. So $\chi(X) = 0$. Thus any torus group action on X has a fixed point even though $\chi(X) = 0$; cf. [Schafer, 1968].

There is an analogue of Theorem (5.4.5) for Stiefel-Whitney classes.

(5.4.9) Theorem
Let M be a topological manifold and let $G = (\mathbb{Z}_2)^r$ be a 2-torus acting on M. Let $M \xrightarrow{i} M_G \to BG$ be the fibre bundle. Suppose that either (i) M is compact or (ii) M and the action of G on M are smooth. Then there is a class $u \in H_G^*(M; \mathbb{F}_2)$ such that $i^*(u) = w(M)$, the total Stiefel-Whitney class of M.

Proof.
The proof in case (ii) is completely analogous to the proof of Theorem (5.4.5). In case (i) one can use the treatment of Stiefel-Whitney classes given in [Spanier, 1966], Chapter 6, Section 10: see especially Theorem 22 there. □

5.5 Golber formulas and other results

Recall, from Theorem (5.3.11), that if G is a torus or p-torus, if $k = \mathbb{Q}$, resp. $\mathbb{F}_p$, if G is acting on a space X so that condition (LT) of the introduction to Section 3.10 is satisfied, if X is a Poincaré duality space over k, and if $X^G \neq \emptyset$, then for any component F of X^G, there is the Borel formula
$$\mathrm{fd}(X) - \mathrm{fd}(F) = \sum_{H \in \theta_1} \{\mathrm{fd}(F(H)) - \mathrm{fd}(F)\},$$
where θ_1 is the set of subtori of corank one, $F(H)$ is the component of X^H which contains F, and for any Poincaré duality space over k, Y, $\mathrm{fd}(Y)$ is the formal dimension of Y.

The Borel formula is very powerful, and it raises the question as to whether there are analogous formulas involving subtori of higher corank

5.5. Golber formulas and other results

which give additional information about the orbit structure. Such higher order formulas, which have only been found under very special conditions, are called Golber formulas. The original Golber formulas are to be found in [Golber, 1971] and [Chang, Skjelbred, 1974]. (See also [Hsiang, 1975]). In this section we shall state without detailed proof the somewhat more general Golber formulas of [Allday, 1979]. We shall state also some related formulas of [Bredon, 1974]. Finally we shall state a theorem of [Duflot, 1983].

Throughout this section we shall follow our usual convention that the cohomology theory and rational homotopy theory are the Alexander-Spanier theory. As usual singular theory suffices for G-CW-complexes.

The first result, from [Allday, 1979], uses rational homotopy, and, hence, applies only to case (0). We shall need the following conditions.

(5.5.1) Conditions

Let G be a torus and let X be a connected G-space. Suppose that condition (LT) is satisfied. Suppose too that $X^G \neq \emptyset$. Suppose that $\dim_\mathbb{Q} \Pi_\psi^*(X) < \infty$. (E.g., if X is a simply-connected CW-complex of finite $\mathbb{Q}$-type, then we suppose that $\dim_\mathbb{Q} \pi_*(X) \otimes \mathbb{Q} < \infty$.) Finally let F be a component of X^G, and suppose that F is full. (See Definition (3.3.13).)

(Since $\dim_\mathbb{Q} \Pi_\psi^*(X) < \infty$, and since the condition (LT) is satisfied, $\dim_\mathbb{Q} H^*(X; \mathbb{Q}) < \infty$, by Remark (4.2.5). So X^G has a finite number of components; and condition (CS) is satisfied automatically.)

(5.5.2) Definitions

(1) Let θ be the set of subtori of G; and for $0 \leq n \leq \dim G$, let θ_n be the set of subtori of G having corank n. For any $L \in \theta$, let $\theta_n(L) = \{K \in \theta_n; L \subseteq K\}$.

(2) For an abelian group A let A^θ be the abelian group of functions $\theta \to A$. For any $L \in \theta$, define homomorphisms $F_n(-, L) : A^\theta \to A$, $n \geq 0$, recursively as follows: for $g \in A^\theta$,

$$F_0(g, L) = g(L) - g(G),$$
$$F_1(g, L) = F_0(g, L) - \sum_{H \in \theta_1(L)} F_0(g, H), \ldots,$$
$$F_n(g, L) = F_{n-1}(g, L) - \sum_{H \in \theta_n(L)} F_{n-1}(g, H).$$

Of course, $F_n(g, L)$ is well defined if and only if both $F_{n-1}(g, L)$ and $F_{n-1}(g, H)$ are well defined for all $H \in \theta_n(L)$, and $F_{n-1}(g, H) = 0$ for all but finitely many $H \in \theta_n(L)$. If $L \in \theta_n$, and if $F_n(g, L)$ is well-defined, then $F_m(g, L) = 0$ for all $m \geq n$.

(3) Under Conditions (5.5.1), let $\beta_i(X) = \dim_{\mathbb{Q}} \Pi_\psi^i(X)$. For $L \in \theta$, let $F(L)$ be the component of X^L which contains F. Let
$$\beta_i(L) = \dim_{\mathbb{Q}} \Pi_\psi^i(F(L)).$$
In particular, $\beta_i(X) = \beta_i(I)$, where I is the trivial subtorus, and $\beta_i(G) = \dim_{\mathbb{Q}} \Pi_\psi^i(F)$.

Note that $\dim_{\mathbb{Q}} \Pi_\psi^*(F(L)) = \dim_{\mathbb{Q}} \Pi_\psi^*(X)$ for all $L \in \theta$, by Corollary (3.3.12) and the assumption that F is full.

(4) Define functions f_n, f_n^0, f_n^1 in $\mathbb{Z}^\theta$ as follows. For each $L \in \theta$, and $n \geq 0$,
$$f_n(L) = \sum_{m=1}^{\infty} \beta_m(L) m^n,$$
$f_n^0(L) = \sum_{m=1}^{\infty} \beta_{2m}(L)(2m)^n$, and $f_n^1(L) = \sum_{m=0}^{\infty} \beta_{2m+1}(L)(2m+1)^n$.
In particular,
$$f_0(L) = \dim_{\mathbb{Q}} \Pi_\psi^*(F(L))$$
and $f_0^0(L) - f_0^1(L) = \chi\pi(F(L))$.

Note that, by Theorem (2.4.3), for any $L \in \theta$, $F(L)$ is a Poincaré duality space over $\mathbb{Q}$, and $\mathrm{fd}(F(L)) = f_0^0(L) + f_1^1(L) - f_1^0(L)$.

Now we have the following theorem from [Allday, 1979].

(5.5.3) Theorem
Under Conditions (5.5.1) and with the notation of Definitions (5.5.2), for any subtorus $L \subseteq G$, for any $n \geq 0$, and for any $m \geq 0$, $F_n(g_m, L)$ is well defined if $g_m = f_m$, f_m^0 or f_m^1; and for $0 \leq m \leq n$,
$$F_n(f_m, L) = F_n(f_m^0, L) = F_n(f_m^1, L) = 0.$$
Furthermore, $F_n(f_{n+1}, L) \geq 0$, $F_n(f_{n+1}^0, L) \geq 0$ and $F_n(f_{n+1}^1, L) \geq 0$.

(5.5.4) Examples
(1) The formulas with $m = n = 0$ follow at once from the fullness of F, and the fact that $\chi\pi(F) = \chi\pi(X)$.

(2) With $m = n = 1$ and $L = I$ one has
$$f_1(I) - f_1(G) = \sum_{H \in \theta_1} \{f_1(H) - f_1(G)\},$$
$$f_1^0(I) - f_1^0(G) = \sum_{H \in \theta_1} \{f_1^0(H) - f_1^0(G)\},$$
and
$$f_1^1(I) - f_1^1(G) = \sum_{H \in \theta_1} \{f_1^1(H) - f_1^1(G)\}.$$

Since $\operatorname{fd}(F(L)) = f_0^0(L) + f_1^1(L) - f_1^0(L)$, and since, by the fullness of F, $f_0^0(L) = f_0^0(I) = \dim_{\mathbb{Q}} \Pi_\psi^{\mathrm{ev}}(X)$ for all $L \in \theta$, by subtracting the second formula from the third, one gets the Borel formula

$$\operatorname{fd}(X) - \operatorname{fd}(F) = \sum_{H \in \theta_1} \{\operatorname{fd}(F(H)) - \operatorname{fd}(F)\}.$$

(3) With $m = n = 2$ and $L = I$, one has, e.g.,

$$f_2(I) - f_2(G) - \sum_{H \in \theta_1} \{f_2(H) - f_2(G)\}$$

$$= \sum_{L \in \theta_2} [f_2(L) - f_2(G) - \sum_{H \in \theta_1(L)} \{f_2(H) - f_2(G)\}].$$

One also has that

$$f_2(L) - f_2(G) - \sum_{H \in \theta_1(L)} \{f_2(H) - f_2(G)\} = F_1(f_2, L) \geq 0,$$

for all $L \in \theta_2$. And one has the corresponding results for f_2^0 and f_2^1 separately.

The next two examples are applications of Theorem (5.5.3).

(5.5.5) Examples
(1) Under Conditions (5.5.1) suppose that $G = T^3$, a torus of rank 3, and that $H^*(X; \mathbb{Q}) \cong H^*(S^3 \times S^5; \mathbb{Q})$. Suppose that the action of G is cohomologically effective. By Theorem (4.2.7) and Remark(4.2.9), X^L is connected for all $L \in \theta$. Furthermore, by the results of Sections 3.3 and 3.4, $\Pi_\psi^*(F(L))$ has a basis consisting of two elements of odd degree. Let us say that L and $F(L)$ have type (m, n) if the basis elements of $\Pi_\psi^*(F(L))$ have degrees m and n with $m \leq n$. So X has type $(3,5)$; and, by the Borel formula, since G is cohomologically effective, $F = X^G$ has type $(1,1)$, any essential subtorus in θ_1 has type $(1,3)$, and any essential subtorus in θ_2 has type $(3,3)$ or $(1,5)$. By the Borel formula there are exactly three essential subtori in θ_1, and hence, by Theorem (5.3.26), at most three essential subtori in θ_2.

If L has type (m, n), then $f_2(L) = m^2 + n^2$. So the left hand side of the second order formula in Example (5.5.4)(3) is $34 - 2 - 3(10 - 2) = 8$. For $L \in \theta_2$ of type $(3,3)$, $f_2(L) - f_2(G) = 16$, and for $L \in \theta_2$ of type $(1,5)$, $f_2(L) - f_2(G) = 24$. And $f_2(H) - f_2(G) = 8$ for any essential $H \in \theta_1$.

If all three essential subtori of corank one conatin one subtorus of corank two, then the right hand side is zero. So there must be three essential subtori in θ_2, the identity components of the pairwise intersections of the three essential corank one subtori. Thus a corank two subtorus of type (3,3) contributes 0 to the right hand side, and a corank two subtorus of type (1,5) contributes 8.

So there is exactly one corank two subtorus of type (1,5) and exactly two corank two subtori of type (3,3).

(2) ([Chang, Skjelbred, 1976]) Suppose that $H^*(X; \mathbf{Q}) = \mathbf{Q}[x]/(x^3)$, where $\deg(x) = 8$. Let G be a torus acting cohomologically effectively on X, and suppose that condition (LT) is satisfied. We shall show that rank$(G) \leq 4$. Since X has the same rational cohomology as the Cayley projective plane, Cay P^2, and since Cay $P^2 \approx F_4/\operatorname{Spin}(9)$, this result is optimal. We shall assume that rank$(G) = 5$ and get a contradiction. Let us say that $L \in \theta$ has the type of Y if $H^*(F(L); \mathbf{Q}) \cong H^*(Y; \mathbf{Q})$ for some component $F(L)$ of X^L.

By repeated applications of the Borel formula and the formula

$$f_1^0(I) - f_1^0(G) = \sum_{H \in \theta_1} \{f_1^0(H) - f_1^0(G)\},$$

it follows that there are corank one subtori having the type of S^2, corank two subtori having the type of S^4, and corank three subtori having the type of S^6. Let $K \in \theta_3$ have the type of S^6: and let F be the component of X^K with $H^*(F, \mathbf{Q}) \cong H^*(S^6; \mathbf{Q})$.

By the formula for f_1^0 applied at F to the action of K on X, and by the cohomological effectiveness, there is exactly one corank one subtorus of K having the type of S^8. If H is any other essential corank one subtorus of K, then we must have $H^*(X^H; \mathbf{Q}) \cong \mathbf{Q}[y]/(y^3)$, where $\deg(y) = 6$. This contradicts the Borel formula for the action of K at F. (See [Chang, Skjelbred, 1976] for much more.)

Before stating a cohomological analogue we shall outline briefly the proof of Theorem (5.5.3) in the following remark.

(5.5.6) Remark

For $L \in \theta$, let $\psi_L : \Pi_G^*(F(L)) \to \Pi_G^*(F)$ be the restriction homomorphism. (We suppose fixed throughout a chosen base point $x_0 \in F$.) Now we define R-modules, $\operatorname{cok}_n(L)$, for $n \geq 0$, where $R = H^*(BG; \mathbf{Q})$ as usual, and homomorphisms $\psi_{n,L}$ recursively as follows. $\operatorname{cok}_0(L) = \operatorname{coker}(\psi_L)$; $\psi_{1,L}$ is the naturally induced homomorphism $\operatorname{cok}_0(L) \to \prod_{H \in \theta_1(L)} \operatorname{cok}_0(H)$; and $\operatorname{cok}_1(L) = \operatorname{coker}(\psi_{1,L})$;

$$\psi_{n,L} : \operatorname{cok}_{n-1}(L) \to \prod_{H \in \theta_n(L)} \operatorname{cok}_{n-1}(H)$$

and $\operatorname{cok}_n(L) = \operatorname{coker}(\psi_{n,L})$. (One can also set $\operatorname{cok}_{-1}(L) = \Pi_G^*(F(L))$ and $\psi_{0,L} = \psi_L$.)

Without assuming that F is full one can show the following.

5.5. Golber formulas and other results

(i) For $n \geq 1$, if $H \in \theta_n$ and if H is not essential at F, then $\cok_{n-1}(H) = 0$. Thus each product, $\prod_{H \in \theta_n(L)} \cok_{n-1}(H)$, above, is just a finite sum. In particular, for any $L \in \theta$ and $n \geq -1$, $\cok_n(L)$ is a finitely generated R-module.

(ii) For $n \geq 0$, if $\cork(L) \leq n$, then $\cok_n(L) = 0$.

(iii) For $n \geq 0$, if $L, K \in \theta$ and $L \subseteq K$, then the restriction homomorphism $\cok_n(L) \to \cok_n(K)$ is surjective.

(iv) For $n \geq -1$, $L \in \theta$ and $\mathfrak{p} \in \Spec(R)$ with $\sigma\mathfrak{p} = PK$, the localized restriction homomorphism
$$\cok_n(L)_{\mathfrak{p}} \to \cok_n(LK)_{\mathfrak{p}}$$
is an isomorphism. And the localization homomorphism
$$\cok_n(K) \to \cok_n(K)_{\mathfrak{p}}$$
is injective. Thus $\cok_n(L)$ is a Chang-Skjelbred module for all $n \geq -1$ and $L \in \theta$.

If F is full, then one can also prove the following.

(v) For $n \geq 0$ and $L \in \theta$, $\psi_{n,L}$ is a monomorphism.

(vi) For $n \geq -1$ and $L \in \theta$, if $\cok_n(L) \neq 0$, then $\cok_n(L)$ is a Cohen-Macaulay module with $\dim_R \cok_n(L) = r - n - 1$ and $\hd_R \cok_n(L) = n + 1$, where $r = \rank(G)$.

(v) and (vi) are proved together by an application of the Cohen-Macaulay trick of [Atiyah, 1974], Lecture 7. See also [Bredon, 1974]. They are the key to the theorem.

The rest of the proof is an easy argument using Poincaré series. Define, g, g^0, and $g^1 : \theta \to \mathbb{Z}[z]$ by
$$g(L) = (1 - z^2)^r P(\Pi_G^*(F(L)), z),$$
$$g^0(L) = (1 - z^2)^r P(\Pi_G^{\ev}(F(L)), z)$$
and
$$g^1(L) = (1 - z^2)^r P(\Pi_G^{\od}(F(L)), z).$$
Then, e.g.,
$$F_n(g, L) = (-1)^{n+1}(1 - z^2)^r P(\cok_n(L), z).$$
Since $\dim_R \cok_n(L) = r - n - 1$ if $\cok_n(L) \neq 0$, $(1 - z)^{n+1}$ is a factor of $F_n(g, L)$ (see Lemma (4.1.3)). Hence
$$\left(\frac{d^n}{dz^n}\right) F_n(g, L)|_{z=1} = 0.$$
So
$$F_n\left(\frac{d^n g}{dz^n}|_{z=1}, L\right) = 0.$$

Now we shall give the cohomological analogue of Theorem (5.5.3) which appears in [Bredon, 1974]. The conditions needed are as follows.

(5.5.7) Conditions

Let G be a torus or p-torus and let $k = \mathbb{Q}$, resp. $\mathbb{F}_p$. Let X be a G-space satisfying condition (LT). Suppose that $\dim_k H^*(X;k) < \infty$, and that X is TNHZ in $X_G \to BG$ with respect to cohomology with coefficients in k.

Now we define $F_n(g, L)$ as before. We also define f_n, f_n^0, f_n^1 just as before except that we put $\beta_i(L) = \dim_k H^i(X^L; k)$ and we add $\beta_0(L)$ into $f_0(L)$ and $f_0^0(L)$: i.e. $f_0(L) = \sum_{m=0}^{\infty} \beta_m(L) = \dim_k H^*(X^L; k)$ and $f_0^0(L) = \sum_{m=0}^{\infty} \beta_{2m}(L) = \dim_k H^{\mathrm{ev}}(X^L; k)$.

Remark (5.5.6) proceeds as before except that $\psi_L = \psi_{0,L}$ is defined to be the restriction homomorphism $H_G^*(X^L; k) \to H_G^*(X^G; k)$. (ii), (iii), and (iv) hold as stated without assuming that X is TNHZ. And, without assuming that X is TNHZ, it follows also that for $n \geq 1$ $\mathrm{cok}_{n-1}(H) = 0$ for all but finitely many $H \in \theta_n$. So again $\mathrm{cok}_n(L)$ is always a finitely generated R-module.

When X is TNHZ (v) and (vi) hold as before. Thus one gets the following cohomological analogue of Theorem (5.5.3).

(5.5.8) Theorem

Under Conditions (5.5.7) and with the notation of Definition (5.5.2), as amended immediately above, for any subtorus $L \subseteq G$, for any $n \geq 0$ and for any $m \geq 0$, $F_n(f_m, L)$ is well defined; and for $0 \leq m \leq n$,

$$F_n(f_m, L) = 0.$$

Furthermore

$$F_n(f_{n+1}, L) \geq 0.$$

In case (0), $F_n(f_m^0, L)$ and $F_n(f_m^1, L)$ are well defined; and for $0 \leq m \leq n$, $F_n(f_m^0, L) = F_n(f_m^1, L) = 0$. Also $F_n(f_{n+1}^0, L) \geq 0$ and $F_n(f_{n+1}^1, L) \geq 0$.

(5.5.9) Remarks

(1) See [Bredon, 1974] for much more along these lines.
(2) Under Conditions (5.5.7), since ψ_L is injective, there is the short exact sequence

$$0 \to H_G^*(X^L; k) \to H_G^*(X^G; k) \xrightarrow{\delta} H_G^*(X^L, X^G; k) \to 0.$$

Thus there is an isomorphism of degree one from $\mathrm{cok}_0(L)$ to

$$H_G^*(X^L, X^G; k).$$

So $\psi_{1,L}$ gives a monomorphism
$$H_G^*(X^L, X^G; k) \to \sum_{H \in \theta_1(L)} H_G^*(X^H, X^G; k).$$
Now let $X_i = \{x \in X : \operatorname{rank}(G_x) \geq i\} = \bigcup_{H \in \theta_{r-i}} X^H$. This is a closed invariant subspace by Proposition (4.1.11); and by the proof of Proposition (4.1.14), $H_G^*(X^H, X_{i+1}^H; k) \neq 0$ for only a finite number of $H \in \theta_{r-i}$. Thus from the Mayer-Vietoris sequence and induction, $H_G^*(X_i, X_{i+1}; k) \cong \sum_{H \in \theta_{r-i}} H_G^*(X^H, X_{i+1}^H; k)$. Thus
$$\sum_{H \in \theta_1(L)} H_G^*(X^H, X^G; k) \cong H_G^*(X_{r-1}^L, X_r; k) = H_G^*(X_{r-1}^L, X_r^L; k).$$
(Here as above $r = \operatorname{rank}(G)$.) Hence $\operatorname{cok}_1(L)$ is isomorphic to
$$H_G^*(X^L, X_{r-1}^L; k)$$
by an isomorphism of degree two.

Continuing inductively it follows that $\operatorname{cok}_n(L)$ is isomorphic to
$$H_G^*(X^L, X_{r-n}^L; k)$$
by an isomorphic of degree $n+1$.

On the other hand, under Conditions (5.5.7), the Cohen-Macauley trick of [Atiyah, 1974], Lecture 7, shows directly that $H_G^*(X, X_i; k)$ is a Cohen-Macauley module with
$$\dim_R H_G^*(X, X_i; k) = i - 1 \text{ and } \operatorname{hd}_R H_G^*(X, X_i; k) = r - i + 1,$$
if $H_G^*(X, X_i; k) \neq 0$. And for each $i \geq 0$ there is an exact sequence
$$0 \to H_G^*(X, X_{i+1}; k) \to H_G^*(X_i, X_{i+1}; k) \xrightarrow{\delta} H_G^*(X, X_i; k) \to 0.$$
(3) For any space Y let $c(Y)$ be the number of components of Y. Thus under Conditions (5.5.7), by the monomorphisms ψ_L and $\psi_{1,L}$ with $L = I$, $\dim_k H_G^1(X, X^G, k) = c(X^G) - c(X)$ and
$$c(X^G) - c(X) \leq \sum_{H \in \theta_1} \{c(X^G) - c(X^H)\}$$
In particular, if X is connected and if X^G has two components, then X^H is connected for some subtorus H of corank one.

We finish with a theorem from [Duflot, 1983]. It is related to the preceding only in that it concerns the subspaces X_i. It is very different in as much as it gives information about the open invariant subspaces $X - X_i$ and the subspaces $X_i - X_{i+1}$. The conditions are as follows.

(5.5.10) Conditions

Let G be a torus or p-torus for p an odd prime, and let $k = \mathbb{Q}$ or $\mathbb{F}_p$, respectively. Let X be a smooth manifold on which G is acting smoothly

with FMCOT. As usual let
$$X_i = \{x \in X : \operatorname{rank}(G_x) \geq i\},$$
and let $r = \operatorname{rank}(G)$. For each $i \geq 0$ suppose that $X_i - X_{i+1}$ has only a finite number of components.

Let $\mathcal{F}_i$ be the kernel of the restriction homomorphism
$$\rho_i^* : H_G^*(X; k) \to H_G^*(X - X_i; k).$$

(5.5.11) Theorem
Under Conditions (5.5.10),

(1) *for each $i \geq 0$, the restriction homomorphism*
$$\phi_i^* : H_G^*(X - X_{i+1}; k) \to H_G^*(X - X_i; k)$$
is surjective; and so, for each $i \geq 0$,
$$\rho_i^* : H_G^*(X; k) \to H_G^*(X - X_i; k)$$
is surjective;

(2) *for each $i \geq 0$, there is an ungraded k-vector-space isomorphism between $\ker(\phi_i^*) \cong H_G^*(X - X_{i+1}, X - X_i; k)$ and $H_G^*(X_i - X_{i+1}; k)$; and*

(3) *for each $i \geq 0$, there is an ungraded k-vector-space isomorphism between $\mathcal{F}_i/\mathcal{F}_{i+1}$ and $H_G^*(X_i - X_{i+1}; k)$.*

See [Duflot, 1983] for the proof and more.

Exercises

In the following exercises, unless stated otherwise, all G-spaces should be assumed to satisfy condition (LT) of the introduction to Section 3.10.

(5.1) Let G be a torus or p-torus, and let $k = \mathbb{Q}$ or $\mathbb{F}_p$, respectively. Let G act on M, a compact topological manifold which is orientable over k. Show that the sequence
$$0 \to H_G^*(M, M - M^G; k) \to H_G^*(M; k) \to H_G^*(M - M^G; k) \to 0$$
is exact.

(5.2) ([Félix, Halperin, 1982b]) Let $G = T^r$, and let X be an almost-free G-space. Show that X is a $\mathbb{Q}$-PD-space of formal dimension n if and only if X/G is a $\mathbb{Q}$-PD-space of formal dimension $n - r$.

(5.3) Let X be a $\mathbb{Q}$-PD-space of formal dimension n. Suppose that there are classes $\omega_1, \ldots, \omega_n \in H^1(X; \mathbb{Q})$ such that $\omega_1 \ldots \omega_n \neq 0$ in $H^n(X; \mathbb{Q})$. Show that the only possible cohomologically effective compact connected Lie group actions on X are almost-free torus actions.

(5.4) Let X be a $\mathbb{Q}$-PD-space of formal dimension $2n$. Suppose that $H^1(X;\mathbb{Q}) = 0$, and that there are classes $\omega_1, \ldots, \omega_n \in H^2(X;\mathbb{Q})$ such that $\omega_1 \ldots \omega_n \neq 0$ in $H^{2n}(X;\mathbb{Q})$. Suppose that $G = T^r$, $r \geq 1$, acts cohomologically effectively on X. Show that X^G is non-empty and non-connected.

(5.5) Let G be a torus or p-torus, and let $k = \mathbb{Q}$ or $\mathbb{F}_p$, respectively. Let X be a k-PD-space, and let G act on X with $X^G \neq \emptyset$. Let F_i, $1 \leq i \leq \nu$, be the components of X^G. Let $\phi_i : F_i \to X$, $1 \leq i \leq \nu$, be the inclusions; and let $\theta_i = (\phi_i)_!(1)$ as in Definition (5.3.13). Show that:

(1) for $1 \leq i \leq \nu$, $\ker \phi_i^* = \{x \in H_G^*(X;k); \theta_i x = 0\}$; and
(2) $x \in H_G^*(X;k)$ is torsional (over R) if and only if $\theta_i x = 0$ for $1 \leq i \leq \nu$.

(5.6) With the same conditions as Exercise (5.5), let
$$H_G(X) = H_G^{\mathrm{ev}}(X;k)$$
in cases (0) and (odd p), and let $H_G(X) = H_G^*(X;k)$ in case (2). Let $\phi : X^G \to X$ be the inclusion. For $1 \leq j \leq \nu$, let I_j be the set of all $x \in H_G(X)$ such that $\phi^*(x)u \in \mathrm{im}(\phi^*)$ for all $u \in H^*(F_j)$. Let $J_j = \phi_j^*(I_j) \subseteq H_G(F_j)$. Show that $J_j = (e_j)$, the principal ideal in $H_G(F_j)$ generated by $e_j := e_G(F_j, X)$.

(5.7) With the same conditions as Exercise (5.5), for $1 \leq j \leq \nu$, let Ω_j be as defined in Definition (5.3.13). For any j, $1 \leq j \leq \nu$, and any $u \in H_G^*(F_j;k)$, show that $\Omega_j \in \sqrt{I_u}$. (Recall that $I_u := \{a \in R; au \in \mathrm{im}(\phi^*)\}$.)

(5.8) (Bredon) Let G be a torus or p-torus, and let $k = \mathbb{Q}$ or $\mathbb{F}_p$, respectively. Let X and Y be G-spaces and k-PD-spaces. Let $f : X \to Y$ be an equivariant map, and suppose that $\deg(f) \neq 0$ in case (0), and $\deg(f) \neq 0$ (modulo p) in cases (2) and (odd p). Show that
$$f^* : H^*(Y^G;k) \to H^*(X^G;k)$$
is injective. In particular, $X^G \neq \emptyset$ if $Y^G \neq \emptyset$.

(5.9) Suppose that $H^*(X;\mathbb{Q}) \cong H^*(\mathbb{H}P^n;\mathbb{Q})$ as graded $\mathbb{Q}$-algebras. With the notation of Definition (5.3.20)(1), show that $T_t^c(X) \leq n+1$. In particular, $T_t(\mathbb{H}P^n) = n+1$.

(5.10) (Skjelbred) Let $G = T^r$; and suppose that G is acting effectively on S^{4n-1} without fixed points, and that G is acting on S^{2n} with a connected fixed point set. Suppose that there is an equivariant map $f : S^{4n-1} \to S^{2n}$ with a non-zero Hopf invariant. Show that $r \leq 3$.

(5.11) ([Browder, 1985]) Let M be a topological manifold, and suppose that M has a non-zero Pontryagin polynomial $P_I(M) \in H^{4i}(M;\mathbb{Q})$.

Suppose that $H^j(M;\mathbb{Q}) = 0$ for all odd $j > 4i$. Show that any circle group action on M has a fixed point.

(5.12) Suppose that $H^*(X;\mathbb{Q}) \cong H^*(S^3 \times S^5 \times S^7;\mathbb{Q})$ as graded $\mathbb{Q}$-algebras. Suppose $G = T^6$ is acting on X cohomologically effectively with $X^G \neq \emptyset$. For any subtorus $L \subseteq G$ say that L has type (a, b, c), where $a \leq b \leq c$, if $H^*(X^L;\mathbb{Q}) \cong H^*(S^a \times S^b \times S^c;\mathbb{Q})$ as graded $\mathbb{Q}$-algebras. Show that there are exactly four corank two subtori of type (1,1,5) and exactly eleven corank two subtori of type (1,3,3).

(5.13) Let G be a torus or p-torus, and let $k = \mathbb{Q}$ or $\mathbb{F}_p$, respectively. Let X be a connected G-space. Suppose that $\dim_k H^*(X;k) < \infty$, X is TNHZ in $X_G \to BG$, and that X^G has exactly ν distinct components. Show that there is a subtorus $K \subseteq G$ of corank at most $\nu - 1$ such that X^K is connected.

(5.14) ([Yau, 1977]) Let M be a connected topological manifold of dimension n. (If M is non-compact, then suppose that $H_*(M;\mathbb{Z})$ has finite type.) Suppose that a torus G can act non-trivially on M with $M^G \neq \emptyset$. Show that for any $\omega_1,\ldots,\omega_{n-1} \in H^1(M;\mathbb{Q})$, $\omega_1\ldots\omega_{n-1} = 0$.

(5.15) ([Yau, 1977]) Let M be as in Exercise (5.14), and suppose that $\chi(M) \neq 1+(-1)^n$. Show that the connected sum $M\sharp T^n$ does not admit any non-trivial compact connected Lie group action.

(5.16) Let M be a connected topological manifold of dimension n, or more generally, let M be a connected integral cohomology manifold of dimension n. Show that $T_t(M) \leq n$; and that $T_t(M) = n$ if and only if M is homeomorphic to T^n. (If T^r is acting effectively on M, then, by the Principal Orbit Type Theorem (see [Borel et al, 1960], Chapter IX), there is a $x \in M$ with trivial isotropy subgroup. Now apply [Borel et al, 1960], Chapter I, Corollary 4.6, to the orbit of x.)

(5.17) Let X be a $\mathbb{Q}$-PD-space, and suppose that X has FDRH (see Definition (2.4.1)(2)). Show that
$$T_t^c(X) \leq \frac{1}{2}\{\mathrm{fd}(X) - \chi\pi(X)\}.$$
(If X is not an agreement space, then FDRH and $\chi\pi(X)$ should be interpreted in terms of the Alexander-Spanier rational homotopy theory of Section 2.6.)

(5.18) ([Gottlieb, 1986]) Let G be a compact Lie group acting smoothly and in an orientation preserving way on a closed connected oriented smooth manifold M of dimension n.

(1) Show that there is a class $\gamma \in H_G^n(M;\mathbb{Z})$ such that $i^*(\gamma) = e(M)$, the Euler class of M, where $i : M \to M_G$ is the inclusion of the fibre as usual.

(2) If $G = (\mathbb{Z}_p)^r$, where p is any prime number, and if the rank of the action is ρ, then deduce that $p^\rho | \chi(M)$; (cf. Exercise (4.11).)

(5.19) ([Gottlieb, 1986]) Let G be a compact Lie group acting smoothly on a closed connected oriented smooth manifold M. For any $j \geq 0$ show that there is a class $\gamma_j \in H_G^{4j}(M;\mathbb{Z})$ such that $i^*(\gamma_j) = p_j(M)$, the Pontryagin class of M of degree $4j$. In particular show also the following.

(1) If G is finite and if the action of G on M is not orientation preserving then all Pontryagin numbers of M are zero.

(2) If $G = (\mathbb{Z}_p)^r$, where p is any prime number, if the action is orientation preserving, and if the rank of the action is ρ, then $p^\rho | \pi$, for any Pontryagin number π of M.

Appendix A

Commutative algebra

In this appendix we summarize briefly some of the algebra needed in the rest of the book. We give proofs in general only for those results which we have not found stated conveniently in one of the main references: otherwise we give a reference to one or more of the standard texts on the subject.

We have included some results which are not used in the rest of the book. In some cases this is intended to put those results which are needed into a more general context. But we have included also a section on Cohen-Macaulay rings and modules: this is intended to make it easier to read certain works on tranformation groups where Cohen-Macaulay theory is used. See, e.g., [Allday, 1979], [Atiyah, 1974], [Bredon, 1974] and [Duflot, 1981]. See also Section 5.5.

On the other hand there are some very useful topics which we have not included. Three which spring to mind are:

(1) Noether's Normalization Lemma: see e.g., [Serre, 1965a], Chapter III, Théorème 2;
(2) The going up and going down (Cohen-Seidenberg) theorems: see e.g., [Matsumura, 1986], §9, or [Serre, 1965a], Chapter III, part A; and
(3) Preparation theorems, Grauert invariants, Gröbner bases, etc.: see e.g., [Chang, Su, 1980].

Unless stated otherwise all rings will be assumed to be commutative with identity, and all modules will be assumed to be unitary.

A.1 Krull dimension

Throughout this section A will denote a commutative ring with identity. For simplicity we shall assume that $A \neq 0$.

(A.1.1) Definitions
(1) Spec(A), the spectrum of A, is the set of prime ideals of A. Max (A) is the set of maximum ideals of A.
(2) For any ideal $\mathfrak{a} \triangleleft A$, let $\sqrt{\mathfrak{a}}$ be the radical of $\mathfrak{a}$: i.e. $\sqrt{\mathfrak{a}} = \{a \in A : a^n \in \mathfrak{a} \text{ for some integer } n \geq 1\}$.
(3) For any ideal $\mathfrak{a} \triangleleft A$, let $V(\mathfrak{a}) = \{\mathfrak{p} \in \text{Spec}(A) : \mathfrak{p} \supseteq \mathfrak{a}\}$. Thus $\sqrt{\mathfrak{a}} = \bigcap_{\mathfrak{p} \in V(\mathfrak{a})} \mathfrak{p}$.
([Matsumura, 1986], §1.)
(4) A chain of prime ideals of A, $\mathfrak{p}_0 \subseteq \mathfrak{p}_1 \subseteq \ldots \subseteq \mathfrak{p}_n$, is said to begin at $\mathfrak{p}_0$ and end at $\mathfrak{p}_n$. The chain is said to have length n if $\mathfrak{p}_i \neq \mathfrak{p}_{i+1}$ for $0 \leq i \leq n-1$.
(5) For $\mathfrak{p} \in \text{Spec}(A)$, the height of $\mathfrak{p}$, denoted ht($\mathfrak{p}$), is defined to be the supremum of the lengths of all chains of prime ideals of A which end at $\mathfrak{p}$.
(6) For $\mathfrak{p} \in \text{Spec}(A)$, the coheight of $\mathfrak{p}$, denoted coht($\mathfrak{p}$), is defined to be the supremum of the lengths of all chains of prime ideals of A which begin at $\mathfrak{p}$.
(7) The Krull dimension of A, denoted dim A, is defined by

$$\dim A := \sup\{\text{ht}(\mathfrak{p}) \; ; \; \mathfrak{p} \in \text{Spec}(A)\} = \sup\{\text{ht}(\mathfrak{m}) : \mathfrak{m} \in \text{Max}(A)\}.$$

(A.1.2) Examples
(1) If k is a field, then $\dim k = 0$.
(2) If A is a principal ideal domain (PID) but not a field, then $\dim A = 1$.
(3) For any $\mathfrak{p} \in \text{Spec}(A)$, $\dim A_\mathfrak{p} = \text{ht}(\mathfrak{p})$ and $\dim A/\mathfrak{p} = \text{coht}(\mathfrak{p})$. (Here, as usual, $A_\mathfrak{p}$ denotes A localised at $\mathfrak{p}$, i.e. A localized with respect to the multiplicative set $A - \mathfrak{p}$.)
(4) For any $\mathfrak{p} \in \text{Spec}(A)$, clearly $\text{ht}(\mathfrak{p}) + \text{coht}(\mathfrak{p}) \leq \dim A$.
The following theorem is crucial.

(A.1.3) Theorem
If A is Noetherian and if $R = A[X_1, \ldots, X_n]$ is a polynomial ring over A in n indeterminates, then

$$\dim R = \dim A + n.$$

([Serre, 1965a], Chapter III, Proposition 13.)

(A.1.4) Definition
For any ideal $\mathfrak{a} \triangleleft A$, let $\operatorname{ht}(\mathfrak{a}) = \inf\{\operatorname{ht}(\mathfrak{p}) : \mathfrak{p} \in V(\mathfrak{a})\}$.

We refer to the next theorem as the Krull Height Theorem. It is also known occasionally as the Krull Altitude Theorem. Many books, however, treat it just as a corollary of Krull's Hauptidealsatz

(A.1.5) Theorem
Suppose that A is Noetherian and that a given ideal $\mathfrak{a} \triangleleft A$ can be generated by n elements: i.e. $\mathfrak{a} = (a_1, \ldots, a_n)$ for some $a_1, \ldots, a_n \in A$. Let $\mathfrak{p}$ be any minimal element of $V(\mathfrak{a})$. Then $\operatorname{ht}(\mathfrak{p}) \leq n$. (And so, in particular, $\operatorname{ht}(\mathfrak{a}) \leq n$.)
([Serre, 1965a], Chapter III, Corollaire 4 of Théorème 1.)

(A.1.6) Theorem
Suppose that A is a finitely generated k-algebra and an integral domain, where k is a field: i.e. there are indeterminates $X_1, \ldots, X_n$ and a prime ideal P of the polynomial ring $k[X_1, \ldots, X_n]$ such that $A \cong k[X_1, \ldots, X_n]/P$. Then, for any $\mathfrak{p} \in \operatorname{Spec}(A)$,

$$\operatorname{ht}(\mathfrak{p}) + \operatorname{coht}(\mathfrak{p}) = \dim A.$$

([Serre, 1965a], Chapter III, Proposition 15.)

(A.1.7) Remarks
(1) With the notation of Theorem (A.1.6), by Theorem (A.1.6), $\dim A = n - \operatorname{ht}(P)$.
(2) Theorem (A.1.6) is valid also if A is any Cohen-Macaulay ring in which all maximal ideals have the same height ([Matsumura, 1986], Theorem 17.4). (See section A.6 for the definition of a Cohen-Macaulay ring.)

The next theorem is sometimes called the Intersection Dimension Formula because of its interpretation in algebraic geometry.

(A.1.8) Theorem
Suppose that $A = k[X_1, \ldots, X_n]$, a polynomial ring over a field k. Let $\mathfrak{p}_1, \mathfrak{p}_2$ be any two prime ideals of A, such that $\mathfrak{p}_1 + \mathfrak{p}_2 \neq A$. Let $\mathfrak{p}$ be minimal in $V(\mathfrak{p}_1 + \mathfrak{p}_2) = V(\mathfrak{p}_1) \cap V(\mathfrak{p}_2)$. Then

$$\operatorname{ht}(\mathfrak{p}) \leq \operatorname{ht}(\mathfrak{p}_1) + \operatorname{ht}(\mathfrak{p}_2).$$

([Matsumura, 1986], Section 33 or [Serre, 1965a], Chapter III, Proposition 17.)

(A.1.9) Remark
Theorem (A.1.8) is also valid more generally when A is any regular ring. ([Serre, 1965a], Chapter V, Théorème 3) (See Section A.4 for the definition of a regular ring.)

A.2 Modules

As before throughout this section A will denote a commutative ring with identity, and we shall assume that $A \neq 0$. All modules will be assumed to be unitary.

(A.2.1) Definitions
Let M be an A-module.
(1) For any $x \in M$, let $\text{ann}(x)$, or, more precisely, $\text{ann}_A(x)$, be defined by $\text{ann}(x) = \{a \in A : ax = 0\}$.
Let $\text{ann}_A(M) = \{a \in A : ax = 0, \text{ for all } x \in M\} = \bigcap_{x \in M} \text{ann}(x)$. ($\text{ann}(x)$ is called the annihilator of x and $\text{ann}_A(M)$ is called the annihilator of M (in A).)
(2) The set of associated prime ideals of M, denoted by $\text{Ass}_A(M)$, is defined to be the set of prime ideals of A which have the form $\text{ann}(x)$ for some $x \in M$. So $\text{Ass}_A(M) = \{\text{ann}(x) : x \in M\} \cap \text{Spec}(A)$.
(3) The support of M, denoted $\text{Supp}(M)$, or, more precisely, $\text{Supp}_A(M)$, is defined to be the set of prime ideals $\mathfrak{p} \triangleleft A$ such that the localized module $M_\mathfrak{p} = A_\mathfrak{p} \otimes_A M \neq 0$. So $\text{Supp}(M) = \{\mathfrak{p} \in \text{Spec}(A) : M_\mathfrak{p} \neq 0\}$
(4) We shall say that an ascending chain of submodules
$$0 = M_0 \subseteq M_1 \subseteq \ldots \subseteq M_n = M$$
is a prime composition series for M if there exist prime ideals $\mathfrak{p}_1, \ldots, \mathfrak{p}_n$ of A such that for each i, $1 \leq i \leq n$, there is an A-module isomorphism $M_i/M_{i-1} \cong A/\mathfrak{p}_i$. We shall say then that $\{\mathfrak{p}_1, \ldots, \mathfrak{p}_n\}$ is a prime composition set for M.
(5) The dimension of M, denoted $\dim_A(M)$, is defined by
$$\dim_A(M) = \dim\left(A/\text{ann}_A(M)\right),$$
where the latter is Krull dimension (Definition (A.1.1)(7)). So
$$\dim_A(M) = \sup\{\text{coht}(\mathfrak{p}) : \mathfrak{p} \in V(\text{ann}_A(M))\}.$$
(6) If E is any set of prime ideals of A, i.e., $E \subseteq \text{Spec}(A)$, then we shall denote by $\min E$, resp. $\max E$, the set of minimal elements, resp. maximal elements, of E with respect to the usual partial ordering provided by set containment. In particular $\text{Max}(A) = \max \text{Spec}(A)$.

(A.2.2) Remark

It is a basic fact that $\operatorname{Max}(A) \neq \emptyset$: one applies Zorn's Lemma to the set of proper ideals of A. On the other hand, the intersection of any descending chain of prime ideals is a prime ideal: and so, applying Zorn's Lemma with the reverse ordering, it follows that $\min \operatorname{Spec}(A) \neq \emptyset$ also.

The following proposition gives some of the basic facts concerning the sets defined above: proofs can be found in [Bourbaki, 1967], Chapter IV, § 1.

(A.2.3) Proposition

Let M be an A-module.

(1) $\mathfrak{p} \in \operatorname{Ass}_A(M)$ *if and only if M contains a submodule isomorphic to $A/\mathfrak{p}$.*
(2) *If $\mathfrak{p} \in \operatorname{Spec}(A)$ and if M is a non-zero submodule of $A/\mathfrak{p}$, then $\operatorname{Ass}_A(M) = \{\mathfrak{p}\}$.*
(3) $\max\{\operatorname{ann}_A(x) : x \in M \text{ and } x \neq 0\} \subseteq \operatorname{Ass}_A(M)$.
(4) *If A is Noetherian, then $\operatorname{Ass}_A(M) \neq \emptyset$ if and only if $M \neq 0$.*
(5) *For any submodule $N \subseteq M$, $\operatorname{Ass}_A(N) \subseteq \operatorname{Ass}_A(M) \subseteq \operatorname{Ass}_A(N) \cup \operatorname{Ass}_A(M/N)$.*
(6) *If M is isomorphic to the direct sum $\bigoplus_{i \in I} M_i$ of a family $\{M_i : i \in I\}$ of A-modules, then $\operatorname{Ass}_A(M) = \bigcup_{i \in I} \operatorname{Ass}_A(M_i)$.*
(7) *If $\mathfrak{p} \in \operatorname{Ass}_A(M)$ and if $\mathfrak{q} \in V(\mathfrak{p})$, then $\mathfrak{q} \in \operatorname{Supp}(M)$.*
(8) *If A is Noetherian and if $\mathfrak{p} \in \operatorname{Supp}(M)$, then there is $\mathfrak{q} \in \operatorname{Ass}_A(M)$ such that $\mathfrak{p} \in V(\mathfrak{q})$.*
(9) *If A is Noetherian, then $\operatorname{Ass}_A(M) \subseteq \operatorname{Supp}(M)$, and $\min \operatorname{Ass}_A(M) = \min \operatorname{Supp}(M)$.*
(10) *If A is Noetherian and if M is finitely generated, then M has a prime composition series.*

(A.2.4) Definition

If $S \subseteq A$ is a multiplicative subset, then we shall let

$$\operatorname{Spec}(A; S) = \{\mathfrak{p} \in \operatorname{Spec}(A) : \mathfrak{p} \cap S = \emptyset\}.$$

It is a basic fact (see [Bourbaki, 1967], Chapter II, § 2) that the function $\mathfrak{p} \to S^{-1}\mathfrak{p}$ gives a bijection

$$\operatorname{Spec}(A; S) \to \operatorname{Spec}(S^{-1}A).$$

And, if M is a finitely generated A-module then

$$S^{-1}\operatorname{ann}_A(M) = \operatorname{ann}_{S^{-1}A}(S^{-1}M).$$

(One always has $S^{-1}\operatorname{ann}_A(M) \subseteq \operatorname{ann}_{S^{-1}A}(S^{-1}M)$, clearly; and the reverse inclusion follows if M is finitely generated.)

(A.2.5) Proposition

Let M be an A-module and let $S \subseteq A$ be a multiplicative subset. Then:

(1) the function $\mathfrak{p} \to S^{-1}\mathfrak{p}$ gives an injection $\mathrm{Ass}_A(M) \cap \mathrm{Spec}(A; S) \to \mathrm{Ass}_{S^{-1}A}(S^{-1}M)$; and

(2) if $\mathfrak{p} \in \mathrm{Spec}(A; S)$, if $\mathfrak{p}$ is finitely generated, and if

$$S^{-1}\mathfrak{p} \in \mathrm{Ass}_{S^{-1}A}(S^{-1}M),$$

then $\mathfrak{p} \in \mathrm{Ass}_A(M)$.

Thus, if A is Noetherian, then $\mathfrak{p} \to S^{-1}\mathfrak{p}$ gives a bijection

$$\mathrm{Ass}_A(M) \cap \mathrm{Spec}(A; S) \to \mathrm{Ass}_{S^{-1}A}(S^{-1}M).$$

([Bourbaki, 1967], Chapter IV, § 1.)

An important theorem is the following, which may also be found in [Bourbaki, 1967], Chapter IV, § 1.

(A.2.6) Theorem

Suppose that A is Noetherian and that M is a finitely generated A-module. Let $\{\mathfrak{p}_1, \ldots, \mathfrak{p}_n\}$ be a prime composition set for M. (See Definition (A.2.1)(4) and Proposition (A.2.3)(10)). Then

$$\mathrm{Ass}_A(M) \subseteq \{\mathfrak{p}_1, \ldots, \mathfrak{p}_n\} \subseteq \mathrm{Supp}(M)$$

and

$$\min \mathrm{Ass}_A(M) = \min\{\mathfrak{p}_1 \ldots, \mathfrak{p}_n\} = \min \mathrm{Supp}(M) = \min V(\mathrm{ann}_A(M)).$$

(A.2.7) Corollary

If A is Noetherian and if M is finitely generated, then

$$\dim_A(M) = \sup\{\mathrm{coht}(\mathfrak{p}) : \mathfrak{p} \in \mathrm{Supp}(M)\}.$$

If, in addition, A satisfies the conditions of Theorem (A.1.6), then

$$\dim_A(M) = \dim A - \inf\{\mathrm{ht}(\mathfrak{p}) : \mathfrak{p} \in \mathrm{Supp}(M)\}$$
$$= \dim A - \inf\{\mathrm{ht}(\mathfrak{p}) : \mathfrak{p} \in \mathrm{Ass}_A(M)\}.$$

(A.2.8) Theorem

Let M and N be A-modules.

(1) If M and N are finitely generated, then

$$\mathrm{Supp}(M \otimes_A N) = \mathrm{Supp}(M) \cap \mathrm{Supp}(N).$$

(2) If A is Noetherian and if M is finitely generated, then

$$\mathrm{Ass}_A(\mathrm{Hom}_A(M, N)) = \mathrm{Supp}(M) \cap \mathrm{Ass}_A(N).$$

([Serre, 1965a], Chapter I, Proposition 10; [Bourbaki, 1967], Chapter IV, §1.)

We turn now to the graded case.

(A.2.9) Definitions

(1) A is said to be graded over the integers, $\mathbb{Z}$, or $\mathbb{Z}$-graded, if there is a direct sum decomposition of A as an abelian group, $A = \bigoplus_{n \in \mathbb{Z}} A_n = \bigoplus_{n=-\infty}^{\infty} A_n$, such that for any $m, n \in \mathbb{Z}$, $a \in A_m$ and $b \in A_n$, $ab \in A_{m+n}$: i.e. $A_m A_n \subseteq A_{m+n}$. The elements of A_n are said to be homogeneous of degree n. In particular the identity $1 \in A$ is homogeneous of degree 0: i.e. $1 \in A_0$.

If $A_m = 0$ for $m < 0$, then we say that A is $\mathbb{N}$-graded, $\mathbb{N} = \{0, 1, 2, \ldots\}$ being the set of non-negative integers.

(2) If A is $\mathbb{Z}$-graded and if M is an A-module, then M is a graded, or $\mathbb{Z}$-graded, A-module if there is a direct sum decomposition of M as an abelian group, $M = \bigoplus_{n \in \mathbb{Z}} M_n = \bigoplus_{n=-\infty}^{\infty} M_n$, such that for any m, $n \in \mathbb{Z}$, $a \in A_m$ and $x \in M_n$, $ax \in M_{m+n}$: i.e. $A_m M_n \subseteq M_{m+n}$.

If $M_m = 0$ for $m < 0$ then we say that M is $\mathbb{N}$-graded.

(3) If A is graded and if $\mathfrak{a}$ is an ideal in A, then $\mathfrak{a}$ is said to be a homogeneous or graded ideal if $\mathfrak{a} = \bigoplus_{n \in \mathbb{Z}} \mathfrak{a} \cap A_n$. So if $a \in A$ and $a = a_1 + \ldots, +a_r$ where $a_i \in A_{n_i}$ for $1 \leq i \leq r$, and integers $n_1 < \ldots < n_r$, and if $\mathfrak{a}$ is homogeneous, then $a \in \mathfrak{a}$ if and only if $a_i \in \mathfrak{a}$ for $1 \leq i \leq r$.

Note that if A is graded and if $\mathfrak{p}$ is a homogeneous ideal of A, then $\mathfrak{p}$ is prime if and only if $ab \in \mathfrak{p}$ implies that $a \in \mathfrak{p}$ or $b \in \mathfrak{p}$ for any homogeneous elements $a, b \in A$.

(4) If A is graded and if M is a graded A-module, then a submodule $N \subseteq M$ is said to be a graded submodule if $N = \bigoplus_{n \in \mathbb{Z}} N \cap M_n$.

(A.2.10) Definition

Suppose that A is graded and that M and N are graded A-modules. A homomorphism $\phi \colon M \to N$ is said to be homogeneous of degree d if $\phi(M_n) \subseteq N_{n+d}$ for all $n \in \mathbb{Z}$.

Now we have the following from [Bourbaki, 1967], Chapter IV, §3.

(A.2.11) Theorem

Suppose that A is graded and that M is a graded A-module.

(1) *For any $\mathfrak{p} \in \mathrm{Ass}_A(M)$, $\mathfrak{p}$ is homogeneous; and $\mathfrak{p} = \mathrm{ann}(x)$ for some homogeneous $x \in M$.*

(2) *For any $\mathfrak{p} \in \mathrm{Ass}_A(M)$, there is a graded submodule $N \subseteq M$ and a homogeneous isomorphism $A/\mathfrak{p} \to N$.*

(3) *If A is Noetherian and if M is finitely generated, then there is a prime decomposition series $0 = M_{(0)} \subseteq \ldots \subseteq M_{(n)} = M$, where each $M_{(i)}$, $0 \leq i \leq n$, is a graded submodule of M; and for each i, $1 \leq i \leq n$, there is a homogeneous prime ideal $\mathfrak{p}_i$ in A, and a*

homogeneous isomorphism
$$\phi_i : A/\mathfrak{p}_i \to M_{(i)}/M_{(i-1)}.$$
Furthermore, if M is $\mathbf{N}$-graded, then each ϕ_i has non-negative degree. (Note that $M_{(i)}$, $0 \leq i \leq n$, is a graded submodule of M, not to be confused with M_i, the set of homogeneous elements of degree i.)

A.3 Primary decomposition

As before, throughout this section A will be a commutative ring with identity, and we shall assume that $A \neq 0$. Again all modules are assumed to be unitary.

We shall state the main results concerning primary decomposition for the general case of submodules, commenting as we go along on the important special case of primary decomposition of ideals.

(A.3.1) Definition
Let M be an A-module. A submodule $Q \subseteq M$ is said to be primary if $\text{Ass}_A(M/Q)$ consists of a single element. If $\text{Ass}_A(M/Q) = \{\mathfrak{p}\}$, then Q is said to be a $\mathfrak{p}$-primary submodule of M.

In particular, if $\mathfrak{q}$ is an ideal of A, then $\mathfrak{q}$ is said to be a primary ideal, or more precisely, a $\mathfrak{p}$-primary ideal if $\text{Ass}_A(A/\mathfrak{q}) = \{\mathfrak{p}\}$.

(A.3.2) Proposition
Suppose that A is Noetherian and that M is an A-module. Then a submodule $Q \subseteq M$ is primary if and only if $Q \neq M$ and, for any $a \in A$, either

(i) $a \in \sqrt{\text{ann}(x)}$ *for all $x \in M/Q$: or,*

(ii) $ax \neq 0$ *for all non-zero $x \in M/Q$; i.e. $a \notin \text{ann}(x)$, for all non-zero $x \in M/Q$.*

Furthermore, if Q is primary, then $\text{Ass}_A(M/Q) = \{\mathfrak{p}\}$, where $\mathfrak{p} = \bigcap_{x \in M/Q} \sqrt{\text{ann}(x)}$.
([Bourbaki, 1967], Chapter IV, § 2.)

(A.3.3) Corollary
If A is Noetherian and if $\mathfrak{q}$ is an ideal in A, then $\mathfrak{q}$ is primary if and only if $\mathfrak{q} \neq A$ and, for any $a, b \in A$, if $ab \in \mathfrak{q}$ and $b \notin \mathfrak{q}$, then $a \in \sqrt{\mathfrak{q}}$.

Furthermore, if $\mathfrak{q}$ is a primary ideal of A, then $\sqrt{\mathfrak{q}}$ is a prime ideal of A, $\mathfrak{p}$, say, and $\mathfrak{q}$ is $\mathfrak{p}$-primary. In particular $\text{Ass}_A(A/\mathfrak{q}) = \{\sqrt{\mathfrak{q}}\}$.

(A.3.4) Remark
The condition in Corollary (A.3.3) is sometimes taken as the definition of a primary ideal. That is an ideal $\mathfrak{q} \triangleleft A$ is often defined to be primary if $\mathfrak{q} \neq A$, and, if, for any $a, b \in A$, if $ab \in \mathfrak{q}$ and $b \notin \mathfrak{q}$, then $a \in \sqrt{\mathfrak{q}}$. This is equivalent to saying that $\mathfrak{q}$ is primary if $\mathfrak{q} \neq A$, and, if, for any $a, b \in A$, if $ab \in \mathfrak{q}$ and $a \notin \sqrt{\mathfrak{q}}$, then $b \in \mathfrak{q}$. It follows that $\sqrt{\mathfrak{q}}$ is prime. And it follows easily that $\mathrm{Ass}_A(A/\mathfrak{q}) = \{\sqrt{\mathfrak{q}}\}$, if $\mathrm{Ass}_A(A/\mathfrak{q}) \neq \emptyset$.

(A.3.5) Proposition
Suppose that A is Noetherian and that M is an A-module. Let $\mathfrak{p} \in \mathrm{Spec}(A)$, and suppose that $Q_1, \ldots, Q_n$ are $\mathfrak{p}$-primary submodules of M, where $n \geq 1$. Then $\bigcap_{i=1}^n Q_i$ is a $\mathfrak{p}$-primary submodule of M.
([Bourbaki, 1967], Chapter IV, § 2.)

(A.3.6) Proposition
Suppose that A is Noetherian and that $S \subseteq A$ is a multiplicative subset. Let $Q \subseteq M$ be a submodule and let $\mathfrak{p} \in \mathrm{Spec}(A)$. Let $i_S \colon M \to S^{-1}M$ be the localization homomorphism. Then

(1) *if $\mathfrak{p} \cap S \neq \emptyset$, i.e. $\mathfrak{p} \notin \mathrm{Spec}(A; S)$, and if Q is a $\mathfrak{p}$-primary submodule of M, then $S^{-1}Q = S^{-1}M$: and*

(2) *if $\mathfrak{p} \cap S = \emptyset$, i.e. $\mathfrak{p} \in \mathrm{Spec}(A; S)$, then Q is a $\mathfrak{p}$-primary submodule of M if and only if $Q = i_S^{-1}(Q')$ for some $S^{-1}\mathfrak{p}$-primary sub-$S^{-1}A$-module $Q' \subseteq S^{-1}M$; and in the latter case $Q' = S^{-1}Q$.*

([Bourbaki, 1967], Chapter IV, § 2.)

(A.3.7) Definitions
(1) Let M be an A-module and let $N \subseteq M$ be a submodule. A primary decomposition of N in M is a finite set of primary submodules of M, $Q_1, \ldots, Q_n$, say, such that $N = \bigcap_{i=1}^n Q_i$.

In particular, a primary decomposition of an ideal $\mathfrak{a} \triangleleft A$ is a finite set of primary ideals of A, $\mathfrak{q}_1, \ldots, \mathfrak{q}_n$, say, such that $\mathfrak{a} = \bigcap_{i=1}^n \mathfrak{q}_i$. In this case, $V(\mathfrak{a}) = \bigcup_{i=1}^n V(\mathfrak{p}_i)$, where $\mathfrak{p}_i = \sqrt{\mathfrak{q}_i}$ for $1 \leq i \leq n$.

(2) A primary decomposition $N = \bigcap_{i=1}^n Q_i$ as in (1) is said to be reduced if

(i) for any j, $1 \leq j \leq n$, $\bigcap_{i=1, i \neq j}^n Q_i \not\subseteq Q_j$; and

(ii) if $\mathrm{Ass}_A(M/Q_i) = \{\mathfrak{p}_i\}$, then $\mathfrak{p}_1, \ldots, \mathfrak{p}_n$ are n distinct prime ideals.

(A.3.8) Proposition
Suppose that A is Noetherian. Let M be an A-module and let $N \subseteq M$ be a submodule. Suppose that $N = \bigcap_{i=1}^n Q_i$ is a primary decomposition of N. Let $\mathrm{Ass}_A(M/Q_i) = \{\mathfrak{p}_i\}$ for $1 \leq i \leq n$. Then this primary

decomposition is reduced if and only if $\mathfrak{p}_1,\ldots,\mathfrak{p}_n$ are n distinct prime ideals, and $\mathfrak{p}_i \in \mathrm{Ass}_A(M/N)$ for $1 \leq i \leq n$. Furthermore, if the primary decomposition is reduced, then

(1) $\mathrm{Ass}_A(M/N) = \{\mathfrak{p}_1,\ldots,\mathfrak{p}_n\} = \bigcup_{i=1}^n \mathrm{Ass}_A(M/Q_i)$;
(2) for any j, $1 \leq j \leq n$, $\mathrm{Ass}_A(Q_j/N) = \{\mathfrak{p}_1,\ldots,\mathfrak{p}_n\} - \{\mathfrak{p}_j\}$; and
(3) if $\mathfrak{p}_i \in \min \mathrm{Ass}_A(M/N)$, then $Q_i = i_{\mathfrak{p}_i}^{-1}(N_{\mathfrak{p}_i})$, where $i_{\mathfrak{p}_i}: M \to M_{\mathfrak{p}_i}$ is the localization homomorphism. ([Bourbaki, 1967], Chapter IV, § 2.)

(A.3.9) Remark

Suppose that A is Noetherian. If M is an A-module, $N \subseteq M$ a submodule, and $N = \bigcap_{i=1}^n Q_i$ a reduced primary decomposition of N in M, then $Q_1,\ldots,Q_n$ are called the primary components of N (in M). If $\mathrm{Ass}_A(M/Q_i) = \{\mathfrak{p}_i\}$, for $1 \leq i \leq n$, then Propostion (A.3.8) (1) shows that $\mathfrak{p}_1,\ldots,\mathfrak{p}_n$ are uniquely determined as the elements of $\mathrm{Ass}_A(M/N)$. Furthermore, by Proposition (A.3.8) (3), for each $\mathfrak{p}_i \in \min \mathrm{Ass}_A(M/N)$, the corresponding primary component Q_i is also uniquely determined by N. If $\mathfrak{p}_j$ is a non-minimal member of $\mathrm{Ass}_A(M/N)$, then $\mathfrak{p}_j$ is said to be embedded.

In the case of an ideal $\mathfrak{a} \triangleleft A$ and a reduced primary decomposition $\mathfrak{a} = \bigcap_{i=1}^n \mathfrak{q}_i$, then the ideals $\mathfrak{p}_i = \sqrt{\mathfrak{q}_i}$, for $1 \leq i \leq n$, are uniquely determined, and so is a primary component $\mathfrak{q}_j$ if $\mathfrak{p}_j$ is minimal. Clearly $\sqrt{\mathfrak{a}} = \bigcap_{i=1}^n \mathfrak{p}_i = \bigcap_{\mathfrak{p} \in \min \mathrm{Ass}(A/\mathfrak{a})} \mathfrak{p}$; and $\min \mathrm{Ass}(A/\mathfrak{a}) = \min V(\mathfrak{a})$.

The main existence theorem for primary decompositions is the following, which is sometimes called the Lasker-Noether Primary Decomposition Theorem.

(A.3.10) Theorem

Suppose that A is Noetherian. Let M be a finitely generated A-module and let $N \subseteq M$ be a submodule. Then, for each $\mathfrak{p} \in \mathrm{Ass}_A(M/N)$, there is a $\mathfrak{p}$-primary submodule $Q(\mathfrak{p}) \subseteq M$, such that $N = \bigcap_{\mathfrak{p} \in \mathrm{Ass}_A(M/N)} Q(\mathfrak{p})$ is a reduced primary decomposition of N in M.

In particular every ideal $\mathfrak{a} \triangleleft A$ has a reduced primary decomposition $\mathfrak{a} = \bigcap_{i=1}^n \mathfrak{q}_i$, where $\{\sqrt{\mathfrak{q}_i} : 1 \leq i \leq n\} = \mathrm{Ass}_A(A/\mathfrak{a})$. (The cases where $N = M$ or $\mathfrak{a} = A$ are considered to be trivial primary decompositions.)
([Bourbaki, 1967], Chapter IV, § 2.)

(A.3.11) Definition

If A is Noetherian and if $\mathfrak{a}$ is an ideal in A, then the elements of $\mathrm{Ass}_A(A/\mathfrak{a})$ are called prime divisors of a. They are either minimal or embedded.

(A.3.12) Example

If A is a PID, then the primary ideals are precisely the ideals of the form (p^n) where p is prime and $n \geq 1$. In this case the primary decomposition of ideals is tantamount to unique factorization. In particular all the prime divisors of an ideal in a PID are minimal.

Once more it is useful to consider the graded case. From [Bourbaki, 1967], Chapter IV, § 3, we have the following two results.

(A.3.13) Proposition

Let A be $\mathbb{Z}$-graded and Noetherian. Let M be a graded A-module and let $Q \subseteq M$ be a graded submodule. Then Q is a primary submodule of M if and only if $Q \neq M$ and conditions (i) or (ii) of Proposition (A.3.2) are satisfied for all homogeneous $a \in A$ and all homogeneous non-zero $x \in M/Q$. Furthermore if Q is primary, then $\mathrm{Ass}_A(M/Q) = \{\mathfrak{p}\}$, where $\mathfrak{p}$ is a homogeneous prime ideal of A which is equal to $\bigcap \{\sqrt{\mathrm{ann}(x)} : x \in M/Q, x \text{ homogeneous}\}$.

In particular, a homogeneous ideal $\mathfrak{q} \triangleleft A$ is primary if and only if $\mathfrak{q} \neq A$, and, for any homogeneous $a, b \in A$, if $ab \in \mathfrak{q}$ and $b \notin \mathfrak{q}$, then $a \in \sqrt{\mathfrak{q}}$.

(A.3.14) Theorem

Suppose that A is $\mathbb{Z}$-graded and Noetherian. Let M be a finitely generated graded A-module and let $N \subseteq M$ be a graded submodule. Let $\mathrm{Ass}_A(M/N) = \{\mathfrak{p}_1, \ldots, \mathfrak{p}_n\}$. Then each $\mathfrak{p}_i$, $1 \leq i \leq n$, is a homogeneous prime ideal of A (by Theorem (A.2.11)(1)) and N has a reduced primary decomposition in M, $N = \bigcap_{i=1}^n Q_i$, where, for $1 \leq i \leq n$, Q_i is a graded $\mathfrak{p}_i$-primary submodule of M.

In particular, any homogeneous ideal $\mathfrak{a} \triangleleft A$ has a reduced primary decomposition, $\mathfrak{a} = \bigcap_{i=1}^n \mathfrak{q}_i$, where each $\mathfrak{q}_i$ is a homogeneous primary ideal of A.

A.4 Homological dimension

As usual, throughout this section A will denote a commutative ring with identity, and we shall assume that $A \neq 0$. All modules will be assumed to be unitary.

(A.4.1) Definitions

(1) Let M be an A-module. Then the homological dimension of M, also called the projective dimension of M, denoted $\mathrm{hd}_A(M)$, is defined to be

the supremum of all integers m such that $\operatorname{Ext}_A^m(M,N) \neq 0$ for at least one A-module N.
(2) The global homological dimension of A, denoted gldh(A), as in [Serre, 1965a], is defined to be the supremum of all integers m such that $\operatorname{Ext}_A^m(M,N) \neq 0$ for at least one pair of A-modules M and N. Thus $\operatorname{gldh}(A) = \sup\{\operatorname{hd}_A(M) : M \text{ an } A\text{-module}\}$.

(A.4.2) Example
If $M \neq 0$, then $\operatorname{hd}_A(M) = 0$ if and only if M is a projective A-module.

(A.4.3) Proposition
$\operatorname{gldh}(A) = \sup\{\operatorname{hd}_A(M) : M \text{ is a finitely generated } A\text{-module}\}$
([Serre, 1965a], Chapter IV, Proposition 16, Corollaire (Auslander).)

(A.4.4) Theorem
Let M be an A-module. Then the following statements are equivalent.
(1) $\operatorname{hd}_A(M) \leq n$.
(2) *For any exact sequence of A-modules $0 \to P_n \to P_{n-1} \to \cdots \to P_0 \to M \to 0$ such that P_i is projective for $0 \leq i \leq n-1$, P_n is projective also.*
(3) *There is a projective resolution of M of length n: i.e. there is an exact sequence $0 \to P_n \to P_{n-1} \to \cdots \to P_0 \to M \to 0$ such that P_i is projective for $0 \leq i \leq n$.*

([Cartan, Eilenberg, 1956], Chapter VI, Proposition 2.1)

(A.4.5) Proposition
Suppose that A is Noetherian and that M is a finitely generated A-module. Then

(1)
$$\operatorname{hd}_A(M) = \sup\{\operatorname{hd}_{A_\mathfrak{p}}(M_\mathfrak{p}) : \mathfrak{p} \in \operatorname{Spec}(A)\}$$
$$= \sup\{\operatorname{hd}_{A_\mathfrak{m}}(M_\mathfrak{m}) : \mathfrak{m} \in \operatorname{Max}(A)\};$$

and

(2)
$$\operatorname{gldh}(A) = \sup\{\operatorname{gldh}(A_\mathfrak{p}) : \mathfrak{p} \in \operatorname{Spec}(A)\}$$
$$= \sup\{\operatorname{gldh}(A_\mathfrak{m}) : \mathfrak{m} \in \operatorname{Max}(A)\}.$$

([Serre, 1965a], Chapter IV, Proposition 19, Corollaire 2.)

(A.4.6) Definition
A is said to be regular if A is Noetherian and if $\operatorname{gldh}(A) < \infty$.

(A.4.7) Theorem
(1) *If A is regular, then $A_\mathfrak{p}$ is a regular local ring for all $\mathfrak{p} \in \operatorname{Spec}(A)$.*

(2) *If A is Noetherian, if $\dim A < \infty$, $\dim A$ being the Krull dimension of A, and if $A_\mathfrak{m}$ is a regular local ring for all $\mathfrak{m} \in \mathrm{Max}(A)$, then A is regular.*
(3) *If A is regular, then $\mathrm{gldh}(A) = \dim A$.* ([Serre, 1965a], Chapter IV, Proposition 23, and Théorème 9, Corollaire 2.)

(A.4.8) Theorem
If A is regular, then the polynomial ring $A[X_1, \ldots, X_n]$ is regular, and $\mathrm{gldh}(A[X_1, \ldots, X_n]) = \mathrm{gldh}(A) + n$.

In particular, if k is a field, then $k[X_1, \ldots, X_n]$ is regular with global homological dimension n.
([Serre, 1965a], Chapter IV, Proposition 25.)

(A.4.9) Remarks
(1) Theorem (A.4.8), especially the second part, is often referred to as the Theorem on Syzygies: in its original form it is due to Hilbert.
(2) If k is a field, then $\mathrm{gldh}(k) = 0$. If k is a PID, then $\mathrm{gldh}(k) \leq 1$.
(3) Let R be a Noetherian local ring with $\dim R = n < \infty$. Let $\mathfrak{m}$ be the maximal ideal of R. Then R is often defined to be regular if $\mathfrak{m}$ can be generated by a set of n elements. It is then a theorem of Serre that this is equivalent to the condition that $\mathrm{gldh}(R) < \infty$, and that when R is regular then $\mathrm{gldh}(R) = n$. See [Matsumura, 1986], Theorem 19.2, or [Serre, 1965a], Chapter IV, Théorème 9. Note that $\dim R < \infty$ for any Noetherian local ring by the Krull Height Theorem (Theorem (A.1.5)).
(4) Any regular local ring is an integral domain. ([Matsumura, 1986], Theorem 14.3, or [Serre, 1965a], Chapter IV, Théorème 9, Corollaire 3.) In fact a theorem of Auslander and Buchsbaum states that any regular local ring is a unique factorization domain. (See, e.g. [Matsumura, 1986], Theorem 20.3.)
(5) Any finite product of regular rings is regular. Hence there are regular rings which are not integral domains. (See, e.g., the concluding remarks of [Kunz, 1985], Chapter VI, § 1.)

A Noetherian local ring, R, with residue field k, is regular if and only if the associated graded ring of R is isomorphic as a graded k-algebra to a polynomial ring over k. ([Matsumura, 1986], Theorem 17.10.) From this one obtains the following.

(A.4.10) Theorem
Suppose that A is Noetherian and $\mathbb{N}$-graded, and that A_0 is a field, k, say. Let $\mathfrak{m} = \bigoplus_{i=1}^{\infty} A_i$ be the unique maximal homogeneous ideal of A. Suppose that $A_\mathfrak{m}$ is a regular local ring, and let $n = \dim A_\mathfrak{m}$.

Then there are homogeneous elements of positive degree, $X_1, \ldots, X_n \in A$, which are algebraically independent over k, such that
$$A = k[X_1, \ldots, X_n],$$
a graded polynomial ring over k.

In particular, if A is $\mathbb{N}$-graded, and if A_0 is a field k, then A is regular if and only if A is a graded polynomial ring over k with a finite number of generators.

(See e.g., [Matsumura, 1986], Exercise 19.1.)

Combining Theorems (A.4.4) and (A.4.8) with the Quillen-Suslin Theorem ([Quillen, 1976] and [Suslin, 1976]; see also [Kunz, 1985]), which says that any finitely generated projective module over a polynomial ring is free, one gets the following.

(A.4.11) Corollary

Let $A = k[X_1, \ldots, X_n]$, where k is a field, and let M be a finitely generated A-module. Then M has a finitely generated free resolution of length at most n: i.e. there is an exact sequence
$$0 \to F_m \to F_{m-1} \to \ldots \to F_0 \to M \to 0$$
where, for $0 \leq i \leq m$, F_i is a finitely generated free A-module, and $m \leq n$.

Corollary (A.4.11) is an old result in the graded case: see, for example, [Zariski, Samuel, 1960], Chapter VII, §13. In the graded case, however, there is a useful refinement of Corollary (A.4.11).

(A.4.12) Proposition

Let $A = k[X_1, \ldots, X_n]$ where k is a field. Suppose that A is $\mathbb{N}$-graded by assigning a positive degree to each X_i, $1 \leq i \leq n$. We do not assume that $X_1, \ldots, X_n$ all have the same degree. Let $\mathfrak{m} = (X_1, \ldots, X_n)$ be the unique maximal homogeneous ideal of A. Let M be a finitely generated $\mathbb{N}$-graded A-module. Then M has a finitely generated free resolution
$$0 \to F_m \xrightarrow{d_m} \ldots F_i \xrightarrow{d_i} F_{i-1} \to \ldots \xrightarrow{d_1} F_0 \xrightarrow{\epsilon} M \to 0$$
such that
(1) $m \leq n$;
(2) each F_i, $0 \leq i \leq m$, is a finitely generated free $\mathbb{N}$-graded A-module; and
(3) the resolution is minimal in the sense that, for $1 \leq i \leq m$, $d_i(F_i) \subseteq \mathfrak{m} F_{i-1}$.

In particular, $\dim_k \operatorname{Tor}_i^A(M, A/\mathfrak{m}) = rk_A(F_i)$, the rank of F_i.

(A.4.13) Remarks
(1) Proposition (A.4.12) follows because, under the given conditions, if a set of homogeneous elements $\{x_1, \ldots, x_n\} \subseteq M$ maps to a k-vector space basis of $M/\mathfrak{m}M$ under the quotient map, then $\{x_1, \ldots, x_n\}$ generates M as an A-module. Indeed M is finitely generated if and only if $\dim_k(M/\mathfrak{m}M) < \infty$; and if M is finitely generated, then $\dim_k(M/\mathfrak{m}M)$ is the minimal possible number of generators of M.

(2) Theorem (A.4.4) and the exactness of localization imply that if A is regular, and if $S \subseteq A$ is any multiplicative subset with $0 \notin S$, then $S^{-1}A$ is regular and $\text{gldh}(S^{-1}A) \leq \text{gldh}(A)$. (Note that if M is any $S^{-1}A$-module, then M and $S^{-1}M_A$ are isomorphic as $S^{-1}A$-modules, where M_A is M viewed as an A-module. So any S_A^{-1}-module is the localization of some A-module. See, e.g., [Bourbaki, 1961], Chapter II, § 2, Proposition 18(ii).)

(3) Corollary (A.4.11) is valid if the phrase 'finitely generated' is omitted throughout. That is, if $A = k[X_1, \ldots, X_n]$ and if M is any A-module, then M has a free resolution of length at most n. This follows as before using a theorem of Bass ([Bass, 1963]), which says that any non-finitely generated projective module over a Noetherian integral domain is free. (Thus one has a Bass-Quillen-Suslin Theorem: If $A = k[X_1, \ldots, X_n]$, where k is any PID, then any projective A-module is free.)

(4) Proposition (A.4.12) gives $\text{Tor}_i^A(M, A/\mathfrak{m}) \cong F_i \otimes_A A/\mathfrak{m}$ under the stated conditions. The Koszul complex, however, gives a more direct way of computing $\text{Tor}_i^A(M, A/\mathfrak{m})$ by providing a simple resolution of the A-module $k \cong A/\mathfrak{m}$. Namely, let $A = k[X_1, \ldots, X_n]$, where k is a field, and let k be viewed as an A-module via the homomorphism $A \to k$ given by $X_i \mapsto 0$ for $1 \leq i \leq n$: i.e. $k \cong A/\mathfrak{m}$, where $\mathfrak{m} = (X_1, \ldots, X_n)$. Now let $s_1, \ldots, s_n$ be indeterminates and let $\wedge(s_1, \ldots, s_n) = \bigoplus_{j=0}^n \wedge^j(s_1, \ldots, s_n)$ be the exterior algebra over k generated by $s_1, \ldots, s_n$. Let

$$F_j = A \otimes_k \wedge^j(s_1, \ldots, s_n).$$

In particular, $F_0 \cong A$ and F_1 is the free A-module generated by $s_1, \ldots, s_n$. Define $d_1: F_1 \to F_0$ by $d_1(s_i) = X_i$ for $1 \leq i \leq n$. Now extend d_1 to be a derivation of $A \otimes_k \wedge(s_1, \ldots, s_n)$. So, for example, for $1 \leq j_1 < \ldots < j_r \leq n$,

$$d_r(1 \otimes s_{j_1} \ldots s_{j_r}) = \sum_{i=1}^r (-1)^{i-1} X_{j_i} \otimes s_{j_1} \ldots \hat{s}_{j_i} \ldots s_{j_r}.$$

It is not hard to see that

$$0 \to F_n \xrightarrow{d_n} F_{n-1} \to \ldots \to F_1 \xrightarrow{d_1} F_0 \to k \to 0$$

is a free resolution of k. Such a resolution is called a Koszul complex. ([Matsumura, 1986], § 16 and Appendix C.)

(A.4.14) Theorem
If A is regular, and if M is a finitely generated A-module, then
$$\mathrm{hd}_A(M) \geq \sup\{\mathrm{ht}(\mathfrak{p}) : \mathfrak{p} \in \mathrm{Ass}_A(M)\}.$$

Proof.
By Theorem (A.4.7)(1) and [Zariski, Samuel, 1960], Chapter VII, Theorem 44, for any $\mathfrak{m} \in \mathrm{Max}(A)$, if $M_\mathfrak{m} \neq 0$,
$$\mathrm{hd}_{A_\mathfrak{m}}(M_\mathfrak{m}) \geq \sup\{\mathrm{ht}(P) : P \in \mathrm{Ass}_{A_\mathfrak{m}}(M_\mathfrak{m})\}.$$
By Proposition (A.2.5),
$\sup\{\mathrm{ht}(P) : P \in \mathrm{Ass}_{A_\mathfrak{m}}(M_\mathfrak{m})\} = \sup\{\mathrm{ht}(\mathfrak{p}) : \mathfrak{p} \in \mathrm{Ass}_A(M) \text{ and } \mathfrak{p} \subseteq \mathfrak{m}\}.$
So the result follows from Proposition (A.4.5)(1). □

(A.4.15) Corollary
Suppose that A is regular, and that all maximal ideals of A have the same height: e.g., $A = k[X_1, \ldots, X_n]$, where k is a field. Let M be a finitely generated A-module. Then
$$\begin{aligned}
\dim_A(M) &= \dim A - \inf\{\mathrm{ht}\,\mathfrak{p} : \mathfrak{p} \in \mathrm{Ass}_A(M)\} \\
&\geq \dim A - \sup\{\mathrm{ht}\,\mathfrak{p} : \mathfrak{p} \in \mathrm{Ass}_A(M)\} \\
&\geq \dim A - \mathrm{hd}_A(M).
\end{aligned}$$

Proof.
We shall see in Section A.6 that any regular ring is Cohen-Macaulay; and so Theorem (A.1.6) applies. The result now follows from Corollary (A.2.7) and Theorem (A.4.14). □

(A.4.16) Remarks
(1) Theorem (A.1.6) implies that all maximal ideals in $k[X_1, \ldots, X_n]$, where k is a field, have the same height.
(2) In [Zariski, Samuel, 1960], homological dimension is called cohomological dimension. This is perhaps the most appropriate name in view of Definition (A.4.1)(1) and the analogous definition of the sheaf-theoretic cohomology dimension of a topological space.

A.5 Regular sequences

Again, throughout this section, A will denote a commutative ring with identity; and we shall assume that $A \neq 0$. All modules will be assumed to be unitary.

(A.5.1) Definitions
Let M be an A-module.

(1) If $a_1, \ldots, a_n$ is a sequence of elements of A, then let $\mathfrak{a}_i$, for $1 \leq i \leq n$, denote the ideal $(a_1, \ldots, a_i) \triangleleft A$.
(2) $a \in A$ is said to be regular on M, or M-regular, if $ax \neq 0$ for all non-zero $x \in M$. So a is regular on M if and only if the homomorphism $\mu_a \colon M \to M$, given by $\mu_a(x) = ax$, is injective.
(3) A sequence of elements $a_1, \ldots, a_n \in A$ is said to be a M-regular sequence, or just a M-sequence, if
(i) a_1 is regular on M;
(ii) for $1 \leq i \leq n-1$, a_{i+1} is regular on $M/\mathfrak{a}_i M$; and
(iii) $\mathfrak{a}_n M \neq M$
(4) An A-regular sequence is just called a regular sequence (in A).

(A.5.2) Proposition
If M is an A-module and if $a_1, \ldots, a_n$ is a M-regular sequence, then
$$a_1^{m_1}, \ldots, a_n^{m_n}$$
is a M-regular sequence for any positive integers $m_1, \ldots, m_n$.
([Matsumura, 1986], Theorem 16.1)

(A.5.3) Definitions
(1) Let $\mathrm{rad}(A)$ denote the Jacobson radical of A: i.e. $\mathrm{rad}(A) = \bigcap_{\mathfrak{m} \in \mathrm{Max}(A)} \mathfrak{m}$, the intersection of all the maximal ideals of A.
(2) Following [Matsumura, 1986] we shall distinguish the following two cases for an A-module M and a sequence $a_1, \ldots, a_n \in A$: (α) M is finitely generated, and $a_i \in \mathrm{rad}(A)$ for $1 \leq i \leq n$; (β) A and M are $\mathbb{N}$-graded, and for $1 \leq i \leq n$, a_i is homogeneous of positive degree. (See Definitions (A.2.9).)

(A.5.4) Theorem
Suppose that A is Noetherian. Let M be an A-module; and suppose that $a_1, \ldots, a_n \in A$ is a M-regular sequence. Then, in cases (α) or (β), any permutation of the sequence $a_1, \ldots, a_n$ is again a M-regular sequence.
([Matsumura, 1986], Theorem 16.3, Corollary; see also Remark (A.6.10).)

Now for $a_1, \ldots, a_n \in A$ we can introduce a Koszul complex as in Remark (A.4.13)(4). Let B be the free A-module generated by indeterminates $s_1, \ldots, s_n$; and let $\wedge_A(s_1, \ldots, s_n)$ denote the exterior algebra of B. So $\wedge_A(s_1, \ldots, s_n) = \wedge B$. (See, e.g., [Matsumura, 1986], Appendix C.) Then $\wedge_A(s_1, \ldots, s_n) = \bigoplus_{j=0}^n \wedge_A^i(s_i, \ldots, s_n)$, $\wedge_A^0(s_1, \ldots, s_n) = A$, and

$\wedge_A^1(s_1,\ldots,s_n) = B = As_1 \oplus \ldots \oplus As_n$. We can define a differential on $\wedge_A(s_1,\ldots,s_n)$, $d_j: \wedge_A^j(s_1,\ldots,s_n) \to \wedge_A^{j-1}(s_1,\ldots,s_n)$, by setting $d_1(s_i) = a_i$, for $1 \le i \le n$, and then extending d_1 to be a derivation of the A-algebra, $\wedge_A(s_1\ldots,s_n)$. For example, for $1 \le i_1 < \ldots < i_j \le n$, and for $a \in A$, $d_j(as_{i_1}\ldots s_{i_j}) = \sum_{r=1}^{j}(-1)^{r-1} a_{i_r} a s_{i_1}\ldots \hat{s}_{i_r}\ldots s_{i_j}$.

We shall denote this Koszul complex by $K(a_1,\ldots,a_n)$. Unlike the situation in Remark (A.4.13)(4), however, $K(a_1,\ldots,a_n)$ is not necessarily a resolution of $A/\mathfrak{a}_n$, as we shall see below.

If M is an A-module, then let

$$\#K(a_1,\ldots,a_n,M) = M \otimes_A K(a_1,\ldots,a_n)$$

with differential $1_M \otimes d$. So now, for $1 \le i_1 < \ldots < i_j \le n$, and for $x \in m$,

$$d_j(x \otimes s_{i_1}\ldots s_{i_j}) = \sum_{r=1}^{j}(-1)^{r-1} a_{i_r} x \otimes s_{i_1}\ldots \hat{s}_{i_r}\ldots s_{i_j}.$$

Clearly the image of

$$d_1 : K_1(a_1,\ldots,a_n,M) = M \otimes_A B \to K_0(a_1,\ldots,a_n,M) = M$$

is $\mathfrak{a}_n M$; we let

$$K_j(a_1,\ldots,a_n,M) = M \otimes_A K_j(a_1,\ldots,a_n) = M \otimes_A \wedge_A^j(s_1,\ldots,s_n).$$

Let $H_j(a_1,\ldots,a_n,M) = \ker d_j/\operatorname{im}(d_{j+1})$, where d_j is the differential on $K_j(a_1,\ldots,a_n,M)$, and $d_j = 0$ if $j > n$ or $j \le 0$. Abbreviate $H_j(a_1,\ldots,a_n,A)$ by $H_j(a_1,\ldots,a_n)$, just as $K(a_1,\ldots,a_n,A) = K(a_1,\ldots,a_n)$.

Note that $H_0(a_1,\ldots,a_n,M) \cong M/\mathfrak{a}_n M$, and $H_n(a_1,\ldots,a_n,M) \cong \{x \in M : a_1 x = \ldots = a_n x = 0\}$.

(A.5.5) Theorem
Let M be an A-module, and let $a_1,\ldots,a_n \in A$.

(1) *If $a_1,\ldots,a_n$ is a M-regular sequence, then $H_j(a_1,\ldots,a_n,M) = 0$ for $j \ge 1$. In this case then $K(a_1,\ldots,a_n,M)$ is a resolution of the A-module $M/\mathfrak{a}_n M$.*

In particular, if $M = A$, and $a_1,\ldots,a_n$ is a regular sequence, then $K(a_1,\ldots,a_n)$ is a finitely generated free resolution of $A/\mathfrak{a}_n$ of length n.

(2) *Suppose that A and M satisfy cases (α) or (β) of Definition (A.5.3) (2), and assume in case (α) that A is local. Under these conditions, if $M \ne 0$ and if $H_1(a_1,\ldots,a_n,M) = 0$, then $a_1,\ldots,a_n$ is a M-regular sequence.*

([Matsumura, 1986], Theorem 16.5.)

(A.5.6) Definition

Let M be an A-module, and let $I \triangleleft A$ be an ideal. A M-regular sequence $a_1, \ldots, a_n \in A$, resp. $a_1, \ldots, a_n \in I$ is said to be maximal, resp. I-maximal, if $a_1, \ldots, a_n, b$ is not a M-regular sequence for any $b \in A$, resp. $b \in I$. The sequence $a_1, \ldots, a_n$ is said to have length n.

(A.5.7) Theorem

Suppose that A is Noetherian, and let M be a finitely generated A-module. Let $I \triangleleft A$ be an ideal such that $IM \neq M$. Then all I-maximal M-regular sequences have the same length, and this length is $\inf\{n : \operatorname{Ext}_A^n(A/I, M) \neq 0\}$

([Matsumura, 1986], Theorem 16.7.)

(A.5.8) Definition

Let A, M and I be as in Theorem (A.5.7). Then the length of a I-maximal M-regular sequence is called the I-depth of M; and it is denoted $\operatorname{depth}(I, M)$. So

$$\operatorname{depth}(I, M) = \inf\{n : \operatorname{Ext}_A^n(A/I, M) \neq 0\}.$$

If A is local with maximal ideal $\mathfrak{m}$, then the $\mathfrak{m}$-depth of M is just called the depth of M, and it is denoted $\operatorname{depth}_A(M)$. In [Serre, 1965a], $\operatorname{depth}_A(M)$ is called the homological codimension of M, and it is denoted $\operatorname{codh}_A(M)$. Similarly if A and M are $\mathbb{N}$-graded, and if A_0 is a field, then the $\mathfrak{m}$-depth of M is just called the depth of M and denoted $\operatorname{depth}_A(M)$ or $\operatorname{codh}_A(M)$, where $\mathfrak{m} = \bigoplus_{i=1}^{\infty} A_i$, the unique maximal homogeneous ideal of A.

(A.5.9) Theorem

Suppose that A is Noetherian and that M is a finitely generated A-module. Let $I \triangleleft A$ be an ideal such that $IM \neq M$. Then

(1) *if $J \triangleleft A$ is any ideal such that $V(J) = V(I)$, then $\operatorname{depth}(J, M) = \operatorname{depth}(I, M)$; and*

(2) *if $I = (a_1, \ldots, a_n)$, then*

$$\operatorname{depth}(I, M) = n - \sup\{j : H_j(a_1, \ldots, a_n, M) \neq 0\}.$$

([Matsumura, 1986], Theorems 16.6 and 16.8)

(A.5.10) Theorem

If A is a regular local ring, and if M is a non-zero finitely generated A-module, then

$$\operatorname{hd}_A(M) + \operatorname{depth}_A(M) = \dim A.$$

([Serre, 1965a], Chapter IV, Prop. 21. See also Remark (A.6.19)(1).)

(A.5.11) Remark
If A is a Noetherian local ring, and if M is a non-zero finitely generated A-module, then
$$\operatorname{depth}_A(M) \leq \inf\{\operatorname{coht}(\mathfrak{p}) : \mathfrak{p} \in \operatorname{Ass}_A(M)\}.$$
([Serre, 1965a], Chapter IV, Proposition 7.)
An elementary result is the following. (See, e.g., [Matsumura, 1986], Theorem 6.1.)

(A.5.12) Proposition
Suppose that A is Noetherian, and let M be a non-zero A-module. Then $a \in A$ is regular on M if and only if a does not belong to any associated prime ideal of M, i.e. $a \notin \bigcup_{\mathfrak{p} \in \operatorname{Ass}_A(M)} \mathfrak{p}$.

(A.5.13) Corollary
Suppose that $A = k[X_1, \ldots, X_n]$ where k is an infinite field. Let M be a finitely generated A-module. Let $\mathfrak{m} = (X_1, \ldots, X_n) \triangleleft A$: and let $L(X_1, \ldots, X_n)$ be the k-linear subspace of A spanned by $(X_1, \ldots, X_n)$; i.e. $L(X_1, \ldots, X_n)$ is the space of homogeneous linear polynomials. Suppose that $\mathfrak{m}M \neq M$, and let $\operatorname{depth}(\mathfrak{m}, M) = r$. Then there is a M-regular sequence $a_1, \ldots, a_r \in L(X_1, \ldots, X_n)$.

Proof.
The proof is by induction on r. If $r \geq 1$, then $\mathfrak{m} \notin \operatorname{Ass}_A(M)$. Since k is infinite, $L(X_1, \ldots, X_n) \not\subseteq \bigcup_{\mathfrak{p} \in \operatorname{Ass}_A(M)} \mathfrak{p}$. (If k in an infinite field, then no k-vector-space is the union of a finite number of proper linear subspaces.) Thus there is a M-regular $a_1 \in L(X_1, \ldots, X_n)$. By Theorem (A.5.7), there is a M-regular sequence $a_1, c_2, \ldots, c_r \in \mathfrak{m}$. Now $c_2, \ldots, c_r$ is a $M/a_1 M$-regular sequence in $\mathfrak{m}$; and so, by induction, there is a $M/a_1 M$-regular sequence $a_2, \ldots, a_r \in L(X_1, \ldots, X_n)$. □

The following observation is useful.

(A.5.14) Proposition
Let M be an A-module and let $S \subseteq A$ be a multiplicative subset. Suppose that $a_1, \ldots, a_n$ is a M-regular sequence and that $S^{-1}(M/a_n M) \neq 0$. Then, for any $s_1, \ldots, s_n \in S$, $a_1/s_1, \ldots, a_n/s_n$ is a $S^{-1}M$-regular sequence.

In particular, if $a_1, \ldots, a_n$ is a regular sequence in A and if $\mathfrak{p}$ is a prime ideal of A such that $\mathfrak{p} \supseteq (a_1,\ldots,a_n)$, then, for any $s_1,\ldots,s_n \in A - \mathfrak{p}$, $a_1/s_1,\ldots,a_n/s_n$ is a regular sequence in $A_\mathfrak{p}$.

A.6 Cohen-Macaulay rings and modules

As before, throughout this section, A will be assumed to be a commutative ring with identity with $A \neq 0$; and all A-modules will be assumed to be unitary.

(A.6.1) Definitions
(1) Suppose that A is a Noetherian local ring, and let M be a finitely generated A-module. Then M is said to be a Cohen-Macaulay A-module if $M = 0$ or $\mathrm{depth}_A(M) = \dim_A(M)$.

If A is a Noetherian local ring, then A is said to be a Cohen-Macaulay local ring if $\mathrm{depth}_A(A) = \dim A$: i.e. if A is a Cohen-Macaulay A-module.

(2) Suppose that A is a Noetherian ring. Then A is said to be a Cohen-Macaulay ring if $A_\mathfrak{m}$ is a Cohen-Macaulay local ring for all $\mathfrak{m} \in \mathrm{Max}(A)$.

Similarly, if A is Noetherian and if M is a finitely generated A-module, then M is said to be a Cohen-Macaulay A-module if $M_\mathfrak{m}$ is a Cohen-Macaulay $A_\mathfrak{m}$-module for all $\mathfrak{m} \in \mathrm{Max}(A)$.

The next theorem follows at once from Theorem (A.5.10).

(A.6.2) Theorem
Any regular ring is Cohen-Macaulay.

(A.6.3) Proposition
If A is Noetherian and M is a Cohen-Macaulay A-module, then $M_\mathfrak{p}$ is a Cohen-Macaulay $A_\mathfrak{p}$-module for all $\mathfrak{p} \in \mathrm{Spec}(A)$.

([Matsumura, 1986], Theorem 17.3(iii).)

(A.6.4) Theorem
Suppose that A is a Cohen-Macaulay local ring with maximal ideal $\mathfrak{m}$. Then

(1) *for any ideal $J \triangleleft A$, if $J \neq A$, then $\mathrm{ht}(J) + \dim A/J = \dim A$; and*
(2) *for any sequence $a_1,\ldots,a_n \in \mathfrak{m}$, the following statements are equivalent:*
    (i) $a_1,\ldots,a_n$ *is a regular sequence;*
    (ii) $\mathrm{ht}(a_1,\ldots,a_i) = i$, *for* $1 \leq i \leq n$;
    (iii) $\mathrm{ht}(a_1,\ldots,a_n) = n$.

([Matsumura, 1986], Theorem 17.4)

(A.6.5) Definition
Suppose that A is Noetherian and that $J \triangleleft A$ is an ideal with $J \neq A$. Then J is said to be unmixed if all the prime divisors of J, i.e. all members of $\mathrm{Ass}_A(A/J)$, have the same height.

(A.6.6) Theorem
Suppose that A is Noetherian. Then A is a Cohen-Macaulay ring if and only if:

(1) *the zero ideal, (0), is unmixed; and*
(2) *for any $n \geq 1$, if $J = (a_1, \ldots, a_n)$ is a proper ideal generated by n elements and if $\mathrm{ht}(J) = n$, then J is unmixed.*

([Matsumura, 1986], Theorem 17.6)

From Theorems (A.6.4) and (A.6.6) and Proposition (A.5.14) one has the following.

(A.6.7) Corollary
Suppose that A is Cohen-Macaulay, and that $a_1, \ldots, a_n \in A$ is a regular sequence. Let $\mathfrak{a} = (a_1, \ldots, a_n)$. Then $\mathrm{ht}(\mathfrak{a}) = n$ and $\mathfrak{a}$ is unmixed.

(A.6.8) Remarks
(1) Theorem (A.6.4)(1) implies Remark (A.1.7)(2).
(2) Under the conditions of Corollary (A.6.7) it follows also that $\mathfrak{a}^m$ is unmixed for any integer $m \geq 1$. Furthermore, $A/\mathfrak{a}^m$ is a Cohen-Macaulay ring for all $m \geq 1$. (See [Matsumura, 1986], Exercise 17.4)

Theorem (A.6.4)(2) gives the converse of Corollary (A.6.7) when A is local. There is also a converse in the graded case, as follows.

(A.6.9) Proposition
Let A be an $\mathbb{N}$-graded Cohen-Macaulay ring such that A_0 is a field. Let $a_1, \ldots, a_n \in A$ be homogeneous elements of positive degree. Suppose that $\mathrm{ht}(a_1, \ldots, a_n) = n$. Then $a_1, \ldots, a_n$ is a regular sequence.

Proof.
Let $\mathfrak{m} = \bigoplus_{i=1}^{\infty} A_i$, the unique maximal ideal of A: and let $\mathfrak{a}_i = (a_1, \ldots, a_i)$ for $1 \leq i \leq n$, and let $\mathfrak{a}_0 = (0)$. If $a_1, \ldots, a_n$ is not regular, then for some i, $1 \leq i \leq n$, a_i is in some associated prime of $A/\mathfrak{a}_{i-1}$; see Proposition (A.5.12). But $A/\mathfrak{a}_{i-1}$ is a graded A-module, and so, by Theorem (A.2.11)(1), all associated primes of $A/\mathfrak{a}_{i-1}$ are homogeneous

and, hence, contained in $\mathfrak{m}$. Furthermore, by Proposition (A.2.5), localization at $\mathfrak{m}$ gives a bijection $\mathrm{Ass}_A(A/\mathfrak{a}_{i-1}) \to \mathrm{Ass}_{A_\mathfrak{m}}((A/\mathfrak{a}_{i-1})_\mathfrak{m})$. It follows that $a_1/1, \ldots, a_n/1$ is not a regular sequence in $A_\mathfrak{m}$.

So, by Theorem (A.6.4)(2), $\mathrm{ht}(a_1/1, \ldots, a_n/1) < n$. (Note that the height is at most n by the Krull Height Theorem.) Thus $\mathrm{ht}(a_1, \ldots, a_n) < n$. □

(A.6.10) Remark

Proposition (A.6.9) is not valid without the assumptions that A is graded and $a_1, \ldots, a_n$ are homogeneous of positive degree. In $k[X, Y, Z]$, let $a_1 = XY - X$, $a_2 = Y$ and $a_3 = YZ - Z$. Then $\mathrm{ht}(a_1, a_2, a_3) = \mathrm{ht}(a_1, a_3, a_2) = \mathrm{ht}(X, Y, Z) = 3$: and a_1, a_2, a_3 is a regular sequence, but a_1, a_3, a_2 is not.

Corresponding to Theorem (A.4.8) there is the following.

(A.6.11) Theorem

If A is Cohen-Macaulay, then the polynomial ring $A[X_1, \ldots, X_n]$ is also Cohen-Macaulay.

([Matsumura, 1986], Theorem 17.7.)

(A.6.12) Lemma

Suppose that A is a regular ring in which all maximal ideals have the same height. Let M be a non-zero finitely generated A-module. Then:
(1) $\dim_A(M) \geq \dim A - \mathrm{hd}_A(M)$;
(2) *if $\dim_A(M) = \dim A - \mathrm{hd}_A(M)$, then M is a Cohen-Macaulay A-module; and, for any $\mathfrak{p} \in \mathrm{Ass}_A(M)$, $\mathrm{ht}(\mathfrak{p}) = \mathrm{hd}_A(M)$; and*
(3) *if M is a Cohen-Macaulay A-module, then*
  (i) $\mathrm{hd}_A(M) = \sup\{\mathrm{ht}(\mathfrak{p}) : \mathfrak{p} \in \mathrm{Ass}_A(M)\}$, *and*
  (ii) *all the associated primes of M are minimal: i.e. $\mathrm{Ass}_A(M) = \min \mathrm{Ass}_A(M)$.*

Proof.

(1) This has already appeared as Corollary (A.4.15).

(2) This follows because $\mathrm{ann}_{A_\mathfrak{m}}(M_\mathfrak{m}) = (\mathrm{ann}_A(M))_\mathfrak{m}$ for any $\mathfrak{m} \in \mathrm{Max}(A)$; and so, if $M_\mathfrak{m} \neq 0$, by Proposition (A.4.5)(1),
$$\dim_A(M) \geq \dim_{A_\mathfrak{m}}(M_\mathfrak{m}) \geq \dim A_\mathfrak{m} - \mathrm{hd}_{A_\mathfrak{m}}(M_\mathfrak{m})$$
$$= \dim A - \mathrm{hd}_{A_\mathfrak{m}}(M_\mathfrak{m}) \geq \dim A - \mathrm{hd}_A(M).$$
Thus, if $\dim_A(M) = \dim A - \mathrm{hd}_A(M)$, then, for any $\mathfrak{m} \in \mathrm{Max}(A)$, either $M_\mathfrak{m} = 0$, or, by Theorem (A.5.10),
$$\dim_{A_\mathfrak{m}}(M_\mathfrak{m}) = \mathrm{depth}_{A_\mathfrak{m}}(M_\mathfrak{m}).$$

The equality of the heights of the associated primes of M follows from Corollary (A.4.15).

To see (3) suppose that M is Cohen-Macaulay, and let $\mathfrak{m} \in \mathrm{Max}(A)$ such that $M_\mathfrak{m} \neq 0$. Then, by Theorem (A.5.10) and the proof of Corollary (A.4.15),

$$\dim_{A_\mathfrak{m}}(M_\mathfrak{m}) = \dim A - \inf\{\mathrm{ht}(\mathfrak{p}) : \mathfrak{p} \in \mathrm{Ass}_A(M) \text{ and} \mathfrak{p} \subseteq \mathfrak{m}\}$$
$$= \dim A - \sup\{\mathrm{ht}(\mathfrak{p}) : \mathfrak{p} \in \mathrm{Ass}_A(M) \text{ and } \mathfrak{p} \subseteq \mathfrak{m}\}$$
$$= \dim A - \mathrm{hd}_{A_\mathfrak{m}}(M_\mathfrak{m}).$$

So
$$\mathrm{hd}_{A_\mathfrak{m}}(M_\mathfrak{m}) = \sup\{\mathrm{ht}(\mathfrak{p}) : \mathfrak{p} \in \mathrm{Ass}_A(M) \text{ and } \mathfrak{p} \subseteq \mathfrak{m}\}$$
$$= \inf\{\mathrm{ht}(\mathfrak{p}) : \mathfrak{p} \in \mathrm{Ass}_A(M) \text{ and } \mathfrak{p} \subseteq \mathfrak{m}\}.$$

The result follows by Proposition (A.4.5) (1). □

(A.6.13) Proposition

Suppose that A is Noetherian and $\mathbb{N}$-graded, and that A_0 is a field. Let $\mathfrak{m} = \bigoplus_{i=1}^\infty A_i$. Then

(1) $\dim A = \mathrm{ht}(\mathfrak{m})$; and
(2) *if A is Cohen-Macaulay and M is a non-zero finitely generated graded A-module, then $\dim_A(M) = \dim_{A_\mathfrak{m}}(M_\mathfrak{m})$.*

Proof.

(1) This follows from [Matsumura, 1986], Exercise 13.6.

(2) To see (2), one has from Theorem (A.6.4)(1),
$$\dim_{A_\mathfrak{m}}(M_\mathfrak{m}) - \dim(A_\mathfrak{m}/(\mathrm{ann}_A(M))_\mathfrak{m}) = \dim A_\mathfrak{m} - \mathrm{ht}(\mathrm{ann}_A(M))_\mathfrak{m}$$
$$= \dim A - \mathrm{ht}(\mathrm{ann}_A(M)).$$
Now let $\mathfrak{p} \in V(\mathrm{ann}_A(M))$ have maximal coheight. Then
$$\dim_{A_\mathfrak{m}}(M_\mathfrak{m}) \geq \dim A - \mathrm{ht}(\mathfrak{p}) \geq \mathrm{coht}(\mathfrak{p}) = \dim_A(M).$$
But, clearly,
$$\dim_{A_\mathfrak{m}}(M_\mathfrak{m}) \leq \dim_A(M).$$

□

(A.6.14) Theorem

Suppose that $A = k[X_1, \ldots, X_n]$, where k is a field, and suppose that A is $\mathbb{N}$-graded by assigning a positive degree to each X_i, $1 \leq i \leq n$. We do not assume that $X_1, \ldots, X_n$ all have the same degree. Let $\mathfrak{m} = (X_1, \ldots, X_n)$. Let M be a non-zero finitely generated graded A-module. Then the following three statements are equivalent.

(1) $\dim_A(M) = \dim A - \mathrm{hd}_A(M)$.
(2) M *is a Cohen-Macaulay A-module.*
(3) $M_\mathfrak{m}$ *is a Cohen-Macaulay $A_\mathfrak{m}$-module.*

Proof.
(1) $\Rightarrow$ (2) by Lemma (A.6.12)(2).
(2) $\Rightarrow$ (3) by definition.
Suppose (3). Then
$$\dim_{A_\mathfrak{m}}(M_\mathfrak{m}) = \dim A_\mathfrak{m} - \mathrm{hd}_{A_\mathfrak{m}}(M_\mathfrak{m})$$
by Theorem (A.5.10). By Proposition (A.6.13),
$$\dim A_\mathfrak{m} = \dim A,$$
and
$$\dim_{A_\mathfrak{m}}(M_\mathfrak{m}) = \dim_A(M).$$
Hence, to show the implication (3) $\Rightarrow$ (1), it is enough to show that $\mathrm{hd}_{A_\mathfrak{m}}(M_\mathfrak{m}) = \mathrm{hd}_A(M)$. Let $h = \mathrm{hd}_A(M)$ and let
$$0 \to F_h \to \ldots \to F_1 \to F_0 \to M \to 0$$
be a minimal free resolution of M as in Proposition (A.4.12). View k as an A-module and an $A_\mathfrak{m}$-module via the isomorphisms $k \cong A/\mathfrak{m} \cong A_\mathfrak{m}/\mathfrak{m}_\mathfrak{m}$. Then, localizing the resolution at $\mathfrak{m}$ one has, $\mathrm{Ext}^h_{A_\mathfrak{m}}(M_\mathfrak{m}, k) = \mathrm{Hom}_{A_\mathfrak{m}}((F_h)_\mathfrak{m}, k) \neq 0$. Since $\mathrm{hd}_{A_\mathfrak{m}}(M_\mathfrak{m}) \leq \mathrm{hd}_A(M)$, the equality follows. $\square$

(A.6.15) Corollary
Under the conditions of Theorem (A.6.14) suppose that M is a Cohen-Macaulay A-module. Then, for any $\mathfrak{p} \in \mathrm{Ass}_A(M)$,
$$\mathrm{ht}(\mathfrak{p}) = \dim A - \dim_A(M) = n - \dim_A(M) = \mathrm{hd}_A(M).$$
(See Lemma (A.6.12)(3).)

A crucial corollary is the following.

(A.6.16) Corollary
Under the conditions of Theorem (A.6.14) suppose that M is a Cohen-Macaulay A-module. Let $N \subseteq M$ be a non-zero, but not necessarily graded, submodule. Then
$$\dim_A(N) = \dim_A(M).$$

Proof.
By Proposition (A.2.3)(5), $\mathrm{Ass}_A(N) \subseteq \mathrm{Ass}_A(M)$. So by Corollary (A.2.7) and Corollary (A.6.15),
$$\dim_A(N) = \dim A - \inf\{\mathrm{ht}(\mathfrak{p}) : \mathfrak{p} \in \mathrm{Ass}_A(N)\}$$
$$= \dim A - \mathrm{hd}_A(M)$$
$$= \dim_A(M).$$
$\square$

(A.6.17) Remark
Under the conditions of Theorem (A.6.14), whether M is Cohen-Macaulay or not, one has, from Theorem (A.5.9)(2) and Proposition (A.4.12),

$$\text{depth}_A(M) = n - \sup\{j : H_j(X_1, \ldots, X_n, M) \neq 0\}$$
$$= n - \sup\{j : \text{Tor}_j^A(A/\mathfrak{m}, M) \neq 0\}$$
$$= n - \text{hd}_A(M).$$

Thus, under these conditions,

$$\dim A - \text{hd}_A(M) = \text{depth}_A(M).$$

Note that in Proposition (A.4.12) one has also that

$$\text{Ext}_A^j(M, A/\mathfrak{m}) = \text{Hom}_A(F_j, k);$$

and so $\dim_k \text{Ext}_A^j(M, k) = rk_A(F_j)$, where $k = A/\mathfrak{m}$.

Note also that under the conditions of Theorem (A.6.14), without loss of generality, M is $\mathbb{N}$-graded: since M is finitely generated and graded there is an integer i_0 such that $M_i = 0$ for all $i < i_0$ and one can raise degrees in M by $-i_0$, if necessary, in order to make M $\mathbb{N}$-graded.

We conclude with a celebrated result and some remarks.

(A.6.18) Theorem
Let $A = k[X_1, \ldots, X_n]$, where k is a field, graded as in Theorem (A.6.14). Let B be an $\mathbb{N}$-graded Cohen-Macaulay ring, and suppose that $B_0 = k$. Let $\phi : A \to B$ be a homogeneous k-algebra homomorphism. Suppose that $\dim B = \dim A = n$; and suppose that B is a finitely generated A-module via ϕ. Then B is a free A-module (and ϕ is injective).

Proof.
Let $\phi(X_i) = b_i \in B$, for $1 \leq i \leq n$. Since B is a finitely generated graded A-module, $\dim_k B/\mathfrak{m}B = \dim_k B/(b_1, \ldots, b_n) < \infty$, where $\mathfrak{m} = (X_1, \ldots, X_n)$. So, by the grading, $\sqrt{(b_1, \ldots, b_n)} = \bigoplus_{i=1}^{\infty} B_i$. Hence, by Proposition (A.6.13)(1), $\text{ht}(b_1, \ldots, b_n) = n$. And so, by Proposition (A.6.9), $b_1, \ldots, b_n$ is a regular sequence in B. Thus $X_1, \ldots, X_n$ is a B-regular sequence.
So $\text{depth}_A(B) \geq n$. Hence, by Remark (A.6.17), $\text{hd}_A(B) = 0$. □

(A.6.19) Remarks
(1) Let A be a Noetherian local ring and suppose that M is a non-zero finitely generated A-module such that $\text{hd}_A(M) < \infty$. Then a theorem

of Auslander and Buchsbaum (see [Matsumura, 1986], Theorem 19.1) states that
$$\mathrm{hd}_A(M) + \mathrm{depth}_A(M) = \mathrm{depth}(A).$$
Thus Theorem (A.5.10) generalizes when A is a Cohen-Macaulay local ring and $\mathrm{hd}_A(M) < \infty$.

(2) Suppose that A is $\mathbb{N}$-graded and Noetherian and let B be a $\mathbb{N}$-graded commutative ring such that $A \subseteq B$. Suppose that $A_0 = B_0 = k$, a field; and suppose that B is a finitely generated A-module. Then, for any finitely generated graded B-module M,
$$\mathrm{depth}_A(M) = \mathrm{depth}_B(M).$$
([Matsumura, 1986], Exercise 16.7)

Clearly, if R is a $\mathbb{N}$-graded Noetherian commutative ring with $R_0 = k$, and if $\phi : R \to A$ is a homogeneous epimorphism of degree 0, then, viewing M as a R-module via ϕ, $\mathrm{depth}_R(M) = \mathrm{depth}_A(M)$, also.

(3) Let A, B, R and M be as in (2). Then
$$\dim_A(M) = \dim_B(M) = \dim_R(M).$$
(See [Matsumura, 1986], § 9, especially Exercise 9.2)

(4) Since Corollary (A.6.16) has important applications, it is useful to see other conditions under which it is valid.

(a) If A is a regular ring in which all maximal ideals have the same height, if M is a non-zero finitely generated A-module, if $\dim_A(M) = \dim A - \mathrm{hd}_A(M)$, and if $N \subseteq M$ is a non-zero submodule, then $\dim_A(N) = \dim_A(M)$.

This follows since $\mathrm{Ass}_A(N) \subseteq \mathrm{Ass}_A(M)$, all associated primes of M have the same height, and so
$$\dim_A(N) = \dim A - \inf\{\mathrm{ht}(\mathfrak{p}) : \mathfrak{p} \in \mathrm{Ass}_A(N)\} = \dim_A(M);$$
see Corollary (A.4.15).

(b) If A is $\mathbb{N}$-graded and Noetherian, if A_0 is a field, if M is a non-zero finitely generated graded A-module, if $\dim_A(M) = \mathrm{depth}_A(M)$, and if $N \subseteq M$ is a submodule such that $N_\mathfrak{m} \neq 0$, where $\mathfrak{m} = \bigoplus_{i=1}^{\infty} A_i$, then $\dim_A(N) = \dim_A(M)$.

To see this, one has from Proposition (A.5.14) and Remark (A.5.11),
$$\mathrm{depth}_A(M) \leq \mathrm{depth}_{A_\mathfrak{m}}(M_\mathfrak{m})$$
$$\leq \inf\{\dim(A_\mathfrak{m}/P) : P \in \mathrm{Ass}_{A_\mathfrak{m}}(M_\mathfrak{m})\}$$
$$\leq \sup\{\dim(A_\mathfrak{m}/P) : P \in \mathrm{Ass}_{A_\mathfrak{m}}(M_\mathfrak{m})\}$$
$$= \dim_{A_\mathfrak{m}}(M_\mathfrak{m})$$
$$\leq \dim_A(M).$$

Moreover
$$\dim_A(N) \geq \dim_{A_\mathfrak{m}}(N_\mathfrak{m}) = \sup\{\dim(A_\mathfrak{m}/P) : P \in \mathrm{Ass}_{A_\mathfrak{m}}(N_\mathfrak{m})\}.$$
So $\dim_A(N) \geq \dim_A(M)$, since $\mathrm{Ass}_{A_\mathfrak{m}}(N_\mathfrak{m}) \subseteq \mathrm{Ass}_{A_\mathfrak{m}}(M_\mathfrak{m})$. But clearly, $\dim_A(N) \leq \dim_A(M)$, since $\mathrm{ann}_A(N) \supseteq \mathrm{ann}_A(M)$.

Note that if $P \in \mathrm{Ass}_{A_\mathfrak{m}}(M_\mathfrak{m})$, then $P = \mathfrak{p}_\mathfrak{m}$ for some $\mathfrak{p} \in \mathrm{Ass}_A(M)$ by Proposition (A.2.5). So if A is a catenary integral domain ([Matsumura, 1986], § 5, p.31), $\dim(A_\mathfrak{m}/P) = \mathrm{ht}(\mathfrak{m}) - \mathrm{ht}(\mathfrak{p}) = \dim A - \mathrm{ht}(\mathfrak{p})$, by Proposition (A.6.13)(1).

By [Matsumura, 1986], Theorem 31.7, Corollary 2, and Theorem 17.9, any finitely generated k-algebra is catenary if k is a field, a PID, or, more generally, any quotient of a Cohen-Macaulay ring.

(5) We have seen that, if k is a field, then $k[X_1, \ldots, X_n]$ is a regular ring in which all maximal ideals have the same height. Let $A = \mathbb{Z}[X_1, \ldots, X_n]$, and let $S \subseteq A$ be the multiplicative subset generated by $X_1, \ldots, X_n$. Then A and $S^{-1}A$ are regular: and all maximal ideals in A and $S^{-1}A$ have the same height, $n + 1$, although this is not obvious.

(6) Let $A = \mathbb{Z}_{(p)}[X]$, where $\mathbb{Z}_{(p)}$ denotes $\mathbb{Z}$ localized at p. Then A is a regular integral domain, but its maximal ideals do not all have the same height. Clearly $\dim A = 2$, but $(pX - 1)$ is a maximal ideal of height 1.

Now let $M = \mathbb{Q}$ viewed as an A-module via the isomorphism $\mathbb{Q} \cong A/(pX-1)$. Then $\dim_A(M) = 0$. But $\mathrm{hd}_A(M) = 1$, from the resolution $0 \to (pX - 1) \to A \to \mathbb{Q} \to 0$. Hence $\dim_A(M) < \dim A - \mathrm{hd}_A(M)$.

A.7 Evaluations and presentations

To begin with we shall let $A = k[X]$ where k is a field, and we shall assume that A is $\mathbb{N}$-graded by assigning a positive degree to X.

(A.7.1) Definitions
(1) For any $\alpha \in k$, let k_α be the field k viewed as an A-module via the homomorphism $k[X] \to k$, $X \to \alpha$. So $k_\alpha \cong k[X]/(X - \alpha)$ as an A-module.

(2) For any A-module M let $M_\alpha = k_\alpha \otimes_A M$. For any A-modules M and N and any homomorphism $f : M \to N$ let $f_\alpha = 1 \otimes_A f : M_\alpha \to N_\alpha$. We shall say that M_α, resp. f_α, is M, resp. f, evaluated at α. Thus evaluation at α is a functor from the category of A-modules to the category of k-vector spaces. Note that $M_\alpha \cong M/(X - \alpha)M$.

(A.7.2) Lemma
If $\alpha \neq 0$ in k, then evaluation at α is an exact functor from the category of $\mathbb{Z}$-graded A-modules to the category of k-vector spaces.

Proof.
From the resolution $0 \to (X - \alpha) \to A \to k_\alpha \to 0$ it follows that $\operatorname{Tor}_1^A(k_\alpha, M) \cong \{y \in M \mid (X - \alpha)y = 0\}$. Clearly this is zero if M is a $\mathbb{Z}$-graded A-module and $\alpha \neq 0$. □

(A.7.3) Lemma
Let M be an $\mathbb{N}$-graded A-module. Then M is finitely generated if and only if $\dim_k M_0 < \infty$. And if M is finitely generated, then $\dim_k M_0$ is equal to the smallest number of elements in any generating set of M.

Let N be another $\mathbb{N}$-graded A-module, and let $f : M \to N$ be a homomorphism (homogeneous of degree 0). Then we also have the following.

(1) *If f is split injective (i.e. f has a left inverse), then f_0 is injective. If f_0 is injective and N is free, then f is split injective.*
(2) *f is surjective if and only if f_0 is surjective.*

Proof.
It is clear from the $\mathbb{N}$-grading that if a set $S \subseteq M$ maps to a basis of the k-vector space $M_0 \cong M/XM$, then S generates M as an A-module. So the first two statements follow easily.

If f is split injective, then clearly f_0 is injective. To prove the rest of (1) suppose that f_0 is injective and that N is free. From the factoring $M \to \operatorname{im} f \to N$ it follows that $M_0 \to (\operatorname{im} f)_0$ is an isomorphism and $(\operatorname{im} f)_0 \to N_0$ is injective. Since $k[X]$ is a PID and N is free, $\operatorname{im} f$ is free, and hence $M \cong \operatorname{im} f \oplus \ker f$. So $(\ker f)_0 = 0$; and thus $\ker f = 0$, since it is $\mathbb{N}$-graded. To show that f is split, it is clearly enough to show that $\operatorname{coker} f$ is free. Since $(\operatorname{im} f)_0 \to N_0$ is injective, and N is free, it follows that $\operatorname{Tor}_1^A(k_0, \operatorname{coker} f) = 0$. But if L is a $\mathbb{N}$-graded A-module such that $\{y \in L \mid Xy = 0\} = 0$, then by considering a minimal generating set for L, it follows easily that L is free: and one need not assume that L is finitely generated for this.

If f is surjective, then clearly f_0 is surjective. If f_0 is surjective, then $(\operatorname{im} f)_0 \to N_0$ is surjective; and so $(\operatorname{coker} f)_0 = 0$. So $\operatorname{coker} f = 0$; and hence f is surjective. □

(A.7.4) Lemma
Let M be a finitely generated $\mathbb{N}$-graded A-module. Let $\alpha \in k$, and suppose that $\alpha \neq 0$. Then

$$\dim_k M_\alpha = \dim_k M_0 - \dim_k \operatorname{Tor}_1^A(k_0, M).$$

Furthermore if N is another finitely generated $\mathbb{N}$-graded A-module, if $f : M \to N$ is a homomorphism (homogeneous of degree 0), if $\alpha \in k$,

and if N is free, then
$$rk(f_0) \leq rk(f_\alpha);$$
rk here stands for the ordinary rank of a k-linear map.

Proof.
Let $0 \to F_1 \to F_0 \to M \to 0$ be a minimal resolution of M in the sense of Proposition (A.4.12). By Lemma (A.7.2),
$$\dim_k M_\alpha = rk_A(F_0) - rk_A(F_1) = \dim_k M_0 - \dim_k \operatorname{Tor}_1^A(k_0, M),$$
by the minimality.

To prove the inequality of ranks, since we are assuming that N is free, there is an exact sequence
$$0 \to \operatorname{Tor}_1^A(k_0, \operatorname{coker} f) \to (\operatorname{im} f)_0 \to N_0 \to (\operatorname{coker} f)_0 \to 0.$$
Hence $rk(f_0) = \dim_k(\operatorname{im} f)_0 - \dim_k \operatorname{Tor}_1^A(k_0, \operatorname{coker} f)$. Since $\operatorname{im} f$ is free, by the first part, $\dim_k(\operatorname{im} f)_\alpha = \dim_k(\operatorname{im} f)_0$. If $\alpha \neq 0$, then $\dim_k(\operatorname{im} f)_\alpha = rk(f_\alpha)$, by Lemma (A.7.2). □

(A.7.5) Remarks
(1) Lemma (A.7.4) implies that $\dim_k M_\alpha = \dim_k M_1$, for any non-zero $\alpha \in k$, and finitely generated $\mathbb{N}$-graded A-module M. Furthermore $\dim_k M_1 = \dim_k M_0$ if and only if M is free.
(2) Let $M = k[X]$, $N = k_0$ and let $q : M \to N$ be the quotient map. Then $q_0 : M_0 \to N_0$ is the identity on k, whereas $q_1 : M_1 \cong k \to N_1 = 0$ is the zero homomorphism. This shows that the assumption that N is free is needed in Lemma (A.7.3)(1) and Lemma (A.7.4).
(3) Lemma (A.7.3) has the following useful generalization. Let k be a field, and let A be a connected $\mathbb{N}$-graded k-algebra. Assume that A is associative but not necessarily commutative. Let $\mathfrak{m} = \bigoplus_{i=1}^\infty A_i$ be the maximal homogeneous ideal of A; and let k_0 be k viewed as an A-module via the isomorphism $A/\mathfrak{m} \cong k$. For left A-modules M, N, and homomorphism $f : M \to N$ define M_0, N_0, and f_0 as before: namely $M_0 = k_0 \otimes_A M$ and $f_0 = 1 \otimes_A f : M_0 \to N_0$. Then Lemma (A.7.3) holds exactly as stated above.

The proof is just the same except for the part which asserts that f is split injective if N is free and f_0 is injective. To see the latter suppose that N is free and f_0 is injective. As before $(\operatorname{im} f)_0 \to N_0$ is injective; and so, since N is free, $\operatorname{Tor}_1^A(k_0, \operatorname{coker} f) = 0$. From [MacLane, 1967], Chapter VII, Lemma 6.2, which holds clearly in this generality, it follows that $\operatorname{coker} f$ is a free A-module. Hence $N \cong \operatorname{im} f \oplus \operatorname{coker} f$, and so $\operatorname{im} f$ is free too. Thus $M \cong \ker f \oplus \operatorname{im} f$; and, since $M_0 \to (\operatorname{im} f)_0$ is an isomorphism, as before, $(\ker f)_0 = 0$. So $\ker f = 0$.

Next we shall review some basic facts concerning commutative graded algebras. Let k be a field and $A = \bigoplus_{i=0}^{\infty} A_i$ be a $\mathbb{N}$-graded k-algebra. We shall assume that A is connected: i.e. $A_0 = k$. We shall assume also that A is commutative in the graded sense: i.e. for any $a \in A_i$, $b \in A_j$, $ba = (-1)^{ij} ab$. (In sections A.1-A.6 all graded rings and algebras were strictly commutative: they were commutative when the grading was forgotten. In the situation here A is strictly commutative if char $k = 2$. And, of course, A is strictly commutative if $A_i = 0$ for all odd i.) Let $M = \bigoplus_{i=1}^{\infty} A_i$, the maximal homogeneous ideal of A.

In the category of connected commutative $\mathbb{N}$-graded k-algebras the free objects are those of the form $k[X_i : i \in I] \otimes \wedge(Y_j : j \in J)$, where $k[X_i : i \in I]$ is the polynomial ring generated by $\{X_i : i \in I\}$ where each X_i, $i \in I$, is homogeneous of positive even degree, and $\wedge(Y_j : j \in J)$ is the exterior algebra generated by $\{Y_j : j \in J\}$ where each Y_j, $j \in J$, is homogeneous of positive odd degree.

(A.7.6) Definition
A is said to be finitely generated if there is a homogeneous epimorphism of k-algebras of degree zero
$$k[X_1, \ldots, X_r] \otimes \wedge(Y_1, \ldots, Y_s) \to A,$$
where each X_i, resp. Y_j, is homogeneous of positive even, resp. odd, degree.

The following is clear from the grading.

(A.7.7) Lemma
A is finitely generated if and only if $\dim_k(M/M^2) < \infty$. Furthermore, if A is finitely generated, and if
$$k[X_1, \ldots, X_r] \otimes \wedge(Y_1, \ldots, Y_s) \to A$$
is an epimorphism as in Definition (A.7.6), such that r and s are minimal, then r, resp. s, is the dimension over k of the even, resp. odd, part of M/M^2.

(A.7.8) Definitions
(1) Suppose that A is finitely generated and let
$$\pi : B = k[X_1, \ldots, X_r] \otimes \wedge(Y_1, \ldots, Y_s) \to A$$
be an epimorphism as in Definition (A.7.6). Let
$$N = (X_1, \ldots, X_r, Y_1, \ldots, Y_s),$$
the maximal homogeneous ideal of B. Let $J = \ker \pi$; J is called the

A.7. Evaluations and presentations

ideal of relations. Clearly $J \subseteq N$. The dimension over k of the even, resp. odd, part of J/NJ is called the number of relations of even, resp. odd, degree.

(2) Somewhat improperly we shall refer to the exact sequence

$$0 \to J \to B \xrightarrow{\pi} A \to 0$$

as a presentation of A. The presentation is said to be minimal if the numbers of generators of B of even and odd degrees are minimal as in Lemma (A.7.7).

The following is also clear from the grading.

(A.7.9) Lemma
A presentation of A

$$0 \to J \to B \to A \to 0,$$

as in Definitions (A.7.8), is minimal if and only if $J \subseteq N^2$.

Now suppose that A is finitely generated, and let

$$0 \to J \to k[X_1,\ldots,X_r] \otimes \wedge(Y_1,\ldots,Y_s) \to A \to 0$$

be a presentation. Then it is easy to see that the number of relations is finite: J is a finitely generated $k[X_1,\ldots,X_r]$-module, and so $\dim_k(J/(X_1,\ldots,X_r)J) < \infty$. We have seen above that the minimal number of generators in a presentation is uniquely determined by A, namely $\dim_k(M/M^2)$: so it would be good to know that the same is true concerning the number of relations. The following proposition verifies this; and it shows that the uniquely determined minimal number of relations occurs in any minimal presentation.

(A.7.10) Proposition
Suppose that A is finitely generated. Let $0 \to J \to B \xrightarrow{\pi} A \to 0$ be a minimal presentation, and let $0 \to I \to C \xrightarrow{\rho} A \to 0$ be any presentation. Let N_B, N_C be the maximal homogeneous ideals of B, C respectively. Then

$$\dim_k(J/N_B J) \leq \dim_k(I/N_C I).$$

Proof.
Since B and C are free there are homogeneous homomorphisms of degree zero $\phi : B \to C$ and $\psi : C \to B$ such that $\rho\phi = \pi$ and $\pi\psi = \rho$. Let $\theta = \psi\phi : B \to B$. Then $\pi\theta = \pi$; and so for any $b \in B$, $\theta(b) - b \in \ker \pi = J$.

Since π is minimal $J \subseteq N_B^2$. Let $B = k[X_1, \ldots, X_r] \otimes \wedge(Y_1, \ldots, Y_s)$ and let $f = f(X_1, \ldots, X_r, Y_1, \ldots, Y_s) \in N_B^n - N_B^{n+1}$, where $n \geq 0$. Then $\theta(f) = f + h$, where $h \in N_B^{n+1}$. Since B is free and f is an arbitrary element of B, it follows that θ is injective. Hence θ is an isomorphism.

Now ϕ and ψ induce homomorphisms $\bar{\phi} : J/N_B J \to I/N_C I$ and $\bar{\psi} : I/N_C I \to J/N_B J$. $\overline{\psi\phi} = \bar{\theta}$ is an isomorphism, since θ^{-1} induces its inverse. So $\bar{\phi}$ is injective; and the result follows. $\square$

(A.7.11) Remark

The minimal number of generators for A, $\dim_k(M/M^2)$, is also equal to $\dim_k \operatorname{Tor}_1^A(k, k)$, where k is viewed as an A-module via the isomorphism $k \cong A/M$. (See, e.g., [Lemaire, 1974], Lemme (1.0.4).) Since the minimal number of relations is also an invariant of A, by Proposition (A.7.10), it is natural to ask if there is a similar intrinsic formula for it. There is: in a certain sense the minimal number of relations is equal to $\dim_k Q \operatorname{Tor}_2^A(k, k)$, where the functor Q gives the space of generators for the algebra structure on $\operatorname{Tor}_*^A(k, k)$. For details see [Moore, Smith, 1968], Proposition 3.1 and Remark following Theorem 3.5.

We shall now generalize the above discussion of presentations to certain other categories of algebras. Let k be a commutative ring with identity and let $\mathcal{A}$ be a category of augmented k-algebras. We shall not assume that the algebras are commutative, and we shall not even assume that they are associative. We shall assume, however, that the algebras are augmented: i.e. each algebra A comes equipped with a homomorphism $\epsilon_A : A \to k$. (In some cases, e.g. Example (A.7.12)(3), the product on k, which makes it an object of $\mathcal{A}$ may not be the same as the ring multiplication.) We shall assume also that $\mathcal{A}$ contains free k-algebras. Recall that an object F of $\mathcal{A}$ is said to be free if there is a set $G \subseteq \ker \epsilon_F$ such that for any object A of $\mathcal{A}$ and any set function ϕ: $G \to \ker \epsilon_A$, there is a unique augmentation preserving homomorphism $\hat{\phi}: F \to A$ such that $\hat{\phi} \mid G = \phi$.

From now on any homomorphism of objects of $\mathcal{A}$ will be assumed to be augmentation preserving, i.e., a morphism of $\mathcal{A}$.

(A.7.12) Examples

(1) As in (A.7.6)-(A.7.11), $\mathcal{A}$ could be the category of connected $\mathbb{N}$-graded commutative (associative) k-algebras, where k is a field. The augmentations here are the obvious homogeneous degree 0 homomorphisms to k concentrated in degree 0. So each object has a unique augmentation.

A.7. Evaluations and presentations

(2) Let R be an object of the category $\mathcal{A}$ in (1) above. Then we could consider the category of $\mathbb{N}$-graded commutative (associative) R-algebras augmented over R. The augmentations here are no longer automatic even for R-algebras which are connected when viewed as k-algebras.

(3) Let k be a field and let $\mathcal{A}$ be the category of $\mathbb{N}$-graded k-Lie algebras with all augmentation homomorphisms being 0.

(4) $\mathcal{A}$ could be as in (1), (2) or (3) above but with grading over $\mathbb{Z}_2$ instead of $\mathbb{N}$.

From now on we shall consider an arbitrary category $\mathcal{A}$ of augmented k-algebras as above.

(A.7.13) Definitions

(1) Let $\bar{A} := \ker \epsilon_A$, the augmentation ideal of A. Let $Q(A) := \bar{A}/\bar{A}^2$, where $\bar{A}^2 = \bar{A}\bar{A}$. For any homomorphism $f : A \to B$ there is induced in the obvious way a homomorphism $Q(f) : Q(A) \to Q(B)$. Thus Q may be viewed as a functor from $\mathcal{A}$ to the category of k-modules. If $\mathcal{A}$ is a category of graded objects then clearly each $Q(A)$ is graded.

$Q(A)$ is called the module of generators. (In some cases it may occur that $Q(A) = 0$ even though A is not 0 or isomorphic to k. For example, let k be a field and let A be the ungraded commutative k-algebra $k[X]/(X - X^2)$ with $\epsilon_A(X) = 0$.)

(2) Suppose that there is a surjective homomorphism $\pi \colon F \to A$ where F is free. Let $J = \ker \pi$. Then we shall say that the exact sequence

$$0 \to J \to F \xrightarrow{\pi} A \to 0$$

is a presentation of A if $Q(\pi)$ is surjective. In this case the k-module $J/(\bar{F}J + J\bar{F})$ will be called the module of relations. For brevity let $R(J) = J/(\bar{F}J + J\bar{F})$.

A presentation as above is said to be minimal if $Q(\pi)$ is an isomorphism.

Now consider two presentations of the same object A:

$$0 \to J \to F \xrightarrow{\pi} A \to 0$$

and

$$0 \to I \to L \xrightarrow{\rho} A \to 0.$$

Since F and L are free, there are homomorphisms $\phi \colon F \to L$ and $\psi \colon L \to F$ such that $\rho\phi = \pi$ and $\pi\psi = \rho$. Let $\theta = \psi\phi \colon F \to F$. So $\pi\theta = \pi$.

Clearly ϕ, ψ, and θ induce homomorphisms (of k-modules), $R(\phi) \colon R(J) \to R(I)$, $R(\psi) \colon R(I) \to R(J)$ and $R(\theta) \colon R(J) \to R(J)$. Clearly $R(\theta) = R(\psi)R(\phi)$.

The following generalizes Proposition (A.7.10).

(A.7.14) Proposition
In the above situation suppose that the presentation $0 \to J \to F \xrightarrow{\pi} A \to 0$ is minimal. Then
(1) $Q(\theta) = 1$, the identity on $Q(F)$. In particular $Q(\psi)\colon Q(L) \to Q(F)$ is surjective.
(2) $R(\theta) = 1$, the identity on $R(J)$. In particular $R(\psi)\colon R(I) \to R(J)$ is surjective.

Furthermore there are homomorphisms $\lambda\colon Q(F) \to R(J)$ and $\mu\colon Q(L) \to R(I)$ such that $\lambda Q(\psi) = R(\psi)\mu$. And
(3) the homomorphism $\mu \mid \ker Q(\psi) \colon \ker Q(\psi) \to \ker R(\psi)$ is injective.

Proof.
First note that $J \subseteq \bar{F}^2$. Clearly $J = \ker \pi \subseteq \ker \epsilon_A \pi = \ker \epsilon_F = \bar{F}$. Suppose there is $f \in J - \bar{F}^2$. Then $f + \bar{F}^2 \neq 0$ in $Q(F)$. Since $Q(\pi)$ is an isomorphism, $Q(\pi)(f + \bar{F}^2) = \pi(f) + \bar{A}^2 \neq 0$, which contradicts $f \in J$.

Also since $\pi\theta = \pi$, $Q(\pi)Q(\theta) = Q(\pi)$; and so $Q(\theta) = 1$.

Now consider $f \in \bar{F}^2$. Let $f = \sum_i a_i b_i$, where $a_i, b_i \in \bar{F}$. Then $\theta(f) - f = \sum_i [\theta(a_i)\theta(b_i) - \theta(a_i)b_i + \theta(a_i)b_i - a_i b_i]$. So $\theta(f) - f \in \bar{F}J + J\bar{F}$, since, for any $h \in F$, $\theta(h) - h \in J$, because $\pi\theta = \pi$.

Hence for any $f \in J$, $R(\theta)(f + \bar{F}J + J\bar{F}) = \theta(f) + \bar{F}J + J\bar{F} = f + \bar{F}J + J\bar{F}$. So $R(\theta) = 1$.

Now define $\lambda\colon Q(F) \to R(J)$ and $\mu\colon Q(L) \to R(I)$ by $\lambda(f + \bar{F}^2) = \theta(f) - f + \bar{F}J + J\bar{F}$, for any $f \in \bar{F}$, and $\mu(g + \bar{L}^2) = \phi\psi(g) - g + \bar{L}I + I\bar{L}$, for any $g \in \bar{L}$. Since, by above, $\theta(f) - f \in \bar{F}J + J\bar{F}$ whenever $f \in \bar{F}^2$, λ is well-defined. And μ is well-defined similarly. Clearly $\lambda Q(\psi) = R(\psi)\mu$, since $\theta\psi = \psi\phi\psi$.

Let $\mu' = \mu \mid \ker Q(\psi) \colon \ker Q(\psi) \to \ker R(\psi)$. Let $h + \bar{L}^2 \in \ker Q(\psi)$. So $\psi(h) \in \bar{F}^2$, and $\phi\psi(h) \in \bar{L}^2$. If $\mu(h + \bar{L}^2) = 0$, then $\phi\psi(h) - h \in \bar{L}I + I\bar{L} \subseteq \bar{L}^2$. Hence $h \in \bar{L}^2$. Thus μ' is injective. □

Note that the conclusion that $R(\psi)$ is surjective immediately implies Proposition (A.7.10). The conclusion that $\mu \mid \ker Q(\psi)$ is injective has an interesting interpretation in terms of complete intersection defects, which are defined as follows.

(A.7.15) Definition
Suppose that the ground ring k is an integral domain. Let $0 \to J \to F \to A \to 0$ be a presentation of the k-algebra A. Call the presentation P. Suppose that $Q(F)$ and $R(J)$ are projective k-modules of finite rank. Then the complete intersection defect of P, cid(P), is defined to

be rank $R(J) - \operatorname{rank} Q(F)$. It is clear from Proposition (A.7.14) that if cid(P) is defined for a minimal presentation P, then it is defined and the same for all minimal presentations. In the latter case the complete intersection defect of A, cid(A), is defined to be cid(P) for any minimal presentation P.

(A.7.16) Corollary
Suppose that k is an integral domain. Let P be a presentation of the k-algebra A. If cid(P) is defined and A has a minimal presentation, then cid(A) is defined; and cid(A) $\leq$ cid(P).

Proof.
Let P be $0 \to I \to L \to A \to 0$ and let $0 \to J \to F \to A \to 0$ be a minimal presentation. Using the notation of Proposition (A.7.14), $Q(\psi)$ and $Q(\phi)$ exhibit $Q(F)$ as a retract of $Q(L)$. Since $Q(L)$ is projective by assumption, so is $Q(F)$. Similarly $R(J)$ is projective.
Let $K_1 = \ker Q(\psi)$ and $K_2 = \ker R(\psi)$. Then $Q(L) \cong K_1 \oplus Q(F)$ and $R(I) \cong K_2 \oplus R(J)$. So K_1 and K_2 are also projective; and, since, by assumption, $Q(L)$ and $R(I)$ are finitely generated, so are $Q(F)$, $R(J)$, K_1 and K_2. Since $\mu \mid K_1$ is injective, rank(K_1) $\leq$ rank(K_2). Hence
$$\operatorname{rank}(Q(L)) - \operatorname{rank}(Q(F)) \leq \operatorname{rank}(R(I)) - \operatorname{rank}(R(J)).$$
□

(A.7.17) Remarks
(1) In most graded situations all the homomorphisms above, e.g. ϕ, ψ, θ and μ, are homogeneous of degree zero. Hence the inequalities above concerning numbers of generators, numbers of relations and complete intersection defects can be split into separate inequalities concerning the same quantities in even degrees and odd degrees.
(2) There is another common way of writing presentations. Let A be a k-algebra as above, and suppose that there are free k-algebras F_1 and F_0 and homomorphisms $\sigma : F_1 \to F_0$ and $\pi : F_0 \to A$ with π surjective. Let $J = \ker \pi$, and suppose that $\sigma(\bar{F}_1) \subseteq J$. Then $\sigma(\bar{F}_1^2) \subseteq J^2 \subseteq \bar{F}_0 J + J\bar{F}_0$. So σ induces a homomorphism of k-modules $\bar{\sigma} : Q(F_1) \to R(J)$. Then the sequence $F_1 \xrightarrow{\sigma} F_0 \xrightarrow{\pi} A$ is called a (free) presentation of A if both $Q(\pi)$ and $\bar{\sigma}$ are surjective. A free presentation $F_1 \xrightarrow{\sigma} F_0 \xrightarrow{\pi} A$ is said to be minimal if $Q(\pi)$ and $\bar{\sigma}$ are both isomorphisms.
Clearly a free presentation, resp. minimal free presentation, gives rise to a presentation, resp. minimal presentation, in the sense used above. Going the other way is not so easy, and depends, of course, on

the existence of sufficiently many free objects among other things. In Examples (A.7.12)(1) and (3) and their $\mathbb{Z}_2$-graded analogues, it is clear that presentations, resp. minimal presentations, give rise to free presentations, resp. minimal free presentations. In Example (A.7.12)(2), however, constructing a minimal free presentation from a minimal presentation $0 \to J \to F \to A \to 0$ would only seem to be possible if $R(J)$ is a free R-module.

(3) Note that if $f : A \to B$ is an epimorphism of augmented k-algebras, then, clearly, $Q(f) : Q(A) \to Q(B)$ is also surjective.

be rank $R(J) - \operatorname{rank} Q(F)$. It is clear from Proposition (A.7.14) that if $\operatorname{cid}(P)$ is defined for a minimal presentation P, then it is defined and the same for all minimal presentations. In the latter case the complete intersection defect of A, $\operatorname{cid}(A)$, is defined to be $\operatorname{cid}(P)$ for any minimal presentation P.

(A.7.16) Corollary
Suppose that k is an integral domain. Let P be a presentation of the k-algebra A. If $\operatorname{cid}(P)$ is defined and A has a minimal presentation, then $\operatorname{cid}(A)$ is defined; and $\operatorname{cid}(A) \leq \operatorname{cid}(P)$.

Proof.
Let P be $0 \to I \to L \to A \to 0$ and let $0 \to J \to F \to A \to 0$ be a minimal presentation. Using the notation of Proposition (A.7.14), $Q(\psi)$ and $Q(\phi)$ exhibit $Q(F)$ as a retract of $Q(L)$. Since $Q(L)$ is projective by assumption, so is $Q(F)$. Similarly $R(J)$ is projective.
Let $K_1 = \ker Q(\psi)$ and $K_2 = \ker R(\psi)$. Then $Q(L) \cong K_1 \oplus Q(F)$ and $R(I) \cong K_2 \oplus R(J)$. So K_1 and K_2 are also projective; and, since, by assumption, $Q(L)$ and $R(I)$ are finitely generated, so are $Q(F)$, $R(J)$, K_1 and K_2. Since $\mu \mid K_1$ is injective, $\operatorname{rank}(K_1) \leq \operatorname{rank}(K_2)$. Hence

$$\operatorname{rank}(Q(L)) - \operatorname{rank}(Q(F)) \leq \operatorname{rank}(R(I)) - \operatorname{rank}(R(J)).$$

□

(A.7.17) Remarks
(1) In most graded situations all the homomorphisms above, e.g. ϕ, ψ, θ and μ, are homogeneous of degree zero. Hence the inequalities above concerning numbers of generators, numbers of relations and complete intersection defects can be split into separate inequalities concerning the same quantities in even degrees and odd degrees.
(2) There is another common way of writing presentations. Let A be a k-algebra as above, and suppose that there are free k-algebras F_1 and F_0 and homomorphisms $\sigma : F_1 \to F_0$ and $\pi : F_0 \to A$ with π surjective. Let $J = \ker \pi$, and suppose that $\sigma(\bar{F}_1) \subseteq J$. Then $\sigma(\bar{F}_1^2) \subseteq J^2 \subseteq \bar{F}_0 J + J \bar{F}_0$. So σ induces a homomorphism of k-modules $\bar{\sigma} : Q(F_1) \to R(J)$. Then the sequence $F_1 \xrightarrow{\sigma} F_0 \xrightarrow{\pi} A$ is called a (free) presentation of A if both $Q(\pi)$ and $\bar{\sigma}$ are surjective. A free presentation $F_1 \xrightarrow{\sigma} F_0 \xrightarrow{\pi} A$ is said to be minimal if $Q(\pi)$ and $\bar{\sigma}$ are both isomorphisms.
Clearly a free presentation, resp. minimal free presentation, gives rise to a presentation, resp. minimal presentation, in the sense used above. Going the other way is not so easy, and depends, of course, on

the existence of sufficiently many free objects among other things. In Examples (A.7.12)(1) and (3) and their $\mathbb{Z}_2$-graded analogues, it is clear that presentations, resp. minimal presentations, give rise to free presentations, resp. minimal free presentations. In Example (A.7.12)(2), however, constructing a minimal free presentation from a minimal presentation $0 \to J \to F \to A \to 0$ would only seem to be possible if $R(J)$ is a free R-module.

(3) Note that if $f : A \to B$ is an epimorphism of augmented k-algebras, then, clearly, $Q(f) : Q(A) \to Q(B)$ is also surjective.

Appendix B

Some homotopy theory of differential modules

For the convenience of the reader we summarize in the first section of this appendix some basic notions and elementary results in the homotopy theory of differential modules. We are actually interested in situations where some additional structure is given, e.g., $\mathbb{Z}$- or $\mathbb{Z}_2$-graded differential modules over a (possibly) graded ring, but the fundamental ideas can be most easily understood in the general framework of differential modules, and the reader, who has become familiar with this situation, should have no trouble with the minor modifications, which are necessary in the different more special cases. A general reference dealing with chain complexes over arbitrary rings (i.e. $\mathbb{Z}$-graded differential modules, such that the differential has degree -1) is [Dold, 1960], which contains much more than we are going to discuss here. An exposition for chain complexes over the integers $\mathbb{Z}$ is given in [Dold, 1980]. See also, e.g., [Brown, K.S. 1982], Chapter I which contains a summary of relevant material from homological algebra, discussing in particular complexes over arbitrary rings.

The general principles in developing the homotopy theory of differential graded algebras (see [Quillen, 1967], [Sullivan, 1977], [Halperin, 1977(1983)], [Bousfield, Gugenheim, 1976]) are very similar to those in the case of differential (graded) modules, but the technical apparatus is much more involved. It therefore seems reasonable first to get acquainted with the homotopy theory of differential modules if one wants to study the homotopy theory of differential graded algebras in detail.

In the second section we consider the case of free cochain complexes over a graded algebra and derive a number of results, which are used in the main part of the book.

B.1 Basic notions and elementary results

A differential module over a ring R is an R-module K together with an R-linear endomorphism $\partial : K \to K$, called the differential or boundary of K, such that $\partial\partial = 0$. We assume here that R is commutative, associative and has a unit.

The homology $H(K;M)$ of K with coefficients in an R-module M is defined in the usual way:

$$H(K;M) := Z(K;M)/B(K;M),$$

where

$$Z(K;M) := \text{kernel of } \partial \otimes \text{id}_M : K \otimes M \to K \otimes M$$

are the cycles, and

$$B(K;M) := \text{image of } \partial \otimes \text{id}_M : K \otimes M \to K \otimes M$$

are the boundaries.

By R-Mod and ∂R-Mod we denote the category of R-modules and the category of differential R-modules, respectively.

For topological spaces many of the fundamental notions in elementary homotopy theory (such as homotopy, homotopy equivalence, fibration, cofibration) can be based on categorical constructions using the interval $[0,1]$ of real numbers (see, e.g., [tom Dieck, Kamps, Puppe, D. 1970]). Here the differential module I, defined by

$$I := R \oplus R \oplus R; \qquad \partial : I \to I, \qquad \partial(r_0, r, r_1) := (-r, 0, r)$$

for $(r_0, r, r_1) \in I$ plays the role of the interval. In fact, the differential module I (equipped with the appropriate grading) may be viewed as the cellular chain complex (with coefficients in R) of the interval $[0,1]$, considered as a CW-complex in the usual way (two 0-cells, one 1-cell). One has two canonical inclusions $i_\nu : R \to I$, $\nu = 0, 1$, $i_0(r) := (r, 0, 0)$, $i_1(r) := (0, 0, r)$, corresponding to the inclusions of the 0-cells into $[0,1]$. Here R is considered as an object in ∂R-Mod having trivial differential.

We give a number of definitions and basic results for ∂R-Mod imitating the standard procedures in the category of topological spaces.

(B.1.1) Definition

Let $f_0, f_1 : K \to L$ be two morphisms in ∂R-Mod. A homotopy from f_0 to f_1 is a morphism $h : I \otimes K \to L$ in ∂R-Mod such that $h(i_\nu \otimes \text{id}_K) =$

B.1. Basic notions and elementary results

$f_\nu : K = R \otimes K \to I \otimes K \to L$ for $\nu = 0, 1$. The tensor product here is taken over R and the differential for the 'cylinder' $I \otimes K$ over K is defined by

$$\partial : I \otimes K = K \oplus K \oplus K \to K \oplus K \oplus K,$$

$$\partial(k_0, k, k_1) = (\partial_K(k_0) - k, -\partial_K(k), \partial_K(k_1) + k),$$

where ∂_K denotes the differential of K.

We say that h starts with f_0 and ends with f_1 and use the following notation $h : f_0 \simeq f_1$ (or just $f_0 \simeq f_1$, if there exists a homotopy, but we are not interested in a particlular one).

It is a very simple exercise (left to the reader) to check that the above definition is equivalent to the usual one.

The standard notions of homotopy equivalence, cofibration, fibration, mapping cone, double mapping cylinder, telescope, etc., are easily carried over from the category of topological spaces to ∂R-Mod, for example:

(B.1.2) Definition
A morphism $f : K' \to K$ in ∂R-Mod is a cofibration if it has the homotopy extension property ('HEP'), i.e. for any pair (g, h') such that $g : K \to L$ is a morphism in ∂R-Mod and $h' : I \otimes K' \to L$ is a homotopy which starts with gf there exists an 'extension' $h : I \otimes K \to L$ of h' ($h(\mathrm{id}_I \otimes f) = h'$) which starts with g.

(Note, though, that there is a problem with defining a 'reasonable' differential on the tensor product of two differential modules in the absence of an additional grading.)

The formal relations between homotopy equivalences, cofibrations, fibrations, etc., in ∂R-Mod are in many aspects similar to the topological situation. Quillen has developed an abstract homotopy theory based on equivalence, certain formal properties of cofibrations and fibrations in a rather general categorial set-up and has used this for his results in rational homotopy theory (see [Quillen, 1967, 1969]). The technical machinery in ∂R-Mod is considerably simpler than, e.g., in the category of differential graded algebras or of topological spaces - mainly because ∂R-Mod is additive (cf. Exercises (B.1)-(B.3)) - but the analogy to the latter may very well serve as a guide. Basic results like the following proposition can be proved in ∂R-Mod by the same formalism as in the category Top of topological spaces (see, e.g., [tom Dieck, Kamps, Puppe, D. 1970] for a detailed presentation in Top), but could also be obtained by arguments which are more specific to the homotopy theory in ∂R-Mod (see below).

(B.1.3) Proposition
Let $f_\nu : K \to L_\nu$, $\nu = 0, 1$ be cofibrations in ∂R-Mod and let $g : L_0 \to L_1$ be a homotopy equivalence and a morphism 'under K', i.e. $gf_0 = f_1$, then g is a homotopy equivalence under K. In particular, if $f : K \to L$ is a cofibration and a homotopy equivalence, then K is a strong deformation retract of L.

Of course, there are results specific to ∂R-Mod like the following characterization of cofibrations, fibrations and homotopy equivalences.

(B.1.4) Proposition
The morphism $f : K \to L$ is a cofibration in ∂R-Mod if and only if f has a left inverse in R-Mod (i.e. there exists a morphism $g : L \to K$ in R-Mod such that $gf = \mathrm{id}_K$).

Proof.
Similar to the situation in Top it follows easily from the definition that $f : K \to L$ is a cofibration if and only if the canonical morphism $s : Z(f) \to I \otimes L$ from the mapping cylinder $Z(f)$ of f, which is the pushout (in ∂R-Mod) of Diagram B.1.1

$$\begin{array}{ccc} I \otimes K & \longrightarrow & Z(f) \\ \uparrow{\scriptstyle i_0} & & \uparrow \\ K & \xrightarrow{f} & L \end{array}$$

Diagram B.1.1

to $I \otimes L$ has a left inverse $r : I \otimes L \to Z(f)$ in ∂R-Mod. As an R-module $I \otimes K$ is just $K \oplus K \oplus K$ and hence $Z(f) \cong L \oplus K \oplus K$ as R-modules. The morphism $s : Z(f) \to I \otimes L$ is given by $\mathrm{id}_L \oplus f \oplus f$.

Assuming that f is a cofibration, the retraction

$$r : I \otimes L = L \oplus L \oplus L \to L \oplus K \oplus K$$

restricted to the last summand in $I \otimes L$ and composed with the projection of $L \oplus K \oplus K$ to the last summand gives the desired left inverse of f in R-Mod.

On the other hand if $f : K \to L$ has a left inverse in R-Mod we can decompose L into a direct sum $L = K \oplus C$ as R-modules. The differential ∂_L of L can be written as

$$\partial_L(k, c) = (\partial_K(k) + \partial_K^C(c), \partial_C^C(c)),$$

where $(k, c) \in K \oplus C$ and ∂_K denotes the differential of K. One now

defines a retraction
$$r : I \otimes L \cong L \oplus L \oplus L \to Z(f) \cong L \oplus K \oplus K$$
by
$$r(0, \tilde{c}, c) = (c, \partial_K^C(c), 0)$$
on $0 \oplus C \oplus C$ using the decomposition
$$I \otimes L \cong L \oplus (K \oplus C) \oplus (K \oplus C)$$
and as the 'identity' on the complementary summand $L \oplus K \oplus K$. Clearly r is a left inverse to s in R-Mod and an easy calculation (using $\partial_L \partial_L = 0$) shows that r is indeed compatible with the differentials, i.e. a morphism in ∂R-Mod. □

We leave the proof of the following proposition (which in a certain sense is dual to (B.1.4)) to the reader.

(B.1.5) Proposition
The morphism $f : K \to L$ is a fibration in ∂R-Mod if and only if f has a right inverse in R-Mod.

(B.1.6) Definitions
(1) A differential module K is called contractible or homotopically trivial ('ht-trivial' for short), if K is homotopy equivalent to 0 (notation: $K \simeq 0$).
(2) A differential module K is called acyclic or homologically trivial ('hl-trivial' for short), if the homology of K vanishes ($H(K) = 0$).

The following proposition is used in the next section, and we applied its consequences in Chapters 1 and 3.

(B.1.7) Proposition
(1) *The morphism $f : K \to L$ is a homotopy equivalence if and only if the mapping cone $C(f)$ of f is ht-trivial.*
(2) *If f is a cofibration (and hence - in particular - injective, (see (B.1.4))), then f is a homotopy equivalence if and only if $L/f(K)$ is ht-trivial.*
(3) *If f is a cofibration and a homotopy equivalence then L is isomorphic to $K \oplus L/f(K)$ as a differential module under K (i.e. the map f corresponds to the inclusion into the first summand) and $L/f(K)$ is ht-trivial.*

For the proof of (B.1.7) we make use of the following characterization of ht-trivial differential modules.

(B.1.8) Proposition
For a differential module N the following assertions are equivalent:
(i) N is ht-*trivial*
(ii) *As an R-module N can be decomposed into a direct sum $N = Z \oplus D$, such that $\partial : N \to N$ is zero on Z and maps D isomorphically onto Z.*
(iii) *As an R-module N can be decomposed into a direct sum $N = N' \oplus N''$, such that $\partial_r'' : N'' \to N'$ is an isomorphism, where ∂_r'' denotes the component of $\partial : N \to N$, which maps to N', restricted to $N'' \subset N$.*

Proof.
We are going to use here the 'classical' definition of (chain) homotopy, which is equivalent to (B.1.1) as remarked above.

(i)$\Rightarrow$(ii): Let $h : N \to N$ be a homotopy which starts with the trivial map and ends with the identity, i.e. $\partial h + h\partial = \mathrm{id} : N \to N$. The short exact sequence $0 \to Z = \ker(\partial) \to N \xrightarrow{\partial} B = \mathrm{im}\,\partial \to 0$ splits. In fact, $h|B : B \to N$ is a right inverse of $\partial : N \to B$ since
$$\partial h \partial(n) = (\partial h + h\partial)\partial(n)$$
for all $\partial(n) \in B$. Hence $N = Z \oplus D$, with $D := h(B)$, and $\partial|_Z = 0$. Since $H(N) = 0$ one has $Z = B$ and $\partial|_D : D \to B = Z$ is an isomorphism.

(ii)$\Rightarrow$(iii): This is obvious (put $N' = Z$, $N'' = D$).

(iii)$\Rightarrow$(i): The differential $\partial : N' \oplus N'' \to N' \oplus N''$ can be written as a 2×2-matrix
$$\partial = \begin{pmatrix} \partial_r' & \partial_r'' \\ \partial_{r'}' & \partial_{r'}'' \end{pmatrix}$$
with $\partial_r' : N' \to N'$, $\partial_r'' : N'' \to N'$ etc. By assumption ∂_r'' is an isomorphism. We define $h : N' \oplus N'' \to N' \oplus N''$ by $h(n', n'') = (0, (\partial_r'')^{-1}(n'))$ for $(n', n'') \in N' \oplus N''$. A straightforward calculation shows that $\partial h + h\partial = \mathrm{id}_N$. Hence $h : N \to N$ gives the desired homotopy from the zero map to the identity. $\square$

Proof of (B.1.7).
One can replace $f : K \to L$ by the inclusion $K \to Z(f)$ of K as the top of the mapping cylinder $Z(f)$ of f, the quotient $Z(f)/K$ being the mapping cone $C(f)$ of f. Since $Z(f)$ is homotopy equivalent to L and $K \to Z(f)$ is a cofibration part (1) of (B.1.7) follows from part (2). In proving (2) we may assume that $f : K \to L$ is an inclusion and L splits into a direct sum $L = K \oplus N$ of R-modules with $N \cong L/K$ (see (B.1.4)). Assume now that $f : K \to L$ is a homotopy equivalence. Then

- by (B.1.3) - K is a strong deformation retract of L, i.e. there is a retraction $r : L \to K$ and a homotopy $h : I \otimes L \to L$, which starts at fr and ends at id_L and which restricted to $I \otimes K$ is 'constant'. Therefore $h : I \otimes L \to L$ induces a homotopy $\overline{h} : I \otimes L/K \to L/K$, which starts at $L/K \to 0 \to L/K$ and ends at $\mathrm{id}_{L/K}$, i.e. L/K is ht-trivial.

For the converse we assume that L/K is ht-trivial and we use the splitting $L = K \oplus N$ (as R-modules) with $N \cong L/K$ and the splitting of $N = Z \oplus D$ as in (B.1.8)(ii). The differential

$$\partial : L = K \oplus Z \oplus D \to K \oplus Z \oplus D = L$$

can be written as

$$\partial(k, z, d) = (\partial_K^K(k) + \partial_K^Z(z) + \partial_K^D(d), \partial_Z^D(d), 0)$$

for $(k, z, d) \in K \oplus Z \oplus D$, where $\partial_K^K : K \to K$ is the differential of K, $\partial_Z^D : D \to Z$ is the 'essential' part of the differential of $N = Z \oplus D$, and $\partial_K^Z : Z \to K$ and $\partial_K^D : D \to K$ are the respective components of the differential of L. We define a map $g : N = Z \oplus D \to L$ by $g(z, d) := (\partial_K^D(\partial_Z^D)^{-1}(z), z, d)$. (Note that $\partial_Z^D : D \to Z$ is an isomorphism (see (B.1.8)(ii))). It is again a simple calculation to check that $g : N \to L$ is a morphism in ∂R-Mod. Since g is a right inverse of the projection $L \to L/K \cong N$ in ∂R-Mod one gets an isomorphism in ∂R-Mod(!) between L and $K \oplus N$ such that $f : K \to L$ corresponds to the inclusion into the first summand. This proves part (3) and, since the projection $K \oplus N \to K$ is a homotopy equivalence (N is ht-trivial), it also finishes the proof of part (2). □

The above argument can be shortened if one uses Exercise (B.1)(c) which implies that any fibration $p : L \to N$ with N ht-trivial splits in the sense that $L \cong \ker p \oplus N$ in ∂R-Mod.

The proof (and the detailed statement) of the following proposition, which is dual to (B.1.7), is left to the reader.

(B.1.9) Proposition
(1) *The morphism $f : K \to L$ is a homotopy equivalence if and only if the 'dual mapping cone' of f is ht-trivial.*
(2) *If f is a fibration (and hence - in particular - surjective (see (B.1.5))), then f is a homotopy equivalence if and only if $\ker(f)$, the kernel of f, is ht-trivial.*
(3) *If f is a fibration and a homotopy equivalence then K is isomorphic to $\ker(f) \oplus L$ as a differential module over L.*

One should compare the Propositions (B.1.7) and (B.1.9) with the following remark, which is an immediate consequence of the long exact homology sequence

$$\ldots \to H(K) \to H(L) \to H(C(f)) \to H(K) \to H(L) \to \ldots$$

(which - more economically - could be written as an exact triangle) for a morphism $f: K \to L$ in ∂R-Mod.

(B.1.10) Remark
A morphism $f: K \to L$ in ∂R-Mod induces an isomorphism in homology if and only if the mapping cone $C(f)$ is hl-trivial ($H(C(f)) = 0$).

It is therefore interesting to compare the two conditions 'ht-trivial' and 'hl-trivial'. Clearly the former implies the latter, but the following proposition shows, that under certain additional assumptions the two conditions are equivalent.

(B.1.11) Proposition
Let P be an object in ∂R-Mod, which is projective as an R-module and let R be of finite global homological dimension (see Section A.4). Then P is ht-trivial ($P \simeq 0$) if and only if P is hl-trivial ($H(P) = 0$).

Proof.
We only have to show that $H(P) = 0$ implies $P \simeq 0$ under the above assumptions. We consider the short exact sequence

$$0 \to Z \to P \xrightarrow{\partial} B \to 0,$$

where $Z = \ker(\partial)$ are the cycles and $B = \operatorname{im} \partial$ the boundaries of P. In order to show that $P \simeq 0$ it suffices to prove that B is a projective R-module. Namely: B projective implies that the above exact sequence splits, i.e. $P = Z \oplus D$ as R-modules with $\partial|Z = 0$ and $\partial|D: D \to B$ an isomorphism. Since $H(P) = 0$, one has that $B = Z$ and therefore P is ht-trivial by (B.1.8).

Let r be the global homological dimension of R. To prove that B is projective one considers the exact sequence (note that $H(P) = 0$)

$$0 \to Z \to P \xrightarrow{\partial} P \xrightarrow{\partial} P \ldots \xrightarrow{\partial} P \to B \to 0$$

containing at least r copies of P. It follows (see (A.4.4)) that Z must be a projective R-module. Since $B = Z$ one has the desired result. $\square$

(B.1.12) Remark
We leave it to the reader as an exercise to show that in the case of chain complexes one can use the grading to replace the assumption on

the homological dimension of R in (B.1.11) by the assumption that the chain complex P is bounded below ($P_q = 0$ for $q < q_0$, $q_0 \in \mathbb{Z}$).

One useful consequence of the above proposition is the following.

(B.1.13) Corollary
Let $f : K \to L$ be a morphism in ∂R-Mod and let K and L be projective as R-modules. If R has finite global homological dimension, then f is a homotopy equivalence if and only if f induces an isomorphism in homology.

Proof.
This follows immediately from (B.1.8), (B.1.10) and (B.1.11). □

If we assume that the homology $H(K)$ of K is projective as a R-module, then there exists a morphism $j : H(K) \to K$ in ∂R-Mod (where $H(K)$ carries the trivial differential), which induces the identity in homology. One just chooses a splitting of the projection $Z \to H(K)$ of the cycles Z of K to their homology classes. Applying (B.1.13) gives the following result.

(B.1.14) Corollary
Let K be an object in ∂R-Mod such that K and $H(K)$ are projective R-modules. If R has finite global homological dimension, then K and $H(K)$ are homotopy equivalent in ∂R-Mod. More precisely, there exists a homotopy equivalence $p : K \to H(K)$ in ∂R-Mod, which restricted to the cycles Z of K is just the projection to their homology classes.

Using (B.1.12) we get corresponding versions of (B.1.13) and (B.1.14).

The class $\mathcal{N}$ of ht-trivial objects in ∂R-Mod is closed under certain categorial constructions.

(B.1.15) Proposition
(1) Let $N = \bigoplus_{\alpha \in \mathcal{A}} N_\alpha$ be a direct sum with $N_\alpha \in \mathcal{N}$ for all $\alpha \in \mathcal{A}$, then $N \in \mathcal{N}$.
(2) Let $f : N' \to N$ be a cofibration with N', $N \in \mathcal{N}$, then
$$\operatorname{coker}(f) = N/f(N') \in \mathcal{N}.$$
(3) Let $f_\nu : N_0 \to N_\nu$, $\nu = 1, 2$ be two morphisms in ∂R-Mod such that N_0, N_1, $N_2 \in \mathcal{N}$, then the double mapping cylinder $Z(f_1, f_2)$ (i.e. the homotopy pushout of Diagram B.1.2)

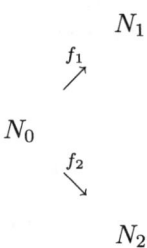

Diagram B.1.2

is in $\mathcal{N}$.

Proof.
Part (1) is obvious and part (2) follows immediately from (B.1.7). Part (3) follows from (1) and (2) since the double mapping cylinder $Z(f_1, f_2) := N_1 \oplus I \otimes N_0 \oplus N_2 / \sim$ (where $\sim$ denotes the equivalence relation generated by: $f_1(n) \sim i_0(n)$ and $f_2(n) \sim i_1(n)$ for all $n \in N_0$) is homotopy equivalent to the cokernel of the cofibration

$$(j_1, -j_2) : N_0 \to Z(f_1) \oplus Z(f_2),$$

where $Z(f_1)$ and $Z(f_2)$ are the mapping cylinders of f_1 and f_2 respectively and $j_\nu : N_0 \to Z(f_\nu)$, $\nu = 1, 2$, is the inclusion of N_0 as the top of the cylinder $Z(f_\nu)$, i.e. in the cokernel of $(j_1, -j_2)$ the two 'tops' are identified. Clearly $(j_1, -j_2)$ is a cofibration by the characterization (B.1.4); and the cokernel of $(j_1, -j_2)$ is nothing but the direct sum of $Z(f_1, f_2)$ and the cone over N_0. Since the latter is ht-trivial, the former is homotopy equivalent to coker$(j_1, -j_2)$. It follows from part (1) (and the fact that $Z(f_\nu)$ is homotopy equivalent to N_ν) that $Z(f_1) \oplus Z(f_2) \in \mathcal{N}$. Hence part (3) is implied by (2). $\square$

The proof of the following proposition, which is dual to (B.1.15), is left to the reader.

(B.1.16) Proposition
The class $\mathcal{N}$ *of* ht-*trivial objects in* ∂R-Mod *is closed under:*

(i) direct products;
(ii) kernels of fibrations $g : N \to N''$ with $N, N'' \in \mathcal{N}$;
(iii) homotopy pullbacks of diagrams such as Diagram B.1.3 with N_0, N_1, $N_2 \in \mathcal{N}$

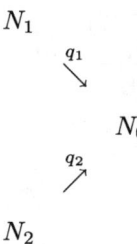

Diagram B.1.3

(B.1.17) Proposition

Let $0 \to K' \to K \to K'' \to 0$ be an exact sequence in ∂R-Mod which splits in R-Mod. Then if any two of the objects K', K, K'' belong to $\mathcal{N}$ so does the third.

Proof.
(1) Assume $K', K \in \mathcal{N}$. Since the sequence splits in R-Mod the morphism $K' \to K$ is a cofibration (see (B.1.4)). Therefore (B.1.15) implies that $K'' \in \mathcal{N}$.
(2) assume $K, K'' \in \mathcal{N}$. This situation is dual to (1), and using (B.1.5) and (B.1.16) gives $K' \in \mathcal{N}$.
(3) Assume $K', K'' \in \mathcal{N}$. Using (B.1.7) one gets that the sequence actually splits in ∂R-Mod, i.e. $K \cong K' \oplus K''$ in ∂R-Mod and hence $K \in \mathcal{N}$. □

(B.1.18) Corollary
(1) Let K be an object in ∂R-Mod which is filtered by subobjects $\mathcal{F}_q(K)$, $q \in \mathbb{Z}$ such that $\mathcal{F}_q(K) = 0$ for $q < q_0$, $q_0 \in \mathbb{Z}$, $\mathcal{F}_q(K) \subset \mathcal{F}_{q+1}(K)$ is a cofibration for all $q \in \mathbb{Z}$ and $K = \bigcup_q \mathcal{F}_q(K)$ (i.e. K is the colimit of $\ldots \mathcal{F}_q(K) \subset \mathcal{F}_{q+1}(K) \ldots$ in the categorical sense). If $\mathcal{F}_{q+1}(K)/\mathcal{F}_q(K) \in \mathcal{N}$ for all $q \in \mathbb{Z}$ then $K \in \mathcal{N}$.
(2) Let $f : K \to L$ be a morphism of filtered objects as in (1):
$$(f(\mathcal{F}_q(K)) \subset \mathcal{F}_q(L) \text{ for } q \in \mathbb{Z}).$$
If f induces homotopy equivalences,
$$\overline{f} : \mathcal{F}_{q+1}(K)/\mathcal{F}_q(K) \to \mathcal{F}_{q+1}(L)/\mathcal{F}_q(L)$$
for $q \in \mathbb{Z}$, then f is itself a homotopy equivalence.

Proof.

(1) By induction - using (B.1.17) - one gets $\mathcal{F}_q(K) \in \mathcal{N}$ for all $q \in \mathbb{Z}$; moreover $\mathcal{F}_{q+1}(K) = \mathcal{F}_q(K) \oplus N_{q+1}$ with $N_{q+1} \cong \mathcal{F}_{q+1}(K)/\mathcal{F}_q(K) \in \mathcal{N}$. Therefore $K = \bigoplus_{q \in \mathbb{Z}} N_q$ is in $\mathcal{N}$.

Part (2) follows from (1) using the mapping cone construction (see (B.1.7)). □

(B.1.19) Remark

Using the standard properties of the homology functor (long exact sequence, additivity etc.) it is very easy to derive analogous properties for the class of homologically trivial objects in ∂R-Mod. If one is only interested in differential R-modules (or chain complexes) which are projective as R-modules, and if R has finite global homological dimension (or the chain complexes are bounded below) then (B.1.11) and (B.1.12) give the possibility to pass from hl-trivial to ht-trivial.

(B.1.20) Corollary

(1) Let Diagram B.1.4

$$\begin{array}{ccccc} K_1 & \xleftarrow{i_1} & K_0 & \xrightarrow{i_2} & K_2 \\ \downarrow{f_1} & & \downarrow{f_0} & & \downarrow{f_2} \\ L_1 & \xleftarrow{j_1} & L_0 & \xrightarrow{j_2} & L_2 \end{array}$$

Diagram B.1.4

be a homotopy commutative diagram in ∂R-Mod. If f_ν, $\nu = 0, 1, 2$ are homotopy equivalences then there exists a homotopy equivalence $f : Z(i_1, i_2) \to Z(j_1, j_2)$ between the double mapping cylinders of the two rows of Diagram B.1.4, which is compatible with the f_νs in the sense that Diagram B.1.5,

$$\begin{array}{ccccccc} K_1 \oplus K_2 & \to & Z(i_1, i_2) & \to & K_0 \\ \downarrow{f_1} \quad \downarrow{f_2} & & \downarrow{f} & & \downarrow{f_0} \\ L_1 \oplus L_2 & \to & Z(j_1, j_2) & \to & L_0 \end{array}$$

Diagram B.1.5

where the horizontal maps are the canonical maps of the double mapping cylinder construction, commutes.

(2) Let Diagram B.1.6

B.1. Basic notions and elementary results

$$\begin{array}{ccccccc}
\cdots \longrightarrow & K_{q-1} & \xrightarrow{i_{q-1}} & K_q & \xrightarrow{i_q} & K_{q+1} & \longrightarrow \cdots \\
& \downarrow f_{q-1} & & \downarrow f_q & & \downarrow f_{q+1} & \qquad q \in \mathbb{Z} \\
\cdots \longrightarrow & L_{q-1} & \xrightarrow{j_{q-1}} & L_q & \xrightarrow{j_q} & L_{q+1} & \longrightarrow \cdots
\end{array}$$

Diagram B.1.6

be a homotopy commutative diagram in ∂R-Mod such that all i_q, j_q are cofibrations. If the f_q, $q \in \mathbb{Z}$ are homotopy equivalences then there exists a homotopy equivalence

$$f : K := \varinjlim(K_q, i_q) \longrightarrow L := \varinjlim(L_q, j_q)$$

such that Diagram B.1.7

$$\begin{array}{ccc}
K_q & \xrightarrow{i} & K \\
\downarrow f_q & & \downarrow f \\
L_q & \xrightarrow{j} & L
\end{array}$$

Diagram B.1.7

is homotopy commutative, where the horizontal maps are the canonical maps into the direct limit (i.e. the categorical colimit).

Proof.
(1) Choosing homotopies $h_\nu : I \otimes K_0 \to L_\nu$, $\nu = 1, 2$ between $f_1 i_1$ and $j_1 f_0$ (resp. $f_2 i_2$ and $j_2 f_0$) one easily obtains a morphism

$$f : Z(i_1, i_2) \to Z(j_1, j_2)$$

which is compatible with the f_νs, $\nu = 0, 1, 2$, in the above sense. Hence part (1) follows from (B.1.18)(2).

For part (2) one can replace L_q, $q \in \mathbb{Z}$, by the mapping cylinder $Z(f_q)$, and $j_q : L_q \to L_{q+1}$ by a map $\tilde{j}_q : Z(f_q) \to Z(f_{q+1})$ (choosing a homotopy $h_q : I \otimes K_q \to L_{q+1}$ between $j_q f_q$ and $f_{q+1} i_q$) such that Diagram B.1.8

$$\begin{array}{ccccccc}
\cdots \longrightarrow & K_{q-1} & \xrightarrow{i_{q-1}} & K_q & \xrightarrow{i_q} & K_{q+1} & \longrightarrow \cdots \\
& \alpha_{q-1} \downarrow & & \alpha_q \downarrow & & \alpha_{q+1} \downarrow & \\
\cdots \longrightarrow & Z(f_{q-1}) & \xrightarrow{\tilde{j}_{q-1}} & Z(f_q) & \xrightarrow{\tilde{j}_q} & Z(f_{q+1}) & \longrightarrow \cdots \\
& \beta_{q-1} \uparrow & & \beta_q \uparrow & & \beta_{q-1} \uparrow & \\
\cdots \longrightarrow & L_{q-1} & \xrightarrow{j_{q-1}} & L_q & \xrightarrow{j_q} & L_{q+1} & \longrightarrow \cdots
\end{array}$$

Diagram B.1.8

strictly commutes, where
$$K_q \xrightarrow{\alpha_q} Z(f_q) \xleftarrow{\beta_q} L_q$$
are the canonical inclusions of the mapping cylinder construction, the map $\beta_q : L_q \to Z(f_q)$ being a homotopy inverse of the projection
$$\pi_q : Z(f_q) \to L_q,$$
and $\pi_q \alpha_q = f_q$. Therefore α_q (together with f_q) and, of course, also β_q are homotopy equivalences. Since i_q and j_q are assumed to be cofibrations so is $\tilde{j}_q$. Using (B.1.18)(2) one gets a commutative diagram, Diagram B.1.9,

$$\begin{array}{ccc}
K_q & \xrightarrow{i} & \varinjlim_q(K_q, f_q) =: K \\
\downarrow{\alpha_q} & & \downarrow{\alpha} \\
Z(f_q) & \longrightarrow & \varinjlim_q(Z(f_q), \tilde{j}_q) \\
\uparrow{\beta_q} & & \uparrow{\beta} \\
L_q & \longrightarrow & \varinjlim_q(L_q, j_q) =: L
\end{array}$$

Diagram B.1.9

such that α and β are homotopy equivalences. Defining $f := \beta'\alpha$, where β' is a homotopy inverse of β gives the desired result. □

(B.1.21) Remarks
(1) The maps $f : Z(i_1, i_2) \to Z(j_1, j_2)$, resp. $f : K \to L$, constructed in (B.1.20) are not unique. In fact, even their homotopy classes may depend on the chosen homotopies h_ν (resp. h_q).
(2) The two parts of (B.1.20) are not formulated in a completely analogous fashion. In (B.1.20)(1) one can replace - up to homotopy equivalence - the double mapping cylinder (i.e. the 'homotopy colimit') by the actual colimit (i.e. the pushout) if one of the maps in each row is a cofibration. On the other hand in (B.1.20)(2) one would use the 'homotopy colimit' (i.e. the telescope construction) instead of the actual colimit if the horizontal maps were not cofibrations.
We leave it to the reader to formulate and prove the 'dual' results.

B.2 Applications to cochain complexes over graded algebras

In this section we apply some of the results of the preceding section to the category δgR-Mod of cochain complexes over R, where R is an (in the graded sense) commutative graded algebra over a field k. We assume that R is connected, i.e. $R^i = 0$ for $i < 0$ and $R^0 = k$, and hence has a canonical augmentation $\varepsilon : R \to k$, being the identity on R^0 and equal to zero otherwise. An object $\tilde{K}$ in δgR-Mod is a graded R-module (in the sense that the map $R \otimes \tilde{K} \to \tilde{K}$, which gives the R-module structure of $\tilde{K}$, preserves the (total) degree) together with an R-linear differential (or coboundary) $\delta_{\tilde{K}} : \tilde{K} \to \tilde{K}$ which raises degrees by one (and fulfils $\delta_{\tilde{K}} \delta_{\tilde{K}} = 0$).

The cochain complexes for the equivariant cohomology of a G-space which are discussed in the main part of the book are objects in such a category, where typically R is a polynomial ring in finitely many variables over a field k. They are actually free as R-modules and carry some additional multiplicative structure.

The results of Section B.1 carry over to the situation at hand without any difficulties.

It is an immediate consequence of the definitions that the functor

$$k \otimes_R - : \delta gR\text{- Mod} \to \delta gk\text{- Mod}, \qquad \tilde{K} \mapsto K := k \otimes_R \tilde{K},$$

preserves homotopies. In particular, if $\tilde{f} : \tilde{K} \to \tilde{L}$ is a homtopy equivalence in δgR-Mod then $f := \mathrm{id}_k \otimes_R \tilde{f} : K = k \otimes_R \tilde{K} \to L = k \otimes_R \tilde{L}$ is a homotopy equivalence in δgk-Mod. Under the additional assumption that $\tilde{K}$ and $\tilde{L}$ as R-modules (i.e. disregarding differentials) are of the form $\tilde{K} = R \otimes K$ and $\tilde{L} = R \otimes L$ (where K and L are graded k-modules and the tensor product is taken over $k = R^0 \subset R$ if not indicated otherwise) one has the following converse.

(B.2.1) Proposition
Let $\tilde{f} : \tilde{K} \to \tilde{L}$ be a morphism in δgR-Mod and suppose that $\tilde{K} = R \otimes K$ and $\tilde{L} = R \otimes L$ as R-modules. Assume that $\tilde{K}$ and $\tilde{L}$ are bounded below. If $f = \mathrm{id}_k \otimes_R \tilde{f} : K \to L$ is a homotopy equivalence in δgk-Mod, then $\tilde{f}$ is a homotopy equivalence in δgR-Mod.

Before we come to the proof of (B.2.1) it should be noted that, although $\tilde{K} = R \otimes K$ and $\tilde{L} = R \otimes L$ as R-modules, neither the differentials of $\tilde{K}$ and $\tilde{L}$ nor the morphism $\tilde{f} : \tilde{K} \to \tilde{L}$ need necessarily be R-linear extensions of corresponding k-linear maps $\delta_K : K \to K$, $\delta_L : L \to L$ and $f : K \to L$ respectively. In fact, in the situation at hand the differentials $\delta_{\tilde{K}} : \tilde{K} = R \otimes K \to \tilde{K} = R \otimes K$, $\delta_{\tilde{L}} : \tilde{L} = R \otimes L \to \tilde{L} = R \otimes L$ and the morphism $\tilde{f} : \tilde{K} = R \otimes K \to \tilde{L} = R \otimes L$ should be considered as 'twisting', 'deformations' or 'perturbations' of $\delta_K = \mathrm{id}_k \otimes_R \delta_{\tilde{K}}$, $\delta_L = \mathrm{id}_k \otimes_R \delta_{\tilde{L}}$ and $f = \mathrm{id}_k \otimes_R \tilde{f}$ (with respect to R as parameter 'space'). To indicate that for example the differential of $\tilde{K}$ is, in general, not the R-linear extension of δ_K we write $\tilde{K} = R \tilde{\otimes} K$ and call this a twisted tensor product (cf. [Brown, E.H., 1959]) and we call $\delta_{\tilde{K}}$ an R-family of deformations of δ_K etc. Since by assumption $\tilde{f} : \tilde{K} \to \tilde{L}$ is R-linear it is completely determined by its restriction

$$\tilde{f}|k \otimes K : k \otimes K = K \to \tilde{L} = R \otimes L$$

which is a k-linear map. The situation with $\delta_{\tilde{K}}$ and $\delta_{\tilde{L}}$ is analogous. In view of the applications in Chapters 1, 3 and 5, we discuss the following special case in somewhat more detail. Let $R = k[t]$ be the polynomial ring over k in one variable t of degree two and let L be bounded below (i.e. $L^j = 0$ for $j \in \mathbb{Z}$ small enough). Then an R-linear map

$$\tilde{f} : \tilde{K} = R \otimes K \to \tilde{L} = R \otimes L$$

corresponds precisely to a sequence $f_p : K^* \to L^{*-2p}$, $p = 0, 1, 2, \ldots$ of k-linear maps in such a way that $f_0 = f$ and

$$\tilde{f}(1 \otimes x) = 1 \otimes f_0(x) + t \otimes f_1(x) + \ldots + t^p \otimes f_p(x) + \ldots .$$

Since L is assumed to be bounded below the above sum is automatically finite for each $1 \otimes x \in k \otimes K \subset R \otimes K$. Hence $f(x) = f_0(x)$ gives the constant term of $\tilde{f}(1 \otimes x)$ considered as a polynomial in t with coefficients in L, and $f_p(x)$ is the coefficient of t^p. For some applications of the concept of (algebraic) deformations to transformation groups on an elementary level see Section 1.3.

Proof of (B.2.1).

Since $f : K \to L$ is assumed to be a homotopy equivalence the mapping cone $C(f)$ is ht-trivial in δgk-Mod (see (B.1.7)). The mapping cone $C(\tilde{f})$ is of the form $R \tilde{\otimes} C(f)$ and hence - again by (B.1.7) - it suffices to show that an object $\tilde{N} = R \tilde{\otimes} N$ in δgR-Mod is ht-trivial if N is ht-trivial in δgk-Mod. Let $N = Z \oplus D$ with $\delta_N|Z = 0$ and $\delta_N|D : D \to Z$ an isomorphism (see (B.1.8)). Then $R \otimes N = R \otimes Z \oplus R \otimes D$. Of course, the differential $\delta_{\tilde{N}}$ need not respect this decomposition, but the part $\delta''_{\tilde{N}}$ of $\delta_{\tilde{N}}$ which maps $R \otimes D$ to $R \otimes Z$ (i.e the projection to $R \otimes Z$ of the restriction of $\delta_{\tilde{N}}$ to $R \otimes D$) can be considered as an R-family of deformations of $\delta_N|D : D \to Z$. Since $\delta_N|D$ is an isomorphism it follows from (A.7.5)(3) that $\delta''_{\tilde{N}}$ is one, too. Hence by (B.1.8) $\tilde{N} = R \tilde{\otimes} N$ is ht-trivial in δgR-Mod. □

(B.2.2) Remark
There is the following analogue of (B.2.1). Let $\tilde{f} : \tilde{K} \to \tilde{L}$ be as in (B.2.1). If f induces an isomorphism in cohomology then so does $\tilde{f}$.

The proof can be given by using a cohomological version of the dual of (B.1.18). This is more or less the same as applying the spectral sequence, obtained by filtering R according to degree. The E_1-term of the spectral sequence (for $\tilde{K}$) is given by $E_1 = R \otimes H(K)$ and the sequence converges to $H(\tilde{K})$. Hence if $H(f)$ is an isomorphism one obtains an isomorphism on the E_1-level, and therefore $H(\tilde{f})$ is an isomorphism, too.

The conditions '$H(f)$ isomorphism' and 'f homotopy equivalence' are equivalent in the case at hand, and under certain conditions on R (e.g., if R is a polynomial ring in finitely many variables); the same holds for $\tilde{f}$ in place of f (see (B.1.11)).

The reader should also compare the above with the topological situation, where one studies maps between fibrations over the same fixed base (see, e.g., [tom Dieck, Kamps, Puppe, D. 1970]).

For the purpose of explicit calculations one often would like to replace a complex by a homotopy equivalent one, which is 'as small as possible'. In certain cases this can be done using the following proposition.

(B.2.3) Proposition
Let $R\tilde{\otimes}K$ be a twisted tensor product in the sense above and let $f : K \to L$ be a homotopy equivalence in δgk-Mod. Assume K and L are bounded below. Then there exists a twisted tensor product $R\tilde{\otimes}L$ and a homotopy equivalence $\tilde{f} : R\tilde{\otimes}K \to R\tilde{\otimes}L$ in δgR-Mod such that $f = \mathrm{id}_k \otimes_R \tilde{f}$.

Proof.
In view of (B.2.1) it is enough to construct $R\tilde{\otimes}L$ and to find a morphism $\tilde{f} : R\tilde{\otimes}K \to R\tilde{\otimes}L$ such that $\mathrm{id}_k \otimes_R \tilde{f} = f$. By Exercise (B.1)(d) we can factor $f : K \to L$ as $K \to K \oplus N' \cong L \oplus N'' \to L$ with N', N'' ht-trivial in δgk-Mod.

(To prove the exercise first replace $f : K \to L$ by the cofibration $(f, \iota_0) : K \to L \oplus CK$, where $N'' := CK$ is the cone over K and $\iota_0 : K \to CK$ is the inclusion of K as the base of the cone. Since ι_0 has a left inverse, so does (f, ι_0) and hence (f, ι_0) is a cofibration (see (B.1.4)). Clearly (f, ι_0) is also a homotopy equivalence because CK is ht-trivial. So by (B.1.7)(3) $L \oplus CK$ can be decomposed as $K \oplus N'$ with N' ht-trivial in such a way that the standard inclusion $K \to K \oplus N'$ corresponds to (f, ι_0). We thus obtain the desired factoriztion of f : $K \to K \oplus N' \cong L \oplus N'' \to L$.)

It is trivial to extend the twisted tensor product structure from $R\tilde{\otimes}K$ to $R\tilde{\otimes}(K \oplus N') = R\tilde{\otimes}K \oplus R\tilde{\otimes}N'$ by just defining the differential on $R\tilde{\otimes}N'$ as the R-linear extension of the differential of N'. Using the isomorphism $K \oplus N' \cong L \oplus N''$ one gets a twisted tensor product structure on $R\tilde{\otimes}(L \oplus N'')$. In general the differential of $R\tilde{\otimes}(L \oplus N'')$ will not respect the given direct sum decomposition $R \otimes (L \oplus N'') = R \otimes L \oplus R \otimes N''$, but one can define a morphism $\tilde{j} : R\tilde{\otimes}N'' \to R\tilde{\otimes}(L \oplus N'')$ in δgR-Mod, where the differential of $R\tilde{\otimes}N''$ is again just the R-linear extension of the differential of N'' such that $j = \mathrm{id}_k \otimes_R \tilde{j} : N'' \to L \oplus N''$ is the standard inclusion and the cokernel of $\tilde{j}$ is isomorphic to $R \otimes L$ in gR-Mod and hence gives a twisted tensor product $R\tilde{\otimes}L$. To define $\tilde{j}$ we choose a splitting $N'' = Z'' \oplus D''$ with $\delta_{N''}|Z'' = 0$ and $\delta_{N''}|D'' : D'' \to Z''$ an isomorphism. One only needs to define $\tilde{j}$ on $R \otimes D''$ in such a way that $\tilde{j}|D'' : D'' \to L \oplus N'' = L \oplus Z'' \oplus D''$ is the standard inclusion. Since $\delta_{R\tilde{\otimes}N''} : R \otimes D'' \to R \otimes Z''$ is an isomorphism of R-modules there is a unique extension of $\tilde{j}|R\otimes D''$ to a morphism $\tilde{j} : R\tilde{\otimes}N'' \to R\tilde{\otimes}(L\oplus N'')$ in δgR-Mod. In particular it follows that $j : N'' \to L \oplus N''$ is the canonical inclusion and (A.7.5)(3) implies that $\mathrm{coker}(\tilde{j})$ is isomorphic to $R \otimes L$ in gR-Mod. One therefore gets an induced differential on $R\otimes L$ which gives the desired twisted tensor product structure $R\tilde{\otimes}L$. The morphism $\tilde{f}$ can be chosen to be the composition $R\tilde{\otimes}K \to R\tilde{\otimes}(K\oplus N') \cong R\otimes(L\oplus N'') \to R\tilde{\otimes}L$, where the first map is the canonical inclusion and the last is the projection to $\mathrm{coker}(\tilde{j})$ (and, in general, not the canonical projection with respect to the given direct sum decomposition). It is immediate from the construction that $\mathrm{id}_k \otimes_R \tilde{f} = f$. □

(B.2.4) Corollary
The twisted product $R\tilde{\otimes}K$ (K bounded below) can, up to homotopy equivalence in δgR-Mod, be replaced by $R\tilde{\otimes}H(K)$.

Proof.
This is an immediate consequence of (B.1.14) and (B.2.3). □

The complex $R\tilde{\otimes}H(K)$ can be viewed as a minimal R-free model (i.e. free as a R-module) for the homotopy type of $R\tilde{\otimes}K$ in δgR-Mod. It is uniquely determined up to isomorphism in δgR-Mod (see Exercise (B.7)).

The constructions in Section 1.3, (cf. also Sections 1.4 and 3.5) lead to twisted tensor products $\tilde{K} = R\tilde{\otimes}K$, where $R = k[t_1, \ldots, t_r]$ is a polynomial ring in r variables $t_1, \ldots, t_r$ over a field k and K is bounded below, e.g., in the case of $(\mathbb{Z}_2)^r$-actions the cochain model (with coefficients in $\mathbb{F}_2$) for the Borel construction is such a twisted tensor product, where $k = \mathbb{F}_2$ and all the variables have degree one. The differential $\tilde{\delta} : \tilde{K} \to \tilde{K}$ applied to an element $1 \otimes x \in R \otimes K$ can be written as a polynomial

$$\tilde{\delta}(1 \otimes x) = 1 \otimes \delta_0(x) + \sum_{i=1}^{r} t_i \otimes \delta_{1,i}(x) + \ldots,$$

B.1. Applications to cochain complexes

where $\delta_{1,i} : K \to K$ is a k-linear map which preserves degree (since $\tilde{\delta}$ increases the total degree by 1). Of course, the relation $\tilde{\delta}\tilde{\delta} = 0$ implies certain conditions on the coefficients $\delta_0(x)$, $\delta_{1,i}(x)$ etc. of the above polynomial, which we write down in a more explicit way for the case $r = 1$. Then $\tilde{\delta}(1 \otimes x) = \sum_{p=0}^{\infty} t^p \otimes \delta_p(x)$ and $\tilde{\delta}\tilde{\delta} = 0$ is equivalent to the following set of equations: $\sum_{i+j=m} \delta_i\delta_j = 0$ for $m = 0, 1, 2, \ldots$ (i.e. $\delta_0\delta_0 = 0$, $\delta_1\delta_0 + \delta_0\delta_1 = 0$, etc.). (Note that compared with the notation in Chapter 1 the order of the factors in the tensor product $R\tilde{\otimes}K$ is reversed, i.e. there we use $K\tilde{\otimes}R$ instead of $R\tilde{\otimes}K$.)

In general we can write $\tilde{\delta} : R\tilde{\otimes}K \to R\tilde{\otimes}K$ applied to an element $1 \otimes x \in R\tilde{\otimes}K$ as a sum $\tilde{\delta}(1 \otimes x) = \sum_{p=0}^{\infty} \tilde{\delta}_p(1 \otimes x)$, where $\tilde{\delta}_p(1 \otimes x)$ is the component of $\tilde{\delta}(1 \otimes x)$ in $R^p \otimes K$. We also denote the R-linear extension of $\tilde{\delta}_p : 1 \otimes K \to R^p \otimes K$ to $R \otimes K$ by $\tilde{\delta}_p$. Again $\tilde{\delta}\tilde{\delta} = 0$ implies certain relations between the $\tilde{\delta}_p$s, i.e. $\sum_{i+j=m} \tilde{\delta}_i\tilde{\delta}_j = 0$ for $m = 0, 1, 2, \ldots$, in particular $\tilde{\delta}_0\tilde{\delta}_0 = 0$ and $\tilde{\delta}_0\tilde{\delta}_1 + \tilde{\delta}_1\tilde{\delta}_0 = 0$. The complex K is homotopy equivalent to its homology $H(K)$, and for any twisted tensor product $R\tilde{\otimes}K$ one can construct a twisted tensor product $R\tilde{\otimes}H(K)$ and a homotopy equivalence $\tilde{f} : R\tilde{\otimes}K \to R\tilde{\otimes}H(K)$ in δgR-Mod, such that $f = \mathrm{id}_k \otimes_R \tilde{f} : K \to H(K)$ is a homotopy equivalence in δgk-Mod. We can choose a decomposition $K = H(K) \oplus N$ such that N is ht-trivial and $f : H(K) \oplus N \to H(K)$ is the standard projection to the first factor. It is clear that the part $\tilde{\delta}_{H,0}$ of the differential $\tilde{\delta}_H = \sum_{p=0}^{\infty} \tilde{\delta}_{H,p}$ of $R\tilde{\otimes}H(K)$ must vanish ($\tilde{\delta}_{H,0}$ is determined by $\delta_H = \mathrm{id}_k \otimes_R \tilde{\delta}_H$, and δ_H fulfils the relation $f\delta_K = \delta_H f$; but δ_K vanishes on elements of the form $(h, 0) \in H(K) \oplus N$).

We want to show that $\tilde{\delta}_{H,1}$ is induced by $\tilde{\delta}_{K,1}$ taking homology with respect to the differential $\tilde{\delta}_{K,0}$. Similar to the presentation of the differentials we can decompose $\tilde{f} : R\tilde{\otimes}K \to R\tilde{\otimes}H(K)$, i.e. $\tilde{f} = \sum_{p=0}^{\infty} \tilde{f}_p$ where $\tilde{f}_p : R\otimes K \to R\otimes H(K)$ is an R-linear map which raises the degree of the first factor by p, but preserves the total degree. The equation

$$\left(\sum_{p=0}^{\infty} \tilde{f}_p\right)\left(\sum_{q=0}^{\infty} \tilde{\delta}_{K,q}\right) = \tilde{f}\tilde{\delta}_K = \tilde{\delta}_H\tilde{f} = \left(\sum_{p=0}^{\infty} \tilde{\delta}_{H,p}\right)\left(\sum_{q=0}^{\infty} \tilde{f}_q\right)$$

implies in particular:

$$\tilde{f}_1\tilde{\delta}_{K,0} + \tilde{f}_0\tilde{\delta}_{K,1} = \tilde{\delta}_{H,1}\tilde{f}_0 + \tilde{\delta}_{H,0}\tilde{f}_1$$

(more generally one has:

$$\sum_{i+j=m} \tilde{f}_i\tilde{\delta}_{K,j} = \sum_{i+j=m} \tilde{\delta}_{H,i}\tilde{f}_j$$

for $m = 0, 1, 2, \ldots$ which is obtained by separating the component of $\tilde{f}\tilde{\delta}_K(1 \otimes x) = \tilde{\delta}_H\tilde{f}(1 \otimes x)$ in $R^m \otimes H(K)$). We know already that

$\tilde{\delta}_{H,0}$ must vanish, hence $\tilde{f}_1\tilde{\delta}_{K,0} + \tilde{f}_0\tilde{\delta}_{K,1} = \tilde{\delta}_{H,1}\tilde{f}_0$. On elements of the form $r \otimes (h,0) \in R \otimes (H(K) \oplus \tilde{N})$ the map $\tilde{\delta}_{K,0}$ is zero. Therefore $\tilde{f}_0\tilde{\delta}_{K,1}(r\otimes(h,0)) = \tilde{\delta}_{H,1}(r\otimes h)$. Since $\tilde{\delta}_{K,1}\tilde{\delta}_{K,0}+\tilde{\delta}_{K,0}\tilde{\delta}_{K,1} = 0$ the element $\tilde{\delta}_{K,1}(r\otimes(h,0))$ is mapped to zero under $\tilde{\delta}_{K,0}$. Hence $\tilde{f}_0(\tilde{\delta}_{K,1}(r\otimes(h,0)))$ is just the homology class of $\tilde{\delta}_{K,1}(r\otimes(h,0))$ with respect to the differential $\tilde{\delta}_{K,0} : R \otimes K \to R \otimes K$.

Exercises

(B.1) Show that all the basic notions of the homotopy theory in ∂R-Mod can be deduced from the class $\mathcal{N}$ of ht-trivial objects by purely categorical constructions, more precisely:

(a) Two morphisms $f_0, f_1 : K \to L$ in ∂R-Mod are homotopic if and only if $f_1 - f_0$ factors through an object $N \in \mathcal{N}$.

(b) A morphism $f : K' \to K$ is a cofibration if and only if for any morphism $g' : K' \to N$ with $N \in \mathcal{N}$ there exists an extension $g : K \to N$ of g' (i.e. $gf = g'$).

(c) A morphism $f : K \to K''$ is a fibration if and only if for any morphism $g'' : N \to K''$ with $N \in \mathcal{N}$ there exists a lifting $g : N \to K$ of g'' (i.e. $fg = g''$).

(d) A morphism $f : K \to L$ is a homotopy equivalence if and only if there exist $N', N'' \in \mathcal{N}$ such that f factors as

$$K \to K \oplus N' \to L \oplus N'' \to L,$$

where the first map in this composition is the canonical inclusion, the second is an isomorphism and the third the canonical projection. In particular: K and L are homotopy equivalent if and only they are '$\mathcal{N}$-stably isomorphic', i.e. there exist $N', N'' \in \mathcal{N}$, such that $K \oplus N'$ and $L \oplus N''$ are isomorphic in ∂R-Mod.

(B.2) Let $\mathcal{C}$ be an additive category which has finite limits and colimits, and let $\mathcal{N}$ be a subclass of objects in $\mathcal{C}$, which is closed under finite sums (biproducts) and retracts. Define cofibrations, fibrations and (weak-)equivalences in $\mathcal{C}$ (with respect to $\mathcal{N}$) by using the characterizations in Exercise (B.1) (b)(c)(d) as definitions, e.g., $f : K' \to K$ is a cofibration if for any morphism $g' : K' \to N$ with $N \in \mathcal{N}$ there exists a morphism $g : K \to N$ such that $gf = g'$, etc.

Assume that $(\mathcal{C}, \mathcal{N})$ has 'enough cofibrations and enough fibrations', i.e. for any object K in $\mathcal{C}$ there exists a cofibration $K \to N'(K)$ and a fibration $N''(K) \to K$ with $N'(K), N''(K) \in \mathcal{N}$.

(a) Show that $\mathcal{C}$ together with the cofibrations, fibrations and (weak-) equivalences defined above form a closed model category in the sense of [Quillen, 1967].

(b) Prove: If $\mathcal{C}$ is the category of chain complexes and $\mathcal{N}$ is the class of ht-trivial complexes then the notion of homotopy (see Exercise (B.1)(a)) derived from the closed model structure of $(\mathcal{C}, \mathcal{N})$ coincides with the usual one.

(c) In the category of chain complexes choose for $\mathcal{N}$ the subclass of complexes which are hl-trivial (instead of ht-trivial), and compare the resulting 'homotopy category' with (b).

(B.3) For any additive category the choice of a class $\mathcal{N}$ of objects which is closed under finite sums and retracts will lead to a notion of cofibrations, fibrations and (weak-) equivalences (see Exercise (B.2)). Try to identify these notions in more common categorical terms in the following examples and check under which additional assumptions they lead to a closed model category (cf. [Hilton, 1965] Chapter 13, [Varadarajan, 1975]):

(a) $\mathcal{C} := R\text{-Mod}$, $\mathcal{N} := \{N \in ob\mathcal{C};\ N \text{ injective}\}$

(b) $\mathcal{C} := R\text{-Mod}$, $\mathcal{N} := \{N \in ob\mathcal{C};\ N \text{ projective}\}$

(c) $\mathcal{C} :=$ full subcategory of finitely generated projective modules in $R\text{-Mod}$, $\mathcal{N} := \{N \in ob\mathcal{C};\ N \text{ free}\}$

(d) $\mathcal{C} := \mathbb{Z}\text{-Mod}$

(α) $\mathcal{N} := \{N \in ob\mathcal{C};\ pN = 0\}$ for a fixed prime p

(β) $\mathcal{N} := \{N \in ob\mathcal{C};\ N \text{ torsion}\}$

(γ) $\mathcal{N} := \{N \in ob\mathcal{C};\ N \xrightarrow{\mu_p} N \text{ isomorphism}\}$, where μ_p denotes multiplication with a fixed prime p ($\mu_p(n) = pn$ for $n \in N$).

(δ) $\mathcal{N} := \{N \in ob\mathcal{C};\ N \xrightarrow{\mu_p} N \text{ isomorphism for all primes } p\}$

(B.4) Let P be an object in $\partial R\text{-Mod}$ which is projective as an R-module. Show that this does not imply that P is a 'projective object' in $\partial R\text{-Mod}$ in the sense that for any surjection $f : K \to L$ any morphism $g : P \to L$ can be lifted to K.

(B.5) Give an example of a morphism $\tilde{f} : R\tilde{\otimes}K \to R\tilde{\otimes}L$ in $\delta gR\text{-Mod}$ which is a homotopy equivalence in $\delta gk\text{-Mod}$ but not in $\delta gR\text{-Mod}$.

(B.6) Give a direct proof of (B.2.4) by constructing inductively pairs of maps $(\tilde{f}|R \otimes K^{q+1}, \tilde{\delta}_H|R \otimes H^q(K))$, where $\tilde{\delta}_H$ is the differential in $R\tilde{\otimes}H(K)$ and $\tilde{f} : R\tilde{\otimes}K \to R\tilde{\otimes}H(K)$ is the desired homotopy equivalence.

(B.7) Let $\tilde{K} = R\tilde{\otimes}K$ be a twisted tensor product as in (B.2.4) (K bounded below). If $\tilde{H} = R\tilde{\otimes}H$ and $\tilde{H}' = R\tilde{\otimes}H'$ are twisted tensor products which are homotopy equivalent to $\tilde{K} = R\tilde{\otimes}K$ in $\delta gR\text{-Mod}$ and if $\delta_H = \text{id}_k \otimes_R \delta_{\tilde{H}}$ and $\delta_{H'} = \text{id}_k \otimes_R \delta_{\tilde{H}'}$ vanish then $\tilde{H} = R\tilde{\otimes}H$ and $\tilde{H}' = R\tilde{\otimes}H'$ are isomorphic in $\delta gR\text{-Mod}$.

References

Adem, A. 1987: $\mathbb{Z}/p\mathbb{Z}$ actions on $(S^n)^k$, *Trans. Amer. Soc.* **300**, 791-809.
Adem, A. 1988: Cohomological restrictions on finite group actions, *J. Pure Applied Algebra* **54**, 117-39.
Adem, A. 1989: Torsion in equivariant cohomology, *Comment. Math. Helv.* **64**, 401-11.
Adem, A. and Browder, W. 1988: The free rank of symmetry of $(S^n)^k$, *Invent. Math.* **92**, 431-40.
Allday, C. 1976: The stratification of compact connected Lie group actions by subtori, *Pacific J. Math.* **62**, 311-27.
Allday, C. 1979: Rational homotopy and torus actions, *Houston J. Math.* **5**, 1-19.
Allday, C. 1985: A family of unusual torus group actions, *Group Actions on Manifolds*, R. Schultz, (Ed.), Contemporary Math. **36**, American Math. Soc., Providence, Rhode Island, 107-11.
Allday, C. and Halperin, S. 1978: Lie group actions on spaces of finite rank, *Quarterly J. Math. Oxford* **29**, 63-76.
Allday, C. and Halperin, S. 1984: Sullivan-de Rham theory for rational Alexander-Spanier cohomology, *Houston J. Math.* **10**, 15-33.
Allday, C. and Puppe, V. 1984: On the rational homotopy of circle actions, *Proc. Aarhus Symposium on Algebraic Topology* (1982), I. Madsen and R. Oliver (Eds.), Lecture Notes in Math. 1051, Springer-Verlag, Berlin-Heidelberg-New York, 533-9.
Allday, C. and Puppe, V. 1986a: Bounds on the torus rank, *Transformation Groups, Poznań* (1985), S. Jackowski and K. Pawałowski (Eds.), Lecture Notes in Math. 1217, Springer-Verlag, Berlin-Heidelberg-New York, 1-10.
Allday, C. and Puppe, V. 1986b: On the rational homotopy Lie algebra of a fixed point set of a torus action, *Trans. Amer. Math. Soc.* **297**, 521-8.
Allday, C. and Skjelbred, T. 1974: The Borel formula and the topological splitting principle for torus actions on a Poincaré duality space, *Ann. of Math.* **100**, 322-5.

Andrews, P. and Arkowitz, M. 1978: Sullivan's minimal models and higher order Whitehead products, *Can. J. Math.* **30**, 961-82.

Assadi, A. 1988: Varieties in finite transformation groups, *Bull. Amer. Math. Soc. (New Series)*, **19**, 459-63.

Atiyah, M.F. 1974: *Elliptic Operators and Compact Groups*, Lecture Notes in Math. 401, Springer-Verlag, Berlin-Heidelberg-New York.

Atiyah, M.F. and MacDonald, I.G. 1969: *Introduction to Commutative Algebra*, Addison-Wesley, Reading, Massachusetts.

Atiyah, M.F. and Bott, R. 1984: The moment map and equivariant cohomology, *Topology* **23**, 1-28.

Back, A.H. 1977: *Involutions on Grassman Manifolds*, Thesis, University of California.

Barnes, D.W. 1985: *Spectral Sequence Constructors in Algebra and Topology*, Memoirs of the Amer. Math. Soc. 317, Providence, Rhode Island.

Bass, H. 1963: Big projective modules are free, *Illinois J. Math.* **7**, 24-31.

Baum, P.F. 1968: On the cohomology of homogeneous spaces, *Topology* **7**, 15-38.

Baumgartner, J.C. 1990: Kohomologie freier $(\mathbb{Z}_p)^r$ - Räume, Dissertation, Konstanz.

Benson, D.J. and Carlson, J.F. 1987: Complexity and multiple complexes, *Math. Zeit.* **195**, 221-38.

Borel, A. 1967: *Topics in the Homology Theory of Fibre Bundles*, Lecture Notes in Math. 36, Springer-Verlag, Berlin.

Borel, A. et al. 1960: *Seminar on Transformation Groups*, Ann. of Math. Studies No. 46, Princeton University Press, Princeton.

Bott, R. 1967: Vector fields and characteristic numbers, *Mich. Math. J.* **14**, 231-44.

Bourbaki, N. 1961: *Algèbre Commutative*, Chapter 2, Hermann, Paris.

Bourbaki, N. 1966: *Topologie Générale*, Chapter 9, Hermann, Paris.

Bourbaki, N. 1967: *Algèbre Commutative*, Chapter 1, Hermann, Paris.

Bousfield, A.K. and Gugenheim, V.K.A.M. 1976: *On P.L. de Rham Theory and Rational Homotopy Type*, Memoirs of the Amer. Math. Soc. 179, Providence, Rhode Island.

Bousfield, A.K. and Kan, D.M. 1972: *Homotopy Limits, Completions and Localizations*, Lecture Notes in Math. 304, Springer-Verlag, Berlin-Heidelberg-New York.

Bredon, G.E. 1960: Orientation in generalized manifolds and applications to the theory of transformation groups, *Mich. Math. J.* **7**, 35-64.

Bredon, G.E. 1967a: *Sheaf Theory*, McGraw-Hill, New York.

Bredon, G.E. 1967b: *Equivariant Cohomology Theories*, Lecture Notes in Math. 34, Springer-Verlag, Berlin-Heidelberg-New York.

Bredon, G.E. 1969: Homotopical properties of fixed point sets of circle group actions, I, *Amer. J. Math.* **91**, 874-88.

Bredon, G.E. 1972: *Introduction to Compact Transformation Groups*, Academic Press, New York.

Bredon, G.E. 1973: Fixed point sets of actions on Poincaré duality spaces, *Topology* **12**, 159-75.

Bredon, G.E. 1974: The free part of a torus action and related numerical equalities, *Duke Math. J.* **41**, 843-54.

Bredon, G.E. 1977: Book review, *Bull. Amer. Math. Soc.* **83**, 711-18.
Browder, W. 1983: Cohomology and group actions, *Invent. Math.* **71**, 599-607; Addendum and Erratum, *Invent. Math.* **75** (1984), 585.
Browder, W. 1985: S^1-actions on open manifolds, *Conference on Algebraic Topology in Honor of Peter Hilton (Saint John's Newfoundland, 1983)*, Contemporary Math. **37**, Amer. Math. Soc., Providence, Rhode Island, 25-30.
Browder, W. 1987: Pulling back fixed points, *Invent. Math.* **87**, 331-42.
Browder, W. 1988: Actions of elementary abelian p-groups, *Topology* **27**, 459-72.
Browder, W. 1989: Talk at the International Conference on Algebraic Topology, Poznań, June 22-27.
Brown, E.H. 1959: Twisted tensor products, I, *Ann. of Math.* **69**, 223-46.
Brown, K.S. 1982: *Cohomology of Groups*, Graduate Texts in Math. **87**, Springer-Verlag, New York-Heidelberg-Berlin.
Carlson, J.F. 1983: The varieties and the cohomology ring of a module, *J. Algebra* **85**, 104-43.
Carlson, J.F. 1984: The variety of an indecomposable module is connected, *Invent. Math.* **77**, 291-9.
Carlson, J.F. 1985: The cohomology ring of a module, *J. Pure Applied Algebra* **36**, 105-21.
Carlson, J.F. 1989: Exponents of modules and maps, *Invent. Math.* **95**, 13-24.
Carlsson, G.E. 1980: On the non-existence of free actions of elementary abelian groups on products of spheres, *Amer. J. Math.* **102**, 1147-57.
Carlsson, G.E. 1982: On the rank of abelian groups acting freely on $(S^n)^k$, *Invent. Math.* **69**, 393-404.
Carlsson, G.E. 1983: On the homology of finite free $(\mathbb{Z}/2)^n$-complexes, *Invent. Math.* **74**, 139-47.
Carlsson, G.E. 1987a: Free $(\mathbb{Z}/2)^3$-actions on finite complexes, *Algebraic Topology and Algebraic K-Theory*, W. Browder (Ed.), Ann. of Math. Studies No. 113, Princeton University Press, Princeton, 332-44.
Carlsson, G.E. 1987b: Segal's Burnside ring conjecture and the homotopy limit problem, *Homotopy Theory, Proceedings of the Durham Symposium 1985*, E. Rees and J.D.S. Jones (Eds.), London Math. Soc. Lecture Note Series 117, Cambridge University Press, Cambridge, 6-34.
Cartan, H. and Eilenberg S. 1956: *Homological Algebra*, Princeton Math. Series No. 19, Princeton University Press, Princeton.
Chang, C.-N. and Su, J.-C. 1980: A theorem on Poincaré algebras, *Bull. Inst. Math. Academia Sinica* **8**, 417-38.
Chang, T. 1976: On the number of relations in the cohomology of a fixed point set, *Manuscripta Math.* **18**, 237-47.
Chang, T. and Skjelbred, T. 1972: Group actions on Poincaré duality spaces, *Bull. Amer. Math. Soc.* **78**, 1024-6.
Chang, T. and Skjelbred, T. 1974: The topological Schur lemma and related results, *Ann. Math.* **100**, 307-21.
Chang, T. and Skjelbred, T. 1976: Lie group actions on a Cayley projective plane and a note on homogeneous spaces of prime Euler characteristic, *Amer. J. Math.* **98**, 655-78.
Conner, P.E. 1960: Retraction properties of the orbit space of a compact topological transformation group, *Duke Math. J.* **27**, 341-57.

Conner, P.E. and Floyd, E.E. 1964: *Differentiable Periodic Maps.* Springer-Verlag, Berlin-Göttingen-Heidelberg.

Cusick, L.W. 1987: Free actions on products of even dimensional spheres, *Proc. Amer. Math. Soc.* **99**, 573-4.

Dade, E.C. 1978: Endo-permutation modules over p-groups II, *Ann. Math.* **108**, 317-46.

Davis, J.F. and Milgram, R.J. 1991: Free actions on products of spheres: the rational case, *Math. Zeit.* **207**, 359-90.

Deligne, P., Morgan, J., Griffiths, P. and Sullivan, D. 1975: The real homotopy theory of Kähler manifolds, *Invent. Math.* **29**, 245-54.

Deo, S. and Tripathi, H.S. 1982: Compact Lie group actions on finitistic spaces, *Topology* **21**, 393-9.

tom Dieck, T. 1970: Bordism of G-manifolds and integrality theorems, *Topology* **9**, 345-58.

tom Dieck, T., Kamps, K.H. and Puppe, D. 1970: *Homotopietheorie*, Lecture Notes in Math. 157, Springer-Verlag, Berlin-Heidelberg-New York.

tom Dieck, T. 1972: Existence of fixed points, *Proc. Second Conf. on Compact Transformation groups*, Part I, Lecture Notes in Math. 298, Springer-Verlag, Berlin-Heidelberg-New York, 163-9.

tom Dieck, T. 1987: *Transformation Groups*, de Gruyter Studies in Math. 8, Walter de Gruyter, Berlin.

Dold, A. 1960: Zur Homotopietheorie der Kettenkomplexe, *Math. Ann.* **140**, 278-98.

Dold, A. 1963: Partitions of unity in the theory of fibrations, *Ann. Math.* **78**, 223-55.

Dold, A. 1980: *Lectures on Algebraic Topology*, Second Edition, Grundlehren der math. Wissenschaften 200, Springer-Verlag, Berlin-Heidelberg-New York.

Dowker, C.H. 1952: Homology groups of relations, *Ann. Math.* **56**, 84-95.

Dranishnikov, A.N. 1988: On homological dimension modulo p, *Math. U.S.S.R. Sbornik* **60**, 413-25.

Duflot, J. 1981: Depth and equivariant cohomology, *Comment. Math. Helv.* **56**, 627-37.

Duflot, J. 1983: Smooth toral actions, *Topology* **22**, 253-65.

Dwyer, W.G. and Wilkerson, C.W. 1988: Smith theory revisited, *Ann. Math.* **127**, 191-8.

Dwyer, W.G. and Wilkerson, C.W. 1991: Smith theory and the functor T, *Comment. Math. Helv.*, 1991 **66**, 1-17.

Dyer, E. 1969: *Cohomology Theories*, W.A. Benjamin, New York.

Eilenberg, S. and Moore, J.C. 1961: Limits and spectral sequences, *Topology* **1**, 1-23.

Evans, E.G. and Griffith, P. 1985: *Syzygies*, London Math. Soc. Lecture Note Series 106, Cambridge University Press, Cambridge.

Félix, Y. 1984: Espaces formels et π-formels, *Soc. Math. de France, Astérisque*, **113-4**, 96-108.

Félix, Y. and Halperin, S. 1982a: Formal spaces with finite dimensional rational homotopy, *Trans. Amer. Soc.* **270**, 575-88.

Félix, Y. and Halperin, S. 1982b: Rational L.S. category and its applications, *Trans. Amer. Math. Soc.* **273**, 1-37.

Félix, Y. and Tanré, D. 1988: Formalité d'une application et suite spectrale d'Eilenberg-Moore, *Algebraic Topology, Rational Homotopy, Proceedings, Louvain-la-Neuve*, 99-123, Y. Félix (Ed.), Lecture Notes in Math. 1318, Springer-Verlag, Berlin-Heidelberg-New York.

Félix, Y., Halperin, S., Jacobsson, C., Löfwall, C. and Thomas J.-C., 1988: The radical of the homotopy Lie algebra, *Amer. J. Math.* **110**, 301-22.

Friedlander, J. and Halperin, S. 1979: An arithmetic characterization of the rational homotopy groups of certain spaces, *Invent. Math.* **53**, 117-38.

Gerstenhaber, M. 1964: On the deformation of rings and algebras, *Ann. Math.* **79**, 59-103.

Gerstenhaber, M. 1966: On the deformation of rings and algebras II, *Ann. Math.* **84**, 1-19.

Gerstenhaber, M. 1974: On the deformation of rings and algebras VI, *Ann. Math.* **99**, 257-76.

Gerstenhaber, M. and Schack, S.D. 1988: Algebraic cohomology and deformation theory, M. Hazewinkel and M. Gerstenhaber, (Eds.), *Deformation Theory of Algebras and Structures and Applications*, Kluwer Academic Publishers, Dordrecht-Boston-London, 11-264.

Godement, R. 1958: *Topologie Algébrique et Théorie des Faisceaux*, Hermann, Paris.

Golber, D. 1971: Torus actions on a product of two odd spheres, *Topology* **10**, 313-26.

Gottlieb, D.H. 1986: The trace of an action and the degree of a map, *Trans. Amer. Math. Soc.* **293**, 381-410.

Greenberg, M.J and Harper, J.R. 1981: *Algebraic Topology, A First Course*, The Benjamin/Cummings Publishing Company, Menlo Park, California.

Greub, W., Halperin, S. and Vanstone, R. 1976: *Connections, Curvature and Cohomology*: Volume III, *Cohomology of Principle Bundles and Homogeneous Spaces*, Academic Press, New York.

Grivel, P.-P. 1977: Suite spectrale et modèle minimal d'une fibration, Thesis, University of Geneva.

Grivel, P.-P. 1979: Formes différentielles et suites spectrales, *Ann. Inst. Fourier* **29**, 17-37.

Hadwiger, H., Debrunner, H. and Klee, V. 1964: *Combinatorial Geometry in the Plane*, Holt, Rinehart and Winston, New York.

Halperin, S. 1977: Finiteness in the minimal models of Sullivan, *Trans. Amer. Math. Soc.* **230**, 173-99.

Halperin, S. 1977 (1983): *Lectures on Minimal Models*, Publications Internes de l'U.E.R. de Mathématiques Pures et Appl. No. 111 (1977), Université des Sciences de Techniques de Lille I; or Mém Soc. Math. France (N.S) No. 9-10 (1983).

Halperin, S. 1978: Rational fibrations, minimal models and fibrings of homogeneous spaces, *Trans. Amer. Math. Soc.* **244**, 199-224.

Halperin, S. 1985: Rational homotopy and torus actions, *Aspects of Topology*, 293-306, London Math. Soc. Lecture Note Series 93, Cambridge University Press, Cambridge.

Halperin, S.: Toral actions and the invariant $\chi_\pi(X)$ for non-simply-connected spaces, preprint.

Halperin, S. and Stasheff, J. 1979: Obstructions to homotopy equivalences, *Adv. Math.* **32**, 233-79.

Hattori, Y. 1985: A note on finitistic spaces, *Questions Answers Gen. Topology* **3**, 47-55.

Hauschild, V. 1983: *Deformationen graduierter Artinscher Algebren in der Kohomologietheorie von Transformationsgruppen.* Habilitationsschrift, Konstanz.

Hauschild, V. 1985: Actions of compact Lie groups on homogeneous spaces, *Math. Z.* **189**, 475-86.

Hauschild, V. 1986: The Euler characteristic as an obstruction to compact Lie group actions, *Trans. Amer. Math. Soc.* **298**, 549-78.

Hazewinkel, M. and Gerstenhaber, M. (Eds.) 1988: *Deformation Theory of Algebras and Structures and Applications*, Kluwer Academic Publishers, Dordrecht-Boston-London, NATO ASI series, Series C: Mathematical and Physical Sciences, Vol. 247.

Heller, A. 1959: A note on spaces with operators, *Illinois J. Math.* **3**, 98-100.

Hilton, P.J. 1965: *Homotopy Theory and Duality*, Gordon and Breach, London.

Hilton, P., Mislin, G. and Roitberg, J. 1975: *Localization of Nilpotent Groups and Spaces*, North Holland Mathematics Studies 15, North Holland/American Elsevier, Amsterdam and New York.

Hirsch, G. 1954: Sur les groupes d'homologie des espaces fibrés, *Bull. Soc. Math. Belg.* **6**, 79-96.

Hoffman, M.E. 1987: Free actions of abelian groups on a Cartesian power of an even sphere, *Can. Math. Bull.* **30**, 358-62.

Hsiang, W.-Y. 1974: Structural theorems for topological actions of $\mathbb{Z}_2$-tori on real, complex and quaternionic projective spaces, *Comment. Math. Helv.* **49**, 479-91.

Hsiang, W.-Y. 1975: *Cohomology Theory of Topological Transformation Groups*, Ergebnisse der Math. und ihrer Grenzgebiete 85, Springer-Verlag, Berlin-Heidelberg-New York.

Hu, S.-T. 1959: *Homotopy Theory*, Academic Press, New York.

Husemoller, D. 1966: *Fibre Bundles*, McGraw-Hill, New York.

Iarrobino, A. and Emsalem, J. 1978: Some zero-dimensional generic singularities; finite algebras having small tangent space, *Compositio Math.* **36**, 145-88.

Illman, S. 1972a: Equivariant algebraic topology. Thesis, Princeton.

Illman, S. 1972b: Equivariant singular homology and cohomology for actions of compact Lie groups, *Transformation Groups. Proc. Conf. Amherst 1971*, Lecture Notes in Math. 298, Springer-Verlag, Berlin-Heidelberg-New York, 403-15.

Illman, S. 1975: Equivariant singular homology and cohomology, *Mem. Amer. Soc.* **156**, 1-74.

Illman, S. 1978: Smooth equivariant triangulations of G-manifolds for G a finite group, *Math. Ann.* **233**, 199-220.

Illman, S. 1983: The equivariant triangulation theorem for actions of compact Lie groups, *Math. Ann.* **262**, 487-501.

Illman, S. 1990: Restricting the transformation group in equivariant CW complexes, *Osaka J. Math.* **27**, 191-206.

Jackowski, S. 1988: A fixed-point theorem for p-group actions, *Proc. Amer. Math. Soc.* **102**, 205-8.

Jacobson, N. 1974: *Basic Algebra* I, W.H. Freeman and Company, San Francisco.

Jänich, K. 1968: *Differenzierbare G-Mannigfaltigkeiten*, Lecture Notes in Math. 59, Springer-Verlag, Berlin-Heidelberg-New York.

Kaplansky, I. 1970: *Commutative Rings*, Allyn and Bacon, Boston.

Kawakubo, K. 1980: Equivariant Riemann-Roch theorems, localization and formal group law, *Osaka J. Math.* **17**, 531-71.

Kirby, S. and Siebenmann, L. 1977: *Foundational Essays on Topological Manifolds, Smoothings and Triangulations*, Ann. of Math. Studies No. 88, Princeton University Press, Princeton.

Kister, J. 1964: Microbundles are fibre bundles, *Ann. Math.* **80**, 190-9.

Kobayashi, M. 1988: Fixed point sets of orientation reversing involutions on 3-manifolds, *Osaka J. Math.* **25**, 877-9.

Ku, H.-T. and Ku, M.-C. 1968: The Lefschetz Fixed Point Theorem for involutions, in P.S. Mostert, (Ed.), *Proceedings of the Conference on Transformation Groups, New Orleans 1967*, Springer-Verlag, Berlin-Heidelberg-New York, 341-2.

Kunz, E. 1985: *Introduction to Commutative Algebra and Algebraic Geometry*, Birkhäuser, Boston.

Lamotke, K. 1968: *Semisimpliziale algebraische Topologie*, Grundlehren der math. Wissenschaften 147, Springer-Verlag, Berlin-Heidelberg-New York.

Lannes, J. 1987: Sur la cohomologie modulo p de p-groupes abéliens élémentaires, *Homotopy Theory, Proceedings of the Durham Symposium 1985*, E. Rees and J.D.S. Jones, (Eds.), London Math. Soc. Lecture Note Series 117, Cambridge University Press, Cambridge, 97-116.

Lee, C.N. 1968: Equivariant homology theories, P.S. Mostert, (Ed.), *Proceedings of the Conference on Transformation Groups, New Orleans 1967*, Springer-Verlag, Berlin-Heidelberg-New York, 237-44.

Lemaire, J.-M. 1974: *Algèbres Connexes et Homologie des Espaces de Lacets*, Lecture Notes in Math. 422, Springer-Verlag, Berlin-Heidelberg-New York.

Lück, W. 1989: *Transformation Groups and Algebraic K-theory*, Lecture Notes in Math. 1408, Springer-Verlag, Berlin-Heidelberg-New York.

MacLane, S. 1967: *Homology*, Springer-Verlag, Berlin-Göttingen-Heidelberg.

Mann, L.N. 1962: Finite orbit structure on locally compact manifolds, *Mich. Math. J.* **9**, 87-92.

Matsumura, H. 1986: *Commutative Ring Theory*, Cambridge Studies in Advanced Mathematics 8, Cambridge University Press, Cambridge.

Matumoto, T. 1971: On G-CW-complexes and a theorem of J.H.C. Whitehead, *J. Fac. Sci. Univ. of Tokyo* **18**, 363-74.

Matumoto, T. 1984: A complement to the theory of G-CW-complexes, *Japan. J. Math.* **10**, 353-74.

May, J.P. 1967: *Simplicial Objects in Algebraic Topology*, Van Nostrand Mathematical Studies no. 11, Van Nostrand, Princeton.

May, J.P. 1982: Equivariant homotopy and cohomology theory, *Contemporary Math.* **12**, 209-17.

Mayer, K.H. 1989: G-invariante Morse-Funktionen, *Manuscripta Math.* **63**, 99-114.

McCleary, J. 1985: *User's Guide to Spectral Sequences*, Math. Lecture Series 12, Publish or Perish, Wilmington.

Miller, H.R. 1984: The Sullivan conjecture on maps from classifying spaces, *Ann. Math.* **120**, 39-87.
Milnor, J.W. 1962: On axiomatic homology theory, *Pacific J. Math.* **12**, 337-41.
Milnor, J.W. 1964: Microbundles, Part I, *Topology* **3**, 53-80.
Milnor, J.W. and Stasheff, J.D. 1974: *Characteristic Classes*, Ann. of Math. Studies No. 76, Princeton University Press, Princeton.
Moore, J.C. and Smith, L. 1968: Hopf algebras and multiplicative fibrations II, *Amer. J. Math.* **90**, 1113-50.
Munkres, J.R. 1975: *Topology, A First Course*, Prentice-Hall, Englewood Cliffs, New Jersey.
Nijenhuis, A. and Richardson, R.W.Jr. 1966: Cohomology and deformations in graded Lie algebras, *Bull. Amer. Math. Soc.* **72**, 1-29.
Oliver, R. 1976: A proof of the Conner conjecture, *Ann. Math.* **103**, 637-44.
Pardon, W. 1980: Mod.2 semicharacteristics and the converse to a theorem of Milnor, *Math. Zeit.* **171**, 247-68.
Petrie, T. and Randall, J.D. 1984: *Transformation Groups on Manifolds*, Marcel Dekker, New York.
Petrie, T. and Randall, J.D. 1986: Finite-order algebraic automorphisms of affine varieties, *Comment. Math. Helv.* **61**, 203-21.
Puppe, D. 1983: Homotopy cocomplete classes of spaces and the realization of the singular complex, *Topological Topics*, I.M. James, (Ed.), London Math. Soc. Lecture Notes Series 86, Cambridge University Press, 55-69.
Puppe, V. 1974: On a conjecture of Bredon, *Manuscripta Math.* **12**, 11-16.
Puppe, V. 1978: Cohomology of fixed point sets and deformation of algebras, *Manuscripta Math.* **23**, 343-5.
Puppe, V. 1984: P.A. Smith theory via deformations, *Soc. Math. de France, Astérisque* **113-4**, 278-87.
Puppe, V. 1988: Simply connected manifolds without S^1-symmetry, *Algebraic Topology and Transformation Groups, Proc. Göttingen (1987)*, T. tom Dieck (Ed.), Lecture Notes in Math. 1361, Springer-Verlag, Berlin-Heidelberg-New York, 261-8.
Quillen, D.G. 1967: *Homotopical Algebra*, Lecture Notes in Math. 43, Springer-Verlag, Berlin.
Quillen, D.G. 1969: Rational homotopy theory, *Ann. Math.* **90**, 205-95.
Quillen, D.G. 1971a: The spectrum of an equivariant cohomology ring: I, *Ann. Math.* **94**, 549-72.
Quillen, D.G. 1971b: The spectrum of an equivariant cohomology ring: II, *Ann. Math.* **94**, 573-602.
Quillen, D.G. 1976: Projective modules over polynomial rings, *Invent. Math.* **36**, 167-71.
Rachid, H.M. 1990: Actions du tore T^n sur les espaces simplement connexes, Thesis, Université Catholique de Louvain.
Raussen, M. 1991: Rational cohomology and homotopy of spaces with circle action, preprint, University of Aalborg.
Raymond, F. 1968: A classification of the actions of the circle on 3-manifolds, *Trans. Amer. Math. Soc.* **131**, 51-78.
Samuelson, H. 1969: *Notes on Lie Algebras*, Van Nostrand Reinhold Math. Studies No. 23, Van Nostrand Reinhold, New York.

Schafer, J.A. 1968: A relation between group actions and index, *Proceedings of the Conference on Transformation Groups, New Orleans 1967*, P.S. Mostert, (Ed.), Springer-Verlag, Berlin-Heidelberg-New York, 349-50.

Schultz, R. (Ed.) 1985: *Group Actions on Manifolds*, Contemporary Math. **36**, Amer. Math. Soc., Providence, Rhode Island.

Segal, G.B. 1968a: Equivariant K-theory, *Publ. Math. Inst. Hautes Etudes Sci.* **34**, 129-51.

Segal, G.B. 1968b: Classifying spaces and spectral sequences, *Publ. Math. Inst. Hautes Etudes Sci.* **34**, 113-28.

Seifert, H. 1932: Topologie dreidimensionaler gefaserter Räume, *Acta. Math.* **60**, 147-238.

Serre, J.P. 1965a: *Algèbre Locale, Multiplicités*, Lecture Notes in Math. 11, Springer Verlag, Berlin-Heidelberg-New York.

Serre, J.P. 1965b: Sur la dimension cohomologique des groupes profinis, *Topologie* **3**, 413-20.

Seymour, R.M. 1982: On G-cohomology theories and Künneth formulae, *Current Trends in Algebraic Topology*, Part 1, Amer. Math. Soc., Providence, Rhode Island, 257-71.

Seymour, R.M. 1983: Some functorial constructions on G-spaces, *Bull. London Math. Soc.* **15**, 353-9.

Singer, W. 1973: Steenrod squares in spectral sequences I,II, *Trans. Amer. Math. Soc.* **175**, 327-36, 337-53.

Skjelbred, T. 1972: Cohomology theory of compact transformation groups, Thesis, University of California.

Skjelbred, T. 1975: Actions of p-tori on projective spaces, *University of Oslo, Institute of Mathematics*, Preprint Series No. 4.

Skjelbred, T. 1978a: Cohomology eigenvalues of equivariant mappings, *Comment. Math. Helv.* **53**, 634-42.

Skjelbred, T. 1978b: Combinatorial geometry and actions of compact Lie groups, *Pacific J. Math.* **79**, 197-205.

Smith, P.A. 1938a: The topology of transformation groups, *Bull. Amer. Math. Soc.* **44**, 497-514.

Smith, P.A. 1938b: Transformations of finite period, *Ann. Math.* **39**, 127-64.

Smith, P.A. 1939: Tranformations of finite period II, *Ann. Math.* **40**, 690-711.

Smith, P.A. 1941a: Transformations of finite period, III Newman's theorem, *Ann. Math.* **42**, 446-58.

Smith, P.A. 1941b: Fixed point theorems for periodic transformations, *Amer. J. Math.* **63**, 1-8.

Smith, P.A. 1941c: Periodic and nearly periodic transformations, *Lectures in Topology*, Univ. of Michigan Press, Ann Arbor, Michigan, 159-90.

Smith, P.A. 1942a: Fixed points of periodic tranformations, Appendix B in Lefschetz, S., *Algebraic Topology*, Amer. Math. Soc., New York.

Smith, P.A. 1942b: Stationary points of transformation groups, *Proc. Nat. Acad. Sci. USA* **28**, 293-7.

Smith, P.A. 1944: Permutable periodic transformations, *Proc. Nat. Acad. Sci. USA* **30**, 105-8.

Smith, P.A. 1945: Transformations of finte period, IV. Dimensional parity, *Ann. Math.* **46**, 357-64.

Smith, P.A. 1959: Orbit spaces of abelian p-groups, *Proc. Nat. Acad. Sci. USA* **45**, 1772-5.

Smith, P.A. 1960: New results and old problems in finite transformation groups, *Bull. Amer. Math. Soc.* **66**, 401-15.

Smith, P.A. 1961: Orbit spaces of finite Abelian transformation groups, *Proc. Nat. Acad. Sci. USA* **47**, 1662-7.

Smith, P.A. 1963: The cohomology of certain orbit spaces, *Bull. Amer. Math. Soc.* **69**, 563-8.

Smith, P.A. 1965: Periodic transformations of 3-manifolds, *Illinois J. Math.* **9**, 343-8.

Smith, P.A. 1967: Abelian actions on 2-manifolds, *Michigan Math J.* **14**, 257-75.

Smith, P.A. and Richardson, M. 1937: Periodic transformation of complexes, *Ann. Math.* **38**, 611-33.

Spanier, E.H. 1966: *Algebraic Topology*, McGraw-Hill, New York.

Steenrod, N.E. and Epstein, D.B.A. 1962: *Cohomology Operations*, Ann. of Math. Studies No. 50, Priceton University Press, Princeton.

Sullivan, D. 1970: *Geometric Topology, Part I: Localization, Periodicity and Galois Symmetry*, mineographed notes, Massachusetts Institute of Technology.

Sullivan, D. 1977: Infinitesimal computations in topology, *Publ. Math. IHES* **47**, 269-332.

Suslin, A. 1976: Projective modules over polynomial rings (in Russian), *Dokl. Akad. Nauk SSSR* 26.

Swan, R.G. 1960: A new method in fixed point theory, *Comment. Math. Helv.* **34**, 1-16.

Tanré, D. 1983: *Homotopie Rationelle: Modèles de Chen, Quillen, Sullivan*, Lecture Notes in Math. 1025, Springer-Verlag, Berlin-Heidelberg-New York.

Thomas, J.-C. 1980: Homotopie rationelle des fibrés de Serre, Thesis, University of Lille I.

Thomas, J.-C. 1982: Eilenberg-Moore models for fibrations, *Trans. Amer. Math. Soc.* **274**, 203-25.

Tomter, P. 1974: Transformation groups on cohomology products of spheres, *Invent. Math.* **23**, 79-88.

Varadarajan, K. 1975: Numerical invariants in homotopical algebra, *Can. J. Math.* **27**, 935-60.

Vigué-Poirrier, M. 1981: Réalisation de morphismes donnés en cohomologie et suite spectrale d'Eilenberg-Moore, *Trans. Amer. Math. Soc.* **265**, 447-84.

Waner, S. 1980: Equivariant homotopy theory and Milnor's theorem, *Trans. Amer. Math. Soc.* **258**, 351-68.

Whitehead, G.W. 1978: *Elements of Homotopy Theory*, Graduate Texts in Math. 61, Springer-Verlag, New York.

Yau, S.-T. 1977: Remarks on the group of isometries of a Riemannian manifold, *Topology* **16**, 239-47.

Yoshida, T. 1979: On the rational homotopy of fixed point sets of circle actions, *Math. J. Okayama Univ.* **21**, 149-53.

Zariski, O. and Samuel, P. 1960: *Commutative Algebra*, Vol. II, Van Nostrand, Princeton.

Index

agreement space 112
Alexander-Spanier pseudo-dual rational homotopy ($\overline{\Pi}^*_\psi(X)$) 112
Alexander-Spanier Sullivan-de Rham algebra, complex ($\overline{A}(X)$) 111
almost-effective 327
almost-free torus actions (see also torus rank) 269ff
associated prime 401
based homotopy 97
belongs 186
bigraded model 118, 121, 122
Borel construction 2, 10ff
Borel formula 370
Borsuk-Ulam Theorem 33, 34, 246
Browder-Gottlieb exponent (see also G-component) 287
cases (0), (2), (odd p) 130, 131
cases (A), (B), (C) 145
cases (B+), (C+) 159
c-connected (CGDA or space) 96
Čech-G-covering 225
Čech nerve 111
centre (of a Lie algebra) 277
c-finite 105
CGDA (commutative graded differential algebra) 93
Chang-Skjelbred module 176ff
Chang-Skjelbred submodule 178
cid (complete intersection defect) 42, 200, 201, 432
cocentre (of a co-Lie algebra) 277
cofibration 437
coformal 128
coheight 399
Cohen-Macaulay (rings and modules) 418ff
cohomologically effective 327
cohomologically separable (CS) 159, 174

co-Lie algebra 102
comparison theorem 8, 9
compatible diagonal 231
complete intersection defect See cid
composition product 232
conditions (A), (B), (C) See cases (A), (B), (C)
condition (CS) See cohomologically separable
condition (LT) 208
connected (CGDA) 96
connecting subtorus 193
connective isotropy types 131
deformation 38ff, 449, 450
depth 416
depth(I, M), depth$_A(M)$ 416
differential module 436
Eilenberg-Moore spectral sequence 125, 126, 127
equivariant canonical map 225
equivariant (co)homology theory 8
equivariant Dowker map 225
equivariant Euler class 366, 367, 372
equivariant Gysin homomorphism 360, 366
equivariant rational homotopy 149ff
equivariant refinement map 226
equivariant Tate cohomology 302, 305, 306, 318
essential (at F) 379
Euler class 397
Euler homotopy characteristic 105
Evaluation Theorem 27, 67 160, 163, 165, 237, 239, 244
faithful covering 317
FDRH (finite dimensional rational homotopy; same as π-finite) 105
fibre number 297 see also G-exponent
filtered model 123
filtration, first, second 303

Index

finite $\mathbb{Q}$-type 102
finitely generated 428
finitely generated Chang-Skjelbred module 178
finitely many connective orbit types (FMCOT) 131
finitely many orbit types (FMOT) 131
finitistic 116, 142
Fitting ideals, Fitting invariants 186-187
formal (CGDA, space) 117
formal dimension 343
formal (homomorphism, map) 123
free actions 249. See also rank and p-rank
free presentation 433
full 156, 157

G-cell of type G/K 3
G-(co)homology theory 8
G-CW-complex 3
generators and relations 42, 200, 337, 425ff
G-exponent 287, 288, 291, 295, 296
global homological dimension (gldh) 409
Golber formulae 387, 389
graded 404
Grivel-Halperin-Thomas Theorem 106, 112, 113
G-variety 314

Harker-Noether Primary Decomposition 407
height 388
Hirsch-Brown model 21
homogeneous 404
homogeneous ideal 404
homological codimension 416
homological dimension 408ff
homotopic 97
homologically trivial, hl-trivial 439
homotopically trivial, ht-trivial 439
homotopy 97, 436, 437
homotopy groups (of an augmented KS-complex) 101
Hopf invariant 395

Hsiang's Fundamental Fixed Point Theorem 189ff
Hsiang-Serre ideal 181ff, 248
hyperformal 120

ideal (of relations) 429
I-depth 416
I-maximal 416
index 366, 375
integration formula 374
intrinsically formal 120
invariant covering 225
isotropy types 131

Kähler space 281
Koszul complex 412, 414
Krull dimension 399
Krull Height Theorem 400
KS-complex 96
KS-extension 97

Λ-extension 97
Λ-minimal Λ-extension 97
length 416
Lefshetz Fixed Point Theorem 249
Localization Condition 176
Localization Theorem 133, 135, 145, 146, 149, 154, 159, 171, 237, 244, 334
local weights 372
Lyusternik-Schnirel'man category 278

minimal CGDA See minimal KS-complex
minimal free presentation 433
minimal Hirsch-Brown model 29, 73ff, 76, 452
minimal KS-complex 96
minimal KS-extension 99
minimal model 98
minimal presentation 429, 433
model 97
model of generators 431
model of relations 431
Monomorphism Condition 176
M-regular sequence, M-sequence 414, 416

nilpotent (space) 100
non-trivial unit 312

$(n, \mathbb{Z})$-pair, $(n, \mathbb{Z})$-space 287
orientation 343
perturbation 38ff, 449
π-finite (same as FDRH) 105
π-formal 128
Π^*_ψ-full See full
Poincaré algebra 343, 344
Poincaré duality 344, 347, 358
Poincaré series 254ff
Pontryagin classes 384, 385, 397
Pontryagin number 385, 393, 395, 405, 406
p-rank (rank$_p$) 80, 286, 288, 291, 295, 296, 316, 318, 319, 321, 323, 324
pre-Chang-Skjelbred module 176
presentation 429
primary (ideal, submodule) 405
primary decomposition 406ff
prime decomposition series, set 401
prime divisors 407
projective dimension 408
pseudo-dual rational homotopy groups (of a space) 101
p-torus ix
pull-backs 106, 108, 113, 127

quasi-isomorphism See weak-equivalence
Quillen's Proposition 142
q-type 328, 330

rank 257. See also p-rank and torus rank
rank variety 314
rational category 278
rational cohomology Kähler space See Kähler space
rational homotopy Lie algebra 102, 104, 336ff
regular local ring 410
regular ring 410
regular sequence 414ff

Serre's Proposition 188
shifted diagonal 232
shifted subgroup 35, 229, 312, 313
P. A. Smith theory ix
special structure 177

spectral sequence, first, second 304
Stiefel-Whitney classes 386
strong Hsiang-Serre ideal 181
Sullivan-de Rham algebra (complex) 92
support (of a prime ideal) 132, 170
support (of a module) 401
Swan's Theorem 308, 340
symmetry 326ff, 376ff

TNHZ (totally non-homologous to zero) 141, 156, 198, 201, 204, 211
Tate G-exponent 318, 319, 321
topological degree of symmetry 327, 330
topological p-toric degree of symmetry 376, 377
topological splitting principal 373
topological symmetry 326ff
topological toric degree of symmetry 245, 327, 376-378
torus rank (rank$_0$) 269-275, 277, 279-281, 285
transfer 217
twisted tensor product 21, 449
type of A^* 39

uniform action 172
unmixed 419
unstable elements 332

Vietoris G-covering 225
Vietoris nerve 111

weak equivalence 97
Weyl group 205ff, 330, 316

Index of Notation

$A(C^*)$ 302
$A(K), A(X)$ 94
$\overline{A}(X)$ 111
$\text{ann}_A(M)$ 401
$A_S(L)$ 95
$\text{Ass}_A(M)$ 401
$B(C^*)$ 302
BG 10
$\beta^G_*(-), \beta^*_G(-)$ 11
$\beta_{G,\alpha}(-)$ 26
$\beta_{G,\tilde{\alpha}}(-)$ 57
$\beta^*_G(C^*)$ 224, 226
$\beta^*_G(X;k)$ 224
$\overline{\beta}^*_G(X;k), \overline{\beta}^*_G(X,A;k)$ 226
$\beta^*_\Gamma(C^*)$ 230
$\overline{\beta}^*_\Gamma(C,A;E)$ 230
$\overline{\beta}^*_{G,\tilde{\alpha}}(X;E)$ 243

$C^{(*)}(-)$ 25
cid 432
$\text{cid}(A^*)$ 42
$\text{codh}_A(M)$ 416
coht 399
$\chi_\pi(X)$ 105
$\chi_{\pi_{\text{top}}}(X)$ 273
$\chi'_\pi(X)$ 274
$\chi''_\pi(X)$ 274
$\partial gk[G]\text{–Mod}, \partial gk[G]\text{–Mod}$ 12
$\partial gR\text{–Mod}$ 448
$\partial R\text{–Mod}$ 436
$\dim A$ 399
$\dim_A(M)$ 401
$d(X)$ 274
$d_{\text{top}}(X)$ 273
$\Delta_0, \Delta_s, \Delta_c$ 231

EG 6
$E(G, C^*)^{*,*}_r$ 227
$e(X;G), e(X,A;G)$ 287
$\hat{e}(X;G)$ 318
$e_G(X,Y)$ 366
$\mathcal{E}^*(G)$ 11, 224

$\mathcal{E}_*(G)$ 10, 223
$\mathcal{E}_*(G;E)$ 229
$f!$ 360
$\text{fd}(A^*)$ 343
FMCOT 131
FMOT 131
$F(G,x), F(x)$ 167
f_n, f^0_n, f^1_n 388
$\mathfrak{g}(A^*)$ 39
$\text{gldh}(A)$ 409
$\text{hd}_A(M)$ 408
$H^{(*)}(-)$ 27, 67
$H^G_*(-), H^*_G(-)$ 11
H^*_G 130
$H_G(X)$ 167
$H_{G,\alpha}(-)$ 26, 66
$H^*_{G,\tilde{\alpha}}(-)$ 53
$\overline{H}^*_{G,\alpha}(X;E)$ 244
$\overline{H}^*_\Gamma(X,A;E)$ 231
$\hat{H}^*_G(C^*)$ 303
$\hat{H}^*_G(X,A;\Lambda), \hat{H}^*_G(X;\Lambda)$ 305
$h^*_G(\ ), h^*_G(\ )$ 8
ht 399

$\mathcal{I}(F)$ 379
$K(\alpha)$ 228
k_α 25
$(K,c)^*$ 167
$k[G]$ 10
$k[t]_{\tilde{\alpha}}$ 52
$L(\alpha)$ 229
$\mathcal{L}(G)$ 49
$\mathcal{L}^*(K), \mathcal{L}^*(X)$ 103
$\mathcal{L}_*(X)$ 104
$\overline{L}^*(X)$ 112
$LW(F)$ 372
$\lambda(K)$ 103
$\text{Max}(A)$ 399
max, min 401
$\mathcal{M}(X)$ 97, 98

$\overline{\mathcal{M}}(X)$ 112
$\mu_G^*(X;k), \mu_{G,\alpha}(X;k)$ 164
$\overline{\mu}_G^*(X;E)$ 243
$\overline{\mu}_{G,\alpha}^*(X;E)$ 243
$\mu d_0(X), \mu d_p(X)$ 377
$n(X)$ 273
$n_T(X)$ 273
$N_t(X)$ 327
$\mathcal{O}_A$ 343
PK 132
$\mathfrak{p}(K,c)$ 167
$\pi^*(K)$ 101
$\Pi^*_{(G,x_0)}(X)$ 149
$\Pi^*_{(G,F)}(X)$ (first appears as $\Pi^*_{(T,F)}(X)$) 155
$\Pi^*_\psi(X)$ 101
$\overline{\Pi}^*_\psi(X)$ 112
$\overline{\Pi}^*_{(G,x_0)}(X)$ 158, 160
$Q(A), Q(f)$ 431
$Q(K)$ 101
QK 167
$Q_{(G,x_0)}(X)$ 149
$Q_{(G,F)}(X)$ (first appears as $Q_{(T,F)}(X)$) 155
R, R_K 130, 131
$R(J)$ 431
$\mathfrak{r}(A^*)$ 42
rank(Φ) 257
$\text{rank}_0(X)$ 269
$\text{rank}_p(X)$ 312
$\text{res}_\alpha X$ 19

R-Mod 436
$S(K)$ 132
$S(K,c)$ 170
$S(V), S(E)$ 93
$\overline{S}^*(X)$ 348
Spec(A) 399
Supp(M) 403
$\sigma\mathfrak{p}$ 132, 170
$T(K)$ 133
$T(K,c)$ 170
$\mathfrak{T}(X)$ 167
$T_t(X)$ 327
$T_{t,p}(X)$ 376
$T_t^c(X), T_{t,p}^c(X)$ 377
$\theta(\Phi)$ 254
$\theta_n, \theta_n(L)$ 370, 387
$\oplus_0, \oplus_s, \oplus_c$ 232
$Un(M)$ 332
$V(\mathfrak{a})$ 399
$W_*(EG)$ 6
X^α 53
X_α 49
X^f, X^S 132
X_G 10
X_i 260
X^K 3
$Z(L)$ (first appears as $Z\mathcal{L}$ and $Z(\mathcal{L})$) 277
$Z_1(X)$ 274
$\zeta(\lambda)$ 277